The Science
of Conjecture

The Science of Conjecture

Evidence and Probability before Pascal

WITH A NEW PREFACE

James Franklin

Johns Hopkins University Press
Baltimore

Johns Hopkins Paperbacks edition, 2002, 2015
Printed in the United States of America on acid-free paper
2 4 6 8 9 7 5 3 1

Johns Hopkins University Press
2715 North Charles Street
Baltimore, Maryland 21218-4363
www.press.jhu.edu

The Library of Congress has cataloged the hardcover
edition of this book as follows:

Franklin, James, 1953–
The science of conjecture : evidence and probability before Pascal /
James Franklin.
p. cm.
Includes bibliographical references and index.
ISBN 0-8018-6569-7 (alk. paper)
1. Probabilities—History. 2. Evidence—History. I. Title.

QA273.A4 F73 2001
519.′09—dc21   00-042429

ISBN 978-1-4214-1880-3

A catalog record for this book is available
from the British Library.

Errata
P. 130 The discovery of the black swan dates to after Burgersdijck's *Logic*.
P. 177 The procedure involving random sampling was a quality-control
method in the Mint, but not the one called the trial of the *pyx*.

Contents

Preface to the 2015 Edition

The first edition of *The Science of Conjecture* reconfigured the field of the history of probability in two ways. The first concerned a wider than usual conception of the nature of probability. The second involved a Renaissance-free view of the history of Western ideas. In the decade since the book's publication, the first of these has gained traction, the second has not.

The book's conception of the subject matter stemmed from an *objective Bayesian* (or logical probabilist) theory of probability. According to that theory, as developed by Keynes in his *Treatise on Probability* and by later authors,[1] the main notion of probability is of an objective logical relation holding between a body of evidence and a conclusion. The body of evidence available in court does or does not make the defendant's guilt highly probable; the known facts do or do not support the theory of global warming, irrespective of any contingent facts about the world or what anyone's opinion is. Logical probability may or may not be numerical; even if it is, qualitative or approximate judgments are often of most importance.

That perspective opened up all kinds of evaluation of uncertain evidence as the natural subject matter of a history of probability. Thus, *The Science of Conjecture* focused on the law of evidence, which, over many centuries of thought, especially in medieval Roman law, had developed evidential concepts like the modern proof beyond reasonable doubt. Moral theory and business were also familiar with concepts of probabilities and risks, mostly quantified only loosely.

During the late twentieth century, debate on "interpretations of probability" largely took the form of pitched battles between frequentists and related schools (who held that probability dealt with relative frequencies or objective propensities) and subjective Bayesians (who took probability to be about degrees of belief, subject to some con-

1. The classic works are J. M. Keynes, *A Treatise on Probability* (London, 1921) and E. T. Jaynes, *Probability Theory: The Logic of Science* (Cambridge, 2003).

straints). But in recent years, a more objective and logical Bayesian interpretation has gradually come to the fore. Statisticians felt the need for some objectivity in prior probabilities, which permitted solid results in a great range of applied areas such as image processing.[2] Legal theorists similarly felt the need for an objective understanding of uncertainty in legal decision making,[3] and there has been extended debate about the use of Bayesian methods in legal cases involving DNA and other identification evidence.[4] Bayesian networks have become a popular method of representing knowledge and making causal inferences in artificial intelligence.[5] Philosophers added objective logical theories to the range of options they considered.[6] It was particularly noticed that probabilistic reasoning works with the confirmation of conjectures in pure mathematics, where there are only logical relations, implying that there must be a purely logical interpretation of probability applicable in those cases.[7] Objective Bayesian approaches to the philosophy of science would seem to be warranted, but have been less popular.[8]

Public understanding of the Bayesian perspective was advanced by Sharon Bertsch McGrayne's semipopular 2011 history, *The Theory That Would Not Die*.[9] It took a triumphalist view of the victory of Bayesian-

2. J. O. Berger, "The case for objective Bayesian analysis (with discussion)," *Bayesian Analysis* 1 (2006): 385–402 and 457–64; J. O. Berger, J. M. Bernardo, and D. Sun, "Objective priors for discrete parameter spaces," *Journal of the American Statistical Association* 107 (2012): 636–48.

3. J. Franklin, "The objective Bayesian conceptualisation of proof and reference class problems," *Sydney Law Review* 33 (2011): 545–61.

4. E.g., P. Roberts and A. Zuckerman, *Criminal Evidence*, 2nd ed. (Oxford, 2010), ch. 4; *Law, Probability and Risk* 11 (4) (Dec 2012), special issue on the *R v T* debate.

5. K. B. Korb and A. E. Nicholson, *Bayesian Artificial Intelligence*, 2nd ed. (London, 2010); L. Bovens and S. Hartmann, *Bayesian Epistemology* (Oxford, 2004).

6. A. Hájek, "Interpretations of probability," *Stanford Encyclopedia of Philosophy* (2002, revised 2009), http://plato.stanford.edu/entries/probability-interpret/; D. H. Mellor, *Probability: A Philosophical Introduction* (London, 2005); J. Franklin, "Resurrecting logical probability," *Erkenntnis* 55 (2001): 277–305; J. Williamson, *In Defence of Objective Bayesianism* (Oxford, 2010).

7. J. Franklin, "Non-deductive logic in mathematics," *British Journal for the Philosophy of Science* 38 (1987): 1–18; later works listed in J. Franklin, "Non-deductive logic in mathematics: the probability of conjectures," in A. Aberdein and I. Dove, eds., *The Argument of Mathematics* (New York, 2013), 11–29.

8. J. Franklin, *What Science Knows: And How It Knows It* (New York, 2009).

9. S. B. McGrayne, *The Theory That Would Not Die: How Bayes' Rule Cracked*

ism, in a more or less objectivist form, over frequentism. The revival of Bayesianism, it argued, was driven by practice more than theory, beginning with Alan Turing's use of Bayes' Theorem in World War II cryptography and given momentum by the availability of sufficient computing power to calculate results of the formula on large databases. That view of the story sees Bayesianism as essentially about a numerical formula, thus neglecting the wider sphere of evidence evaluation, such as in law and scientific hypothesizing. It thus confines the story to recent times and to more mathematically technical fields.

The global financial crisis of 2008 also brought to the fore the problems of the relation between numerical formulas for probabilities and the true chance of events happening. Nassim Nicholas Taleb's *The Black Swan* was the most successful of several works that explained an inherent difficulty in predicting rare events. Because they are rare, there is little relevant evidence, so purely inductive methods are unreliable and must be supplemented with expert opinion to situate the event in a context; unfortunately, expert opinion is chronically unreliable, especially with very small probabilities.[10]

The recognition of imprecise or fuzzy probabilities within Bayesianism has advanced, but it is still a minority interest.[11] Nevertheless, it is recognized as important that statistical methods should be robust, that is, only slightly sensitive to errors in the data. Thus, imprecision is allowed for in such statistical methods. Imprecision is important in applications like safety science, where there may be insufficient evidence to ground precise probabilities.[12]

Perhaps the most surprising recent development has been the grow-

the Enigma Code, Hunted Down Russian Submarines, and Emerged Triumphant from Two Centuries of Controversy (New Haven, 2011).

10. N. N. Taleb, *The Black Swan: The Impact of the Highly Improbable* (New York, 2007, 2nd ed., 2010); J. Franklin, "Operational risk under Basel II: A model for extreme risk evaluation," *Banking and Financial Services Policy Report* 27 (10) (Oct 2008): 10–16; the unreliability of expert opinion shown in, e.g., P. E. Tetlock, *Expert Political Judgment: How Good Is It? How Can We Know?* (Princeton, 2005).

11. P. Coolen-Schrijner, F. P. A. Coolen, M. C. M. Troffaes, and T. Augustin, "Imprecision in statistical theory and practice," *Journal of Statistical Theory and Practice* 3 (2009): 1-9; Society for Imprecise Probability: Theory and Practice (http://www.sipta.org).

12. T. Augustin and R. Hable, "On the impact of robust statistics on imprecise probability models: A review," *Structural Safety* 32 (2012): 358–65; D. R. Insua and F. Ruggeri, eds., *Robust Bayesian Analysis* (New York, 2000).

ing realization among developmental psychologists that babies are powerful Bayesian reasoners. Six-month-old infants show surprise if a box which they know contains mostly pink and a few yellow balls reveals a sample that is mostly yellow, whereas they are not surprised when the sample is mostly pink. That is, they know implicitly that the composition of a sample is likely to approximately match the composition of the population.[13] Twelve-month-olds have much more sophisticated abilities to integrate such frequency perceptions with other knowledge, such as expectations about object motions.[14] Regarding infants as "Bayesian ideal observers" has proved to be predictive of baby behavior in such studies. Since these abilities are acquired even before infants have learned to speak, it is clear that humans have pre-linguistic abilities to respond to and reason about probabilities, confirming the view of *The Science of Conjecture* that the story of probability is one of bringing to consciousness existing but implicit probabilistic knowledge.

The second main idea behind *The Science of Conjecture* was to take a pro-medieval, anti-Renaissance approach to the early modern history of ideas. The scholastics and legal writers of the late middle ages developed the main concepts of the evaluation of uncertain evidence as well as of commercial risk. The development of those ideas slowed in the "Renaissance," to be revived and driven forward in the scientific revolution of the seventeenth century. That will appear as a strange tale, until it is recognized that much the same happened for nearly all areas of intellectual thought, especially the more abstract ones. A similar thesis was maintained for physics and related sciences by Pierre Duhem a century ago and has recently been revived convincingly in James Hannam's *God's Philosophers*.[15] The implication that the medieval contribution was con-

13. S. Denison, C. Reed, and F. Xu, "The emergence of probabilistic reasoning in very young infants: evidence from 4.5- and 6-month-olds," *Developmental Psychology* 49 (2013): 243–49; similar in F. Xu and V. Garcia, "Intuitive statistics by 8-month-old infants," *Proceedings of the National Academy of Science of the U.S.A.* 105 (2008): 5012–15; C. A. Lawson and D. H. Rakison, "Expectations about single event probabilities in the first year of life: the influence of perceptual and statistical information," *Infancy* (2013), doi: 10.1111/infa.12014.

14. E. Téglás, E. Vul, V. Girotto, M. Gonzalez, J. B. Tenenbaum, and L. L. Bonatti, "Pure reasoning in 12-month-old infants as probabilistic inference," *Science* 332 (2011): 1054–59.

15. J. Hannam, *God's Philosophers: How the Medieval World Laid the Foundations of Modern Science* (London, 2009), published in the United States as *The Genesis of Science: How the Christian Middle Ages Launched the Scientific Revolution* (Washington, DC, 2011).

fined to or especially prominent in (what we now call) science is incorrect. On the contrary, science strictly speaking was not the natural bent of the scholastics, who were much more at home in disciplines that rely on conceptual analysis, such as economic theory and law. That has not yet been generally accepted. While there are some older texts on particular topics, such as Marjorie Grice-Hutchinson's *The School of Salamanca* and James Gordley's *The Philosophical Origins of Modern Contract Doctrine*,[16] they have not been followed up and situated in an overall story. There remains a gap in the market. A substantial book needs to be written on the full extent of the scholastic contribution to modern thought.[17]

Updates

On the subject matter of *The Science of Conjecture* itself, the development of ideas about evidence and probability up to 1650, there have not been any major new works. The second edition of Hacking's *The Emergence of Probability* reprinted the first edition, prefaced with brief comments on more recent work.[18] It will suffice to review briefly some studies of particular topics.

Gabbay and Koppel review the Talmudic rules on "follow the majority" (for example, to declare a piece of meat of unknown origin to be kosher if a majority of nearby butchers are kosher). They conclude that the rules do not involve probabilistic thinking but are strictly rules of action to determine an outcome.[19]

16. M. Grice-Hutchinson, *The School of Salamanca: Readings in Spanish Monetary Theory, 1544–1605* (Oxford, 1952); J. Gordley, *The Philosophical Origins of Modern Contract Doctrine* (New York, 1993).

17. An initial attempt and an overview of what needs to be done is in J. Franklin, "Science by conceptual analysis: the genius of the late scholastics," *Studia Neoaristotelica* 9 (2012): 3–24.

18. I. Hacking, *The Emergence of Probability: A Philosophical Study of Early Ideas about Probability, Induction and Statistical Inference*, 2nd ed. (New York, 2006); his more positive review of *The Science of Conjecture* in *Isis* 95 (2004): 460–64. There was little attempt to cover the pre-Pascalian period in D. M. Gabbay, S. Hartmann, and J. Woods, eds., *Handbook of the History of Logic*, vol. 10, *Inductive Logic* (Oxford, 2011). Summaries of the story as it now are in R. Schüssler, "Probability in medieval and Renaissance philosophy," *Stanford Encyclopedia of Philosophy*, 2014, http://plato.stanford.edu/entries/probability-medieval-renaissance/, and J. Franklin, "Pre-history of probability," in *Oxford Handbook of Philosophy and Probability*, ed. C. Hitchcock and A. Hájek, forthcoming, 2015.

19. D. M. Gabbay and M. Koppel, "Uncertainty rules in Talmudic reasoning," *History and Philosophy of Logic* 32 (2011): 63–69.

Some studies of the moral doctrine of probabilism in the early modern period have clarified its nature and shown how widespread it was in education in Catholic countries.[20] Juan Caramuel Lobkowitz, the most extreme of the probabilist moral theologians—presented in *The Science of Conjecture* as something of a figure of fun—has been defended by Julia Fleming on the basis of his later works.[21]

One of the main findings of *The Science of Conjecture* was that late medieval canon lawyers, reflecting on business practice with contracts like insurance, annuities, and options that depend on chance outcomes, made great strides in understanding the pricing of risk. That has been confirmed by the detailed studies of Giovanni Ceccarelli on Olivi and other canonistic thinkers.[22] These studies show that late medieval thinkers had a deep understanding of the nature of risk, although the quantification involved is imprecise (as is indeed appropriate to the approximate frequencies and multiple sources of information involved in estimating risks such as in marine insurance).

20. R. A. Maryks, *Saint Cicero and the Jesuits: The Influence of the Liberal Arts on the Adoption of Moral Probabilism* (Aldershot, 2008); R. Schüssler, "On the anatomy of probabilism," in J. Kraye and R. Saarinen, eds., *Moral Philosophy on the Threshold of Modernity* (Dordrecht, 2005), 91–113. R. Schüssler, "Scholastic probability as rational assertability: the rise of theories of reasonable disagreement, *Archiv für Geschichte der Philosophie* 96 (2014): 151–284.

21. J. Fleming, *Defending Probabilism: The Moral Theology of Juan Caramuel* (Washington, DC, 2006); also A. Solana-Ortega and V. Solana, "Morality principles for risk modelling: needs and links with the origins of plausible inference," in *Bayesian Inference and Maximum Entropy Methods in Science and Engineering, 2009* (American Institute of Physics Conference Proceedings, vol. 1193), 161–69.

22. G. Ceccarelli, "Risky business: theological and canonical thought on insurance from the thirteenth to the seventeenth century," *Journal of Medieval and Early Modern Studies* 31 (2001): 607-58; G. Ceccarelli, "The price for risk-taking: marine insurance and probability calculus in the Late Middle Ages," *Journal Electronique d'Histoire des Probabilités et de Statistique* 3/1 (2007); G. Ceccarelli, "Stime senza probabilità: assicurazione e rischio nella Firenze rinascimentale," *Quaderni storici* 45 (2010): 651–702; also E. D. Sylla, "Business ethics, commercial mathematics, and the origins of mathematical probability," *History of Political Economy* 35 (2003), annual supplement: 309–37; S. Piron, "Le traitement de l'incertitude commerciale dans la scolastique médiévale," *Journal Electronique d'Histoire des Probabilités et de Statistique* 3/1 (2007). Some clarifications on the meaning of Oresme's contemporary work in N. Meusnier, "À propos d'une controverse au sujet de l'interprétation d'un théorème 'probabiliste' de Nicole Oresme," *Journal Electronique d'Histoire des Probabilités et de Statistique* 3/1 (2007).

On lotteries and games of chance, there have been a few studies but no new major findings.[23] There remains the possibility that intense research in obscure vernacular literatures may yet show that games of chance were much better understood in earlier times than is as now apparent.

The chapter on probability in religious argument showed the beginnings of a divergence between English and continental thought related to "reasonableness," with English works following Hooker's *Laws of Ecclesiastical Polity*, emphasizing the need for reasonableness and moderation in matters of religion and politics. Recent work has shown that England really has had a unique history related to probability and reasonableness. Although English thinkers were not prominent in the development of probabilistic thinking up to 1650, John Graunt's work of 1662 on inferences from mortality tables was a unique departure, representing a large first step in drawing conclusions from statistical data.[24] It was followed up in England especially. At a deeper cultural level, Anna Wierzbicka's *English: Meaning and Culture* describes how John Locke's writings on probability, reasonableness, and moderation became ingrained in the English language. The words "reasonable" and "probably" appear in modern English with a frequency and range of application very much higher than their cognates in other European languages, as do words indicating hedges relating to degrees of evidence such as "presumably," "apparently," "clearly." These phenomena indicate an Anglophone mindset that particularly values attention to the uncertainties of evidence.[25]

Our understanding of the history of probability has reached a stable state. It is time to use that history as a resource for deepening our insights into some of the most conceptually tangled but crucial concepts of modern life: risk, evidence, and uncertainty.

23. G. Ceccarelli, "Le jeu comme contrat et le risicum chez Olivi," in A. Boureau and S. Piron, eds., *Pierre de Jean Olivi (1248–1298): Pensée scolastique, dissidence spirituelle et société* (Paris, 1999), 239–50; N. Meusnier, "Le problème des partis peut-il être d'origine arabo-musulmane?" *Journal Electronique d'Histoire des Probabilités et de Statistique* 3/1 (2007); E. Welch, "Lotteries in early modern Italy," *Past and Present* 199 (2008): 71–111.

24. J. Franklin, "Probable opinion," in P. Anstey, ed., *Oxford Handbook of British Philosophy in the Seventeenth Century* (Oxford, 2013), ch. 15; further background in B. J. Shapiro, "Testimony in seventeenth-century English natural philosophy: legal origins and early development," *Studies in History and Philosophy of Science A* 33 (2002): 243–63.

25. A. Wierzbicka, *English: Meaning and Culture* (Oxford, 2006), esp. chs. 4 and 8.

Preface

"Probability is the very guide of life," in Bishop Butler's famous phrase.[1] He does not mean, of course, that calculations about dice are the guide of life but that real decision making involves an essential element of reasoning with uncertainty. Humans have coped with uncertainty without the benefit of advice from mathematicians, both before and after Pascal and Fermat's discovery of the mathematics of probability in 1654. And they have talked and written about how to do so. So there is a history of probability that concerns neither mathematics nor anticipations of mathematics.

This is a history of rational methods of dealing with uncertainty. It treats, therefore, methods devised in law, science, commerce, philosophy, and logic to get at the truth in all cases in which certainty is not attainable. It includes evaluation of evidence by judges and juries, legal presumptions, balancing of reasons for and against scientific theories, drug trials, and counting shipwrecks to determine insurance rates. It excludes methods like divination or the consulting of oracles, which are substitutes for reasoning about uncertainty.

Three levels of probabilistic reasoning are distinguishable:

1. Unconscious inference, the reactions to uncertain situations that the brain delivers automatically at a subsymbolic level. This system of actions—the cloud out of which, so to speak, talk about uncertainty condensed—can be studied by psychology, but it is not history. (A review of what is known about it, and its relation to conscious inference, is given in chapter 12.)

2. Ordinary language reasoning about probabilities. It is this middle level that is the main subject of this book. It may avoid numbers entirely, as in "proof beyond reasonable doubt" in law or the nonnumerical judgments of plausibility that scientists and detectives make in evaluating their hypotheses. Or it may involve rough numerical estimates of probabilities, as in racecourse odds and guesses about the risks of rare events.

3. Formal mathematical reasoning of the kind found in textbooks of
   probability and statistics.

The higher levels may be more noble and perfect, but they are so at
a cost: they are less widely applicable.

The theme of this book, then, must be the *coming to consciousness* of
uncertain inference. The topic may be compared to, say, the history of
visual perspective. Everyone can see in perspective, but it has been a difficult and long-drawn-out effort of humankind to become aware of the
principles of perspective in order to take advantage of them and imitate nature. So it is with probability. Everyone can act so as to take a
rough account of risk, but understanding the principles of probability
and using them to improve performance is an immense task.

There is one further essential distinction to be made. Probability is
of two kinds.[2] There is *factual* or *stochastic* or *aleatory* probability, dealing with chance setups such as dice throwing and coin tossing, which
produce characteristic random or patternless sequences. Almost always
in a long sequence of coin tosses there are about half heads and half tails,
but the order of heads and tails does not follow any pattern. On the
other hand, there is *logical* or *epistemic* probability, or *nondeductive logic*,
concerned with the relation of partial support or confirmation, short of
strict entailment, between one proposition and another. A concept of
logical probability is employed when one says that, on present evidence,
the steady-state theory of the universe is less probable than the big bang
theory or that an accused's guilt is proved beyond reasonable doubt
though it is not absolutely certain. How probable a hypothesis is, on
given evidence, determines the degree of belief it is rational to have in
that hypothesis, if that is all the evidence one has that is relevant to it.

It is a matter of heated philosophical dispute whether one of these
notions is reducible to the other.[3] In any case the surface distinction is
clear and provides an orientation in the history of the subject. In the period covered by the present study, logical probability was the main focus
of interest, and the word *probability* was reserved solely for this case. The
little study of factual probability there was—concerned with dice and insurance—was not seen as connected with logical probability.

Consequently, the book opens with three chapters on the law of evidence, in which there has been the most consistent tradition of dealing explicitly with evidence that falls short of certainty. Conscience, conceived as a kind of internal court of law, could also be in doubt; the rule
of probabilism concerning it is the subject of the fourth chapter. The
fifth chapter describes the (not very successful) attempts by rhetoricians

and logicians to give some account of uncertain reasoning. Evidence for scientific theories (understanding *science* widely) is considered in the next two chapters, followed by two chapters on probability in philosophy and religion, dealing largely with inductive arguments and design arguments for the existence of God. The tenth chapter describes commercial and legal thought on the nature of aleatory contracts (agreements like insurance, annuities, and bets whose fulfillment depends on chance). One aleatory contract, gaming with dice, has outcomes that can be exactly evaluated mathematically: It is the subject of chapter 11.

The reader with an average familiarity with received ideas on intellectual history is asked to make a small number of reorientations, at least provisionally.

The first concerns probability specifically. Two points should be made, to avoid perceptions that early writers are indulging merely in confused "anticipations" of later mathematical discoveries. The first is that the process of discovering the principles of uncertain reasoning is far from over. It can sometimes appear that, beginning with Fermat and Pascal's success with dice in 1654, there has been a successful colonization of all areas of uncertain reasoning by the mathematical theory of probability. As in so many areas, the arrival of the computer has shown that previous knowledge about thinking processes was not nearly as precise as had been thought—not precise enough, in particular, to allow a complete mechanical imitation of them. More is said about this in the epilogue; suffice it to say here that there is no agreement on, for example, how to combine evidence for conclusions in computerized expert systems for medical diagnosis. The disagreements are fundamental and are about quite simple issues that have occupied thinkers about uncertain inference for two thousand years: how to decide the strength with which evidence supports a conclusion, how to combine pieces of evidence that support each other, and what to do when pieces of evidence conflict.

The second point is that while the probability of outcomes of dice throws is essentially numerical, and advances in understanding are measured by the ability to calculate the right answers, it is otherwise with logical probability. Even now, the degree to which evidence supports hypotheses in law or science is not usually quantified, and it is debatable whether it is quantifiable even in principle. Early writers on probability should therefore be regarded as having made advances if they distinguish between conclusive and inconclusive evidence and if they *grade* evidence by understanding that evidence can make a conclusion "almost certain," "more likely than not," and so on. Attempts to give numbers to those grades are not necessarily to be praised. One should not give in to the easy assumption that numbers are good, words bad.

The other requested reorientations concern two features of the history of ideas generally. They will seem strange to anyone even slightly familiar with the usual portrayals of the rise of modern science. The template antiquity > medieval decline > renaissance > scientific revolution does not fit the history of probability; certainly not the history of logical probability. In particular, it is not possible to read the story with the medieval Scholastics as "them" and the men of the seventeenth century as "us." The Scholastics made many advances in the clarification and deepening of concepts necessary to understand probability. And contrary to the myths put about by their many enemies, they explained themselves perfectly clearly.

Finally, the reader is asked to regard it as normal to find many ideas developing in legal contexts. Like the Scholastics, lawyers are often thought of as pursuing esoteric interests of little consequence for the outside world and as, by and large, enemies of scientific progress. It is argued that the prominence of both Scholastics and lawyers is not unique to probability but that their contributions to the development of modern ideas generally have been substantially underrated. A brief overview of their wider importance in the history of ideas is given in chapter 12, in order to situate the development of probability in its appropriate context.

It is useful to keep a few questions in mind while reading the detailed history. Researchers in the field have wondered why the development of probability theory was so slow—especially why the apparently quite simple mathematical theory of dice throwing did not appear until the 1650s. The main part of the answer lies in appreciating just how difficult it is to make concepts precise, especially when mathematical precision is asked for in an area that seems at first glance to be imprecise by nature. Mathematical modeling is always difficult, as is evident in contemporary parallel cases such as the mathematization of continuity that led to the calculus of Newton and Leibniz. The very idea of a "geometry of chance," as Pascal put it, is revolutionary. An evaluation of this and alternative explanations of the slowness of the rise of probability is given in chapter 12. It is suggested that, nevertheless, some mystery remains.

The book has an unusually high proportion of quotation. It is in the nature of the material that, once a small amount of context has been supplied, the authors can be allowed to speak for themselves. Paraphrase is pointless. The book is to be read in only one place at a time: the notes contain references only, the purpose of which is solely to increase the reader's degree of belief in the statements referenced. It is written in only one language, except for occasional words from the original language of texts, included to indicate that there is no overinterpretation through tendentious translation.

The purpose of history may not be to teach us lessons, but the story told here does have a certain contemporary relevance, even though it ends in 1660. The last century of the old millennium saw a gradual waning of faith in the objectivity of the relation of uncertain evidence to conclusion. In the philosophy of science, Popper, Kuhn, and their schools denied that observational evidence could make scientific theories more probable, and attention in the field moved to sociological and other nonevidential influences.[4] Postmodernism, presuming rather than arguing for the absence of objective methods of evaluating theories, offered a number of other reasons—or rather causes—of actual beliefs, such as the demands of "power."[5] The situation is not so bad in law, which has largely retained a commitment to the objectivity of evidence, but even there, theory is not as robust as practice.[6]

The past is a counterweight to these febrile inanities of pygmies who stand on the shoulders of giants only to mock their size. Just as one who feels battered by the relentless enfant terriblisme of "modern" art or music can revive his spirit by communion with Vermeer or Mozart, so the friend of reason can draw comfort from the achievements left by the like-minded of the past. The story of the discovery of rational methods of evaluating evidence can serve as a point of reference and can supply material for the defenses of rationality that will have to be undertaken.

I OWE A GREAT DEBT TO David Stove, who introduced me to the subject of logical probability and provided constant encouragement as well as reading and commenting on the manuscript.

For advice on various matters, I thank James Brundage, David Burr, Hilary Carey, Michael Cowling, Anthony Garrett, Paul Gross, Grahame Harrison, Michael Hasofer, Jamie Kassler, Roger Kimball, Joe McCarthy, Br. Michael Naughtin, Carlos Pimenta, Margaret Sampson, Sandy Stewart, Jenny Teichman, E. H. Thompson, Stephen Voss, John O. Ward, George Winterton, and the Institute of Medieval Canon Law, Berkeley.

The support and competence of Robert J. Brugger, Kimberly Johnson, and Glen Burris, of the Johns Hopkins University Press, as well as copyeditor Diane Hammond, have been invaluable.

For financial support, I am grateful to Enid Jenkins and the University of Sydney's Thyne Reid Memorial Fellowship.

CHAPTER ONE

The Ancient Law of Proof

T HE LAW OF EVIDENCE is the central thread in the history of probability. The reason is that the law must decide disputes. To do so,
it must often decide who or what to believe. What witnesses are credible? Which documents? What is to be done about conflicting testimony? Is evidence of what a witness heard another person say to be admitted? And on what basis should one make the final decision on whose
evidence is to be believed? It is possible to leave all these questions to
the discretion of the judge, but a developed system of law will tend to
introduce rules of evidence, so that this part of the law, like others, will
be aboveboard and seen to be so. The end users will thereby receive the
desired impression that judges are the instruments of a higher reality,
universal and omnicompetent, while legal insiders, who know that
judges are only human, will have some framework for constraining fellow judges and producing a consistent system.

A consistent system is also one in which disputes once settled stay
settled and in which present decisions can be compared with previous
ones. The record keeping necessary to achieve these results is another
reason for the centrality of law: law keeps its conceptual achievements,
and they remain accessible.

Egypt and Mesopotamia

"Civilization is the art of living in towns of such size that everyone does
not know everyone else."[1] With anonymity comes litigation, and with
litigation comes conflict of evidence.

An Egyptian document of the Sixth Dynasty of the Old Kingdom
(about 2200 B.C.) deals with a conflict between documentary evidence
and witnesses. It is among the earliest records of human thought. Sebek-
hotep claims that the will of Ouser has given him the right of use over
Ouser's goods and has named him guardian of Ouser's children. Taou,
the eldest son of Ouser, claims his father made no such will and that the
document produced by Sebek-hotep is forged. The tribunal gives judgment as follows: "If Sebek-hotep brings forward three honorable witnesses, in whom one can have confidence, who repeat [the oath], 'Thy

power be against him [Taou], O God! Since the truth is that this document is in conformity with what Ouser said on this matter,' then the goods will remain in the house of Sebek-hotep, after he has brought forward these witnesses in whose presence these things were said, and Sebek-hotep will have the right of use over the goods. But if he cannot produce three witnesses in whose presence the words were said, in that case, the goods of Ouser will not stay with him but will be given to his [Ouser's] son, the intimate of the king and director of the caravan, Taou."[2]

In later times there are brief mentions of procedures in Egyptian courts that are recognizable as the kind of methods of evaluating evidence that later legal systems have regarded as normal. A foreman of works who is alleged to have spoken offensively of the pharaoh has his case evaluated by eight of his peers. Courts question the claimants on both sides of cases and may summon witnesses, order beatings to elicit confessions, and order suspects to visit the scene of the crime to confirm their confessions.[3]

At the same time, there existed a system of evaluating evidence by mentalities not now accessible. A stele of Abydos records a judgment, much later than that of Sebek-hotep, by the god-king Ahmose I in a dispute among priests. Pasar claims a field of which Tai is now proprietor. "The priest Pasar came with the priest Tai to bring an action before Nebpehtire [Ahmose I]. The priest Pasar came, saying, 'That this field belongs to Tai, son of Sedjemenef, and to the children of Haiou.' The god remained motionless. He turned again to the god, saying, 'That it belongs to the priest Pasar, son of Mes.' The god nodded assent very strongly, in the presence of the priests of Nebpehtire."[4] The pharaoh delivering the judgment has at this point been dead some three hundred years, and reference is being made to what would now be called a statue of him. It is quite unclear what was really happening in such cases. Oracular judgments are not clearly earlier in time than ones by a more normal procedure, nor do they seem to have been reserved for a different class of cases. They were not necessarily even accorded greater credibility. In one case, a god is applied to twice and is for acquittal both times, while in another witnesses are brought forward against the god's decision. Attempting to manipulate an oracle was a recognized offense.[5]

Some faith in divine intervention is revealed also in the *Code of Hammurabi* and other ancient law codes, which appeal to oaths and ordeals when evidence is lacking. Two consecutive provisions in the code are "If the husband of a married woman has accused her but she is not caught lying with another man, she shall take an oath by the life of a god and return to her house"; and "If the finger has been pointed at a married woman with regard to another man, and she is not caught lying with the other man, she shall leap into the river-god for her husband."[6] It is natural to think that the juxtaposition of the two parallel cases is intended to mark some distinction—perhaps that the first case involves weaker evidence and is thus more lenient on the suspect. Nothing like this is said, however, nor is there a sign of any language in which it could be said. All that is visible is a method of deciding cases where there is a lack of evidence; the gods and their earthly legal system fill the gap. Methods of these kinds can be rational as ways of preventing crime, given the beliefs to be expected in a population in those days before the oracles fell silent. As procedures for actually evaluating evidence, they are not so much rational methods as the context out of which rational methods eventually arose. As they do not (at least avowedly) rely on any evaluation of evidence by a rational agent, it is correct to say that "procedure was not essentially deliberation, but action."[7]

On the other hand, there is a point in these procedures at which rational evaluation of evidence may have taken place. Many Babylonian cases, and a few Egyptian ones, conclude with a decision like, If X takes the oath, Y will make compensation. While this appears to refer the decision to the gods, it appears that the judge had some discretion in deciding which party should be given the opportunity of completing the case by oath.[8] Much later, medieval legal theorists believed that the point of the procedure was to give the party with the stronger case the chance of divine confirmation, and it may be that ancient judges too often hid rational thought behind their formulaic utterances.

Early Greek law, too, shows the usual reliance of primitive law on methods of "proof" like ordeals and oaths.[9] Greek thought of the classical period of course had a great deal to say about reasoning in legal cases—but more from the point of view of rhetoric than in a strictly legal context. (The rhetorical tradition is discussed in chapter 5.)

The Talmud

Late antiquity is normally regarded as "decadent" in such intellectual fields as science and mathematics, literature, perhaps philosophy. By contrast, law is one of the few areas of thought in which the legacy of antiquity consists almost exclusively of the productions of the last centuries of the ancient world. It consists, in particular, of the two great compilations of ancient law, the Jewish Talmud and the Roman *Corpus of Civil Law*.

Jewish oral traditions, some going back many centuries, were codified in the Mishnah in the early third century A.D. In the next two centuries, the interpretation of the Mishnah texts in the academies of Babylon was elaborated into a much larger work, the Babylonian Talmud, completed in the early fifth century. During the same centuries, the Roman jurists developed their law, mostly as opinions on particular cases. The classical period of Roman law is that of Ulpian and other jurists of the period around 150–250 A.D. Excerpts from their works were collected, and sometimes adjusted, by a commission set up by the Emperor Justinian. This collection, the *Digest* or *Pandects*, was declared law in 533; all other writings of the jurists were declared to be no longer law and have almost entirely disappeared. The *Digest* forms the largest part of Justinian's complete codification of Roman law, the *Corpus of Civil Law*. The Talmud and the *Corpus* are almost independent, except that a few of the older precepts of Jewish law were inserted into Roman law by the Christian emperors. What is in these collections is of crucial importance for the history of thought. In contrast to, for example, most ancient scientific and philosophical works, which were largely lost for many centuries, every word of the Talmud and the *Corpus of Civil Law* was minutely examined, compared with others, and commented on for century after century.

On matters of evidence, the Talmud and the *Digest* share recognizably similar ideas on three matters: rules concerning witnesses, the high standard of evidence required in criminal cases, and reasoning from "presumptions."

It is natural to adopt rules excluding certain classes of people from giving evidence or about the number of witnesses (e.g., requiring a minimum number to establish a fact or accepting the greater number of witnesses in case of conflict). They may be reasonable rules if they are not applied too rigidly. In Jewish law, some categories of witnesses were not admitted, including relatives of the parties, aliens, slaves, and dice players, but there was little discussion of the reasons for excluding them.[10] Of much greater importance was the two-witness rule, an attempt to ensure a minimum standard of evidence before anyone was condemned to punishment in a criminal case. It is not unique to Jewish law: there are versions of it in the most ancient Cretan, Greek, and Roman codes.[11] Nevertheless, because of the influence of the Bible, the Jewish law is much the most important source of the rule. From ancient times, Jewish law required "two or three" witnesses to establish wrongdoing, and the New Testament repeated the rule.[12] Its divine sanction caused the rule to be taken very seriously in later centuries, sometimes with extreme rigidity.

The contact between Jewish and Greek thought produced the first treatises on the theme so central to medieval philosophy, the harmony of faith and reason. Of the many themes of Aquinas and others that can be found in the works of Jesus' contemporary Philo of Alexandria, one of the most characteristic is the "reasons for the laws": what seems in the Old Testament to be purely an expression of God's will is often justified rationally by Philo and later philosophers. The two-witness rule is reasonable, he says, because a single witness should not be preferred against one or more others: "Against more than one, because their number makes them more worthy of credence than the one; against one, because the witness has not got preponderance of number, and equality is incompatible with predominance. For why should the statement of a witness made in accusation of another be accepted in preference to the words of the accused in his own defense? Where there is neither deficiency nor excess it is clearly best to suspend judgment."[13] In the Talmud itself, the demand for a high standard of evidence in criminal cases developed into a prohibition of any uncertainty in evidence:

> Witnesses in capital charges were brought in and warned: perhaps what you say is based only on conjecture, or hearsay, or is evidence from the mouth of another witness, or even from the mouth of an untrustworthy person: perhaps you are unaware that ultimately we shall scrutinize your evidence by cross-examination and inquiry? Know then that capital cases are not like monetary cases. In civil suits, one can make restitution in money, and thereby make his atonement; but in capital cases one is held responsible for his blood and the blood of his descendants till the end of the world . . . whoever destroys a single soul of Israel, scripture imputes to him as though he had destroyed a whole world . . . Our Rabbis taught: What is meant by "based only on conjecture"?—He [the judge] says to them: Perhaps you saw him running after his fellow into a ruin, you pursued him, and found him sword in hand with blood dripping from it, whilst the murdered man was writhing. If this is what you saw, you saw nothing.[14]

This is the earliest instance of a requirement similar to the rule in English law that in criminal cases proof must be "beyond reasonable doubt."

An unusual feature of Jewish law was that it always refused to admit confessions as evidence in criminal law. It is not clear why. Perhaps a confession was not admitted on the grounds that there is something wrong with anyone who would willingly bring evil on himself, and hence he cannot be believed.[15] Or perhaps it was not admitted simply because, being his own relative, he is disqualified in his own case. A happy result of this prohibition is that Jewish legal procedure was always free of torture.

It was understood that various lower standards of evidence were applicable in civil cases, in which it is not a question of determining wrongdoing but of deciding between the claims of two parties. The two-witness rule is not then generally applicable. Gamaliel the Elder, the teacher of St. Paul, laid it down that a woman is allowed to remarry on the testimony of a single witness that her husband is dead. This case soon became surrounded by elaborate rules, with evidence being admitted from those whose evidence was normally inadmissible, such as slaves, women, and Gentiles.[16] One witness was also allowed in some cases involving ritual prohibition, if there was no prima facie reason to doubt the evidence.[17]

There is also in the Talmud a good deal of reasoning from presumption (*hazakah*), that is, something that is taken as true unless there is reason to think otherwise. For example, it is presumed that what was true beforehand remains true unless there is evidence of change. "If the daughter of an ordinary Israelite is married to a priest and her husband goes abroad, she goes on eating the *terumah* dues on the presumption that he is still alive."[18] Another kind of presumption arises from "what happens mostly" (a kind of reasoning that we see also in Aristotle, Cicero, and the Greek medical writers). Decisions can be based on such well-known generalizations as "One does not ordinarily pay a debt before term" (though exceptions are recognized as common enough), "Most women's pregnancies last nine months," "Most children will turn out to be fertile."[19] Such presumptions may conflict, in which case a distinction needs to be drawn to see which is applicable.[20] Certain presumptions are taken as equivalent to certainty, such as "A scholar will not let go out of his hand something that is not ritually prepared."[21] There are also a few cases in which there is in effect equal probability on opposing sides, because of symmetry in the evidence (though the text does not actually speak about anything like probability): "A house collapsed on a man and his wife. The husband's heirs say, 'The woman died first and the husband afterward'; the woman's heirs say, 'The husband died first and the wife afterward.' . . . Bar Koppara [2d century A.D.] taught: 'Since these are heirs and those are heirs, they divide [the inheritance].'"[22]

Roman Law: Proof and Presumptions

Roman law too was rather explicit about its rules of evidence, though, as in other areas, in a rather haphazard way.[23] The concept of "onus of proof," though not itself probabilistic, was basic to further developments. The fundamental rule in Roman law, and since, was that "proof is incumbent on the party who affirms a fact, not on him who denies it."[24] This was developed into the rule that is the ancestor of modern "proof beyond reasonable doubt" laws. Ulpian wrote that "no-one should be convicted on suspicion; for the Divine Trajan stated in a Rescript to Assiduus Severus: 'It is better to permit the crime of a guilty person to go unpunished than to condemn one who is innocent.'"[25] The charge must be "supported by suitable witnesses, or established by the most open documents, or brought to proof by signs (*indiciis*) that are indubitable and clearer than light."[26] From there, the Roman lawyers were prompted by various particular cases to lay down a number of rules as to what evidence was to count and what kinds of evidence were better than others. The tendency of Roman law was always to keep very close to the cases rather than to generalize; this causes some of the rules to be, to our eyes, overspecialized ("Whether a hermaphrodite can witness a will depends on his sexual development").[27] But there are a few rules of thumb concerning the weight to be attached to evidence: desirable characteristics in a witness include rank, honorable life, wealth ("lest he swear falsely for the purpose of gain"), and lack of friendship or enmity with the accused.[28] Certain classes of witnesses were not to be believed: slaves were not competent in cases against their masters, while those convicted of writing libelous poems were not competent in any case whatever.[29]

Nevertheless such remarks were rare in Roman law. The Romans had enough confidence in their judges—and one might say enough contempt for the rabble—to feel little need to restrict judges' evaluation of evidence by rules. A rescript of Hadrian says that "it is impossible to define strictly the amount and mode of proof needed on each issue. The truth can often but not always be found without recourse to public records. Sometimes the number of witnesses, sometimes their dignity and authority, at others common knowledge settles the truth of the matter in issue. In short, all I can reply to you is that a judicial inquiry should not be tied at once to a single mode of proof. You must judge from your own conviction what you believe and what you find not proved."[30] An emphasis on the inner conviction of the judge is characteristic of such later absolutist regimes as those of Robespierre and Stalin.[31]

Another passage argues against anything like arithmetical rules that might hamper the evaluation of evidence by judges: "If the witnesses are all of the same honest reputation and the circumstances and the inclination of the judge agrees with them, their evidence should be followed. But if some, though fewer, disagree, that evidence should be accepted which fits the circumstances and is not tainted by suspicion of favor or enmity. The judge confirms his personal view from the arguments and evidence that seem more appropriate and closer to the truth. What is decisive is not numbers, but sincere and reliable testimony which illuminates the truth."[32] But in the later empire there was some tendency to move away from free evaluation and to constrain the judge by rules on various evidential matters.[33] The later emperors adopted a two-witness rule: one witness was not enough, even one of senatorial rank, while certain cases required three witnesses.[34]

Torture, especially of slaves, was an integral part of Roman criminal procedure. The attitude of the Roman mind is summed up in Cicero's chilling phrase, *Facit etiam necessitas fidem* (Necessity produces faith).[35] The use of torture was restricted by various rules that, in effect, allowed it only when there was a sufficient weight of other evidence; another rescript of Hadrian says that "slaves are to be subjected to torture only when the accused is suspected, and proof is so far obtained by other evidence that the confession of the slaves alone seems to be lacking."[36] The connection between torture and the sufficiency of evidence is especially clear in the requirement that judges investigating public crimes should not begin with torture but should use whatever "likely and probable arguments" (*argumentis verisimilibus probabilibusque*) were already available.[37] It was recognized that confessions under torture were not always reliable, and again it was left to the judge to decide in particular cases.[38]

As in Jewish law, it was sometimes found necessary to weaken the rules in cases of a sexual and marital nature, where evidence was harder to come by. For example, torture of slaves to obtain evidence against their masters could be allowed. Given the Roman political system, it is not surprising to find that anyone could be tortured in a case of high treason, "if circumstances demand it."[39]

The concept of the probable (earlier expressed by Cicero in the synonyms *probabilis* and *verisimilis* in the rhetorical contexts treated in chapter 5) does active work in the *Digest*. Indeed, it is the first place in which there is an easy use of the day-to-day concept of the probable. Many of the occurrences of these words involve making a reasonable estimate of someone's intention, for example, the intention of someone who has made a will and is now dead: "It is very likely (*verisimilius*) that the testator had intended rather to point out to his heirs where they could readily obtain forty aurei . . . than to have inserted a condition in a trust"; "It is not likely (*verisimile*) that she intended the benefit to be enjoyed by the substitutes sooner than it could have been by her son"; and the like.[40] That these words really do mean "probable" is confirmed by the fact that such arguments can be overcome by proofs to the contrary: "It is likely (*verisimile*), however, that in this instance also the party who gave the dowry had a view to his own interest; for he who made the gift on account of the marriage can, if the marriage is not performed, bring an action for recovery as if on the ground of want of consideration, unless the woman should be able to show by the most evident proofs that he did this rather for her benefit than for his own advantage."[41] There is one instance in which the uncertainty is about an event rather than an intention. Considering a case where it is known only that a woman and child died in a shipwreck, the *Digest* argues that, "since it was more likely that the infant died before its mother, it was decided that the husband could keep a part of the dowry."[42]

An explicit link appeared between probability and presumptions. Sometimes presumptions were, as in Jewish law and the modern presumption of innocence, methods of reaching decisions under uncertainty, irrespective of whether the decision was probable or not. The Carbonian Edict declared that if there was a doubt as to whether a child was in fact the issue of someone who had left money in a will, the presumption was in favor of the child.[43] That is, the question was settled for moral reasons, without considering how probable the outcome was. But at other times, a presumption clearly does mean what is probable: "It is not likely (*verisimile*) that a person would pay in a city, under compulsion and unjustly, something which he did not owe, if he showed that he was of illustrious rank; since he could invoke the public law, and apply to someone in authority who would forbid his being treated with violence. The most open proofs of violence must be given to oppose a presumption of this kind."[44] Here and elsewhere the essential idea of a presumption is seen to consist in what is now called default reasoning: A presumption is what is to be taken as true, unless and until there are reasons to the contrary.[45] That there can be grades of presumption is never explicitly stated but is suggested by the mention at one point of a "not light" presumption.[46]

It must be emphasized that Roman law is very untheoretical compared to modern law. While probability and presumptions are used in reasoning about particular cases, there is no account of what probability or presumptions are or any general rules about them. Only in the later empire, in the time of Justinian, does a little more generality appear with the collection of rules of law. Legal historians emphasize that even these were not meant as generally as they appear to a modern eye. The single one relevant to probability is, "In obscure things it is usual to consider what is more likely (*verisimilius*) or what happens for the most part (*quod plerumque fieri solet*)."[47] This is a quotation from the jurist Paulus, of about 220 A.D. Its isolation as a rule by Justinian gives it a certain generality, but it appears that the "obscure things" are actually only obscure meanings of words, so that the rule means something like, "When words are ambiguous, their ordinary meanings should be adopted."

Indian Law

The Talmud and the *Digest* are rational law codes that developed virtually independently but simultaneously. Their similarities, beneath obvious surface differences, are remarkable. Even more remarkable is that the phenomenon occurred not twice but three times. The Sanskrit Laws of Manu (of perhaps the first century B.C.) and the more developed *Code of Narada* (of perhaps the fifth century A.D.) show no signs of Western influence, except for the single word *dinara* (*denarius*) in *Narada*.[48] Yet it has been rightly remarked that "a [Roman] jurisconsult would immediately feel at home in running over these briefly formulated definitions and precepts [in the *Narada*]."[49] The similarity in method with Roman law is clear. Internal laws of development of law codes generally could explain the common outcome but hardly the simultaneity.

The older *Laws of Manu* include a three-witness rule for certain cases involving debt,[50] and for resolving conflict of testimony they lay down the rule: "On a conflict of witnesses the king shall accept the majority; if equal in number, those distinguished by good qualities; on a difference between distinguished [witnesses] the best among the twice-born."[51] But much of this code's section on evidence deals with the problem of false testimony; penalties are severe, offenders to be "firmly bound by Varuna's fetters, helpless during one hundred existences."[52] An equally ancient code says, "The king shall not convict on suspicion,"[53] but there is not enough context to make very clear what this means.

The *Code of Narada* begins with matters of procedure especially concerning evidence. The author seems to regard the evaluation of evidence as the most difficult part of law; the reason is the argument from illusion: "Liars may have the appearance of veracious men, and veracious men may resemble liars. There are many different characters. Therefore it is necessary to examine [everything]. The firmament has the appearance of a flat surface, and the fire-fly looks like fire. Yet there is no surface to the sky, nor fire in the firefly."[54] Proof is divided into human (documents and witnesses, the former being preferred) and divine (ordeals, used only when human proof is lacking). As in Roman law, the onus of proof is on the plaintiff, with exceptions in special circumstances.[55] When the authenticity of documents is in question, it may be established "by examining the handwriting, the tenor of the document, peculiar marks, circumstantial evidence, and the probabilities of the case." A standard commentary on this text says that investigating probabilities involves asking such questions as, "How has he got hold of this document?" or, "Is he nervous?"[56] There should generally be three unimpeachable witnesses to guilt, but various weakenings of this rule are permitted if there are other indications of crime, if the crime is grave, or if the parties agree to a single witness.[57] In case of a balance in the number of witnesses on either side, those not liable to suspicion are preferred; if there is still a balance, those with a superior memory.[58] The list of disqualified witnesses is based on similar principles to that in the *Digest*, while the complicated rules that apply when a husband is missing abroad are close to those in the Talmud.[59]

The ancient law codes all provide examples of how to evaluate evidence in cases of doubt and conflict. By and large, they do so reasonably. But they are almost entirely devoid of discussion of the principles on which they are operating.

CHAPTER TWO

The Medieval Law of Evidence: Suspicion, Half-proof, and Inquisition

MEDIEVAL LAW IS BUILT on ancient law. Continental medieval law actually claims to be the same as ancient (Roman) law, with a few minor changes. It is not. The difference is that the medievals and their successors are always explaining principles, comparing texts, arguing about the conflict of one text or principle with another. In the course of doing so, they are forever drawing distinctions. Among them are distinctions among the grades of proof.

The medieval lawyers were the first to consider explicitly the grading of the degrees of proof, with some discussion of how fine it should be; the combination of different pieces of evidence for the same conclusion; and the conflict of different pieces of evidence bearing on the same conclusion. The resulting theory is a coherent one. It is not numerical, and there is no reason to think that it would have been improved if it had been numerical. On the contrary, since modern (English) law has a similar theory, and insists on keeping it nonnumerical,[1] there is every reason to believe the medievals were correct in avoiding numbers.

The historical importance of the medieval legal theory of evidence is that it is the foundation of what "probability" meant at the time of the scientific revolution. As we shall see, when a seventeenth-century writer evaluates scientific evidence, or discusses conflicting claims in religion, he turns to the language of the law of evidence. Consequently, this is the central chapter in the book. The material is generally unfamiliar, and in view of its importance, a certain amount of context and detail are needed in order to demonstrate exactly how the medievals' reflections on Roman law resulted in a method of discovering truths about probability.

Dark Age Ordeals

The story of anything intellectual in the centuries between the time of Justinian and the Gregorian Reform amounts to occasional survivals, at the most minimal level, of what was done in antiquity. Nevertheless, the common picture of the monks blindly copying ancient manuscripts they did not understand is something of a caricature of the way ancient thought survived. Among advanced concepts that survived in actual use, legal ones were among the most successful. Much of the surviving documentary record of the period from 500 to 1000 is in the form of legal documents of one form or another; the survival of Roman concepts is substantial, and the use of them is far from blind.[2]

The legal systems of the time are combinations of late imperial provincial law with the practice of the Germanic tribes. The most famous feature of them is the rituals of proof by ordeals (mostly by hot iron or water, sometimes by lot), trial by battle, and oaths (of both parties and "compurgators" such as their kin). The legal system of the tribal world was not as irrational as it seems at first. The more picturesquely irrational methods seem to have been largely reserved for doubtful cases, as proofs of last resort, when the common knowledge of the neighborhood failed to produce an agreed outcome.[3] A Carolingian instruction on procedure of 816 says that "if a party suspects the witnesses brought against him, he can put forward other, better, witnesses against them. But if the two groups of witnesses cannot agree, let one man be chosen from each group to fight it out with shields and spears. Whoever loses is a perjurer and must lose his right hand."[4] Here it appears that the battle is reserved for breaking the deadlock, when evidence is more or less in balance. There is certainly a case for saying that if evidence really is balanced, any method of breaking the deadlock is as rational as any other. And while one is no doubt right to be shocked by the combination of irrationality and violence in proof by battle and ordeal, that very combination seems to have acted in many cases to concentrate the minds of the prospective participants on weighing the evidence in order to avoid the pain of proceeding to the unpredictable outcome of the final step. The records also show a pragmatic and not unreasonable combining of evidence from documents and witnesses with oaths and ordeals.[5]

During periods of relative calm some collections of laws were promulgated that reveal at least a modest acquaintance with the concepts of the Roman law of evidence. In these the words *probabilis* and *verisimilis* are occasionally used of witnesses and their testimony.[6] The Catholic Church, in particular, generally preferred proof by witnesses to the tribal methods.[7] Church decrees also preserved the rudiments of the ancient concepts, speaking here and there of "likely suspicions," "likely indications," and a "probable defense."[8]

The only novel thing concerning probability is the following remarkable text, which appears in the *False Decretals*, an influential mixture of old papal letters, quotations taken out of context, and outright forgeries put together somewhere in Western Europe about 850. The passage itself may be much older. "A bishop should not be condemned except with seventy-two witnesses . . . a cardinal priest should not be condemned except with forty-four witnesses, a cardinal deacon of the city of Rome without thirty-six witnesses, a subdeacon, acolyte, exorcist, lector, or doorkeeper except with seven witnesses."[9] It is the world's first quantitative theory of probability. Which shows why being quantitative about probability is not necessarily a good thing.

The Gregorian Revolution

The periods immediately following the Dark Ages are known as the Gregorian Reform and the twelfth-century "Renaissance." These weak terms entirely fail to express the magnitude of what happened.

To explain why probability developed in the contexts it did, it is necessary to sketch the general historical background a little more fully than usual, since it is somewhat unfamiliar outside the circle of professional historians. If developments in probability suddenly appear in fifth-century B.C. Athens or seventeenth-century France, no one is surprised, since the general outline of events in those times and places is well known, and it is generally understood why those were periods of rapid intellectual change. It may be less expected that twelfth-century France and Italy should be a hotbed of probability theories, since modern perceptions of that epoch have been distorted by propaganda connected with the myth of the Renaissance, which requires the Renaissance to contrast with an earlier period of darkness and slow change. A series of books in recent decades has convinced the majority of medieval historians that the period 1050–1200 was the real turning point in Western history that created modern Europe.[10] The general public, including historians in other fields, has kept to its belief in the gradualness of historical change and the possibility of finding renaissances anywhere[11] and has resolutely refused to change its large-scale picture of history.

In barest summary, the story is this. Camped in the ruins, civilization survived but "by the skin of our teeth."[12] The small advances made in the time of Charlemagne and his successors had petered out, and the Europe that feared the passing of the year 1000 showed little sign that it was incubating events of great moment. Then in the late eleventh century there was an outburst of energy and confidence seen before only in Periclean Athens. The suddenness of the change must be emphasized: "It all happened in a single lifetime."[13] Change was everywhere. Economically, it "affected all the graphs of social activity."[14] By 1100, cities, merchants, and cloth manufacturing were suddenly in rapid growth. French-speaking armies conquered England in 1066, Sicily in 1091, Jerusalem in 1099. The story is clear from the material remains still visible. The few remaining buildings of the Dark Ages are almost all room-sized piles of rubble, but Durham Cathedral (begun 1093) is one of the world's great buildings, and many other Romanesque cathedrals are still admired. As enduring as these structures in stone are ways of organizing people and ways of organizing thought. The Domesday Book of 1086 is a symbol both of the use of writing in administration and of the manorial system of rural life, which provided the economic base for the expansion of northwest Europe. Different but equally successful were the merchant-controlled republics of Italy. The central individual of the eleventh century was the monk Hildebrand, later Pope Gregory VII; the Catholic Church as a centralized bureaucracy of educated celibate clergy, independent of local political control, is largely his creation. The university, still the leading means of organizing intellectual life, descends from the University of Bologna, which grew up in the late eleventh century.

For the story of this book, it is important that the Gregorian revolution was led on the intellectual front by logic and law, the fields in which probability had been developed in antiquity. These two subjects achieved a relative importance that they have never had before or since.

The Glossators Invent Half-proof

Law was the means by which all the other advances were put securely into place. The very names of Europe's sacred books, the Old and New Testaments, are reminders of how full they are of laws, covenants, judgments, witnesses, and trials. The Jewish heritage thus paved the way for the sudden impact of the other law-soaked ancient civilization, the Roman. Though all the major ancient thinkers—Aristotle, Ptolemy, Euclid, Plato—had been rediscovered and translated by 1200, the first great find was Justinian's *Digest*, discovered in an Italian library about 1070, possibly by agents of Gregory VII looking for texts to support papal claims.[15]

It would be too much to attribute the dramatic change in mentality to a single chance event, for the first readers of the rediscovered *Digest* already possessed ways of thinking that were (just) sufficient for them to make sense of the text. The essential idea that one applies reasoning to texts to understand them (in drawing deductions from one's own and one's opponents' views of the text, in trying to reconcile apparent contradictions, in considering emendations of passages that seem against reason) had been developed by the school of commentators on Lombard law in Pavia by about 1050. The laws they discuss are primitive, but the issues they take up and their reasoning are not:

"If the bearer wished to validate a charter that had been challenged, and the notary and all witnesses are dead, how ought it to be done?" Bonifiglio answered him: "By custom the bearer of the charter should validate it with twelve compurgators and with two other charters." . . . Lanfranc: "Then custom is against law; for it is this custom that the prologue to the legislation of Otto I has in mind where it says that 'a detestable and dishonest custom has grown up in Italy.'" . . . Guglielmo: . . . "Otto said a dishonest and detestable custom has grown up not in respect to the aforesaid custom but respecting this, that certain greedy men were drawing up false charters of alienation and defending the charters by perjury, thus acquiring the goods of others. Thus Otto gave the challenger of the charter the choice of battle or letting the bearer swear."[16]

On such matters of evidence, the Lombard commentators already show the characteristic medieval tendency to generalize much more than the Roman lawyers and to inquire into abstract principles. How can a law legislate against those who plot against the king's life, when only God can know anyone's thought? "[The question] is solved in this way: it is known through indications, for example if someone is discovered in the king's chambers after hours having a naked sword under his cloak, or with a knife in his sleeve, or if the cupbearer of the king while near him is seen to prepare poison."[17] The theme of proof by indications, later called circumstantial evidence, has been one of the most vexed themes in the law of evidence.[18] It is not essentially a legal question but one involving judgments of probability from facts to facts.

With that preparation, the *Digest* was immediately recognized as a classic, and the first university, that of Bologna, was set up solely to study it. There were said to be ten thousand students at Bologna by 1200.[19] The world of the Glossators, as the first generations of commentators were called, was circumscribed yet immensely rich. It was based essentially on a single text, but that text contained so much and was so completely known in all its intricacies and interrelations that an almost complete worldview could arise from the interpretations that the text admitted. What the Glossators did has a unique importance for modern thought, in that the tradition of thinking they began is still alive. Whereas understanding the *Digest* itself is like interpreting fossil remains, reading the Glossators is not unlike reading a modern legal judgment. The reason is simply that modern law, both English and Continental, descends from the Glossators in an unbroken tradition.

The Glossators were at once faced with a problem. Their initially reverent and literal approach to the text had to coexist with a number of inconsistencies in the text. Of this dilemma was born the Scholastic method characteristic of the whole of medieval thought. A topic is discussed by collecting authoritative texts and arguments that are or seem to be contradictory and then resolving the apparent contradiction by looking for suitable general principles and distinctions. It is the principles, vocabulary, maxims, and concepts of great generality produced by this method that constitute the most lasting contribution of medieval thinking to later thought.[20] The method survives intact, of course, in law, but much of the argument in all the social sciences still proceeds in effectively the same way. The method in due course led naturally to considerations of logical probability in weighing the arguments on both sides.

Since the *Corpus* almost wholly lacks any explanation of general principles, a certain amount of creative interpretation was necessary. The originality of a medieval juristic proposal may often be judged by the irrelevance of the texts cited to support it. This greatly incensed the Renaissance humanists, who complained at length about the medieval jurists' barbaric accretions on the original Roman law ("nothing but filth and villainy," says Rabelais),[21] but from the point of view of developing concepts, as opposed to preserving the purity of a more primitive era, one's sympathies may well lie with the Glossators.

The Glossators all show a taste for very general maxims, and many of them relate to the strength of evidence and to presumptions. As in Roman law, the level of proof needed in criminal cases was very high: "In sum it is to be noted that in accusations there is need for proofs clearer than light, whether by indubitable documents, or by more than twenty adult witnesses, as per *Code* 4.19.25."[22] The earliest of the Glossators, before 1150, perceived that Roman law occasionally mentions pieces of evidence that do not constitute full proof but that are not worthless either. These pieces of evidence are either a single witness or private papers, of which Roman law says, respectively: "If someone says that he does not recognize his seal, that is not, indeed, a reason for not opening the will, but in other respects it becomes suspect"; and "private papers, that is to say, instruments executed in the presence of witnesses, or notes, if they are not supported by other testimony, are not alone sufficient as evidence."[23] The early Glossators place these together in a single category and say they are valid as presumptions and may be used when there is more than one.[24] Presumption is said to be "not properly proof but standing in place of proof" and "standing until the contrary be proved."[25] Some examples of presumptions are from Azo of Bologna, writing about 1205: "Ignorance is presumed, unless knowledge is proved"; "Such a presumption transfers the onus of proof onto the defendant"; and "The law presumes her a slave, who has been a slave, unless by some supervening circumstance the contrary is proved."[26]

The last example illustrates reasoning by presumption from the past to the present, in which legal reasoning comes close to induction.[27] Of course such reasoning needed a few restrictions in order to avoid unwanted results: "It would seem unnecessary to prove he was a minor at the time of the contract, because it is certain that he was a minor at one time, therefore he should be presumed to be still a minor."[28] The problem is solved with the obvious distinction. A presumption particularly rich in practical consequences, especially in an age in which documentary evidence of sale was only just becoming the norm, is "He who is in possession is regarded as owner."[29] Azo saw the need for a classification of presumptions, arising because certain of the presumptions in Roman law were unrebuttable. Azo calls them *presumptiones iuris et de iure*. The more normal presumptions, which do admit evidence to the contrary, came to be called *presumptiones iuris* and were distinguished from "mere" presumptions "of man or of fact," or those not specifically mentioned in the law but drawn by the judge at his discretion.[30]

Further reflection on the kinds of evidence short of certainty led to a word that expressed the most significant and original idea of the Glossators for probabilistic argument: half-proof (*semiplena probatio*). In the 1190s, this word was invented for the class of items of evidence that were neither null nor full proof.[31] The word expresses the natural thought that, if two witnesses are in theory full proof, then one witness must be half. Azo writes: "It would seem this does not hold, because either the plaintiff proves, or not. If he proves, the defendant should be condemned. If not, he is acquitted, according to the rule. . . . I reply that although according to Aristotle it would seem to be an exhaustive division, it is not so according to the laws. There is a medium, namely, half-proof. Say therefore that in such a case proof has been less than full. Then the judge gives the right to complete the case by oath to the plaintiff. Otherwise if he has not proved at all, the defendant is acquitted, and an oath is not allowed, except in cases where the law gives the defendant the right to clear himself by oath. . . . The right to an oath is therefore given to the plaintiff if he half proves, either by private documents, or by the flight of the defendant . . . or by one witness. . . . Full proof is by two witnesses, therefore half proof by one."[32]

Another kind of half-proof admitted by Azo, on the basis of an obscure passage in the *Code*, is the comparison of handwriting in different documents.[33] More dangerous, and contrary to a general reluctance to admit hearsay evidence, was the admission of reputation, or common knowledge (*fama*) as a half-proof, on the basis merely of a phrase in the *Digest* to the effect that "reputation when in agreement confirms" the truth of a matter.[34]

The passage of Azo just quoted explains what to do when there is a single half-proof. But the point is normally to have two and to add them: "Two half-proofs make a full one—what the individual ones are insufficient for can be done with the help of many." Half-proofs of different kinds may be added: one may add a single witness to private documents to determine the validity of a contract. The same idea is expressed also in the language of presumptions: "Presumptions do not suffice, but they perfect, as one witness with a just suspicion as well . . . one presumption can be overcome by another presumption." But one may not subtract: "proof overcomes presumption."[35]

The important matter of the addition of different pieces of uncertain evidence is handled also by one of the later Glossators, William of Drogheda, in his *Summa aurea* of about 1239. It is an interpretation of a passage from the Roman law. The original merely says that "if anyone . . . should produce a will that has not been canceled or annulled and is not defective in any respect but appears in its original form without alteration and is fortified by the attestation of the legal number of witnesses, he shall be placed in possession."[36] William notices the abstract fact that there are here different pieces of evidence relevant to the same conclusion. His comment is that "it was safer . . . and better, in proving a matter of meaning and intention, that it should be so proved with more and thus with better. . . . That is firmer, that is linked with more chains."[37]

If there are half-proofs, can there be quarter-proofs and eighth-proofs? Voltaire says there were certain "Visigoths" in the Toulouse legal profession who used them.[38] But even if he means to be taken literally, there appear to be no such things in medieval law. The accusations that the "formal" Continental system of proofs had a "fairly elaborate tariff of gravity" or "strict mechanical rules" should be taken with a grain of salt.[39] The system later had two powerful enemies, the philosophers of the Enlightenment and the English common lawyers, both with an interest in caricaturing it. The medieval treatises nowhere give the impression of asking judges to depart from normal rational standards in the interest of keeping to rigid rules.

Nevertheless, it is clear that the system had not as yet graded proof finely enough; it was not satisfactory to count everything between no proof and full proof as half-proof. The problem of what to say about minor pieces of evidence that did not reach half-proof was left to the next century to solve.

Presumptions in Canon Law

The period of the Gregorian reform was characterized by its variety of legal systems. Besides the rediscovered Roman law, the church, kings, emperors, cities, and merchant guilds all developed systems that existed alongside tribal and feudal custom.[40] Under pressure from one another, and from the increasing complexity of life, all of them developed conceptually, but the law of the church was the one that commanded the greatest share of the available intellectual resources, such as people who could read and write.

The law of the church could not be allowed to lag behind the law of the emperors. Canon law had first to be provided with an authoritative text, and in the early twelfth century much work was done in collecting the decrees of the popes and councils. The project culminated in the collection known as the *Decretum* of Gratian, produced in Bologna about 1140. It orders the papal decrees according to subject matter, with short comments by Gratian, especially in cases in which the contradictions between neighboring decrees are particularly flagrant. His method is again to explain away problems by bringing to bear general principles. The section on evidence begins with the biblical two-witness rule but immediately qualifies it. As in Roman law, witnesses must be "suitable." There follows the rule from the *False Decretals* quoted above about the need for seventy-two witnesses to convict a bishop. Gratian in effect says it is nonsense, being contrary to the two-witness rule. He insists that two witnesses of obviously good life are sufficient to condemn anyone; it is just that the "improbity" of some makes their testimony worthless.[41] The rule continued to be taken seriously by those likely to benefit from it. Aquinas quotes it with a straight face and remarks simply that those who are appointed bishops should have such sanctity as to make them more credible than many witnesses.[42]

The crucial grading of presumptions according to strength appears in the earliest commentaries on Gratian. A husband could legally separate from his wife after accusing her of adultery, either manifest or suspected. "But it is to be known that suspicion from a medium presumption is different from that from a violent one. A medium one is if, say, he sees his wife frequently exchanging lewd nods with a lustful young man; a violent one is if, say, he finds them in bed together at night, even if he does not see them coupling, and the like. From the former presumption, she cannot be sent away; from the latter, proved by suitable witnesses, she can be sent away before the church."[43] There is a Roman law text on which this is based, but it only distinguishes between knowing an adultery certainly and knowing it merely on suspicion.[44] The importance of this text is the statement in the abstract that there are degrees of rational belief. The division is close to the Glossators' divisions of proofs. It is not unlike some remarks of Quintilian (see chapter 5), but the relevant book of Quintilian was apparently unknown in the Middle Ages.[45] Its source seems instead to be the usual creative interpretations of scattered remarks in Roman law. The phrase "a light presumption" does occur in the *Digest*, as do "a light suspicion" and "a rash indication."[46]

It is especially unclear where the grade of "violent presumption" came from. The only passage alleged in its support that has any relevance merely says that "if a deed is canceled . . . the debtor is presumed discharged."[47] Another suggestion links the grading of proofs or suspicions to a brief remark by the Emperor Hadrian in the *Digest*, where it simply says that "slaves are to be subjected to torture only when the accused is suspected and proof is so far obtained by other evidence that the confession of the slaves alone seems to be lacking."[48] The grades of suspicion arise by separating the clauses of this sentence, according to a later writer on canon law: "Proof has its grades. . . . This is proved by [the *Digest* title] On Tortures l.1 there: 'when the accused is suspected,' that is the first grade; 'by other evidence,' that is the second, 'that the confession of the slaves alone seems to be lacking,' that is the third."[49] These ingenious jurists found a theory of grades of proof in the *Digest*, but plainly they had a good idea of what they wanted to find before they went looking for it.

The next challenge was to find a theory of probable proof in a text even more authoritative than the *Digest*. There was only one such text, the Bible. It was no easy task; the books of the Bible on law deal with an almost tribal society, while the later books are about quite other things. Still the challenge was met. The single text that proved useful was the famous story of the judgment of Solomon on two women who both claimed to be the mother of a certain child. Solomon said, "Divide the child in two, and give half to one, and half to the other." The first woman agreed while the second asked that the living child be given to the other woman. The second was of course judged the real mother. John of Salisbury in 1159 reports lawyers using this as an example of a probable argument, and in the next century it entered canon law as a support for the proposition, "From a violent presumption a definitive [sentence] can arise."[50]

The threefold division of presumptions into rash, probable, and violent was repeated by all standard legal authors for centuries.[51] Aquinas, for example, divides suspicion into light, probable, and violent, emphasizing that these qualities are properties of the evidence rather than subjective matters of someone's opinion.[52]

Speculations and fine conceptual distinctions by academic lawyers are one thing, implementation that affects practice is another. The man most responsible for making the intellectual apparatus of canon law part of everyday practice is Innocent III, pope from 1198 to 1216 and the most powerful pope in history. His rule marks a dividing line in medieval history. Medievalists divide into two camps according to their attitudes to the times before and after this date. Thomists and others see the age before 1200 as essentially a prelude to the building of great structures, Aquinas's *Summa theologiae*, the high Gothic cathedrals, and the *Divine Comedy*. The opposite view is that a bright new dawn was hijacked by careerist bureaucrats, lawyers, dry Scholastics, and inquisitors. These views agree at least that the world became more organized, and writings even of the time reflect a recognition that everything from then on was controlled by and for "the system," that is, the Papal Curia.[53]

Innocent had at least some training in canon law, at Bologna, and his decrees on canon law show a special concern for questions in which there are doubts about evidence. In the first year of his reign, he excepted himself from the two-witness rule: the pope's testimony counts as two.[54] He formalized the principle that witnesses are to be preferred to documents and required higher standards of evidence for the authentication of relics and for canonizations.[55] Finding the modes of proof in Roman law insufficient, he adopted a new one: if a crime is "notorious" in a district it may be regarded as fully proved, without the usual formalities of witnesses or confessions.[56] The crime giving rise to this novelty was clerical concubinage, a major target of the Gregorian reform; it tended to scandal when "well known" but was not easily proved in court.[57]

But the most serious difficulties with evidence tended to arise in marital questions, especially in the very common cases of clandestine marriages and questions of consanguinity. The church, though of course demanding that marriages be celebrated publicly, accepted that a couple's consent and their living together was sufficient to produce a valid marriage without a ceremony. In particular, the church authorities were unwilling to allow annulment and remarriage on the grounds that the first marriage had not been celebrated publicly. The strict rules forbidding marriage within a number of degrees of consanguinity produced prob-

lems in villages in which many people were related but no one could remember exactly how. It was feared that couples "discovered" common
ancestors when they wished to divorce. The marital problems of kings
were equally complicated, leading in Innocent's reign to war between
France and England and an interdict on France.

In this context a papal decree of 1209 laid down what became the
classic statement of the grading of presumptions: "We believe it should
be distinguished whether the spouse knows for certain the impediment
to the marriage, and then he may not engage in carnal intercourse without mortal sin, even though he could not prove it [the impediment] before the church; or whether he does not know the impediment for certain but only believes it. . . . In the second case, we distinguish, whether
his conscience is thus from a light and rash belief or a probable and discreet one (*ex credulitate levi et temeraria, an probabili et discreta*). . . . When
his conscience presses his mind with a probable and discreet belief, but
not an evident and manifest one, he may render the marriage debt but
ought not to demand it."[58] Equally difficult problems of evidence were
posed by clandestine marriages and their proofs, to which Gratian de-
votes a section.[59] By Innocent's time it was declared that the testimony
of seven examining women was to be accepted as to the virginity of a
wife, the husband's oath notwithstanding.[60] Given the extreme difficulty
of proof of carnal knowledge, the judge was to be careful to confirm
even a violent presumption of it, and records show the ecclesiastical
courts applying very high standards of proof in such cases.[61]

Innocent was also much exercised by the spread of heresy. His first
line of response was to dispatch the crusade that wiped out the Albi-
gensians in southern France, but almost equally important was his decision to increase the number of trials, or inquisitions, for heresy. He
recognized that heresy was a particularly difficult crime to collect reliable evidence for; his instructions of 1206 say, "Inasmuch as we do not
wish anyone condemned for such a grave crime solely on suspicion,
however vehement that may be, we order that to achieve such security
you should, besides requiring his oath, coerce him by fear of corporal
punishment, while enjoining him to real penance; from which it will
surely appear whether he walk in darkness or in the light, whether he be
truly penitent or falsely converted."[62] The alarming feature of this passage, to have grave consequences later, is the nexus between the admirable
desire not to condemn when there is any chance of innocence and the
recommendation of coercion to supply any defects in the evidence.

The reader will suspect that I am not choosing examples at random
but am acting like filmmakers who view the Middle Ages as a setting in

which one may be particularly free with examples of sex and violence.
That is not true.

Grades of Evidence and Torture

"The fear of corporal punishment" brings us again to the unpleasant
matter of torture and confessions, which were part of the major reforms
in secular as well as ecclesiastical law of the time. Innocent III had a hand
in these matters too. The Roman law studied at Bologna was not at first
in force anywhere. In southern Europe, a much simplified system of
Roman law had partly survived, but in most of Europe the law up to
about 1200 remained largely the customary law of the Germanic tribes.
The methods of proof by ordeal did not survive the rationalist mood of
the twelfth century. Popular belief waned as ordeal by battle was sometimes replaced by contests of hired champions,[63] a procedure whose
faults are especially evident, though the concept is not entirely unknown
to modern law. Civil and canon lawyers, trained in Roman legal thought,
were scandalized by the whole barbaric business—and perhaps exaggerated the pagan and popular aspects of procedures in which the
learned had, after all, colluded.[64]

The propaganda of Peter the Chanter around 1200 was particularly
effective in eroding the last stages of belief, using some well-chosen an-
ecdotes of executions for murder after trial by ordeal, followed by the
supposed victim's return from pilgrimage.[65] He expresses the new mood
using the language of reason and probability: "No one ought to tempt
God when he has rational courses of action. . . . If the miracles the Lord
promised in the Gospel . . . are not guaranteed (*nec sunt in necessitate*)
how can these ordeal miracles be guaranteed to happen or have their result? . . . In doubtful cases sentence should be withheld, especially in capital cases, where there should be no proceeding merely conjecturally or
probably but only judging with reasons transparent and clearer than
light."[66] When Innocent III called the Fourth Lateran Council of 1215,
one of its main decrees forbade ecclesiastics participating in ordeals.
Without divine sanction, the ordeals convinced no one, and they soon
disappeared. Compurgation is not strictly irrational, provided the sanctity of oaths retains its meaning. Though it was retained in some places
as a useful second line of defense after some rationalization,[67] it was
clearly unsuited as a main method of deciding criminal cases.

With one law of evidence on its way out, a replacement was
needed.[68] The most natural option would have been to adopt the system found in the *Digest* of more or less free evaluation of evidence by
the judge. If this had happened, there would have been little develop-

ment in law of rules of probability. That it did not reflects the fact that there was no longer a Roman Empire. The emperor and his judges had the power and prestige to act as they thought fit; there was appeal to the emperor but no appeal against the system. But King John and the barons were not Pilate and the Sanhedrin. By 1200, no authority was absolute. People clamoring about their ancient rights and privileges had to be taken seriously. Competing courts—ecclesiastical and royal, urban, feudal, and merchant—had to look to their "image," since there was often the possibility of moving a case from one to the other. Evaluation of evidence had to be seen to be fair.

There were two choices available, and the different choice made by English and Continental law remains the most prominent difference between the two to the present day. The English option stemmed from the fact that Henry II instituted a working system of evidence evaluation. This was the jury, in the form later known as the grand jury, or jury of presentment. Its original purpose was to decide whether the evidence was strong enough to proceed to the ordeal. The evidence was to be evaluated more or less freely, not by the judge but by a panel of men to whom the accused could not reasonably object, namely his peers ("who would understand"). Glanvill's treatise on English law, written about 1190, explains that men may thus preserve their rights while avoiding the doubtful outcome of the duel.[69]

It appears from Glanvill that there was a mixed system of evaluation of evidence in criminal cases, with juries inquiring into crimes in their locality and referring to the ordeal only those cases in which the evidence against the accused was strong. The language used in explaining the procedure is that of the Glossators: "[In treason cases] if no specific accuser appears but the accusation is based only on public notoriety (*fama*) . . . then the truth of the matter will be investigated by many and varied inquests and interrogations before the justices and arrived at by considering the probable facts and possible conjectures both for and against the accused (*ex verisimilibus rerum indiciis et coniecturis nunc pro eo nunc contra eum*), who must as a result be either absolved entirely or made to purge himself by the ordeal. . . . [In cases of fraudulently concealing a treasure trove] no one shall be purged by ordeal unless it has previously been proved against, or admitted by, him in court that metal of some other kind was found and recovered from the place in question. If this much has been proved against the accused, then the presumption being against him, he is bound to purge himself by the ordeal."[70]

Cases of the time show that juries did normally refer for the ordeal only those cases in which the accused was suspected, with the evidence mentioned being in fact reasonably substantial; cases in which the accused successfully passed the ordeal afterward seem to have been common, even normal, thus relieving the jury of total responsibility in doubtful cases.[71] When the ordeals were forbidden, England thus had a working system of evaluating evidence capable of replacing them. By 1220 a system of juries (petty juries, distinct from grand juries) was enforced for final verdicts in cases in which guilt was neither manifest nor only lightly suspected.[72] Since the evaluation of the evidence was still effectively free, however, there was little call for rules concerning it, so English law has not contributed as much to the discussion of probability as has Continental law. Another effect was that English law never accepted torture to obtain evidence.

The solution adopted on the Continent was that the judge evaluated the evidence but was bound by strict rules that were publicly known and seen to be fair. The two-witness rule was a beginning. Actually reducing the evaluation of evidence completely to rules was not contemplated, however; the possibility was denied in the Roman law texts, and it has not proved possible since. The upshot was that Continental criminal procedure came to rely more and more on the one mode of proof that could be accepted as certain, the "queen of proofs": confession. Confessions, though, do not just happen—and if they do, they are probably false. They usually need to be made to happen. The desire for certainty, the wish not to condemn anyone "solely on account of suspicion, however vehement," as Innocent put it, pushed the legal system toward a policy of trying and trying again until a confession was achieved. Torture was used more and more as time went on.[73]

Torture was foreign to European custom at the time. So its use was confined by strict rules, at least in the beginning. The original idea, from Roman law, was that torture should be used only when the evidence of guilt was already strong (the Glossators insisted that the *Digest* allowed torture only when there was so much suspicion that only confession was lacking)[74] and that the judge should then confirm the facts confessed to (for example, if the accused confessed to burying the murder weapon under a certain tree, the judge should send someone to dig it up).[75] This is by no means entirely unreasonable. As compared with the jury system, it can indeed work to prevent condemnation purely on suspicion. It has been remarked that the English jury could convict on less evidence than the Glossators required for torture,[76] and it can hardly be doubted that the jury system avoided torture at a high cost in unjust sentences of death. The most serious problems with the European system arose for those suspected of particularly heinous crimes that left no traces—like rape, treason, heresy, and witchcraft.

In the meantime, the problem was to produce some system that could publicly—that is, by fixed rules—evaluate the intermediate grade of suspicion on which a suspect could be tortured though not yet proven guilty.[77] For this purpose, the Glossators' theory of evidence was elaborated into a theory of considerable complexity. From the conceptual point of view, there is very little in the later theory that is not already in Azo's comments on half-proof, but as a practical tool it considers a great many more cases in detail.

As Aquinas's *Summa* dominated theology, so law was dominated for centuries by the great collections and commentaries of the thirteenth and early fourteenth centuries. The major canon law collection, the *Decretals of Gregory IX*, or *Liber Extra*, of 1234, includes a section on presumptions, similar to the treatment in the Glossators, and examples, such as "A stronger proof is required from one who wishes to prove what is not likely (*verisimile*)."[78] By the time of Guillaume Durand's *Speculum judiciale* of about 1275, all the concepts have been tabulated, defined, and classified into kinds. Half-proof, sufficient for summary procedures, is divided into the same six kinds given by the Glossators. But doubts arose, as well they might, as to whether reputation really amounted to half-proof, and it was sometimes regarded instead as merely a light (minor) proof or as a support (or quasi-presumption).[79] Durand defines presumption, one of the twelve varieties of proof, as "an argument to the belief in one fact, arising from the proof of another: for example, from the cohabitation of suspect persons is presumed coitus." He divides it into four grades: rash, probable, violent, and necessary.[80]

Hostiensis, the most respected commentator on the *Decretals*, discusses the conflict of different numbers of witnesses, especially of an equal number of equally credible witnesses. In such a case, an object in dispute may be divided, "except that perhaps if a woman is claimed as wife by two who prove equally, the woman is given the choice, if she consents to it." In cases in which witnesses differ in credibility, "authority is to be preferred to numbers. . . . For what if there are a hundred sinners, notorious bawds, and perjurers on one side and on the other side two bishops or honest religious?"[81]

With a hierarchy of grades of proof in place, it made sense conceptually to vary the effects of the different grades. The laws of Italian city-states did not always coincide with Roman law in this respect. Bologna did forbid torture when there were only indications of guilt, allowing it only in case of violent presumptions. On the other hand, some city-states granted their magistrates full and free discretion. According to Thomas of Piperata, a Bolognese jurist of the 1270s, this meant that they could order torture on a single indication and could convict when there were undoubted indications, short of a confession or two witnesses. The standard seems equivalent to the modern "proof beyond a reasonable doubt." The examples given are of a man seen coming out of a room with only one exit, pale, and carrying a bloody sword; immediately afterward, a man slain with a sword is found in the room. (This example plainly derives from the Talmud, though how it got from there into Western law is unknown.) Or again, "Titius is killed in a vineyard, and Seius is accused of the crime. No witnesses testify that they saw Seius kill Titius, but there is proof of the following: Seius was Titius's enemy; Seius once threatened Titius with a sword; Seius fled from the vicinity of the crime; and there is general belief and public outcry that Seius killed Titius." Thomas holds that no two of these pieces of evidence constitute sufficient proof of guilt but that all four together do.[82]

These developments passed by English law almost completely. Bracton, the chief thirteenth-century writer on English law, does briefly deal with presumption from half-proof, holding that trial by duel or jury may be dispensed with in case of "violent presumption . . . as when he is arrested over the body of the dead man with the knife dripping blood."[83] But English law did not incorporate the concepts at the time. Only after 1600 did English lawyers find it desirable to use these concepts. Bracton was then found convenient as an English authority to refer to, allowing the pretense to be maintained of an English law safe from Continental influences.[84] A moderately clear standard of suspicion or sufficient cause for arrest is visible in Bracton and some medieval cases, and the Statute of Westminster (1276) provides for bail if the accused has been arrested for a serious crime on light suspicion (*leger suspecio*).[85] Even if one avoids talking about proof by leaving the jury as a black box evaluator of guilt, there is still the problem of explaining the level of probability of guilt needed for pretrial actions like arrest and indictment.[86]

The Postglossators Bartolus and Baldus: The Completed Theory

By the fourteenth century, law was well established—far too well established, according to some. It had less need to look to its own prestige. There was much quoting of a maxim of Innocent III, "It is in the public interest that crimes not go unpunished," to justify whittling away of the rights of defendants.[87] The Roman law theorists of this time, the Postglossators, were thus able to adopt a somewhat looser interpretation of the ancient texts and adapt them more freely to the extra complexities of the day, such as the realities of international commerce. There was less need for judges to be confined by artificial rules of evidence. They write, for example, that it is the function of the judge, not the witness, to draw conclusions from testimony; thus the witness does not testify that Peter is the owner of a garment but that he saw Peter buying the wool out of which he then made the garment.[88] In drawing conclusions, the judge must be unfettered by presumptions and other rules; he must be free to weigh up the evidence, especially that of contradictory statements, using the appearance of witnesses in court.[89] The requirement of proof by the testimony of unimpeachable witnesses, or documents, or the genuine confession of the accused could only be weakened in the most difficult cases. For example, if a crime is notorious, this could be proved by sufficient witnesses. In cases of adultery, for which direct evidence is difficult to obtain, the judge may admit such probable presumptions as hearsay, but these have only corroborating value, and the judge need not adhere to them. Less-than-unimpeachable witnesses may need to be admitted concerning crimes in brothels.[90]

The mental operations the judge performs are detailed by Bartolus of Sassoferato, the most distinguished of the Postglossators (c. 1350). The judge must arrive at conviction (*credulitas*) that facts have been proved, a condition between knowledge and ignorance. To begin with, the judge is in doubt as to the facts alleged by the prosecutor: "Doubt is when he does not apply his mind more to one side than to the other." After some evidence the judge may have *opinio* or *suspicio*: "The judge's mind begins to incline to one side, though doubt remains." He is to take the facts as proved only if the argument no longer permits of any other reasonable conclusion. Thus if the accused was seen fleeing with drawn sword from the house in which the victim was found dead, that is not enough for a conviction of murder, since other conclusions can reasonably be drawn from the evidence.[91] When suspicions are supported by "likely and probable" arguments, torture is called for. Though there are no rules for evaluating signs, there are certain restrictions. The incriminating circumstance should itself be proved by the testimony of two witnesses, except that one completely reliable witness to guilt itself is sufficient for torture. Thus the statement of one witness that he saw the defendant at night near the house that was broken into would be insufficient to justify torture, but two witnesses to the same effect would.[92] Since the point of torture is not punishment but the finding of the truth,

one tortures first those who will divulge the truth most readily. Therefore "a woman is tortured first, rather than a man, because a man is of
greater constancy and so will confess more slowly. The woman will confess faster, because she has a movable and unstable heart."[93] A judge is
not liable for the death or permanent injury of the tortured, since he is
presumed to have acted out of zeal for justice.[94]

Bartolus recognizes that reaching half-proof is a matter of balancing
positive and negative evidence; if both are present, the positive must
overbalance the negative.[95] The Postglossators are very clear that there
must be a finer grading of evidence between the extremes of no proof
and full proof, instead of calling all such things half-proof. Sarcasms
abound in their works on the idea that private documents constitute
half-proof, since that would allow anyone to fabricate half a proof for
himself. The flight of the defendant might be perfectly reasonable and,
although giving rise to some suspicion, might not amount to a half-
proof against him. Reputation was especially regarded as unreliable and
far from constituting a half-proof. What category, then, should be set
up for such pieces of evidence that lie between worthless suspicion and
half-proof? A word of rather obscure meaning from Roman law, *indicium* (indication) names this category; as well as the items of evidence
just mentioned, it includes such things as threats made before a crime
and the extrajudicial confession of the accused.[96]

The most developed form of the theory is found in the late fourteenth-century commentaries of Baldus de Ubaldis. He is the most
philosophical of the medieval legal writers; his knowledge of Aristotle
was considerable, and he sees law on the model of an Aristotelian science, with theorems derivable from the abstract notion of justice.[97] His
especially abstract point of view suited the ideas of a later age, and on
evidence, as on many topics, he is the authority most quoted by textbook
writers of the sixteenth century. Since his work is at once the most the-
oretically developed—indeed, unsurpassed—and entirely unknown in
modern discussions of probability, it is worth quoting at some length.
The fact that it is still easy to understand is a tribute to the close connection between medieval law and modern everyday concepts of reasoning.

> Before knowledge is arrived at, a man goes through many grades and
> media, and little by little, part by part, comes to knowledge of the truth.
> Thus I define the following: first, what is suspicion? I answer that sus-
> picion is an application of the mind to something with vehement hesi-
> tation. That suspicion is a certain motion of the mind toward something

with vehement hesitation is proved by *Digest* 26.2.17[?] and other laws that mention suspicion—though the definition does not apply to the case of the suspect tutor or other administrator, because that is not simple suspicion but suspicion mixed with conjectures, presumptions, and other likelihoods, on which grounds he may be removed from office (*Digest* 26.10.7). I ask, What is presumption? I say that presumption of man is a certain concept caused in the mind by some probable conjecture (*Digest* 45.1.137). Presumption of law is of two kinds: one is simple presumption, which is thus defined: A simple presumption of law is a certain likeness sufficient, in a doubtful matter, to make the believable believed. This is proved by *Code* 8.37(38).1 and *Digest* 2.14.7 etcetera. Presumption *iuris et de iure* is thus defined: It is a conjecture without doubt, established by law; as proved by *Digest* 27.7.4. A fiction is thus defined: It is a falsehood accepted as true, for a special and just reason expressed in law, as is clear from *Digest* 4.2.23, and everything on fictions. Indication is twofold, either half-full and doubtful or full and undoubted. A half-full indication is a presumption strongly moving the mind to believe or disbelieve something, as proved by *Digest* 22.3.7. A full indication is a demonstration of a thing through sufficient signs, in which the mind rests as in something fully apparent, feeling no need for further investigation, as per *Code* 3.32.19 and the like. Argument is thus defined: It is a proposition drawn from certain facts, tending to show or conclude to a result. Hence when several indications or propositions, or several presumptions, or several witnesses are joined together in proof of some conclusion, the combination is called argumentation or argument, which is the collecting of several things toward one conclusion, as per *Digest* 22.5.3 and 48.3.6. A support is twofold, either vehement or not vehement. A vehement support is thus defined: It is a supplement to a defect in proof, which does not have effect alone but only when working with another proof, as proved by *Decretals* 3.27.3. A nonvehement support is some confirmation of a probable thing, or some confirmation of truth tending to overcome a defect in proof; such supports do not have effect, even when there are several of them, as per *Digest* 27.1.2.1. Conjecture is thus defined: It is an accepting or regarding of something as true, from some other thing that it is likely is designed to show it, as from a signboard we conjecture a shop, or from style of dress a prostitute, as per *Digest* 47.10.15.[98]

On indications Baldus writes more fully: "Sometimes indications are light and by a great space of reason distant from necessary belief in truth, such as flight alone, being seen talking together, and the like. In those cases indications are less than half-proof, as per *Digest* 22.3.7. But what is the theoretical basis of this distinction? I answer by the theory of the gloss on this law. For if one thing occurs frequently and easily without another, then there is a light indication, as per *Code* 4.19.12 and 6.2.2. But if they frequently and easily occur together, then there is a weighty indication, which is equal to a presumption of law, as per Cynus on *Code* 2.19(20).9. This applies whether it is a question of proving a human act or of proving intention, as per *Digest* 48.8.1 and *Decretals* 5.12.16."[99] On the grades of presumptions his divisions and examples are much the same as those of other writers, but he urges care on their addition: "Note that two imperfect kinds of proof, and two presumptions of different sorts, and one witness to truth, and two to reputation do not make full proof in a criminal case; for there is only the single witness to truth."[100] He mentions a kind of presumption called circumferent, which proves only when collected with others, apparently the origin of modern "circumstantial evidence."[101]

The concepts thus set out are used to explain the difference between the standards of proof in civil and criminal cases: "In criminal cases proof ought to be superior in three ways. First, in admissibility, because the witnesses should be questioned with greater subtlety, as per *Novels* 90.5 and *Digest* 22.5.4. Then in demonstrative quality, because more weighty and mature indications are needed in criminal than in civil cases; and the reason is that fuller proof is required where greater sentence is involved, as said in *Code* 4.1.3. Then in the probative quality of witnesses, because more suitable witnesses, called in the laws unexceptionable, are required in criminal cases but not in civil ones, as per *Digest* 22.5.4, *Code* 4.20.6. So that is the difference: for the extra proof is required by reason of the greater matter, so the proofs should reach certainty, so that there can be no persuasion to the contrary, or at least no probable persuasion (*Code* 9.47.16)."[102]

Having made all these distinctions, Baldus allows indications of less than half-proof to suffice for torture and vehement indications to suffice for condemnation.[103] But he does consistently refuse to allow the adding together of extremely weak indications for these purposes. In this context he invents what seems to be the smallest nonzero quantum of proof considered by the medieval lawyers, the "indicationlet" (*indiciolum*). If in a murder case a witness testifies that the brother of the accused was an enemy of the deceased, "an indication is certainly not thereby proved, although it makes an *indiciolum* that is very small and of no strength." A number of such things cannot be added together to produce evidence sufficient for torture. Baldus mentions having himself seen executed on weak evidence people whose innocence afterward came to light.[104] On the addition of presumptions, Baldus approves the collecting of those of different kinds; there results "an abnormal proof, just as from a horse and an ass there results an abnormal kind of animal that does not breed, that is, a mule, but that is very effective at work." He also countenances a kind of subtraction of half-proofs, allowing that three witnesses might defeat a document.[105]

An idea of how the rules worked in practice can be gathered from the advice Baldus gives in the following case. An important deed of contract, about eighteen years old, is alleged to be forged. It bears the signatures of four witnesses, of whom two have died and two deny all knowledge of it. Should the notary who drew up the deed be condemned? It might seem that he should, as there are two witnesses against him. But this is not so, according to Baldus, because the two witnesses are not really testifying to the same fact, as they cannot give evidence that the other failed to sign the deed. Nor is there even enough evidence to justify torturing the notary, since the half-proof against him seems to be reduced by indications to the contrary, such as the fact that the deed has been in the continuous ownership of one party for the whole eighteen years without any question having been raised about its authenticity. The judge might even consider torturing the two witnesses, especially if they are of low degree and suspect and the notary is a man of authority. While much will depend on what the judge thinks, in general two witnesses do not defeat a public document.[106] Notaries are members of the legal profession, which explains the bias in their favor. One understands why Wat Tyler, Baldus's contemporary who led the Peasants' Revolt, greatly desired to kill all lawyers.[107]

Canon law too exhibited a slowing of development. Without a leading figure like Baldus, it reached a fullness without originality, describable according to taste as either maturity or ossification. By the fourteenth century, the theory of evidence in canon law had become static. There are at most a few new examples: "When someone has studied negligently in canon law, it is presumed that he will study negligently in civil law, and conversely."[108] But if the main body of canon law underwent little growth, there was one small shoot showing extreme vigor.

The Inquisition

By the time of the Postglossators, there was a new organization with a special professional interest in evidence, its acquisition and evaluation.

The medieval church regarded heresy in much the same way that a modern state regards terrorism, with the added fear that the success of heresy would destroy not only civil society but also immortal souls. As with terrorism, it proved particularly difficult to obtain evidence, since heretics lied on each others' behalf, were protected by sympathetic populations, and merged into other heretical groups "by whatever names they are called, having many faces but intertwined in their tails."[109] The famous inquisitor, Bernard Gui, explains in his *Manual* (c. 1323) why the inquisitor cannot afford to be squeamish about evidential difficulties: "It is, indeed, all too difficult to bring heretics to reveal themselves when, instead of frankly avowing their error they conceal it, or when there is not sure and sufficient testimony against them. Under these circumstances difficulties rise on all sides for the investigator. On the one hand his conscience will torment him if he punishes without having obtained a confession or conviction of heresy; on the other hand, all that repeated experience has taught him of the falseness, guile and malice of such people will cause him still greater anguish. If they escape punishment owing to their fox-like craftiness it is to the great harm of the faith, for they become even stronger, more numerous and more wily than before. Moreover, lay persons devoted to the faith find it scandalous that an inquisitorial case, once begun, should be abandoned more or less for lack of method."[110]

Consequently, the inquisitors were granted certain concessions in their evaluation of evidence. Unfortunately for heretics, this idea found support in Roman law, where it was said that the category includes those found by a "light argument" to deviate from the true faith.[111] Medieval lawyers did not wish to go to that extreme but made a number of carefully defined adjustments. The most radical change was that the Inquisition was allowed to take evidence in secret and so confront the accused with charges without revealing who had made them.[112] The aim was to prevent intimidation of witnesses in areas in which heresy was rife. As a check on abuses, the inquisitors were required to discover whether the accused had any mortal enemies likely to give false evidence against him. In fact, the *Manual for Inquisitors* (1376) of Nicolau Eymerich, inquisitor general of Aragon and Catalonia, suggests not relying fully on witnesses at all: even the evidence of two witnesses does not necessarily justify a conviction, and the inquisitor, if he remains dubious, should check such evidence by torture.[113]

No doubt the inquisitors were genuinely concerned to avoid false convictions, but of course the general trend was to relax the rules to make conviction easier. Witnesses not normally competent—such as his wife and family, his accomplices, perjurers, excommunicated persons and other heretics—could give evidence against the accused but not for him.[114] The inquisitor's most extreme fury was reserved for relapsed heretics, since the principal aim of the whole process was not so much conviction of offenders as their repentance. If repentance was genuine, even leaders of heretical sects could be treated leniently and uses found for them (for example, as inquisitors; who better?).[115] But relapse meant that repentance had not been genuine and was punished with the utmost ferocity. To ensure that heresy was really destroyed, a remarkable weakening of the rules of evidence was countenanced in this case, which was not allowed in any other legal context: "One accused or suspected of heresy against whom there has arisen great and vehement suspicion of this crime and who has abjured heresy in court, if he afterward falls into it should be regarded by a legal fiction as relapsed, even though before his abjuration of heresy the crime was not fully proved against him. But if there was only a light and moderate suspicion against him, he may be more severely punished for that reason but should not be punished as a relapsed heretic."[116]

Here the theory of grades of evidence receives an application; without it, the rule could not even have been formulated. The developed state of the theory of proof allowed great precision in stating exactly how far it was that concessions would extend. The language of probable signs and vehement suspicions appears throughout the writings of the inquisitors.[117]

Again, though the heretic could not actually be convicted on suspicion, if suspicion did remain after all inquiries had been completed, the suspect could not expect to simply go free. If the suspicion is light, Gui says, a routine of standing daily outside the inquisitor's house might be sufficient: "When an accused is strongly suspect and in all likelihood and probability guilty, and when the inquisitor is thoroughly convinced thereof, in such a case, when the person is obdurate in his testimony and persists in his denials, as I have observed time and time again, he should not be released for any reason whatever but should be held for a number of years in order that his trials may open his mind. Many have I seen who, thus subjected for a number of years to this regime of vexations and confinement, end by admitting not only recent but even long-standing and old crimes, going back thirty and forty years or more."[118]

Naturally, the Inquisition was concerned to allow torture on lower levels of evidence than was usual. Eymerich gives rules that allow torture if the witness vacillates or has against him one witness and a bad public reputation but do not allow torture if there is only a bad reputation, a single witness, or a single indication. The latter is a real restriction, but zealous inquisitors had wide discretion in deciding which indications were "grave." "It belongs to the inquisitor to decide on the value of this kind of evidence from witnesses when, disagreeing in some details, they agree essentially."[119] (One of the legal subtleties produced by the two-witness rule was the problem of witnesses who agreed essentially but disagreed on details. This matter had been treated extensively in Jewish law, and in Christian thinking the rule "Diversity of time and place is not taken to vitiate testimony" was applied to the Gospel accounts of the Crucifixion, one of which placed that event at the third hour and the other at the sixth.)[120]

The working effect of these rules of evidence when under pressure can be seen in the trials of the Templars of 1308–14. All the knights of the Templar order in France were suddenly arrested on royal orders in 1307 and charged with heresy, sodomy, urinating on the cross, worshipping idols, and a variety of other astonishing crimes. Severe torture produced many confessions. The masters of theology of the Sorbonne, consulted by the king, reasoned that the confessions so far obtained created a vehement suspicion of like offenses, or at least of concealing knowledge of like offenses, against remaining members of the order, so that there were grounds for an inquisition against the whole order. The pope, at first admitting that "it did not seem likely or credible that men so religious" should do such things, was eventually convinced by the number of witnesses.[121] At one point certain members of the order put to their accusers that there were "good presumptions" on their behalf, "against which proofs to the contrary ought not to be received," since holy men joined the order to save their souls, not lose them, and so would have revealed any shameful practices.[122] The defense had no effect. An incident from the trial of the Templars in Paris recalls Bartolus on the need for judges to inquire into the source of witnesses' knowledge. A knight testifies to the common report that initiation into the order involved the ceremony of kissing the anus. "Asked what he means by common report, he replies: what is publicly discussed in different places by different persons. Asked if he knows the origin of the said report, he replies no, but it was said by good and serious men."[123] The mass burnings revealed a presumably unintended consequence of the rules of evidence. Confessions under torture had to be repeated later "freely"; and heretics were executed only if they relapsed. These two rules, individually, were obviously designed to make the last stage of punishment harder to reach. But together, as appeared in the case of the Templars and later in that of Joan of Arc, the rules implied that someone who retracted a confession made under torture could be regarded technically as a relapsed heretic, and burned.

The Inquisition's ability to produce confessions of whatever it wanted was legendary. Even at the time it was said that if St. Peter and St. Paul had appeared before the inquisitors, they would have been found to be heretics.[124] The results were seen again in the leper scare of 1321, in which Bernard Gui was involved. The French king, finding that "public knowledge and the course of experience" had shown that lepers had plotted to poison the wells, ordered an extensive investigation and the harshest penalties. A widespread plot involving lepers, Jews, and Moslems was uncovered, and the usual burnings followed.[125] The same happened again when confessions of clandestine poisonings were extracted from German Jews at the time of the Black Death, though secular more than inquisitorial law was involved.[126]

Law in the East

Before going on to later developments in European law, this is a convenient place to survey briefly some developments in other legal systems. Maimonides, the leading medieval Jewish thinker, lived mostly in Egypt around 1200, wrote in Arabic and Hebrew, and presumably had never heard of any contemporary writer in Latin, but his ideas on legal evidence, based on the Talmud and Aristotle, are recognizably close to those of the Glossators. Commenting on the Talmud's assertion that the discovery of someone standing with bloody sword over a just-dead body is insufficient evidence for conviction, Maimonides argues that allowing conviction on such a strong presumption would have led to a gradual weakening of standards, since there are continuous grades of evidence. "For among contingent things some are very likely, other possibilities are very remote, and yet others are intermediate. The 'possible' is very wide. Had the Torah permitted punishment to be carried out when the possibility is very likely—such that it is almost a necessity, as in the example we have given—some might inflict punishment when the chances are more distant than that, and then when they are even further still, until they would punish and execute people unjustly on slight presumptions according to the judge's imagination."[127]

Maimonides' discussion of the laws of marriage contains much about the conflict of testimony. The cases are complicated rather than realistic, such as this one on symmetry in evidence: "If one woman A, and two men, B and C, arrive from another country, and B says, 'A is my wife and C my slave,' while C says, 'A is my wife and B my slave,' whereas A says, 'Both B and C are my slaves,' A is permitted [in marriage] to any man.

For though both B and C claim that she is married, yet inasmuch as each one of them has testified to his own advantage, neither one is to be believed."[128]

One might ask how much difference it would have made if European law had developed without the apparatus of theory described above. From the victims' point of view, not much, if a comparison with Chinese law is any indication. Generally, Chinese law, though developed independently of Roman law and its derivatives, used witnesses and torture in a similar way but discussed them in much less precise terminology. Han law, contemporary with the early Roman Empire, produced "clear" proof by using witnesses and torture, and there was some recognition that torture could produce false confessions.[129] The Confucian approach to disputes emphasized mediation to restore harmony rather than adjudication of rights.[130] Hence there was nothing like a presumption of innocence, anyone who appeared in court being regarded as already a disturber of the harmony of the state and so liable to arbitrary torture.[131] Mongol Chinese law, contemporary with the Postglossators, says the following in connection with cases in which the evidence is unclear (it is as far as Chinese law went in grading evidence): "If a case is reasonably suspicious and evidence and other indications are clear and yet [the offender] nevertheless conceals and does not confess, then [the official] shall establish with other participating officials [of the joint conference] a case to confer together and, in accordance with law, impose torture for eliciting a confession. If neither an accusation is clear nor evidence is reliable [the official] shall first use reason to analyze and surmise and shall not impose abruptly any torture."[132]

Chinese law thereafter changed little until the impact of Western forms of thought in the nineteenth century. Confession was regarded as almost always necessary, and torture was ordered when guilt was already certain and clear but the accused refused to confess. Early Portuguese visitors to China found torture routinely used on suspects against whom there was the least evidence and on witnesses who disagreed with one another. There was no space in Chinese law for a legal profession, and hence for any formal science of law, and so for a forum for discussion of legal questions like the strength of evidence.[133]

Russia meanwhile retained the system that the West had before the twelfth-century reforms, with oaths and ordeals normal practice for the final decision of doubtful cases. The Russian Church preferred witnesses and documents but did not succeed in converting secular practice to its own habits, as happened in the West.[134]

Islamic law made an attempt to achieve certainty in verdicts through a system of fixed procedural rules. A plaintiff, whether in a civil or a criminal case, needed to prove through the eyewitness testimony of two witnesses (adult, male, Muslim, and of unimpeachable character). There might be witnesses to the character of other witnesses. Documents and circumstantial evidence were not admitted, in principle, though there were minor exceptions (fifty confirmatory oaths might suffice to confirm a strong suspicion, and some jurists thought that a smell of alcohol on the breath might establish the crime of drinking alcohol). Confessions were permitted, but torture to obtain them was not. When the evidence was weaker, the defendant was acquitted on swearing an oath to his innocence; the judge had no discretion to evaluate the evidence. The inevitable result of the demand for certainty in religious courts was the evolution of a system of administrative courts unconstrained by these rules and permitting circumstantial evidence, conviction on the basis of character and previous offenses, and the extortion of confessions.[135]

CHAPTER THREE

Renaissance Law

W hat is interesting about proof in Renaissance law is not so much developments in theory, of which there are almost none, as the insight that the extensive records of real cases provide into people's familiarity with concepts of evidence. In case after case, the medieval language of probability and presumptions is used by many, not just lawyers, to discuss the worth of evidence.

Henry VIII Presumed Wed

On November 14, 1501, Arthur, Prince of Wales, married Catherine of Aragon, daughter of Ferdinand and Isabella of Spain. The happy event promised a long period of harmony between the emerging Atlantic powers, the one recently recovered from civil war, the other now ethnically cleansed and beginning its conquest of the New World. The fifteen-year-old couple were often seen together in bed, royal domestic arrangements not being conducive to privacy, and the marriage was (or according to a competing later theory, was not) consummated. Five months later, Arthur was dead. In view of the importance of the alliance, Catherine was married to Arthur's younger brother Henry, the pope having purported to grant (respectively, validly granted) a dispensation from the Old Testament prohibition against marrying one's brother's wife. In 1527, Henry VIII, moved by a troubled conscience (respectively, by lust for his latest concubine, Anne Boleyn), began to doubt the validity of his marriage to Catherine and instituted a series of canon law proceedings to seek an annulment. Much depended on whether Catherine's earlier marriage to Arthur had been consummated.

The paradigm of a violent presumption in law was the inference from being seen naked together in bed to intercourse; that was exactly how things stood in the case in question. But Catherine consistently maintained on oath that the marriage had never been consummated, and available evidence to the contrary was certainly flimsy. The imperial ambassador, a doctor of civil and canon law, secretly organized witnesses both to Catherine's character, in support of her oath, and to Arthur's impotency.[1] Whether proof of the alleged consummation was actually necessary was canvassed widely among the most learned theological and legal faculties of Europe, with mixed results. The University of Salamanca replied that while "daily marital cohabitation" did produce a violent presumption of consummation, proofs to the contrary, such as the queen's oath, could certainly be heard.[2] On the other relevant legal issues, Oxford and Cambridge Universities were persuaded, under pressure, to declare that it was "more probable, valid, true, and certain" that it was against divine law to marry one's brother's wife—but only provided consummation had taken place with the brother. Cranmer proved simultaneously that proof of consummation was unnecessary for annulment and that Catherine could be "violently presumed" to have lost her virginity through at least one of the technical possibilities.[3]

Tudor Treason Trials

For English courts, the most detailed accounts of sixteenth-century trials come from the state trials. Sir Thomas More and Bishop John Fisher were each condemned on the evidence of a single witness; in More's case, the witness, Rich, was by reputation an inveterate liar. More asked the jury, "Can it therefore seem likely unto your honorable lordships, that I would, in so weighty a cause, so inadvisedly overshoot myself as to trust Master Rich, a man of me always reputed for one of so little truth . . . that I would unto him utter the secrets of my conscience touching the King's Supremacy? . . . A thing which I never did, nor never would, after the statute thereof made, reveal either to the King's highness himself or to any of his honorable counselors. . . . Can this, in your judgments, my lords, seem likely to be true?"[4]

The paranoia about treason became institutionalized in Elizabeth's reign and was reflected in its exceptional legal position.[5] The "black poison of suspect" spread through all circles of society with any connection to the court. The secret police uncovered plot after plot, and the niceties of the law of evidence were not allowed to stand in the way of repression. Lines in the first of the great Elizabethan dramas, Kyd's *Spanish Tragedie*, give a sense of the atmosphere of suspicion in England, besides giving evidence that the language of presumptions could be understood even by a theater audience: "For I suspect, and the presumptions great. . . . We are betraide."[6]

It is remarkable that everyone involved in the state trials seems able to use the language of proofs and presumptions to discuss the state of the evidence. The orders issued by the monarch's council for torture regularly assert the existence of "vehement presumptions" against the suspect.[7] The Jesuit Edmund Campion, conducting his own defense at his trial for treason in 1581, argued—correctly but needless to say fruit-lessly—that there was only "a naked presumption (who seeth it not) and nothing vehement nor of force against me," and that "the wisdom and providence of the laws of England, as I take it, is such as proceedeth not to the trial of any man for life or death by shifts of probabilities or con-jectural surmises, without proof of the crime by sufficient evidence and substantial witnesses . . . be the theft but of an halfpenny, witnesses are produced, so that probabilities, aggravations, invectives, are not the bal-ance wherein justice must be weighed, but witnesses, oaths, &c. All that is yet laid against us, be but bare circumstances, and no sufficient argu-ments to prove us Traitors, in so much that we think ourselves very hardly dealt with, that for want of proof we must answer to circum-stances."[8]

In his final speech to the jury, Campion's oratory went further: "The speech and discourse of this whole day consisteth, first, in Presumptions and Probabilities; secondly, in matters of Religion; lastly, in Oaths and Testimonies of Witnesses. The weak and forceless Proof that proceedeth from conjectures are neither worthy to carry the Verdict of so many, nor sufficient evidence for trial of man's life. The constitutions of the realm exact a necessity, and will that no man should totter upon the hazard of likelihoods; and albeit the strongest reasons of our accusers have been but in bare and naked Probabilities, yet are they no matters for you to rely upon who ought only to regard what is apparent. Set circumstances aside, set presumptions apart, set that reason for your rule which is war-ranted by certainty."[9]

Modern lawyers will recognize here a "golden thread" speech ("Throughout the web of the English Criminal Law one golden thread is always to be seen, that it is the duty of the prosecution to prove the prisoner's guilt. . . . If, at the end of and on the whole of the case, there is a reasonable doubt . . . the prisoner is entitled to an acquittal")[10]—that is, a general purpose appeal to the burden of proof that the defendant uses as a last resort. Campion had been a professor of rhetoric and phi-losophy before his mission to England and so was presumably more fa-miliar with such concepts than the average defendant. (Speeches of the same general tenor were known in ancient rhetoric, as described in chapter 5.) But he clearly expects to be understood, and others involved in the trial use some of the same language. The jury found him guilty "directly and by the most sufficient and manifest evidence," and a gov-ernment tract aimed at calming public disquiet over the torture of Cam-pion and others claimed that no one had been put to the rack "but where it was first knowen and evidently probable by former detections, con-

fessions and otherwise, that the partie so racked or tortured, was guylty
. . . and the racke was never used to wring out confessions at adventure
upon uncertainties."[11]

Mary Queen of Scots was convicted of plotting against Elizabeth's
life on the evidence of just two witnesses, her secretaries; she maintained
she had no knowledge of the treasonable contents of the letters they had
written to conspirators. Some comment was occasioned by the Star
Chamber's accepting these witnesses without allowing her to confront
them.[12] The petitions to Elizabeth that urged her to overcome her reluctance, or pretended reluctance, to proceed to Mary's execution are
full of what would now be called decision-theoretic considerations. The
queen's safety is at peril and cannot be allowed to depend on "the event
of the like miraculous discoveries" as that of the conspiracy[13] (which had
almost certainly been in fact overseen by government agents all along,
though Elizabeth probably did not know that). When the queen still inquired whether there were no alternatives to proceeding to Mary's execution, it was argued, in language that recalls the inquisitors, that "there
was no probable hope of anie conversion, but rather great doubt and
feare of relapse and recidivation, forasmuch as she stood obstinatlie in
the denial of matter most evidentlie prooved."[14]

At a lower rung of the legal order, Elizabethan justices of the peace,
responsible for legal and police action at the lowest level, were expected
to possess considerable legal expertise, including some understanding of
Bracton on suspicion and reputation and the distinction between a suspicion with some basis and a "bare surmise" without.[15]

Continental Law: The Treatises on Presumptions

In the three hundred years from Baldus to Leibniz, there were no conceptual developments of any importance in the legal theory of evidence.
The reason is that, if the law is to keep to a nonnumerical approach to
evidence, there is essentially no development possible. Baldus's theory
is complete and is in no important respect improved on by modern
treatments of evidence in law.

It remains, then, merely to survey briefly the theory of the law of evidence from the late Middle Ages to the time of Pascal. In discussing
Renaissance law and its impact, it should be kept in mind that the law
was then much less an esoteric specialization than it is today. Such unlikely people as Alberti and Copernicus were actually doctors of canon
law, while almost all of the founders of mathematical probability had
some legal connection: Fermat was a professional lawyer, Cardan and
Pascal were the sons of lawyers, Huygens was a doctor of civil and canon

law, and Leibniz, who (as described in the epilogue) saw the legal theory of evidence as a logic of probability, was first trained as a lawyer. (A
fuller account of the place of law in the early modern development of
ideas is given in chapter 12.)

Despite humanist attacks, the elaborate medieval developments in
Continental law remained generally intact. In terms of quantity, if not
quality, the high point of academic legal thought on probability and presumptions is reached in the three massive volumes of Mascardi's *On
Proofs*, of 1584, and the two of Menochio's *On Presumptions, Conjectures,
Signs, and Indications*, of 1587. Though the production of large Latin
tomes on evidence did not cease after that time,[16] the two were always
taken to be the definitive treatments. A hundred years later, Leibniz remarks that to learn what the lawyers know about these matters, it would
be desirable to excerpt Menochio and Mascardi, and the treatment of
presumptions in Scots law is based on Menochio.[17] The principal books
on evidence that appeared in the United States in the nineteenth century refer to them enthusiastically.[18]

Unfortunately, what quality of thought there is in these authors is all
secondhand. Their idea of creating a theoretical synthesis is to quote extensively from Baldus, then to quote less extensively from other authors
whether or not they agree with Baldus, then to stop. Menochio identi-
fies presumption of law with "likely proof, also called violent presumption by Baldus" and also with a "probable conjecture." A "presumption
of man" is "called probable by a Venetian statute" and is less than half-
proof; it is to be noted that Baldus's division of this kind into major and
minor does not create a difference in species. According to Aristotle's
Rhetoric to Alexander, the *Rhetoric to Herennius*, and Quintilian, presumption includes argument, sign, and example.[19] A vast number of
references are adduced in discussing whether and how presumption, indication, conjecture, sign, suspicion, and support differ, but the incon-
sistency of the references prevents any sensible conclusion being
reached.[20] Some theory is attempted on such questions as how to decide
which presumptions are stronger than others (for example, the more
special is stronger than the more general) and when two half-proofs can
be put together to make a whole proof (generally, in unimportant
cases).[21] The bulk of the work is taken up with a large number of particular presumptions.

Mascardi's volumes have much the same organization. Conjecture is
"a reasonable vestige of hidden truth, whence is born the opinion of the
wise." An indication, "a notable sign of some crime or other thing," is
less than half-proof, as Baldus says. "From the little garden of half-proof

we can pluck four kinds of minor and pallid flowers."[22] These are one witness, reputation, comparison of handwriting, and private documents; the four hundred years since the Glossators have not led to the discovery of any new kinds. (The attitude of these authors to new ideas is expressed in Menochio's view that innovations generally are a sign of fraud.)[23] Mascardi continues at length on the proof of particular facts. Is adultery proved by the woman's wandering around at night outside the husband's house? It is, unless done from a just cause, such as having been thrown out of the house by the husband, or unless done openly with the husband's knowledge. Adultery is also proved from a man's being found in the home of a beautiful woman, especially a young and poor one, although it must be taken into account that tastes in these matters differ. The age of rustics will have to be proved by any probable reason available, such as what age they appear to be. Gamesters are excluded from giving testimony, except, for example, when it is necessary to prove that other gamesters have been playing with false dice. "If one presumption concerns what happens for the most part, and another what happens rarely, that from rarity is defeated by that from what happens frequently."[24]

A substantial proportion of these academic subtleties spread into law as actually practiced. Roman law was "received" in Germany, replacing local customary law. The code known as the Carolina, of 1532, was intended as a summary in German of all the new criminal law, usable by officials without special legal training. Nevertheless, the law of evidence in it contains a great deal of the medieval apparatus of half-proof, indications, and presumptions. As usual, half-proof, such as one witness or an article owned by the suspect being found at the scene of the crime, is sufficient for torture. Less than half-proof are sufficient indications, such as bad reputation, previous criminal record, flight, or enmity toward the victim. One indication is insufficient for torture, but two are sufficient if the judge thinks that together they amount to half-proof.[25]

Textbooks on torture continued to discuss the level of evidence required. On the whole, they were concerned to make torture a little more easily available than the medieval lawyers had recommended. Francisco Bruni's *On Indications and Torture*, of 1493, does not agree with Baldus in forbidding torture when a number of indications have only one witness each for them: "A single witness makes some kind of presumption, so if there are many witnesses each testifying to different indications, there are said to be many presumptions, which make one indication for torture. . . . Say if one witness says he saw Seius wounded and lying on the ground, and when asked who wounded him, he replies, Titius; this is one witness testifying to an indication. Another says he saw Titius coming straight out, fleeing into a wood, and hiding. Certainly the individual witnesses testify to different indications, but no one of sound mind will deny that these presumptions make an indication sufficient for torture."[26]

The debate was a live issue in real cases. In the trial of Mari Gonsáles Panpán before the Spanish Inquisition in 1483–84 on the charge of being a relapsed Jew, the accused has confessed to some minor Jewish practices, saying she was forced into them by her husband. Several neighbors testify to her observance of Christian practices outside her house but say they have no knowledge of what has gone on inside. A number of people testify that her husband was a Jewish butcher. A former servant girl says that during the time she lived with the family they all observed the Jewish Sabbath. One neighbor claims to have seen them wearing clean clothes on Saturdays but working on Sundays. Counsel for the accused argues that there is no evidence for anything she has not already confessed to: the evidence for the more serious charges is either hearsay or given by only a single witness for each charge. The accused was found guilty and burned at the stake; on the evidence, it does seem she was probably "guilty."[27]

Grillandus's *On the Question and Torture* distinguishes between doubtful or half-full indication, full indication, reputation, rumor, four types of presumption, argument, vehement and nonvehement support, conjecture, the likely, and the notorious, before going on to detail with equal learning the five degrees of torture.[28] Damhouder's *Practice in Criminal Cases*, the standard manual of criminal procedure in the later sixteenth century, admits that reputation is not in general sufficient for torture but asserts that it may be if vehement or joined with other supports. A single indication may sometimes be sufficient for torture, notwithstanding the plural language of the laws. Too much torture does, however, produce false confessions.[29] Other manuals of practice admit that, strictly according to law, half-proof is necessary for torture and that public fame against an individual must be proved before questioning of witnesses against him—but cheerfully maintain that usage in the courts dispenses with these pettifogging restrictions.[30]

In one of the rare instances in which statistics are available, it appears that in Florence in 1425–28, when the legal system was comparatively well run and perhaps less subject to panics than later, 21 percent of convictions were obtained by confessions extracted under torture. This does not count confessions in which the threat of torture was sufficient.[31] At least for hidden crimes like conspiracy, torture was permissible even in the absence of indications. Despite a schedule of standards of proof required for conviction for various crimes, such as two witnesses of sight plus two of public fame, the discretion of the judge was in practice wide.[32]

On the other hand, rules about evidence could be used to impede political demands for convictions, if a court so desired. Aragon created difficulties for the Spanish crown by preserving various ancient rights, including strict standards of evidence in court. Military action was necessary, in the end, to suppress them.[33]

The Witch Inquisitors

So far the story of probability is one of progress. Progress with interruptions and losses, certainly, but still a movement without notable regressions. Now for something completely different.

The learned of the early Middle Ages regarded belief in witches as a vulgar superstition, a point of view expressed in the only reference to the subject in Gratian's *Decretum*. But gradually, fear of sorcery increased, the interpretation of biblical passages about witches became more literal, and the forces of legal repression became more organized.[34] The fourteenth century saw little actual attention to witches, but the increased use of torture had the capacity to produce occasional bizarre confessions. In a case in which a witch confessed to adoring the devil and killing children by spells, Bartolus advised that she be burned.[35] The fifteenth century, the century of the burning of Joan of Arc, of Bluebeard, and Dracula, of the pictures of *danses macabres* everywhere, saw a sharp increase in fears in general and in fear of witchcraft in particular.[36] The outcome can be read at firsthand in the *Malleus Maleficarum*, or *Hammer of Witches*, written by the inquisitors Kramer and Sprenger in 1487.

The book is the most successful of any written in the early years of printing. The authors are plainly disturbed, and disturbed in a way usually associated more with fin de siècle Vienna than with fifteenth-century Germany. They adduce many authorities and widely reported experiences in support of such propositions as that incubi exist, that an olive tree planted by a harlot will not bear fruit, that hardly a hamlet lacks a midwife who is a witch, and that witches can by magic make the male organ seem to disappear.[37] Who can believe, they ask, that witches do not have congress with devils, when the inquisitor of Como burned forty-one witches in the year 1485 alone, "who all publicly affirmed, as it is said, that they had practised these abominations with devils. Therefore this matter is fully substantiated by eye-witnesses, by hearsay, and the testimony of credible witnesses."[38] Learned debates proceed on whether the fact that most witches are female is due more to the insatiable lusts of the female sex or to its greater credulity[39] and on "whether the relations of an Incubus devil with a witch are always accompanied by the injection of semen . . . no infallible rule can be stated as to this matter, but there is this probable distinction: that a witch is either old and sterile, or she is not. . . . But if it is asked whether he is able to collect the semen emitted in some nocturnal pollution in sleep, just as he collects that which is spent in the carnal act, the answer is that it is probable that he cannot, though others hold a contrary opinion."[40]

The second part of the *Malleus* deals with obtaining and classifying the evidence for witchcraft. The authors argue that witchcraft is in effect a form of heresy, so they take over all the theory of evidence and its grades from the handbooks for inquisitors. Thus because of the gravity of the crime, two witnesses may not be enough if the accused is of good reputation; the judge has discretion when witnesses conflict on details; and accomplices, servants, notorious criminals, close relatives, and other witches may be admitted as witnesses (for the prosecution only) but not mortal enemies. Less serious degrees of enmity need not disqualify, since women are always quarreling, and their depositions can be believed if supported by other proofs.[41]

The *Malleus* gives more details and examples of actual trial procedure than earlier inquisitors' works. One has in reading the *Malleus* a greater feel for what the judges considered good and bad evidence. There is an outline questionnaire for examining witnesses: "The witness, N., of such a place, was called, sworn, and questioned whether he knew N. (naming the accused). . . . Further, he was asked whether any of the accused's kindred had formerly been burned as witches, or had been suspected, and he answered. . . . Asked further how he could distinguish the accused's motive, he answered that he knew it because he had spoken with a laugh. This is a matter which must be inquired into very diligently; for very often people use words quoting someone else, or merely in a temper, or as a test of the opinions of other people."[42] Then there are some suggested questions to be put to the suspected witch or wizard:

> Then he must be asked whether he believes that there are such things as witches, and that such things as were mentioned could be done, as that tempests could be raised or men and animals bewitched. Note that for the most part witches deny this at first; and therefore this engenders a greater suspicion than if they were to answer that they left it to a su-

perior judgement to say whether there were such or not. So if they deny
it, they must be questioned as follows: Then are they innocently condemned when they are burned? And he or she must answer.[43]

In the end the accused will allege that the informer has spoken out
of enmity; but when this is not mortal, but only a womanish quarrel, it
is no impediment. For this is a common custom of witches, to stir up
enmity against themselves by some word or action, as, for example, to
ask someone to lend them something or else they will damage his garden. . . . Asked why she touched a child, and afterwards it fell sick, she
answered. Also she was asked what she did in the fields at the time of a
tempest, and so with many other matters. Again, why, having one or two
cows, she had more milk than her neighbours, who had four or six.
Again, let her be asked why she persists in a state of adultery or concubinage; for although this is beside the point, yet such questions engen-
der more suspicion than would be the case with a chaste and honest
woman who stood accused. And note that she is to be continually questioned as to the depositions which have been laid against her, to see
whether she always returns to the same answers or not.[44]

The matter of cows is important in evaluating just how much the inquisitors' canons of evidence led to wrong conclusions. Though real
witches did not exist, we cannot dismiss all the confessions as false, since
there were no doubt many people who believed they were witches, especially after attending drug-assisted orgies.[45] Again, psychosomatic illness being what it is, it is only too likely that many people became sick
as a result of being threatened by someone with the reputation of being
a witch. On the other hand, there seems to be no way of actually be-
witching a cow. (Animals are also an indicator of how rational some of
the judicial practice of the day was. A rooster was tried, convicted, and
burned at Basel in 1474 for laying eggs, the law of the state thus making
up for what was defective in the law of nature. Trials of animals peaked
in the sixteenth century. Pigs were probably the worst treated, possibly
because of their near-human squeals.[46] Combination witch-animal trials were also possible, as numbers of werewolves were discovered,
mostly in Germany. It is a relief to be able to report similar legal
processes against vampires, who had the good fortune to be dead already.)[47]

What if the accused continues to deny everything, and the judge has
to evaluate the evidence himself?

The Judge has three points to consider, namely, her bad reputation, the
evidence (*indicium*) of the fact, and the words of the witnesses; and he

must see whether all these agree together. And if, as very often is the case, they do not altogether agree together, since witches are variously accused of different deeds committed in some village or town; but the evidences of the fact are visible to the eye, as that a child has been harmed by sorcery, or, more often, a beast has been bewitched or deprived of its milk; and if a number of witnesses have come forward whose evidence, even if it shows certain discrepancies (as that one should say she had bewitched his child, another his beast, and a third should merely witness to her reputation, and so with the others), but nevertheless agree in the substance of the fact, that is, as to the witchcraft, and that she is suspected of being a witch; although those witnesses are not enough to warrant a conviction without the fact of the general report, or even with that fact, as was shown above at the end of Question III, yet, taken in conjunction with the visible and tangible evidence of the fact, the Judge may, in consideration of these three points together, decide that the accused is to be reputed, not as strongly or gravely under suspicion (which suspicions will be explained later), but as manifestly taken in the heresy of witchcraft.[48]

Meanwhile the accused's house should be searched for instruments of witchcraft and her servants and companions imprisoned and questioned. It is advisable to remove a witch from her house in a basket or on a plank of wood so that she cannot touch the ground, as "we know from experience and the confessions of witches that when they are taken in this manner they more often lose the power of keeping silence under examination" (not, be it noted, to prevent her from working harm on the captors and judges, which is already impossible, since a witch loses her power when she falls into the hands of public justice).[49]

How difficult an acquittal is may be gathered from the following hypothetical case:

Katharina's child, or she herself, is bewitched, or she has suffered much loss in her cattle; and she suspects the accused because her husband or brothers had previously brought an unjust accusation against her own husband or brother. Here the cause of enmity is twofold on the part of the deponent, having its root both in her own bewitchment and in the unjust accusation brought against her husband or brother. Then ought her deposition to be rejected or not? From one point of view it seems that it should, because she is actuated by enmity; from another point of view it should not, because there is the evidence of the fact in her bewitchment. We answer that if in this case there are no other deponents, and the accused is not even under common suspicion, then her depositions cannot be allowed, but must be rejected; but if the accused is rendered suspect, and if the disease is not due to natural causes but to witchcraft (and we shall show later how this can be distinguished), she is to be subjected to a canonical purgation. . . .

But when the Advocate assumes the second line of defence, admitting that the accused has used such words against the deponent as, "You shall soon know what is going to happen to you," or "You will wish soon enough that you had lent or sold me what I asked for," or some such words; and submits that, although the deponent afterwards experienced some injury either to his person or his property, yet it does not follow from this that the accused was the cause of it as a witch, for illnesses may be due to various different causes. Also he submits that it is a common habit of women to quarrel together with such words, etc.

The Judge ought to answer such allegations in the following manner. If the illness is due to natural causes, then the excuse is good. But the evidence indicates the contrary; for it cannot be cured by any natural remedy; or in the opinion of the physicians the illness is due to witchcraft, or is what is in common speech called a Night-scathe. Again, perhaps other enchantresses are of the opinion that it is due to witchcraft. Or because it came suddenly, without any previous sickening, whereas natural diseases generally develop gradually. Or perhaps because the plaintiff had found certain instruments of witchcraft under his bed or in his clothes or elsewhere, and when these were removed he was suddenly restored to health, as often happens.[50]

Torture is then described (those administering it to do so "not joyfully, rather appearing to be disturbed by their duty"). An inability to weep is a strong sign of guilt; a suspect who weeps should be released if the other evidence is not very strong. Judges should beware of being bewitched by the voice or look of the accused and losing their anger.[51] The authors do not recommend frequent resort to torture and are somewhat skeptical of its value, "for some are so soft-hearted and feeble-minded that at the least torture they will confess anything, whether it be true or not. Others are so stubborn that, however much they are tortured, the truth is not to be had from them. . . . Others are bewitched, and make use of the fact in their torture, so that they will die before they confess anything."[52] There is no point in continuing torture unless there is some indication of its probable success.[53] A better method of securing confessions may be to falsely promise mercy: "Let some honest women visit her and promise that they will set her entirely at liberty if she will teach them how to conduct certain practices."[54]

One question deals at length with the grades of suspicion. They are divided, as usual, into light, probable, and vehement, with supporting quotations from various canons on heresy and the commentaries of canon lawyers. Light suspicion in cases of witchcraft is illustrated by the congregating of people in remote woods or forming friendships with suspected witches, "since it is proved that heretics often act in this manner."[55] The results of each grade of suspicion have become harsher. In particular, "a violent suspicion is sufficient to warrant a conviction, and admits no proof to the contrary."[56] One is surprised to find, in fact, that the grades of suspicion are also grades of verdict, that is, that there are *verdicts* of lightly, strongly, and vehemently suspect of heresy or witchcraft. The schedule of punishments, or more strictly "penances," is as follows. "Lightly suspect of heresy: a formal swearing that one has never believed in that heresy. Strongly (that is, probably) suspect: a formal swearing and some penance such as a term in prison or a pilgrimage. Gravely suspect: imprisonment for life or a long period."[57] Almost all of what has been quoted so far would tend to make convictions for witchcraft easier. The only rule likely to limit convictions concerns the harsh penalties for those who confess to making false accusations.[58]

The witch-burning craze continued unabated through the sixteenth century and only lessened toward the middle of the seventeenth. Estimates of the dead do not start below fifty thousand. The *Malleus* figured prominently; editions rolled off the presses. Sylvester Prierias (noted in the next chapter in connection with probabilism in moral theology) was also an inquisitor in Lombardy and was one of the first to quote the *Malleus* with approval. In the spirit of that work, he asserted that witches really do sometimes fly bodily from place to place, "for otherwise the nearly infinite processes of the inquisitors would necessarily be false, and it would deny the evidence of the senses, for much has been found out concerning this which can no more be rationally denied than that I am writing—which some one might deny and say I am deluded in seeming to myself and others to be writing."[59] The *Malleus* was certainly mentioned by real inquisitors, and there are verbatim reports that follow the procedure there described.[60] More mainstream legal works are saner—but only slightly.[61]

Outside pressures on the law may be illustrated by Martin Luther's remark, when told that a little girl had been forced by a witch to weep tears of blood: "Such a woman ought to be promptly punished with death. The lawyers want too much evidence and they despise these open and flagrant proofs. I have had today before me a matrimonial case; the woman had tried to poison the man, and he vomited up lizards. When she was questioned on the rack she answered nothing; for such sorceresses are dumb, they despise punishment; the devil will not let them speak. Such deeds, however, are evidence enough that for the sake of frightening others they ought to be made an example."[62] And Jean Bodin (noted in a later chapter for his reasonable views on the evaluation of histories) says of witches that "one accused of being a witch ought never to be fully acquitted and set free unless the calumny of the accuser is clearer than the sun, inasmuch as the proof of such crimes is so obscure and so difficult that not one witch in a million would be accused or punished if the procedure were governed by the usual rules."[63]

How the evaluation of evidence could work in practice can be seen in a rather later case. In 1672 the town of Balingen in Württemberg was largely destroyed by fire, and three women were soon arrested on suspicion of witchcraft. The magistrates consulted the legal faculty of Tübingen but were disappointed to learn that there were insufficient indications to torture even the prime suspect. They wrote to the Superior Council in Stuttgart explaining that the townspeople were not happy with the decision: "They would not only kill them but would place the magistrates in great personal danger. . . . In order that we might give the vulgar a sop and silence their mouths, we have been moved to petition your princely grace graciously to allow us to consult a different faculty, like Strasbourg, since it is well known the legal faculty of Tübingen are much too lenient in criminal matters and especially in hidden crimes and are always inclined to the more gentle."[64] The torture permitted by Strasbourg on the main suspect failed to produce a confession, and banishment was ordered. While being led through the streets, she was set upon by the mob and stoned so badly that she died a few days later.

But at times when there was no witch hysteria in train, the legal rules of evidence could act to obstruct popular pressures for witch burnings. In 1536 in Saxony, a village community demanded action against a seventy-year-old woman accused as a sorcerer. At an initial interrogation, she confessed and named two accomplices but then retracted the confession. The governor believed she should be released but, under further pressure from the commune, allowed further torture. The woman insisted on her innocence and was released, despite the commune's threatening to appeal to the elector of Saxony. Another case in a Saxon village in 1581 also resulted in the acquittal of an old woman after torture, followed by a long-drawn-out claim by her family for compensation on the grounds that she had been tortured too early and to excess, given the amount of evidence against her. This despite the fact that she had admitted all along to a little sorcery and refused to attend the sacraments.[65]

The witch-hunting craze did not reach England in earnest until Elizabeth's reign, apparently being brought back from the Continent by the Reformers. Some thousand witches were executed in England in the course of the next century, a much smaller figure than for Germany, France, or Scotland. Reginald Scot's *Discoverie of Witchcraft* of 1584 attacked the credulity of the *Malleus* and Bodin, including their reliance on presumptions and suspicions to an extent not acceptable with other crimes.[66] Scot was answered by King James VI of Scotland in his *Demonologie*, a book based on the king's own interrogation of witches who confessed to trying to kill him. James thinks the accused against whom there is only moderate evidence will all be eventually found guilty of at least something, "so jealous is God, I say, of the fame of those that are innocent in such causes."[67] One of James' first acts on becoming king of England was to order all copies of Scot's book burned, but thereafter his interest in witchcraft waned—though not before Shakespeare, nonjudgmental as ever, seized the opportunity to pander to the royal obsession with his portrait of the witches in *Macbeth*.[68] James even gained a reputation as an exposer of fraudulent claims of witchcraft. Sir Walter Raleigh's *History of the World*, written in the years between his conviction for treason and his execution on the king's orders, writes of "uncertain report only, which his majesty truly acknowledgeth for the author of all lies." The reference is to the *Demonologie*.[69]

In England, as later in New England, there was some association between Calvinism and the persecution of witches.[70] The Puritan William Perkins' *Discourse of the Damned Art of Witchcraft*, of 1608, says, "I say not that a bare confession is sufficient, but a confession after due examination taken upon pregnant presumptions. For if a man examined without any grounds or assumptions should openly acknowledge this crime, his act may be justly suspected, as grounded upon by-respects. But when proceeding is made against him at the first, upon good probabilities, and hereupon he be drawn to a good confession, that which he hath manifested thereby cannot but be a truth."[71]

But one of the most intelligent examinations of the inadequacy of evidence against witches was by a nonconformist minister in Essex, George Gifford. He states clearly the decision-theoretic argument that it is "much better that some should be put to death wrongfully, than to leave only one witch, which might kill and destroy many" but replies that scripture requires us "to condemn none but upon sure ground, and infallible proofs, because presumptions shall not warrant or excuse them before God if guiltless blood be shed." He argues especially that accusations by witches about other witches are worthless, as Satan would like to have the godly condemned and often works by planting evil suspicions. Since curdled cream, dead cows, and the like are acts of Satan, their occurrence is not evidence against the person suspected of causing them, so that common fame is unreliable. "How can that jury answer before God," he asks, "which upon their oath are not sure, but that so proceeding they may condemn the innocent, as often it cometh to pass?"[72]

Seventeenth-century English books on witchcraft were among the main routes by which the Continental language of the grading of evidence became familiar in England.[73] The typically Continental phrase, "Not legally guilty . . . but just ground of vehement suspicion," occurs in a New England witch trial of 1673.[74]

Among the forces restraining the witch craze was one that might not be expected in this context, the Spanish Inquisition. Though zealous against heretics and crypto-Jews, the Spanish authorities tended not to believe reports of witches, and Spain remained largely free of witches. The Inquisition warned its judges against believing everything in the *Malleus*, even if the author "writes about it as something he himself has seen and investigated."[75] It worked against an incipient outbreak in Navarre in 1526, ruling that there should be no arrests solely on the basis of accusations by confessed witches.[76] When the same region saw another scare in 1610, it was investigated by the inquisitor Salazar, whose report is a synthesis of the legal theory of evidence with a solid grasp of the realities of evidence. He proclaimed the usual Period of Grace, in which those who confessed would receive light punishments, and received 1,802 applicants, 1,384 of them aged twelve to fourteen. He granted them secrecy for their confessions and warned them of the danger of perjury. He took the trouble to check confessions that could be checked, finding that some had confessed to being at Sabbaths when others saw them at home, that several girls who had confessed to intercourse with demons were virgins, and that reputed magical ointments had no effect when tested on animals. Various people retracted their confessions, saying they had confessed only under torture; others admitted to being bribed to make accusations. Salazar concluded that 1,672 persons were known to have been falsely accused and that there was not sufficient evidence for even one genuine act of witchcraft.[77] He reported:

> I have not found even indications from which to infer that a single act
> of witchcraft has really occurred, whether as to going to aquelarres [Sab-

baths], being present at them, inflicting injuries, or other of the asserted facts. This enlightenment has greatly strengthened my former suspicions that the evidence of accomplices, without external proof from other parties, is insufficient to justify even arrest. Moreover, my experience leads to the conviction that, of those availing themselves of the Edict of Grace, three-quarters and more have accused themselves and their accomplices falsely . . . in the diseased state of the public mind, every agitation of the matter is harmful and increases the evil. I deduce the importance of silence and reserve from the experience that there were neither witches nor bewitched until they were talked and written about. This impressed me recently at Olague, near Pampeluna, where those who confessed stated that the matter started there after Fray Domingo de Sardo came there to preach about these things. So, when I went to Valderro, near Roncesvalles, to reconcile some who had confessed, when about to return the alcaldes begged me to go to the Valle de Ahescoa, two leagues distant, not that any witchcraft had been discovered there, but only that it might be honored equally with the other. I only sent there the Edict of Grace, and, eight days after its publication, I learned that already there were boys confessing.[78]

Salazar's report was upheld by the *suprema* of the Inquisition, and there were no more witch crazes in Spain.

The Italian Inquisition was also generally skeptical of witchcraft, perhaps partly because of its strict rules on accepting evidence. The testimony of witnesses of poor reputation was not admitted; alleged participants in Sabbaths were not allowed to name accomplices; implausible confessions were regularly deemed invalid; failure of the accused to show emotion during interrogation was not regarded as significant; the accused was given a defense lawyer, who was provided with a copy of the trial proceedings; torture was permitted only after the defense was heard.[79] In 1582 a minor judge in the papal states was condemned to seven years in the galleys for the unjust prosecution of four women "on the pretext that they were witches."[80]

Montaigne attacked witch trials, again on the basis of the insufficiency of evidence:

I have my ears battered with a thousand such flim-flams as these: "Three persons saw him such a day in the east; three, the next day in the west; at such an hour, in such a place, in such habit"; in earnest, I should not believe myself. How much more natural and likely (*vraisemblable*) do I find it that two men should lie, than that one man in twelve hours' time should fly with the wind from east to west? How much more natural that

our understanding should be carried from its place by the volubility of
our disordered minds, than that one of us should be carried by a strange
spirit upon a broomstick, flesh and bones that we are, up the shaft of a
chimney? Let us not seek illusions from without and unknown, we who
are perpetually agitated with illusions domestic and our own. Methinks
one is pardonable in disbelieving a miracle, at least, at all events where
one can elude its verification as such, by means not miraculous; and I am
of St. Augustine's opinion, that "tis better to lean towards doubt than
assurance, in things hard to prove and dangerous to believe." . . . It is
true, indeed, that the proofs and reasons that are founded upon experience and fact, I do not go about to untie, neither have they any end; I
often cut them, as Alexander did the Gordian knot. After all, 'tis setting
a man's conjectures at a very high price, upon them to cause a man to be
roasted alive.[81]

Such a quick and general way of dismissing experience hardly carries the
conviction of Salazar's comprehensive questioning of the actual evidence. Montaigne's attack on judicial torture suffers from the same defect.[82] On the other hand, the passage about witches is preceded by some
discussion of the spread of rumors, which does persuasively explain how
false reports come to be widely believed.

The best known of later attacks on witchcraft was that of Friedrich
von Spee, the German Jesuit, poet, and canon lawyer. His *Cautio Criminalis*, published anonymously in 1631, must be one of the most useful
works of a canon lawyer, if only in restraining the excesses of other
canon lawyers.[83] Like almost everyone opposed to witch trials, Spee
avoids discussing whether witches really exist and concentrates instead
on abuses of the law of evidence. He represents the psychological realities of witch finding as distorting the law of evidence so as to lead to inevitable conviction, not only for the original accused but, by a chain reaction, for a widening circle of those named under torture. What he says
is based on long experience as a confessor to the condemned:

> 9. If now some utterance of a demoniac or some malign and idle
> rumor then current (for proof of the scandal is never asked) points es-
> pecially to some poor and helpless Gaia, she is the first to suffer.
> 10. And yet, lest it appear that she is indicted on the basis of rumor
> alone, without other proofs, as the phrase goes, lo a certain presump-
> tion is at once obtained against her by posing the following dilemma:
> either Gaia has led a bad and improper life, or she has led a good and
> proper one. If a bad one, then, they say, the proof is cogent against her;
> for from malice to malice the presumption is strong. If, however, she has

led a good one, this also is none the less a proof; for thus, they say, are witches wont to cloak themselves and try to seem especially proper.

11. Therefore it is ordered that Gaia be haled away to prison. And lo now a new proof is gained against her by this other dilemma: Either she then shows fear or she does not show it. If she does show it (hearing forsooth of the grievous tortures wont to be used in this matter), this is of itself a proof; for conscience, they say, accuses her. If she does not show it (trusting forsooth in her innocence), this too is a proof; for it is most characteristic of witches, they say, to pretend themselves peculiarly innocent and wear a bold front. . . .

22. She is, however, tortured with the torture of the first degree, i.e. the less severe. This is to be understood thus: that, although in itself it is exceeding severe, yet, compared with others to follow, it is lighter. Whereupon, if she confesses, they say and noise it abroad that she has confessed without torture. . . .

31. For it is never possible to clear herself by withstanding and thus to wash away the aspersion of crime, as is the intention of the laws. It would be a disgrace to her examiners if when once arrested she should thus go free . . . for this were a disgrace to the zeal of Germany.

36. These in their turn are forced to accuse others, and these still others, and so it goes on: who can help seeing that it must go on without end?

37. Wherefore the judges themselves are obliged at last either to break off the trials and so condemn their own work or else to burn their own folk, aye themselves and everybody: for on all soon or late false accusations fall, and, if only followed by the torture, all are proved guilty.

38. And so at last those are brought into question who at the outset most loudly clamored for the constant feeding of the flames.[84]

English Legal Theory and the Reasonable Man

English law certainly had reasons for feeling superior to the system across the Channel, and English lawyers then and later have taken every opportunity to stress its superiority, especially in the treatment of evidence. The truth is perhaps not so black and white. The issue as to whether the Continental or the English system of evidence was better at reaching the truth was taken up in Sir John Fortescue's *Praises of the Laws of England*, of about 1470. A jury of twelve men of the neighborhood under oath, he says, is more reasonable and effective than a system using two suitable witnesses, who may be false.[85] In England, the accused can only be condemned by a jury, whose membership he can challenge.

"Who, then, in England can die unjustly for a crime," he asks, "when he can have so many aids in favour of his life, and none save his neighbours, good and faithful men, against whom he has no manner of exception, can condemn him? I should, indeed, prefer twenty guilty men to escape death through mercy, than one innocent to be condemned unjustly. Nevertheless, it cannot be supposed that a suspect accused in this form can escape punishment, when his life and habits would thereafter be a terror to them who acquitted him of his crime. In this process nothing is cruel, nothing inhuman; an innocent man cannot suffer in body or members. Hence he will not fear the calumny of his enemies because he will not be tortured at their pleasure. Under this law, therefore, life is quiet and secure."[86] These words were written in the middle of the Wars of the Roses, England's most lawless period for several centuries; Fortescue wrote during his enforced exile in France after the defeat of his party. It was also true that the Englishman was still subject to Continental rules in the ecclesiastical courts, and in the Court of Admiralty dealing with maritime matters, which had explicitly incorporated Continental rules of procedure into its practice.[87] He might in addition find himself tortured as an administrative measure by the king.[88] Selden, the legal historian of the early seventeenth century, claims that "the rack is used nowhere as in England. In other countries 'tis used in judicature, when there is a *semi-plena probatio*, a half-proof against a man; then to see if they can make it full, they rack him if he will not confess. But here in England they take a man and rack him I do not know why or when; not in time of judicature but when some body bids."[89] Significantly for the transmission of ideas, Continental influence was also evident in the large numbers of lawyers trained in Continental law at the English universities.[90]

Fortescue's criticism of the Continental system is over the sufficiency of two witnesses for establishing guilt. But the original point had been as much the necessity of two witnesses. A statute of Henry VIII complains that under Admiralty rules it is too hard to obtain convictions against pirates, because the trials are conducted by Continental rules, which require confessions or proof by disinterested witnesses, which is unlikely to be available.[91] The necessity of two witnesses was also a rule in which the survivors of Henry VIII's purges discerned virtue. As we saw, both More and Fisher had been condemned on the evidence of one witness. Early in Edward VI's reign, a two-witness rule was instituted for treason trials only.[92] In Mary's reign, a new statute declared that treason trials "shall be had and used only according to the due order and course of the common laws of this realm and not otherwise."[93] This ap-

parently contradicted the earlier statute, which was however not re-
pealed. Confusion ensued, especially after a further statute of Elizabeth
required two witnesses for treason of certain kinds. The English rule
came to prevail.

Sir Walter Raleigh believed himself safe from charges of treason in
1603, since there was only one witness against him, but he found himself fatally mistaken. He argued, "You try me by the Spanish Inquisition
if you proceed only by the circumstances, without two witnesses,"[94] and
quoted the statutes, Fortescue, Jesus, and St. Paul on the matter. But his
judges told him that, "by the Common Law, one witness is sufficient,
and the accusation of confederates or the confession of others is full
proof. . . . By law, a man may be condemned upon presumption and circumstances, without any witness to the main fact. As, if the King—
whom God defend—should be slain in his chamber, and one be shown
to have come forth of the chamber, with his sword drawn and bloody.
Were not this evidence, both in law and opinion, without further inquisition?"[95] Later the Puritans in England and the American colonies
made some attempts to institute the Biblical two-witness rule for serious crimes, but nothing permanent came of it.[96]

During the seventeenth century, English law recovered the auton-
omy that had been threatened by absolute monarchy, with its unconstrained use of torture and its unconstitutional doctrine of the divine
right of kings. The main mover was Sir Edward Coke, who defended
the independence of the judiciary against James I, when absolutism was
at its height. Although by and large a jealous guardian of English traditions in law, in matters of evidence Coke tended to adopt Continental
concepts, since English law had virtually nothing to say on the subject.
His classification of the grades of evidence, much quoted later, is that of
medieval Roman law: "Bracton saith, there is a *probatio duplex, viz. viva*,
as by witnesses *viva voce*, and *mortua*, as by deeds, writings and instruments. And many times juries, together with other matter, are much induced by presumptions; whereof there be three sorts, viz. violent, probable and light or temerary. *Violenta praesumptio* is manie times *plena
probatio*; as if one be runne throw the bodie with a sword in a house,
whereof he instantly dieth, and a man is seen to come out of that house
with a bloody sword, and no other man was at that time in the house.
Praesumptio probabilis moveth little but *praesumptio levis seu temeraria* not
at all. So it is in the case of a charter of feoffment, if all the witnesses to
the deed be dead (as no man can keep his witnesses alive, and time
weareth out all men) then violent presumption which stands for a proof
is continuall and quiet possession."[97]

On the standard of proof, Coke held that two witnesses or an un-forced confession constituted proof in cases of treason. In a cause so criminal, where *probationes oportent esse luce clariores* [proofs should be clearer than light]."[98] Coke's use of the medieval Latin phrases confirms that the English law of evidence descends from the medieval. He regu-larly quotes medieval Latin maxims on evidence and presumptions: "The law supposes a local to know the deeds of a local." "One eye-wit-ness is worth ten by hearsay." "A single voice induces neither proof nor presumption." "When witnesses depose in equal numbers, the more worthy are believed." "On a crime committed in a brothel, frequenters of brothels may be witnesses." "Since as knowledge grows, doubts grow with it . . . the authority of philosophers, doctors, and poets is to be held to in cases." "The manifest requires no proof." "No-one is presumed heedless of his eternal salvation, especially at the point of death."[99] Coke takes over the medieval distinction between presumptions which admit proof to the contrary and those which do not: "So if a man be within the four seas, and his wife hath a child, the law presumeth that it is the child of the husband, and against this presumption the law will admit no proof."[100] Even Ockham's Razor gets a mention, in the form, "It is in vain done by more, which can be done by fewer."[101] Coke's authority made these into maxims of English law as well, thus giving English law access to the conceptual achievements of the medieval lawyers.

On argument from authority itself, Coke quotes a Latin saying that is important for understanding the nature of legal reasoning, and espe-cially its difference from argument in, for example, science. He writes: "Our booke cases are the best proofes what the law is, *Argumentum ab auctoritate est fortissimum in lege* [Argument from authority is the strongest in law]."[102] All the old problems about balancing the number and weight of witnesses, and of authorities, recurred with argument from precedents. When one criticizes Scholastics and others for relying on arguments from authority, it is well to remember that there are dis-ciplines, like law and theology, in which such arguments are indeed ap-propriate.

Coke, like so many others, became embroiled in controversy with the Jesuits. They had, he said, "slandered" the English common law on the grounds that the prosecution could have counsel and examine wit-nesses but the defendant could not. Coke replied that "the law of Eng-land is a law of mercy . . . the Judge ought to be for the King, and also for the party indifferent, and it is far better for the prisoner to have a Judge's opinion for him, than many counsellors at the Bar."[103] The Jes-uits no doubt remained skeptical of the "indifference" of English judges, especially those Jesuits personally tortured by Coke.[104] Some further confusions in the English law of evidence were exposed by Slade's case of 1602. The facts were the simplest: Slade, the plaintiff, alleged an unwritten contract between himself and the defendant; there was a straightforward "your word against mine" situation, with very little evidence on either side to weigh. Coke appeared for Slade, Bacon for the defense. Hitherto, the defendant could normally win such a case by compurgation, but Coke's argument, ultimately accepted, was that a simple matter of fact was at issue, so that a trial by jury was appropriate. Since English law did not admit the parties as competent witnesses, the jury was effectively being asked to decide the case on zero evidence. The resulting misgivings helped concentrate legal minds on questions of evidence, and the question was partly resolved by the Statute of Frauds of 1677, which required a written contract in such cases.[105]

On another front of the battle between the English and Continental systems, Coke advanced the breathtaking argument that English law is more certain because it is derived from cases: "Upon the text of the civil law there be so many glosses and interpretations and again upon those so many commentaries and all these written by doctors of equal degree and authority, and therein so many diversities of opinions; as they do rather increase than resolve doubts and uncertainties, and the professors of that noble science say that it is like a sea full of waves . . . our expositions or commentaries upon Magna Carta and other statutes are the resolutions of judges in courts of justice in judicial courses of proceeding, either related and reported in our books, or extant in our judicial records, or in both, and therefore being collected together shall (as we conceive) produce certainty."[106] English law has certainly continued to emphasize decided cases over learned treatises on theory, though without its advantage over Continental law with regard to certainty becoming apparent.

The survival of Continental ideas on evidence is still clear at the end of the seventeenth century in the use of the two-witness rule in certain cases[107] and in the remark of the leading authority, Sir Matthew Hale: "The evidence at Law which taken singly or apart makes but an imperfect proof, *semiplena probatio*, yet in conjunction with others grows to a full proof, like *Silurus* his twigs, that were easily broken apart, but in conjunction or union were not to be broken."[108]

But subtle distinctions among grades of presumptions, and fractions of proof, were ill adapted to explanation to juries. Eventually all probabilistic concepts in English law were reduced to one word, *reasonable*. The common understanding that the standard of proof in criminal trials should lie somewhere between suspicion and complete certainty came to be expressed solely in the formula "proof beyond reasonable doubt."[109] Gradually, any question on the evidential relation between facts became expressible in terms of what the reasonable man would think. The relevant sense of *reasonable*, and the phrase *reasonable doubt*, appeared in the sixteenth-century Scholastic, Suarez,[110] while the *reasonable man* has an ancestor in the *steady man* of medieval canon law: If a man and woman are surprised by her father and compelled to marry, the man's consent is null and the marriage void if the fear inflicted on him is such as would coerce a "steady man."[111] The reasonable man himself appears in Hooker's defense of conservatism in religion, the *Laws of Ecclesiastical Polity*, where he prefers papists to Turks, follows the definitive sentence of church councils even when he disagrees with their decisions and, most characteristically, inclines toward the opinion of the most learned divines when there is no reason to favor the opposite opinion.[112]

The use of the formula "proof beyond reasonable doubt" to express the standard of proof in criminal cases became established in English law around 1800.[113] A brief overview of later developments in probability in the law of evidence is given in the epilogue.

CHAPTER FOUR

The Doubting Conscience and Moral Certainty

If rules for evaluating evidence in court seem natural to us, the idea that there should be rules for the deliberations of the internal forum of conscience is a very strange one. The modern attitude, at least before the recent widespread formation of ethics committees in medical research and the like, has been that there are no experts in morals. Everyone not warped by an unhappy childhood is presumed to be able to decide on what is right by an immediate intuition. Or if that is too absolutist a view of morals, it is regarded as the right of all to decide their own paths. In either case, experts are not invited to give their views.

Things used to be otherwise. In the heyday of Catholic casuistry, from about 1200 to about 1650, manuals of an increasing level of detail appeared giving advice on "cases of conscience." "Cases" had of course been dealt with in the Talmud, but the Catholic point of view is different, in a way that is important for the history of probability. There is a greater emphasis on the intention of the individual, leading to rules on what to do when the conscience is in doubt. It was in this context that discussion of "moral certainty" and "probabilism" appeared. The word "probability" in the early seventeenth century mostly refers to the doctrine of probabilism in moral theology, and Pascal's attacks on probabilism in the *Provincial Letters* did more than anything else to discredit casuistry and replace it with the modern attitude, in which each is his own judge of morality.

There is the inevitable origin in Aristotle. Perhaps the most read of all Aristotle's works was the *Nicomachean Ethics*, and consequently the remarks about probability in the introduction are the most quoted and influential opinions on the subject, without exception:

> Our discussion will be adequate if it has as much clearness as the subject-matter admits of; for precision is not to be sought for alike in all discussions, any more than in all the products of the crafts. Now fine and just actions, which political science investigates, exhibit much variety and fluctuation, so that they may be thought to exist only by convention, and not by nature. And goods also exhibit a similar fluctuation be-

cause they bring harm to many people; for before now men have been undone by reason of their wealth, and others by reason of their courage. We must be content, then, in speaking of such subjects and with such premises to indicate the truth roughly and in outline, and in speaking about things which are only for the most part true and with premises of the same kind to reach conclusions that are no better. In the same spirit, therefore, should each of our statements be *received*; for it is the mark of an educated man to look for precision in each class of things just so far as the nature of the subject admits; it is evidently equally foolish to accept probable reasoning from a mathematician and to demand from a rhetorician demonstrative proofs.[1]

The effect of this passage was that probability came to be associated mainly with ethics, or "practical reasoning," rather than with mathematics or science.

Penance and Doubts

The story resumes, as do so many in the history of ideas, with the twelfth-century Renaissance. The period was one of the discovery of the inner life and the origin of European individualism.[2] Antiquity does not have autobiography, in the modern sense, and the meaning of the word "classical" attests to a certain impersonality that everyone can sense in ancient art. The Gospels' ethical emphasis on the intention of the heart, developed by Augustine's *Confessions*, provided the foundation for a new orientation.[3] The twelfth century concentrates on the personal in sculpture and stained glass, in the autobiographies of Abelard and others, in the poetry of the troubadours and Chrétien de Troyes, in the piety of confession and devotion to Mary, and in Abelard's theory that sin is essentially a matter of internal consent to evil.[4] The involvement of all classes of society in civilization, symbolized by the craft-built cathedrals and the mystery plays and vernacular poetry,[5] is in marked contrast to most ancient civilizations. Writings on penance appeared when this background, of a presumed importance of the inner life of individuals, met the stream of theory coming out of canon law.

Law, and especially canon law, is today a subject rather remote from the daily round of life. One would not expect technical concepts arising in that discipline to become general knowledge quickly or even to become known to thinkers in other fields. That the opposite is true of the thirteenth century is due in large part to one of Innocent III's initiatives. The most far-reaching of the decrees of his Fourth Lateran Council made canon law part of the life of every person in Western Europe: every Christian was commanded to go to confession at least once a year. Now it is no accident that *confession* is the common name for the sacrament of penance as well as that of the "queen of proofs" in law. In effect, Innocent and the church hierarchy saw the confessional as a miniature, though secret, court of canon law. This is clear in the *Penitentials*, the books that soon appeared in great numbers as guides for confessors; and it is through these books that the concepts of canon law, including the theory of evidence, reached the hearts and minds of the people. An earlier example of this genre says that the "judge," meaning confessor, will elicit by subtle questioning "what the sinner himself does not know, or because of shame will wish to hide."[6] A *Liber Poenitentialis* of about 1210 again speaks in language hardly appropriate to private confession at all: "If both [spouses] do not confess [impotence] and the woman adduces more probable reasons (*probabiliores rationes*), she is to be believed; otherwise the man is to be believed, because he is the head of the woman. If the man will not or cannot prove, the woman proves by her bodily aspect and public reputation and the sworn testimony of witnesses, the number of whom is at the discretion of the judge."[7]

The decree requiring confession was soon obeyed almost everywhere, and the effect on the European soul was profound.[8] The mere effort to classify a year's sins was a greater demand for abstract thought than the common man or woman had experienced before. Guilt flourished, though without as much diminution as might have been hoped in things to be guilty about. Sin and conscience became prominent topics of study, and it is here that much of the subsequent discussion of doubts and their resolution occurs. Canon lawyers were encouraged to debate the resulting questions also because of a kind of converse of the miniaturization of the canon law court: sins that required restitution could have the necessary action enforced by canon law (or in England by the "equitable remedies" in the Court of Chancery, originally sometimes called a Court of Conscience).[9] Significant too was a connection in the opposite direction between conscience and law: judges and juries were commanded to make up their minds "according to conscience."[10] Evaluation of evidence was thus a moral matter, while conscience had a cognitive role.

When evaluating cases of doubt from a moral point of view, there are requirements other than that of simply reaching the most probable conclusion. The Middle Ages had inherited certain authoritative statements on the matter from antiquity. From Roman law there was the rule that in cases of doubt, for example about the interpretation of wills, the more benign or humane interpretation was to be preferred as safer and more just.[11] Another was the demand that proofs of guilt in criminal cases be "clearer than light." Augustine had asserted a principle of charitable interpretation, "Doubts are to be interpreted in the better part," as a general moral requirement, adding that the apparently contrary principle of inference in the New Testament, "By their fruits you shall know them," referred only to manifest fruits.[12] Innocent III again had a part to play, in enunciating another of the general maxims so favored by medieval lawyers: "In doubtful matters the safer path is to be chosen" (*In dubiis via est tutior eligenda*).[13] This much-discussed axiom, sane enough on the surface, stands in need of some interpretation. What does *safer* mean? Does it mean more probable? Or morally safer? Can these two meanings conflict? The more natural interpretation is morally safer, as explained by Peter the Chanter: in doubtful cases in which doctors disagree and no decision can be found in scripture or in papal decrees, "it is safer to incline to that part in which it is certain there is no sin."[14] But this moral doctrine, called tutiorism, is a very harsh one if taken literally. Is every minor doubt or scruple to force us to choose the "safer" path? Can a lack of probability in the doubt somehow balance the danger? Clearly some distinctions are in order.

The problem was taken up by Stephen Langton, archbishop of Canterbury, leader of the barons in the struggle for the Magna Carta. Here is a man eminently at the intersection of the theoretical and the practical. He deals with cases in which one is in doubt as to whether a proposed course of action is sinful; for example, can a cleric celebrate mass if he suspects he has been promoted to his office corruptly? Langton begins with the tutiorist answer: "When he doubts, he is held to abstain; because he should not commit himself to disaster in doubtful matters." But what counts as doubt? "It is one thing to be unaware that he is worthy of promotion, it is another thing to doubt. If he doubts, he ought not to celebrate; but not if he is unaware; for to doubt is to have reasonable proofs (*rationabiles probationes*) for each side."[15] One writer goes so far as to require that doubt is "what has equal reasons for and against": "To doubt is to consent equally to either side of a contradiction."[16] But that is surely an excessively strong requirement. Plainly, what is needed to sort out the confusion is some explicit understanding of *gradations* in how much reasons can support a proposition, that is, in probability. Then it will be possible to say that one doubt is supported by weighty reasons, and must be taken seriously, while another is supported by light reasons, and can safely be dismissed.

The matter is excellently put by Albert the Great (c. 1240). He is clear first on the ambiguity of *safer*: "On the rule that doubts are to be interpreted on the more secure side, it would seem that the more secure side is the one which has more reasons. . . . To the last point, it should be said that 'more secure' has two senses, namely further from falsity and further from danger. The rule is taken in the second sense."[17] But there are distinctions to be drawn in what counts as doubt: "Doubt is the indeterminate movement of the reason on either side of a contradiction. Ambiguity is a movement of the reason between either side of a contradiction by equal means [in a later text, "for equal reasons"]. Opinion is the acceptance of one side with fear of the other [in the later text, "when there is no or weak reason to fear the opposite"], when there is no express reason for the side for which there is fear, but the fear is only because of the weakness of the reasoning for the side accepted. Faith is a complete persuasion of one side through many probables. Knowledge is what is understood through its cause so that it could not possibly be otherwise."[18] The relevance to morality is that conscience has greater or lesser obligations according to which of these states the mind is in; at least opinion is needed before any action is required.

These distinctions were applied to the burning moral questions of the day. What is one to do, asks Aquinas, if he has vowed to enter religion but has not been accepted by the order of his choice? Is he bound to enter another order? He must consider whether his vow was to enter religion in general or one order in particular. But if he is in doubt as to what his vow was, "he must choose the safer path lest he commit himself to disaster." What of a married man who is converted to the faith? If there is "some probable hope" that his wife will also convert, he should not take a vow of continence. And when giving alms, how are we to decide which of our possessions are necessary and which superfluous? According to "what happens probably and for the most part."[19]

Of more intrinsic interest, as moral problems go, are those connected with just wars. Though the early church had strong pacifist tendencies, Augustine insisted on the justice of some wars and on the obligation on the citizen to fight in them. It is the business of the ruler to decide when a war is just, and his subjects are obliged to follow his decision. But medieval canon lawyers were much exercised by the possibility that subjects might be commanded to kill contrary to the law of God, which of course overrode the law of princes. Many therefore granted subjects a right to refuse to fight in a war that was "manifestly unjust." A borderland of subtle cases was immediately created, leading to some exceptionally tangled discussion. Stephen Langton, certainly an expert on disobedience to princes, advised the individual French knight commanded to fight in an unjust war against England to go to the war but to withdraw at the beginning of battle and refuse to fight. There was scope for asking how manifest the injustice of a war had to be, and there were occasional discussions of the case of a subject in doubt as to the justice of a war. In this case it was generally held that the subject had to fight. One canon lawyer held that a condition for a war to be just was that the enemy should deserve to be punished by war "or, if he does not deserve it, there should be just presumptions that he does." Another wrote that when the whole body of the king's men knew the king's attack was unjust, they must disobey, for God must be obeyed before men. A solitary doubter might recognize, however, that the justice of wars is hard to discern, and might then obey the king, but should not exert himself too hard in the field.[20] The matter of doubt about the justice of wars was taken up more vigorously in the sixteenth century, as described below.

The Doctrine of Probabilism

In hard times, there are always demands that academic research should be "relevant." Political theorists are funded to justify the tyrannies of their patrons, mathematicians to perfect weaponry and devise tax minimization schemes. He who pays the piper calls the tune. Fifteenth-century paintings are full of donors, smaller indeed than the sacred figures occupying the center of the paintings but often more realistic, and never awed by their illustrious company. In law, the push to the applied resulted in manuals for hunting heretics and witches. The outcome in theology was the production of manuals for confessors. It was these that eventually produced the theory known as probabilism.

Particularly influential were the various works on morals by Jean Gerson, chancellor of the University of Paris around 1400. Gerson seems to have been the first to introduce the term, occasionally still heard in English, "moral certainty" (*certitudo moralis*) to mean a very high but not complete degree of persuasion. One of the contexts in which this notion applies is the case of a priest worried that he may be unworthy to celebrate mass because of "nocturnal pollution." Gerson advises against being overscrupulous in these cases. No one has complete certainty of his own worthiness at any time. If "evident certainty" were required, "it is clear that then no one, however pure and just, would be able to celebrate; it would not be possible unless a special revelation were made to him. Therefore there suffices the kind of certainty we usually look for and accept in moral matters. This certainty can be called moral or civil certainty; of which the Philosopher [Aristotle] speaks in *Ethics* I ch. 3, saying: Our discussion will be adequate if it has as much certainty as the nature of the subject admits."[21] In general, one should have moral certainty that a proposed course of action is right before doing it. To acquire moral certainty, one should consider what usually happens, what authorities say, and what one's own learning suggests.[22]

Gerson obviously did not mean to suggest by "moral certainty" any more than the content of the Aristotelian remark that less certainty is to be found in morals than in mathematics. It is a dangerous phrase nevertheless, tending to suggest as it does that something that is not certain *is* certain. Just as a suspected criminal is not a kind of criminal, so moral certainty is not a kind of certainty. But there is an inevitable tendency to think it is. The problem is the same as in some modern philosophies of science that describe a theory against which there is (merely) statistical evidence as "refuted"—it confuses deductive reasons with nondeductive ones.[23] Nider, a follower of Gerson, actually says that "certainty taken loosely does not remove all improbability or opinion of the other side, although it leans more to one side than the other."[24] Nider introduced what proved to be the fateful consideration of the numbers of doctors to be found on either side of an opinion and how they can be used to excuse the penitent's choice: "One can with a good conscience hold to one side of an opinion, and act according to it, at least without scandal, when that side has in its favor notable or very notable doctors, provided it is not contrary to the express authority of holy scripture or the determination of the holy catholic church, and provided the contrariety of those opinions does not cause one to doubt, but rather forms in one conscience or faith in the more probable side—especially when one has diligently applied oneself to inquiring into the matter and has found nothing that sufficiently moves one to think it illicit."[25]

Another major figure in the intellectual life of the time was Saint Antoninus, the reforming archbishop of Florence in the time of Cosimo de Medici, the patron of Fra Angelico, and the author of works on economics and morals, subjects then taken to be related. It is characteristic of the times that his major work is a *Summa Moralis*.[26] The road to laxism is traversed a step further by his fixing on another ambiguity in the word "safer": "To choose the safer way [when in doubt] is a counsel, not a precept; otherwise, many ought to enter religion, in which it is safer to live than in the world. Therefore it is not necessary to choose the safer way; another way which is also safe can be chosen." To oppose "safer" to "safe" in this way is to change the meaning of the tutiorist axiom, since in the original "safer" had been contrasted instead with "more dangerous." The saint's purpose is more to excuse than to bind also in his treatment of the rule that one who is in doubt whether something is a mortal sin, and does it, commits a mortal sin. This is so, he

says, only "taking 'doubt' properly and strictly, namely as having the reasons equally weighty on either side and not inclining more to one than
the other"; but if he acts against a light or scrupulous doubt, "he does
not sin while he adheres to the opinion of some doctor and has probable reasons for that opinion more than the opposite."[27]

These distinctions and their consequences were the received opinions of the next hundred years. They can be seen in the handbook of
Sylvester Prierias of about 1515, the standard manual for confessors for
most of the sixteenth century.[28] The topics are in alphabetical order,
here—as usually—a sign that one is hearing well-digested received ideas.
Doubt is "the indifferent movement to either side of a contradiction or
. . . the equality of contrary reasons"; opinion lies between doubt and
scruple; a scruple is "a movement to one side of a contradiction from
light and very weak conjectures, which is also called suspicion or pusil-
lanimity of spirit . . . the devil attacks the scrupulous secretly, urging that
in doubtful matters the safer way is to be chosen."[29] The entry for *probable* is, "'Probable' is used in two ways. First as the opposite of hidden,
that is what is proved by witnesses (see 'Notorious'). Second, as what
pertains to opinion. And this in two ways. First, the object of a believed
opinion; thus Aristotle (*Topics* I) says that the probable is what seems to
be to all, or most, or the wise . . . according to the Chancellor [Gerson],
what is thus probable is called morally certain . . . it is equally vicious for
the mathematician to seek the persuasive as for the moralist the demonstrative."[30]

This happy, even smug, synthesis was disturbed by one who combined the high moral seriousness of tutiorism with the authority of the
papal throne. Adrian VI was pope in 1522–23, the last non-Italian pope
before the present one. He denied that a more probable opinion always
prevails over a lesser but safe one and even made an attempt at combining the forces of lesser danger and lesser probability (*mathématique
assez subtile*, as a modern author says).[31] While tutiorism in the hands
of the medievals had been an instrument for directing the faithful along
the paths of approved conduct, Adrian drew a consequence that was certainly logical but not such as to commend itself to the powers that be.
This was an age, it will be recalled, when the powers were more than
usually powerful. The period of the Renaissance and the Reformation
discovered the individual to be worthless in the sight not only of God
but also of the state. Western Europe took centuries to recover from the
diversion of political development away from medieval constitutionalism to primitive divine right absolutism. As the historian Tawney remarked, "Sceptical as to the existence of unicorns and salamanders, the

age of Machiavelli and Henry VIII found food for its credulity in the worship of that rare monster, the God-fearing Prince."[32] Into this un-receptive milieu Adrian dropped his tutiorist corollary: if a soldier has doubts as to the justice of the prince's war, he must not serve in it, this being the safer course (morally safer, of course, rather than physically).[33]

This argument was dealt with rather neatly by Francisco de Vitoria, founder of the Thomist School of Salamanca. Before describing his an-swer, let us look at how he used probability in something more central to his work. According to many, Vitoria was the real founder of inter-national law.[34] Professor of theology at the University of Salamanca from 1526 to 1546, he was noted especially for his work *De Indis*, which reported, at the request of the king of Spain, on the morality of colo-nization in America. Vitoria's conclusion is not entirely clear. He agrees that the Indians have title to their land and cannot be forcibly converted to Christianity but concedes to Spain a right to protect the innocent from cannibalism and human sacrifice and perhaps to act as trustees while the Indians acquire civilization.[35] As for the Indians, he thinks they do not in general have enough evidence for Christianity to place them under any obligation to believe it:

> It is foolhardy and imprudent of anyone to believe a thing without being sure it comes from a trustworthy source, especially in matters to do with salvation. But the barbarians could not be sure of this, since they did not know who or what kind of people they were who preached the new re-ligion to them. This is confirmed by St Thomas, who says that things which are of faith visibly and clearly belong to the realm of the credible; the faithful man would not believe them unless he could see that they were credible, either by palpable signs or by some other means (*ST* II-II. 1. 4 ad 2, 1. 5 ad 1). Therefore where there are no such signs nor any other persuasive factor, the barbarians are not obliged to believe. A fur-ther confirmation is that if the Saracens were to preach their own sect in this simple way to the barbarians at the same time as the Christians, it is clear that the barbarians would not be obliged to believe the Sara-cens. Therefore, since they would not be able or obliged to guess which of these two was the truer religion without some more visible proof of probability on one side or the other, the barbarians are not obliged to believe the Christians either, unless the latter put forward some other motive or persuasion to convince them. To do so would be to believe too readily, like the "light-headed man" (*Ecclesiasticus* 19:4).

Nor are the Indians in a particularly good evidential situation: "They are not bound to believe unless the faith has been set before them with persuasive probability. But I have not heard of any miracles or signs, nor of any exemplary saintliness of life sufficient to convert them. On the contrary, I hear only of provocations, savage crimes, and multitudes of unholy acts."[36] To Adrian's argument Vitoria answers with a balancing of risks: "If subjects fail to obey their prince in war from scruples of doubt, they run the risk of betraying the commonwealth into the hands of the enemy, which is much worse than fighting the enemy, doubts notwithstanding."[37] He interestingly compares the case with other standard cases of action under uncertainty: an officer of the law must carry out the judge's sentence, even if he doubts its justice. Vitoria discusses another case in which doubts can conflict. After agreeing that doing something while in doubt as to whether it is sinful is itself sinful, that to act against an improbable opinion when there is a probable one is wrong, and that if two conflicting opinions are probable one may follow either, he considers the following: "Take the case of one who is in complete doubt whether this woman is his wife or not. The woman seeks the [marriage] debt. It would follow that he is in perplexity. This is clear. If he renders the debt, he does wrong, since he acts against a doubt, so he sins if he renders it. If he does not render it, it would seem that he also sins, since he does injury to this woman, of whom he is in doubt whether she may be his wife and asking licitly. Therefore he does her injury, since he is not certain that she seeks wrongly. A similar case can be given: suppose there is a woman who is in doubt whether it is permitted to lie in defence of her husband. Can she do so? If she lies, she sins. If she tells the truth, she also sins, because she does her husband injury, when she is in doubt whether she must tell the truth."[38] His resolution of the cases is that one must choose the lesser of two evils (in the latter case, this is lying). If the evils were exactly equal, either action would be permissible, "but such cases certainly never happen."

Vitoria was succeeded in his chair at Salamanca by his student Melchior Cano, later bishop of the Canary Islands and famous for heated controversies with the Jesuits, the pope, and a long list of other enemies. In pursuing the same questions of morality in cases of doubt, Cano constructs a case that, though far-fetched in itself, really does have the evidence on both sides exactly equal: "Fifth rule: If everything is in doubt, so that the sins and dangers are in equilibrium, the choice of action is free. . . . Against this it could be argued that these rules cannot be satisfied in particular cases. Take first a woman who, hearing of the death of her first husband from seven reliable witnesses, contracts marriage with a second. After a long time, there appear seven witnesses of equal trustworthiness who assert that the first husband is alive. It seems such a woman cannot deny the marriage debt to the second husband, since danger of injustice would arise from not rendering it. But neither could she render it when he asks, since that would expose her to the danger of fornication."[39] There follows a long discussion justifying the decision that the woman's choice is free.

A number of Cano's other opinions also tend to allow doubt to generate freedom of choice in action: when there are different probable opinions of doctors on the licitness of contracts, either choice is safe (otherwise "all human contracts would cease, provisions necessary to human life would not be made, and republics and provinces would be laid waste"); one may act against one's own opinion if the opposite course is still safe, though the first course may be safer.[40] Indeed in some cases "it is more probable that a man not only can, but should, act contrary to his own opinion when he considers the other probable." A confessor, for example, should absolve a penitent whose opinion is probable but contrary to the confessor's. (Cano's opinions on the probability of histories are described in chapter 7.)

Bartolome de Medina, another student of Vitoria, was celebrated as the author of probabilism, the doctrine that in moral matters one may follow an opinion that is probable, even if the opposite is more probable. This proposition appeared in his commentary on the *Summa* of Aquinas, published in Salamanca in 1577. It is worth quoting his views at some length, since they summarize everything that has gone before, and they are the classic statement of what probability normally meant in the century before Pascal.

> This rule [that in doubt the safer part is to be chosen] is not always true. . . . For if someone seeks from me a farm that I legally own, and I begin to doubt whether it is mine, I am not bound to yield it to him, for "in doubtful matters the condition of the possessor is better," although in yielding it I would be certain not to sin. So, when the master rule says that in doubtful matters the safer part is to be chosen, it is understood as: when from following the safe part, I do not suffer great detriment. Otherwise it is not true. . . .
>
> When doubt is equal, and on either side the danger is equal, the doubter may choose either side, for there is no more reason why he should be held to choose one side rather than the other. . . . When the doubt is not equal on either side, and there are more urgent reasons for one side than the other, that part is to be preferred that is confirmed by more arguments; for the excess of reason makes that part preponderate. . . .
>
> Opinions are of two kinds: those are probable, which are confirmed

by great arguments and the authority of the wise (such as that one may charge interest for delayed payment), but others are completely improbable, which are supported neither by arguments nor the authority of many (such as that one may hold a plurality of benefices). . . .

The second conclusion: When both opinions, one's own and the opposite, are equally probable it is licit to follow either indifferently.

From this arises a great question: Whether we are bound to follow the opinion that is more probable, rejecting a probable one, or whether it is enough to follow some opinion that is probable. [There follow standard arguments for tutiorism.]

Lastly, the judge who passes sentence against him who adduces more probable testimony, certainly sins mortally and commits an injustice, as is clear; therefore one who follows a probable opinion, rejecting the more probable, without doubt sins.

Medina's originality comes in his reply to these arguments:

These arguments certainly seem very good, but it seems to me that if an opinion is probable it is licit to follow it, though the opposite be more probable.

For, a probable opinion in speculative matters is one that we can follow without danger of error and deception; therefore a probable opinion in practical matters is one that we can follow without danger of sin.

Secondly, a probable opinion is called probable from the fact that we can follow it without criticism and vituperation; therefore it would involve a contradiction if something were probable but it were not possible to follow it licitly. The antecedent is proved: for an opinion is not called probable because there are apparent reasons adduced in its favor, and it has assertors and defenders (for then all errors would be probable opinions), but that opinion is probable that wise men assert, and very good arguments confirm, to follow which is nothing improbable. This is the definition of Aristotle, *Topics* bk 1 ch 1 and *Ethics* bk 1 ch 4.

Thirdly, a probable opinion is in conformity with right reason and the consideration of prudent and wise men; therefore to follow it is not sinful. The consequence is clear, and the antecedent is thus proved: For if it is contrary to right reason, it is not a probable opinion but a manifest error.

It could be argued against this that it is indeed in conformity with right reason, but since the more probable opinion is more in conformity and safer, we are obliged to follow it. Against this is the argument that no one is obliged to do what is better and more perfect: it is more perfect to be a virgin than a wife, to be a religious than to be rich, but no one is obliged to adopt the more perfect of those.

Fourthly, it is licit to teach and propound a probable opinion in the schools, as even our opponents concede; therefore it is licit to counsel it. In confirmation of this: it is licit to assent to this conclusion internally, therefore it is licit to propound it externally.

Also, a confessor cannot require a penitent to follow the more probable opinion, therefore, there is no obligation to follow the more probable opinion. The antecedent is clear from the above.

Lastly, the opposite assertion would overburden timorous souls, for they would always have to establish which is the more probable opinion, which timorous men never do.[41]

Medina is a powerful arguer, but plainly something has gone very wrong. He is trying to think of *probable* and *more probable* by analogy with *safe* and *safer* and also sees *probable* as meaning something like *approvable*—as its etymology might suggest. But the case he is dealing with is that of a *probable* opinion and, its opposite, which is *more probable*. This is completely impossible on a usual understanding of *probable*, since if the opposite of a proposition is more probable than it, then the proposition itself is not probable but improbable. But this is only so if *probable* is taken to mean, as it should, probable on one's total evidence, that is, after consideration of all reasons for and against. It is natural not to do this, and to regard *probable* as meaning "supported by some good reasons" (irrespective of the reasons there may be to the contrary). It is Medina's conflation of "supported by good reasons" with "supported by the total of all reasons relevant to it" that has led to his assertion of probabilism.[42]

Suarez: Negative and Positive Doubt

By far the best-known of the School of Salamanca was Francisco Suarez, the most famous philosopher of the sixteenth century and of the Jesuit order. He is regarded by modern followers of Aquinas as the foremost of the "decadent Scholastics," to whose perversions they attribute the decline of medieval thought. His taste for novelty (as we would say, originality) in opinions, while staying within the Scholastic framework of questions, is expressed in his treatment of obligation and doubt. In Suarez, tutiorism has almost disappeared. While admitting that the certain is to be preferred to the probable, and the more probable to the less, in matters of fact, it is otherwise in matters of doubtful obligation. He assures one who is in doubt as to whether he is obliged by a vow that he is not so obliged. His reason is a legal one. Since "the obligation of law is in itself onerous and in a way odious," a doubtfully obliging law may be regarded as not sufficiently promulgated and hence not binding. Finally, "whenever there is a probable opinion that this action is not evil or prohibited or advised against, one can form a certain or practical conscience conforming to the opinion."[43] What was once an obligation to follow the safer course has become a practical certainty that the less safe course is permitted.

Suarez is not so cavalier in allowing dispensations from the law of the state. While obedience to human laws that are contrary to God's is not permitted, Suarez asks how much certainty one must have of the evil of a law in order to disobey it. The standard he requires is moral certainty. "For if the matter is doubtful, a presumption must be made in favour of the lawgiver . . . otherwise the subjects would assume an excessive licence to disregard the laws, since the latter can hardly be so just that it is impossible for them to be called in doubt by some for apparent reasons."[44]

Suarez' main contribution to theory was a clear distinction between negative and positive doubt. Doubt is negative when there are no reasons for either side, positive when there are reasons but there is some balance between them. The distinction is relevant to Suarez' treatment of the case of common soldiers in doubt whether the war of their prince is just: "If the doubt is purely negative, it is more probable that they can go to war, without examining the fact, and leaving the whole burden to the Prince to whom they are subject; whom however I suppose to be of good reputation among all. This is clearly taught by Vitoria and agreed to by all Thomists. But if the doubt is positive, and there are probable reasons on either side, I believe they are perhaps obliged to inquire into the truth. If they cannot arrive at that, they should follow what is more probable, and aid him who more probably is in the right; for when there is doubt on a matter of fact, and one which concerns the harm of our neighbour or the defence of the innocent, one must follow what appears more probable."[45]

While Suarez is of his time in his optimism concerning the intentions of princes of repute, his distinction between negative and positive doubt points to a much later era. The debates of the time of Laplace on the principle of insufficient reason never satisfactorily resolved whether the probability of a coin's landing heads is a half because the coin is symmetrical, and hence there is no reason to prefer heads to tails, or because many throws of coins have been observed to produce about half heads and half tails. Keynes' chapter on the weight of evidence shows that we are still no closer to explaining the difference between the probability of a hypothesis in the two cases in which there is little evidence with a certain probability on balance and in which there is a great deal of evidence with the same probability on balance.[46] Are there two dimensions of probability, one giving the total probability and the other the firmness or weight with which that probability can be held?

Having left the decision on war mainly to the prince, Suarez proffers advice to that potentate wise enough to take the advice of theologians: Where there is probability on both sides (of for example a territorial dispute), the king may assert his rights only if the probability is more on his side, while if there is equal probability, or certainly equal doubt, the side in possession is to be preferred.[47]

Suarez made another and more lasting contribution to debate on the just war. Twentieth-century Catholic just war theory has been prepared to consider that some wars might be just if they satisfy certain stringent conditions. The list of these is not fixed but is roughly the following: the war must be fought for a just cause, involving a grave injury to the state; it must be declared by a legitimate authority—in particular only states may declare war; it should be conducted by legitimate means for a limited end, in proportion to the injury; and there must be a reasonable chance of winning. The first three of these stem from medieval thinking on the subject, as elaborated in Vitoria's *Law of War*, but the Middle Ages barely had the conceptual apparatus even to think of the last. It reflects an appreciation of probability too often absent from discussions of right and wrong. Wisely followed by the South Vietnamese in 1975, the precept could well have been used to advantage by the Argentine garrison on the Falklands. The initiator of the reasonable chance condition was Suarez, in the following passage, which asserts that there ought to be some reasonable proportion between the probability of victory and the harms involved:

> [Cajetan says] that for a war to be just, the Prince ought to know he has enough power to be morally certain of victory: first because otherwise he exposes himself to the manifest danger of imposing on his state greater harm than is fair. In the same way, [he says] a judge acts badly in exacting a penalty from a defendant whose guards are so many that it is certain they cannot be overcome. . . . But this condition does not seem to me absolutely necessary. First because in human life it is almost impossible. Secondly because it is often for the common good of the state not to wait upon such certainty, but rather to make the attempt, even when there may be some doubt whether the enemy can be over-

come. Thirdly, because otherwise the less powerful king would never be permitted to make war on the more powerful, because the certainty that Cajetan requires could not be achieved. Consequently it should be said that the Prince is indeed required to achieve the maximum certainty of victory that he can; he ought also compare the hope of victory with the danger of harm, and see whether, all things weighed, hope prevails. If he cannot achieve so much certainty, he should at least have a more probable hope of victory, or doubt equally balanced as to defeat or victory, according to the necessity of the state and of the common good. If the probability of hope is less, and the war is aggressive, it should almost always be avoided; if defensive, undertaken, since the former is of choice, the latter of necessity.[48]

Grotius, Silhon, and the Morality of the State

The wreck of Scholastic casuistry in the mid-seventeenth century left among its detritus the theory of international relations. The survival of this fragment owes something to the distributed nature of international power: There is no overarching legislative or legal power, so individual nations are thrown back on their own casuistical resources to justify their positions. It also owes something to the individual success of the last of the Scholastic theorists and first of the moderns, Grotius.

In the era of Rembrandt, Vermeer, Huygens, Snell, Spinoza, and Tasman, the leading contributor from Holland in the area of social science was Hugo Grotius. His work was centered on, though far from confined to, law. His chief work, *The Law of War and Peace*, is regarded as the founding classic of international law; its legal topic is probably responsible for its status as the last influential work of fully Scholastic cast. Grotius's learning was vast, and besides his political and diplomatic activities he wrote extensively on philology, history, and especially religious controversy, where his moderate views were attacked by extremists on both sides. (His views on probability in religious controversy are considered in chapter 9.)

The *Law of War and Peace* is a descendant of the works of Vitoria and Suarez on the same subject, and on the matter of doubts about the justice of war his treatment is much the same, though he does take a somewhat stricter view of what is required in cases of doubt.[49] His treatment here is a summary of that in a youthful work, *On the Law of Booty*. Written in 1605 in reply to attacks on the Dutch East India Company's policy of piracy on the high seas, the work remained unpublished, apparently because of the vanishing of criticism on distribution of the company's profits.[50] There he had written a passage that is unusually revealing about the intellectual ancestry of ideas on probability, as perceived in the seventeenth century:

> Again, very few facts are discernible through the senses, since we cannot be in more than one place at one particular time, and since the senses perceive only those things which are very close at hand. Yet there is no other way of attaining to true knowledge. Impelled thus by necessity, human reason has fashioned for itself certain rules of probability, or *ton eikoton*, for passing judgement in regard to facts. These rules consist of various *prolepseis*, or (to use the Latin term) *presumtiones*, which are not fixed and unchangeable like scientific rules but rather of a character considered concordant in the greatest possible degree with nature; that is to say, on the basis of what commonly occurs, conclusions of a similar trend may be drawn. For, among the proofs which we accept in forming judgements, there is not one that is necessarily conclusive; on the contrary, all of them are derived from the aforesaid presumptions *hōs epi to poly*, from what happens for the most part [he refers to Aristotle, the *Digest*, the *Decretals* on presumptions, and Aquinas].[51]

Grotius is clear that political affairs are bound by the same moral principles as any other actions. At the opposite extreme was the position of Machiavelli, or at least the one popularly attributed to him, that there is no connection between morality and political action. While Machiavellianism was universally abhorred, there was a school of thought that sought a middle way, holding that actions not normally licit, like lying, might be justified in political affairs by *reason of state*—"a mean between that which conscience permits and affairs require."[52] The notion was justified in terms of a primitive decision theory involving considerations of risk. It is all very well for the prince to keep to high-minded moral principles like keeping his word, but ought he do so when the safety of his subjects is at stake? An individual only risks his own when he voluntarily suffers loss in order to do noble deeds, but for princes or their ministers to do the same is not noble but imprudent. "They are unjust if they sacrifice that which is not theirs and has been placed in their hands as a sacred trust."[53] Thus says Jean de Silhon in his *The Minister of State*, a justification of Cardinal Richelieu's many deviations from accepted principles of rectitude in pursuit of the interests of the state. The safety of the king's realm is threatened by heretic princes and the like whose own word cannot be trusted, who have spies and agents everywhere, and who plot with the enemies of the state, both internal and external. "Although it is always condemned in speech and conversation . . . let us conclude that on these occasions, distrust is the mother of security, and in order to avoid being deceived, one must be prepared as though one might be."[54]

Silhon is aware of the usefulness of probabilism in the justification of doubtful opinions and applies it to reason of state: "In doubtful cases, he [the minister of state] will always choose what is safest and most advantageous to his master even though the least probable, provided that it is truly probable. In this, he combines two maxims, one of conscience and the other of prudence. Conscience permits us to select from two probable opinions that which we prefer, while prudence directs us . . . to choose that which is the most profitable. The reason for conscience is that although we are always obliged to adhere to known truths, this is so difficult a quest and falsehood imitates the appearance of truth so artfully that it is often least present where it seems to be. The reason for prudence is so natural that it is known to all."[55]

Hobbes and the Risk of Attack

Hobbes' political theories were, in a sense, entirely founded on the concept of risk, or danger, although he does not exactly express himself in those terms. Or from another point of view, his position can be seen as the just war theory writ small, in which the actors are individual persons instead of states, though he does not use precisely those terms either. The famous phrase in *Leviathan* about the life of man in a state of nature being "nasty, brutish, and short" occurs in a paragraph that attributes that condition to lack of certainty about the effects of actions. "In such condition, there is no place for industry; because the fruit thereof is uncertain: and consequently no culture of the earth, no navigation . . . no arts, no letters, no society; and, which is worst of all, continual fear, and danger of violent death; and the life of man, solitary, poor, nasty, brutish and short."[56] The remedy is the removal of just suspicions of being attacked. That is a main reason for replacing the state of nature with a compact of civil society:

> If a covenant be made, wherein neither of the parties perform presently, but trust one another; in the condition of mere nature, which is a condition of war of every man against every man, upon any reasonable suspicion, it is void: but if there be a common power set over them both, with right and force sufficient to compel performance, it is not void. For he that performeth first, has no assurance the other will perform after; because the bonds of words are too weak to bridle men's ambition, avarice, anger, and other passions, without the fear of some coercive power; which in the condition of mere nature, where all men are equal,

and judges of the justness of their own fears, cannot possibly be supposed. And therefore he which performeth first, does but betray himself to his enemy; contrary to the right, he can never abandon, of defending his life, and means of living.

But in a civil estate, where there is a power set up to constrain those
that would otherwise violate their faith, that fear is no more reasonable;
and for that cause, he which by the covenant is to perform first, is
obliged so to do.[57]

The more elaborate *De Cive* at the corresponding place emphasizes
even more the concept of the reasonableness of the evaluation of risk:
"For it suits not with reason, that any man should perform first, if it be
not likely that the other will make good his promise after; which,
whether it be probable or not, he that doubts it must be judge of."[58]
Hobbes' phrase "just suspicions" alludes to the casuists' use of this concept. It descends from the fear that the steady man of medieval canon
law was to evaluate. The casuists define the just fear that is sufficient to
nullify a contract as one that would disturb a steady man.[59]

The same reasoning, according to Hobbes, is behind the law's forbidding of preemptive first strikes: "Nature gave a right to every man to
secure himself by his own strength, and to invade a suspected neighbour,
by way of prevention: but the civil law takes away that liberty, in all cases
where the protection of the law may be safely stayed for."[60] Hobbes is
less than clear on what is to be done when one does reasonably fear that
the sovereign lacks the power to protect or when the protection of the
law may not be "safely stayed for."[61] A sovereign without power, it
seems, has no claim to allegiance, a consequence that was the source of
not a little ill feeling between Hobbes and the Royalist exiles in Paris
during the Commonwealth. The situation was all the more awkward because the Commonwealth's propagandists were using exactly the same
argument. Marchamont Nedham, first a defender of Parliament, then
a Royalist, later a Cromwellian, in 1650 rested the case for the Commonwealth on two theses: "That the Sword is, and ever hath been, the
Foundation of all Titles to Government" and that there was a great improbability of any of the Commonwealth's adversaries, whether Royal-
ists, Scots, Presbyterians, or Levellers, ever succeeding in their designs.
He estimates it at ten to one that any Royalist attempt to regain power
will fail. The second edition of Nedham's book includes an appendix de-
lightedly quoting Hobbes, whose arguments about the need for a strong
ruler, Nedham suggests, could hardly apply to the "King of Scots" or
"any power beside the present."[62] When Nedham's ten to one chance

came off at the Restoration, his career as a consistent defender of whatever regime happened to be in power proved to be far from over. The connection between cold-blooded calculation with probabilities and lack of moral fiber was beginning to look close.

The Scandal of Laxism

In the eighty years after Bartolome de Medina proposed it, the doctrine of probabilism had an astonishing success among Catholic theologians, as author after author strove to attenuate the obligation of law in cases of a doubting conscience. Moral theology was the concern of the age, and the print expended on that subject far exceeded that on science, for which the period in question is now chiefly known.[63]

It was not really true, as Pascal's *Provincial Letters* claim, that the entire Jesuit order was committed to the thesis that an opinion was probable, and hence permitted, if there was a single doctor who asserted it. But the truth is not so distant as to make the caricature unrecognizable. The Jesuit Thomas Sanchez, in his widely read *Moral Work on the Precepts of the Decalogue*, of 1611, certainly asserts that the authority of a single "worthy and learned" doctor is enough to render an opinion probable and that one might counsel a probable opinion even if contrary to one's own. He also distinguishes explicitly between the extrinsic probability of an opinion (that is, the weight of authorities in favor of it), and its intrinsic probability (the reasons for it). Another Jesuit excuses holders of academic chairs from having to teach the opinions that appear to them more probable; what appeared probable yesterday might seem less so today, so it suffices to expound opinions that have some probability and are free from scandal.[64] Perhaps wise counsel in the years after Galileo. The situation was compared to cases in ordinary life in which one prefers the weight of the opinion of others against one's own: "If I probably think a man I see at a distance is Titius, but on the contrary others who are with me think he is not Titius, I can prudently assent to the belief of the others against my own opinion, either because they are many, or because I think they can more certainly and clearly perceive at a distance; therefore also in other things I can prudently embrace the opinion of others, contrary to my own."[65] Yet another Jesuit advises that one need not be *certain* of the probability of an opinion one adopts; it is enough that it be "probably probable."[66] Escobar, later made famous by the attacks on him in the *Provincial Letters*, rejoices in the diversity of moral opinions among doctors, by which the yoke of the Lord is made lighter.[67] These doctors naturally praise one another extravagantly, giving their opinions even more extrinsic probability. The expression *Barockmorale* is unavoidable.[68]

Opposition to these lax tendencies was remarkably late in coming. Perhaps the first publication devoted exclusively to the subject of probabilism is the *Disputation on the Practice of Opinions*, of 1642, by the Jesuit Bianchi. Interestingly, he sees the problem as a mistake in the logic of probability: "It is therefore an equivocation and the cause, I believe, of error, to take as the same two things which are quite different: to judge both contradictories probable, and to judge that probably both contradictories are true. For the first can happen, but not the second. And the first was not enough, even according to what our adversaries say elsewhere, for forming an opinion on any proposition and following it in practice."[69] Especially important is Arnauld's attack on probabilism in his *Moral Theology of the Jesuits* of 1643 (about which more will be said in connection with Pascal). Opinion at the highest level of the church remained unconcerned until 1656.[70]

English Casuists Pursue the Middle Way

Churches other than the Roman did not develop casuistry to the same degree but could hardly avoid altogether the giving of moral advice to their flocks. The point of view that sees moral choice in terms of cases of conscience was fundamental to Elizabethan thought and evident in its literature.[71] But the determination of mainstream Anglicanism to maintain a position between Catholic and Calvinist created special problems in moral theology. A church in schism could hardly have the same respect for authorities, especially recent ones, as Catholics. So English casuistry avoided probabilism in the sense of reliance on the opinions of doctors and gave greater weight to the reasons for and against opinions. ("Trust neither me, nor the adverse part, but the Reasons," says John Donne.)[72] The Anglicans tended toward probabiliorism, the position between probabilism and tutiorism that holds that the more probable opinion must be followed. Donne complained that probabilism indulges the human propensity to intellectual laziness: "To which indisposition of ours the casuists are so indulgent, as that they allow a conscience to adhere to any probable opinion against a more probable, and do never bind him to seek out which is more probable, but give him leave to dissemble and to depart from it, if by mischance he come to know it."[73]

The word "dissemble" here alludes to the notorious views of the Jesuits on the permissibility of equivocation and mental reservation. The Jesuits were committed to the view that an outright lie was intrinsically wrong and hence never permissible. But they did regard it as allowable, for sufficient cause, to give answers that were deliberately ambiguous or to give only part of an answer, mentally reserving a continuation of the sentence that would change its meaning. The case for the permissibility of these practices was argued in the *Treatise of Equivocation* of Henry Garnett, the superior of the English Jesuits in the dangerous years around 1600. Admitting that not all doctors approve of mental reservation, he calls the doctrine of probabilism in his support: where there are two probable opinions, "a man may without sinne follow either, if it may be done without preiudice of our neighbour; and if one be lesse probable than the other, yet so long as it is within ye compasse of probability, wch it is if it have 2 or 3 grave autours (as ours hath very many), then may a man be bound under sinne . . . to chuse ye lesse probable in case a superior comaund or our neighbour may be otherwise notably [text damaged]."[74] He means that one may balance a lesser probability of the doctrine of mental reservation against the harm that will follow from answering truly such questions as whether there is a priest hidden in one's home.

Father Garnett evaded capture for many years but fell into the hands of the authorities after the Gunpowder Plot. Coke achieved his conviction of complicity in the plot, claiming to offer, in the Continental phrase, proofs *luci clariores* (clearer than light) of his guilt; he was almost certainly not involved, though he may have learned of the plot through confession. According to the government report of his execution, he said on the scaffold, "I am sorie that I did dissemble . . . but I did not think they had had such proof against me, till it was shewed me."[75]

Who was lax and who strict became curiously reversed in the case of the Oath of Allegiance controversy. After the Gunpowder Plot, James I imposed the Oath to distinguish between loyal Catholics and "such other Papists as in their heartes maintained the like violent bloody *Maximes*, that the Powder-Traitours did."[76] Many Catholics complied although forbidden to do so by the pope, and the Jesuits urged non-compliance. Donne suggests that Catholics might like to apply probabilism to the case and to conclude that obedience to the king may be legitimately preferred to obedience to the pope: "The reasons upon which *Carbo* builds this Doctrine of following a *probable* opinion, and leaving a more *probable*, which are, *That no man is bound, Ad melius et perfectius* [to be better and more perfect], *by necessity, but as by Counsell*: And that this Doctrine hath this Commoditie, *that it delivers godly men, from the care and sollicitude, of searching out, which is the more probable opinion*, shew evidently, that these Rules give no infallible direction to the conscience, and yet in this matter of Obedience, considering the first native

Certaintie of subjection to the King, and then the damages of the re-
fusall to sweare it, they encline much more to strengthen that civill obedience, than that other obedience which is plainly enough claimed, by
this forbidding of the Oath."[77]

The uncompromising biblicism of the more extreme Puritans left no
place for the subtle jugglings of casuistry. Francis Bacon criticized the
Calvinists for hollow exhortation in generalities in place of real moral
direction. "The word," he says, "[the *bread of life*] they toss up and down,
they break it not. They draw not their distinctions down *ad casus consci-*
entiae; that a man may be warranted in his particular actions whether
they be lawful or not."[78] At least, this was so until the 1640s, at which
point the Parliamentarians experienced some difficulty with the text of
Romans 13:2, "Whosoever therefore resisteth the power, resisteth the
ordinance of God: and they that resist shall receive to themselves
damnation." The Puritan casuists whom this necessity called forth were
with justice called by an opponent "perfect Jesuits in their principles,
and resolutions concerning Kings." He remarked that the doctrine of
allegiance to kings had always been counted "a principall head of difference between Protestants and the worst of Papists"[79] and the papacy's
denial evidence of its being Antichrist. History granted Cromwell the
last word on textual interpretation: that the text of Romans was "a Ma-
lignant one; the wicked and ungodly have abused it very frequently, but
(thanks bee to God) it was to their Ruine; yet their abuse shall not hin-
der us from making a right use of it."[80]

There was, however, a less extreme strand of Puritanism that did
admit casuistry of a kind, including considerations of probability.
William Perkins, around 1600, adopted the Scholastic view of conscience as a largely intellectual faculty, concerned with "practical syllogisms," and spoke of a "satisfied conscience" standard, which seems to
be equivalent to moral certainty.[81] However, as we saw, his *Discourse of*
the Damned Art of Witchcraft suggests his own conscience was rather
easily satisfied.

The high point of Anglican casuistry actually appeared after Pascal's
Provincial Letters, and bears no evidence of being affected by Pascal's
ridicule of casuists. It is Bishop Jeremy Taylor's *Ductor dubitantium*, of
1660.[82] His praise of probable arguments is colorful:

> Probable arguments are like little stars, every one of which will be use-
> less as to our conduct and enlightening; but when they are tied together
> by order and vicinity, by the finger of God and the hand of an angel,
> they make a constellation, and are not only powerful in their influence,

but like a bright angel, to guide and to enlighten our way. And although the light is not great as the light of the sun or moon, yet mariners sail by their conduct; and although with trepidation and some danger, yet very regularly they enter into the haven. This heap of probable inducements, is not of power as a mathematical and physical demonstration, which is in discourse as the sun is in heaven, but it makes a milky and a white path, visible enough to walk securely.

And next to these tapers of effective reasoning, drawn from the nature and from the events, and the accidents and the expectations and experiences of things, stands the grandeur of a long and united authority; the understanding thus reasoning, That it is not credible that this thing should have escaped the wiser heads of all the great personages in the world, who stood at the chairs of princes, or sat in the ruler's chair, and should only appear to two or three bold, illiterate, or vicious persons, ruled by lusts, and overruled by evil habits.[83]

On the issue of the doubting conscience, Taylor is firmly for probabiliorism, laying down the rule, The greater probability destroys the less. His argument against probabilism is that "the safety [of an opinion] must increase consequently to the probability, it is against charity to omit that, which is safer, and to choose that, which is less safe."[84]

Another interesting point is a brief decision-theoretic consideration concerning the need for a quick decision in some cases: "One case more happens, in which a small probability may be pursued, viz. when the understanding hath not time to consider deeply, and handle the question on all sides; then that which first offers itself, though but mean and weak, yet if it be not against a stronger argument at the same time presented, it may suffice to determine the action."[85]

The following case of conscience, the case of a woman in Brescia, illustrates both the level of detail the casuists were prepared to consider and the way in which circumstances or consequences could sometimes make up for lack of probability. The description is a long one because the point of the case is to consider the effects of gradually taking into account extra pieces of information:

Her husband had been contracted to a woman of Panormo, "per verba de praesenti"; she taking her pleasure upon the sea, is, with her company, surprised by a Turk's man of war, and is reported, first to have been deflowered, and then killed. When the sorrow for this accident had boiled down, the gentleman marries a maid of Brescia, and lives with her some years; after which she hears that his first spouse was not killed, but alive and in sorrow in the isle of Malta, and therefore that herself lived in a state of adultery, because not she, but the woman in Malta, was the true wife of her husband. In this agony of spirit, a mariner comes to her house, and secretly tells her, that this woman was indeed at Malta, but lately dead, and so the impediment was removed. The question now arises, whether, upon the taking away this impediment, it be required that the persons already engaged should contract anew? That a new contract is necessary, is universally believed, and is almost certain (as in its proper place will be made to appear); for the contrary opinion is affirmed but by a very few, and relies but upon trifling motives, requiring only the consent of either of the parties as sufficient for renewing the contract. But this being but a slender probability, ought not to govern her; she must contract anew by the consent of her husband, as well as by her own act. But now a difficulty arises; for her husband is a vicious man, and hates her, and is weary of her, and wishes her dead; and if she discover the impediment of their marriage, and that it is now taken away, and, therefore, requires him to recontract himself, that the marriage which was innocently begun, may be firm in the progression, and legally valid, and in conscience; she hath great reason to believe that he will take advantage of it, and refuse to join in a new contract. In this case, therefore, because it is necessary she should, some way or other, be relieved, it is lawful for her to follow that little probability of opinion which says, that the consent of one is sufficient for the renovation of the contract. And in this case, all the former inconveniencies mentioned above do not cease: and this is a case of favour in behalf of an innocent marriage, and in favour of the legitimation of children, and will prevent much evil to them both. So that although this case hath but few degrees of probability from its proper and intrinsical causes, yet by extrinsical and collateral appendages, it is grown favourable, and charitable, and reasonable: it is almost necessary, and, therefore, hath more than the little probabilities of its own account.[86]

Juan Caramuel Lobkowitz, Prince of Laxists

The story of probability in moral theology concludes with two opposed figures, Caramuel and Pascal. They were opponents and in many ways exact opposites, though, as is the way with these things, in certain respects identical. Since Caramuel is as obscure as Pascal is famous, some basic facts on his biography will be helpful.[87]

Juan Caramuel Lobkowitz was born in Madrid in 1606. After a career as a child genius, including the publication of astronomical tables at the age of twelve, he entered the Cistercian order, which conferred on him the titles "Abbot of Melrose" and "Vicar-general in England,

Scotland and Ireland," though local religious conditions did not permit him to visit those places. He taught at the University of Louvain for some years, thus being on hand at the outbreak of Jansenism, of which he was one of the earliest opponents. His contribution to the fight included his *Mathesis Audax* of 1642, in which he claims to have resolved the major problems of logic, physics, and theology, especially those touching grace and free will, by ruler and compass construction. His *Rational and Real Philosophy* is one of the most rococo works of late Scholastic philosophy; it contains an unusually detailed treatment of arguments from authority, indicating the direction of Caramuel's thought.[88] In 1648 he was in Prague, then besieged by the Swedish army. He commanded a troop of ecclesiastics in the successful defense of the city, his courage receiving recognition. On the advent of peace, he "converted a great number of heretics," no doubt assisted by the absence of the Swedes.

His writing continued undiminished and ultimately included works on theology, philosophy, mathematics, grammar, heraldry, civil and military architecture, the Cabala, ecclesiastical hierarchy, the rights of various monarchs to various thrones, astronomy, musical theory, cryptography, and many other subjects. It was said by his contemporaries that if God were to allow all the knowledge in all the universities of the world to perish, but preserved Caramuel, he would by himself be able to reconstitute it to its present state.[89] The author of a book called *Anticaramuel* wrote, "Caramuel has the ability of eight, the eloquence of five, the judgement of two," the relative ordering of which seems reasonably accurate.

There was a center to this huge vortex of activity. It was moral theology, especially the consequences of probabilism. The first work to make him famous was his *Regular Theology*, or *Commentary on the Rule of St Benedict*.[90] Especially notorious was its Disputation 6, "On probable opinion." Almost any proposition is probable, he says. "Since we are not angels, but men, we have evident knowledge of scarcely four things, so we are required to act according to probable opinions." While intrinsic probability is relevant to judges in law, who are not versed in theology, it is not a useful criterion in general, as it is possessed by any proposition that is not against faith and has enough reasons in its favor to confuse objectors; in fact the only really improbable opinions are self-contradictory ones. Fortunately, extrinsic probability is better known, and "morally speaking" superior. So how many doctors suffice to give an opinion (extrinsic) probability? Some have thought a minimum of four, but this opinion itself implies that the minimum necessary is one, since there exist four doctors who think that one is sufficient. "Nevertheless," he says, "on my principles authorities are not to be reduced to mathematical computation, but to geometrical proportion. There are some authorities who prevail over many others." One must consider the authorities against an opinion, as well as those for, and balance them. If there is one authority for an opinion, how many against are sufficient to remove the resulting probability? Caramuel is not ashamed to give numbers: "If you have one authority on your side, against 7 of the same quality, you can say your side is still probable; but not against 9." In comparing authorities of different qualities, one of "the greatest reputation" may count as equivalent to five "of great name." Still, in practice, those who do not have expertise in comparing doctors may regard them as all "morally equal" and simply count them.[91]

What of the saying, "Happy the primitive church, which did not labor under so many opinions of doctors?" It is not true, Caramuel replies, that the primitive church lacked men of learning with contrary opinions. He continues with the expressions that caused the gravest scandal of all: "To say that probable opinions are burdens and diseases of the Church Militant is a manifest error, for they are signs of an easier and more excellent salvation. For the Church is not unhappy for having Doctors giving opinions constantly, but happier, since thus more mercifully and easily do its own advance to the Empyrean Laurel. Many would be damned, whom the probability of opinions saves. Thus should the Church be called happier, rejoicing in those most holy and learned men, who introduced benign opinions. . . . You ask, why are there now found so many different opinions in morals? The answer is that of the same Joannes Sancius *In. select. disp.* 44 no. 70, in which he concurs with the learned and distinguished Caramuel in his *Tract. super Regulam S. Benedicti* no. 60: that providence thereby shines the more. For by the variety of opinion the yoke of Christ is more lightly sustained."[92] Not only does Caramuel refer to himself here; the very passage refers to itself. In effecting this postmodern play, Caramuel must be either joking or revealing himself unable to distinguish between joke and nonjoke.

Though it hardly seems necessary, Caramuel goes on to apply his principles to various moral cases arising in the life of a monk. What if he wishes to follow a lax but probable opinion when his abbot orders him to follow a stricter one? He may do as he pleases ("give or take the scandal and other extrinsic inconveniences that commonly occur in such cases"). The abbot should not confuse his role with that of a tyrant, and those who, to keep control, teach that it is sinful to follow opinions that are in fact probable are wolves to their flocks, not shepherds. "It is the ruin of religion to invent obligations where there are none; for true obligations cannot be multiplied."[93]

He repeats these opinions in his *Fundamental Moral Theology* of 1652, the work from which the propositions attacked by Pascal and others were taken. The section on probability begins with the assertion that, while there are grades in probability itself, "all probable opinions in themselves are equally safe: the less harsh, even if less probable, are *per accidens* the more safe."[94] Of special interest is a passage in which he argues against the rigorist position that holds there can be no merely venial sexual sins on the grounds that, because of human frailty, there is always a risk that a small offense will lead to a greater one. This, Caramuel objects, is like saying that it is always a mortal sin to travel by sea on the grounds that there is some danger of shipwreck. One must consider also the size of this risk (here Caramuel develops ideas of earlier casuists on risk, which are described in chapter 10): "I say therefore that there can be motives which obviously move and influence us in grave matters, but which do not always move gravely, but often lightly. Plainly, to go by ship is a grave thing, and there is certainly a risk, but a light one. Why light? Because of a hundred ships that ply our seas, two or three at most perish, as the experience of the Dutch merchants shows. For there are men in their ports called Insurers, who go surety for Sea and Wind; if you pay them five in the hundred, they contract to pay back the full amount, if the ship happens to be wrecked. Therefore in the opinion of these men, of the two propositions, 'This ship will perish' and 'It will not perish,' the first is probable as five (or perhaps as two or three, for these Insurers profit and grow rich) and the second probable as a hundred."[95] It is the same in moral matters. One must evaluate the propositions, "I will sin" and "I will not sin" in terms of one's own and others' experience.

He speaks of evaluating probabilities also in connection with the morality of loans. In lending there is always a risk of losing one's money completely, especially in times like the recent German war. "Where it is equally probable that one will lose and recover the whole loan, it seems certain that one can in law receive 100, when the capital is 100. For otherwise, who would be willing to expose himself to an equally probable fortune of gaining 10, and an equally probable fortune of losing 100? . . . What if the loss arising [from advancing a loan] is not certain but probable? There ought to be a charge, more or less according to the quantity of the probability."[96]

The appearance he (and his enemies) sometimes give that he is forever counting doctors is balanced by remarks to the effect that manifest experience must be preferred to the opinions of ignorant doctors, who sometimes "do not dare to say that the sun is light, snow white, pitch black, or fire hot, without the support of the testimony of some old philosopher." And what was improbable yesterday may become probable today, as the existence of the Antipodes did after Columbus's voyage.[97]

Caramuel appears briefly in Pascal's sixth and seventh *Provincial Letters.* The purpose of the *Letters* is to expose the laxism of the Jesuits, so it is not entirely to the point to attack Caramuel, a Cistercian, but obviously Pascal had found things in Caramuel's works too useful for his purposes to leave out altogether. For Caramuel, after asserting that priests may under certain circumstances kill slanderers, considers the question, May the Jesuits kill the Jansenists? To which his answer is, no, "inasmuch as the Jansenists can no more obscure the glory of the Society than an owl can eclipse that of the sun; on the contrary, they have, though against their intention, enhanced it."[98] Pascal draws attention quite correctly to Caramuel's concentration on extrinsic probability, quoting Caramuel as saying "that the great Diana has rendered many opinions probable which were not so before, and that, therefore, in following them, persons do not sin now, though they would have sinned formerly."[99] Pascal's purpose is to portray almost all Jesuits as forever excusing, forever using probability to make morality easy. Caramuel was perhaps the only person who did genuinely fit the description. Probably the most scandalous of his doctrines was one complained of by the curés of Paris in their continuation of Pascal's campaign. Caramuel admits as probable that a man of religion may kill a prostitute who threatens to reveal his relations with her. The curés attribute to him a principle that they claim underlies the excesses of laxism: "What is required for a proposition to be probable to reason? That it be not evidently false."[100] While Caramuel was lax enough, this particular claim is not correct and is a prime example of quotation out of context or, as the Jesuits would have called it, mental reservation. What Caramuel had actually said was: "That it be not evidently false, and nevertheless be supported by a weighty basis capable of being defended by a learned man, and the opposite of which could be capably attacked by him."[101]

Let us return to more abstract subjects. The *Pandoxion Physico-Ethicum*—a typical Caramuel title—includes the results of his meditations on the insufficiency of Aristotle's logic, thoughts that, he claims, date from his student days in Spain. As Leibniz does later, he regards the flaw in Aristotle as his dealing only in strictly universal propositions. His logic is thus inapplicable to matters of fact in law and ethics, in which universal propositions, like "All men tell the truth," or "Caramuel never hallucinates," are not to be had. So he proposes a logic with more quantifiers that treats such propositions as morally universal, or most vehement: for example, "Almost all mothers love their sons"; and ones of usual force: "Around half of mothers love their sons."[102]

Caramuel was known to the scientific world mainly for his work in astronomy.[103] He has a characteristically unique view on the probability of the Copernican hypothesis: When the heliocentric theory was confined to the circle of philosophers, its probability had not been in dispute, but when the Scriptural passages were used in the controversy, it became a matter of faith, requiring intervention by the pope. Galileo, in defending the speculative probability of a thesis that had been rendered practically improbable by the earlier condemnation, showed himself disobedient. Nevertheless, the condemnation did not, strictly speaking, remove all speculative probability from the Copernican theory. He remarks also that God could put Mercury, or Jupiter, as the immobile center of the universe without it making any observable difference.[104]

Caramuel's views had been investigated by the authorities before Pascal's attack. He was called to Rome to answer for his lax views and also because his summaries of atheist and libertine objections to Christianity were thought unnecessarily sympathetic. He was called "the Carneades of this age" and "the eternal dishonor of our age and of the Church."[105] Opposition in print came with the *Observations on Caramuel's Theologia Fundamentalis* of 1656, written by the Spanish Dominican Martinez de Prado. Among propositions of Caramuel found particularly objectionable were that heresy is almost always excusable in Germany because of invincible ignorance; that some articles of faith are "accidental," and in them the church is not infallible; that one may read the books of heretics; that the obligation of abstaining from meat on Friday is not grave; and that opinions, for example in astronomy, which are attacked by the cardinals do not become improbable.[106] The expected condemnation did not eventuate. Instead, a longtime friend and colleague of Caramuel, whom he had assisted in the anti-Jansenist struggle in northern Europe, became pope. Caramuel "astonished the pope with the facility of his responses."[107] His *Fundamental Theology* escaped condemnation, though important corrections were required. He was made bishop of a small Italian diocese; one perquisite of the office was a printing press, of which he made good, or at least frequent, use. (His work on dice which appeared from this press is considered in chapter 11.)

A full official answer to the excesses of probabilism did not come until Prospero Fagnani's rigorist *Commentary on the Five Books of Decretals* in 1661. Fagnani first sets out ten reasons in favor of the proposition that it is permitted to follow probable opinions in morals. On the surface these seem fair summaries of the probabilists' reasonings, but he repeatedly says that probability is *indifferent* to truth or falsity, thus denying any grading in probability. Then he replies with twenty-six arguments to the contrary, all of which amount to saying that one must act only on the truth, which is attainable with moral certainty. He admits, referring to the legal authorities Bartolus and Baldus, that to attain moral certainty one might sometimes have to put together sufficiently many weaker arguments. He then insists that, if moral certainty is not attained, one may never risk following the laxer opinion.[108]

These views were excerpted for more popular consumption in the author's *Tract on Probable Opinion* of 1665; Caramuel's views are quoted and singled out for special attack. Caramuel's reply, the *Apology for the Most Ancient and Universal Doctrine of Probability*, was placed on the Index of Forbidden Books in 1664, together with a book by one of his few defenders, Verde's *Select Positions of Caramuel's Fundamental Theology Wrongly Accused of Novelty, Singularity, and Improbability*.[109]

Pascal's *Provincial Letters*

The story of the Jansenists and their struggle with the Jesuits is well known. It is surely one of history's best remembered storms in a teacup. The Thirty Years' War, with casualties running into millions, has left less impression on the collective memory, at least outside Germany, than the entirely bloodless harassment of two French convents. Its renown is almost solely due to Pascal, whose *Provincial Letters* and *Pensées* are deservedly acknowledged as classics of French literature. The publication of modern accounts of the affair with high literary merit, notably Sainte-Beuve's *Port Royal*, Bremond's *Histoire littéraire du sentiment religieux en France*, and Ronald Knox's *Enthusiasm*, has only added to the fame of this rather peripheral episode.

In the previous century, controversy had arisen over the difficulty of reconciling the theses that humans can act only with God's help ("grace"); that God knows future human decisions; and that the human will is free. The Jesuit Luis de Molina, in his *Concordance of Free Will with the Gift of Grace* of 1588, defended the freedom of the will, arguably at the expense of the other propositions. After the making of all the necessary and unnecessary distinctions, disagreement remained over the question, If God gave two persons identical graces, in total, could they make different decisions? Molina's answer, "With the same grace given

to many, one man is converted, another is not," enraged the Dominicans on the grounds that it envisaged God giving grace that turned out
to be ineffective. The matter was examined by the Inquisition in Spain,
whereupon Molina took the precaution of accusing his opponents of
Lutheranism, on account of their apparent denial of free will. A commission in Rome examined the question exhaustively over some ten
years. Though a decision against Molina several times seemed imminent, and rumors to that effect were prematurely celebrated in Spain,
the pope finally decreed that the views of both sides were permissible.
Each side was forbidden to call the other heretical, pending a final decision from the Holy See. That decision is still awaited.[110] Controversy
died down for a time but was reopened by the publication in 1640 of the
large book *Augustinus,* by Cornelius Jansen, bishop of Ypres. The book
quoted St. Augustine at length in support of various strict opinions on
grace, free will, and predestination. The work was condemned in general terms by Urban VIII, the same pope who had condemned Galileo.

In France, the influence of Jansen's work became attached to an already existing dispute between, on the one hand, certain fervent souls
associated with the convent of Port-Royal, near Versailles, and on the
other, the Jesuits, then enjoying the powerful support of Cardinal Richelieu. The strict Augustinianism of Jansen appealed to Port-Royal, and
with Richelieu's death in 1642, controversy was let loose. The defense
of Jansen was entrusted to Antoine Arnauld, brother of the formidable
Mère Angélique, who had reformed Port-Royal while still in her teens.
Arnauld was distinguished, even in that argumentative century, for his
lifelong devotion to controversies, apologies, and refutations. ("Day by
day he went into his study, sharpened his pen and attacked someone—or
defended someone, or refuted an answer, or answered a pretended refutation, or wrote a *Premier Ecrit pour la defense de la seconde lettre.*" "As Bremond says, to be a Jansenist you must always be writing *against* somebody.")[111] Arnauld's *Apologies for Jansen* enjoyed considerable success, but
he made rather little impression with his small *Moral Theology of the Jesuits, Faithfully Extracted from Their Works* of 1643. The opening words
of this work are "There is almost nothing the Jesuits do not permit to
Christians by reducing all things to probabilities, and teaching that one
may put aside the more probable opinion, which one believes true, to
follow the less probable, and maintaining further that an opinion is
probable as soon as two doctors teach it, or even one."[112] The Jesuits did,
however, sustain a minor defeat in this area over the possible consequences of lax views on the excusability in certain circumstances of regi-
cide—a fraught topic in the decade of the English Civil War.[113]

The matter was again little heard of for several years but was revived in 1649 by the appearance of the "five propositions" said to summarize the principal heresies to be found in Jansen.[114] The Jansenists maintained that the propositions as they stood were ambiguous and wrongly summarized Jansen's opinions. In view of the length and unreadability of Jansen's *Augustinus*, it was difficult to say whether this was a reasonable contention or not, and the focus of controversy tended to move to the question of whether the five propositions represented the true Jansen. *Le tout Paris* being scandalized by the apparent denial of free will by the Jansenists, the matter was referred to Rome, where opposing delegations conducted the usual intrigues. The result was the papal bull *Cum occasione* of 1653, condemning the five propositions and attributing them to Jansen.

In Paris, the Jansenists hastened to proclaim their full submission to the bull but continued to maintain that the pope had not condemned the propositions in the sense in which Jansen had intended them. Arnauld went further in the use of arguments of the kind commonly called "Jesuitical" in his *Memoir on the Fallibility of Popes and General Councils on Decisions of Fact*. Here the logic of uncertainty was first deployed in the controversy. He failed to prevent a further condemnation of Jansenist opinions by the French bishops.[115] On the other side, the Jesuit confessor to the king, Père Annat, published his *Chicaneries of the Jansenists*, a work that lacked most virtues but did have vigor. The Jesuits at this point maintained an advantage in the propaganda war, since Arnauld's works did not have the literary qualities to make them read in the salons, where the controversy now raged as fiercely as in the Sorbonne. Annat did however lay himself open to attack in the future by rashly committing himself to the assertion that the five propositions were to be found *word for word* in *Augustinus*.

By the winter of 1655–56, Jansenist fortunes were at a low ebb. The new pope was Alexander VII, the colleague of Caramuel. Arnauld faced a renewed conflict with the Sorbonne, with almost certain defeat at the end. Police action to disperse the community at Port-Royal was talked of. At this point there appeared Pascal's first *Provincial Letter*, dated January 23, 1656. Seventeen more letters appeared anonymously between then and May 1657, despite some minor police harassment of the printers. It may be conjectured that, though great serious literature survives, humor and propaganda are cumulative arts, like mathematics, in which each age improves on the techniques of its predecessors. Hence the indulgence needed when reading Shakespeare's or Dickens' humor. If this is so, then the *Provincial Letters* are some centuries ahead of their time.

The brevity, the timing, the always potent combination of irony and moral indignation are as effective as on the day of printing.

The first four letters are about grace and need not concern us here. The letters gained fame with each new publication, and the fifth created a completely new sensation. In it, Pascal widened the front to attack the entire edifice of Jesuit moral theology. (This plan was suggested, according to tradition, by the Chevalier de Méré.)[116] Pascal claims that the source of Jesuit views on grace is their moral laxism. And laxism is caused by probabilism. The attack of the fifth *Letter* is concentrated on the "doctrine of probable opinions, which is at once the source and the basis of all this licentiousness."[117] The letter is structured as a dialogue between a naive Everyman and a Jesuit father who explains to him the doctrines of his Society, illustrated with quotations from the Jesuit moral theologians. The father quotes an opinion from the casuists Ponce and Bauny that "we may seek an occasion of sin directly and designedly when our own or our neighbor's spiritual or temporal advantage induces us to do so." Pascal omits to mention that only Bauny is a Jesuit and that his quoted work was put on the Index of Forbidden Books in 1640, while his condemnation by the Sorbonne was only prevented at the last moment by Richelieu.[118] The dialogue continues:

"Think you that Father Bauny and Basil Ponce are not able to render their opinion *probable?*"

"Probable won't do for me," said I, "I must have certainty."

"I can easily see," replied the good father, "that you know nothing of the doctrine of *probable* opinions. If you did, you would speak in another strain. Ah! my dear sir, I must really give you some instructions on this point; without knowing this, positively you can understand nothing at all. It is the foundation—the very A, B, C of our whole moral philosophy."

Glad to see him come to the point to which I had been drawing him on, I expressed my satisfaction and requested him to explain what was meant by a probable opinion.

"That," he replied, "our authors will answer better than I can do. The generality of them, and among others the four-and-twenty elders, describe it thus: 'An opinion is called probable, when it is founded upon reasons of some consideration. Hence it may sometimes happen that a single *very grave doctor* may render an opinion probable.' The reason is added: 'For a man particularly given to study would not adhere to an opinion unless he was drawn to it by a good and sufficient reason.'" [The quotation is from Escobar.]

"So," I observed, "a single doctor may turn consciences round about and upside down as he pleases, and yet always land them in a safe position."[119]

Pascal's nonsequitur is obvious. If an opinion needs reasons of some consideration to be probable, a doctor cannot make an opinion probable "as he pleases." It is little excused by the ensuing recital of the lax opinions of a few Jesuit doctors. The real source of the disagreement is that Pascal believes there are no experts in morals and that anyone (or at least the saved) can work out his opinions in morals for himself. The modern world has followed Pascal in this—in morals, though not in science; no one would deny that if a noted scientist asserts a scientific opinion, that would be a reason for the layman to believe it, since the scientist can be expected to know the sufficient reasons for the opinion. The casuists' assumption that the same can be done in morals renders them fossils indeed. We, like Pascal, are shocked even when the casuist is dealing with genuinely disputable cases in which reasoning and experience with similar cases might seem to have some relevance. Pascal has the father say, "Now the authority of a learned and pious man is entitled to very great consideration; because (mark the reason), if the testimony of such a man has great influence in convincing us that such and such an event occurred, say at Rome, for example, why should it not have the same weight in a question of morals?" Pascal's answer is, "An odd comparison this . . . between the concerns of the world and those of conscience!"[120] It marks a sudden break between medieval and modern thought. Before, one supports one's opinion by adducing the doctors and their reasonings in its support. Afterward, one claims one's opinion is supported by one's bare conscience. Which method is likely to lead, in practice, to greater laxity of behavior, Pascal does not inquire.

The letter, and later letters, continue with further lax opinions quoted from various authors. Father Bauny appears again, and Escobar is held up for special ridicule. Whether these authors are fairly represented by Pascal has been the subject of dispute, with the tide of opinion running firmly against Pascal.[121] The quotations are selective and misrepresent the authors as lax in toto, which they are not. After the success of the *Letters*, the aged Escobar became a minor tourist attraction and was given to mildly protesting to visitors that his views had been exaggerated and that there were others more lax.[122] Escobar's work actually has only half a page directly on probabilism, and it is by no means extreme. He writes that "an opinion is called probable from there arising reasons for it of some moment. Thus sometimes a single weighty doctor can render an opinion probable because a learned man specially trained would not adhere to any proposition unless led to it by a solid or sufficient reason. Is it permitted to follow a probable reason, leaving aside the more probable? It is, and it is even safer to do so, unless there is a threat of some peril that ought, in prudence or justice or charity, to be avoided by choosing the opposite opinion. Can I choose a probable opinion, leaving aside those that seem to me more probable and safer? Indeed so, nor do I act against my conscience in doing so, provided I think the opinion I follow is probable."[123]

As Escobar's work is so rarely printed, let us allow him at least one reply in his own words on the strictness of opinions, in a passage not quoted by Pascal: "When a problem arises about which opinions are equally divided, as regards the number, the testimony, and the authority of the doctors who have expressed them, I choose and give greater approval to the opinion which is more favourable to religion, piety and justice. . . . In the matter of vows, oaths and testamentary dispositions, my approval goes to the opinion which seems to agree better with the nature of such actions, and also to that which tends more to the protection of orphans, widows, strangers from foreign parts, and other persons called in law 'miserabiles.'"[124] This passage gives a little more insight into the *purpose* of laxism.

Pascal's story is—and this is the weakest part of his case—that the Society of Jesus deliberately maintained both lax and strict casuists so that they would be able to make themselves agreeable to, and hence ensnare, people of all kinds, and so aggrandize the Society. Did Pascal really believe this? The motive of men such as Escobar seems rather to be to adapt centuries of undoubtedly rather strict Christian moral teaching to the diversity of circumstances. And diversity of circumstances means, in the first instance, borderline cases in which there are conflicting rules of morality. It is perhaps Pascal's omission of the context of conflicting considerations in the cases he extracts that is the most consistent index of his bad faith. As a modern commentator writes, "Pascal deliberately chose to give the impression that an odd collection of out-of-context opinions that were at best occasionally permissible deviations from a quite different norm represented a coherent morality."[125]

Since the casuists went so far as to treat particular cases in detail, they inevitably made fools of themselves more than occasionally, as the law sometimes makes an ass of itself in having to pronounce on the myriad combinations of unlikely occurrences. In many cases, the casuists criticized by Pascal reach conclusions that the average person today might agree with, though he would pretend his conscience had told him the answer directly rather than admit he had indulged in reasoning. For example, in the ninth *Letter*, the jovial "good father" speaks of "good cheer, which is accounted one of the greatest pleasures of life, and which Escobar thus sanctions in his *Practice according to our Society:* 'Is it allowable for a person to eat and drink to repletion, unnecessarily, and solely for pleasure? Certainly he may, according to Sanchez, provided he does not thereby injure his health; because the natural appetite may be permitted to enjoy its proper functions.'"[126] One is supposed to be shocked by the laxity of this and similar opinions of Escobar, such as that a woman may adorn herself "merely to gratify a natural inclination to vanity." (What Escobar actually said was that adornment for the sake of exciting lust is a mortal sin but for attracting a husband or wife it is not a sin at all, while doing so from a natural inclination to vanity is either a venial sin or no sin at all.)[127] Modern opinion has sided with Escobar rather than with Pascal and the Puritans on such issues. At least, it has sided with him in practice, while refusing to commit itself in theory.

The subsequent fortune of the campaign may be briefly described. The renown of the *Letters* continued to grow, though not all on the Jansenist side were entirely happy. Mère Angélique commented, with some justice, on "the eloquence which amuses more people than it converts."[128] Between the fifth and sixth *Letters*, when, according to Racine, "Port Royal was in consternation and the Jesuits at the height of their joy," help for the Jansenist cause appeared from an outside source. Pascal's niece, about to undergo an operation for a weeping fistula of the eye, was miraculously cured by the Holy Thorn, a relic said to be from Jesus' Crown of Thorns and kept at Port-Royal. Jansenist apologetic came to rely heavily on this miracle, necessitating attacks on the certainty of the miracles of their enemies.[129] On the other hand, a papal bull arrived declaring that the five propositions were in Jansen and had been condemned in the sense in which Jansen meant them. Following the eighteenth and last *Letter*, the campaign against laxism was continued by the parish clergy of France, some of whose statements on the question were written by Pascal. Jesuit replies to the *Letters* appeared but were of poor quality.

The matter seemed about to expire by the end of 1657 when there appeared the Jesuit *Apology of the Casuists against the Calumnies of the Jansenists.* It does contain a few shrewd replies on the subject of probability: "And we say it belongs only to proud spirits who presume to know all truths, or to misguided souls who persuade themselves they have revelations of everything, to blame probable opinions and say that a probable opinion is not enough for acting prudently and exempting from sin one who follows it."[130] Unfortunately, rather than taking the obvious line that Pascal's selection of quotations from the casuists was biased, the *Apology* defended all the most excessive of the casuists' statements, causing the scandal to revive.[131]

Both laxist and Jansenist statements were condemned by popes before the end of the century, though not by the same pope in each case. A list of lax propositions condemned in 1679 included one taken almost word for word from Caramuel, to the effect that fornication is not intrinsically evil and is only condemned because of the disorders it could introduce into society.[132] After certain doctrinal accommodations, Port-Royal enjoyed in the 1660s a period of comparative peace and even prosperity with "the coming and going of great ladies in carriages."[133] But it was never entirely free of persecution, actual or threatened, and the community was eventually dispersed on the king's order.

CHAPTER FIVE

Rhetoric, Logic, Theory

W hose business is it to explain the *theory* of probability? What discipline owns the theory and the right to evaluate applications of it in other fields? The modern answer has been mathematics. It is an answer that has brought a great deal of clarity into obscure matters and has, for example, saved millions of lives by finding accurate methods of testing drugs for their curative powers. Yet it biases the theory toward aspects of probability that can be given precise numbers. Dice are studied intensively; how to take into account context and background information is neglected. The ancient answer to the same question was that the theory of probability belonged to rhetoric. There are some advantages in that answer too, especially in directing attention to a wide body of data—all the arguments that are found to be persuasive in practice. The concomitant disadvantage is a persistent tendency to blur the distinction between what ought to persuade, because it is probable, and what can be made to persuade by giving it a color of plausibility. That is, the logical and the psychological are easily confused. It is true that the distinction is clearly stated by Aristotle and others and also that the ancient sense of "rhetorical" did not emphasize the *merely* persuasive as much as the modern sense does, but the focus on what actually persuades was enough to prevent an autonomous theory of nondemonstrative argument from being seriously developed in logic. As a consequence, the present chapter, potentially the most exciting because it concerns the *theory* of probability most directly, is in fact something of a disappointment.

It is for this reason that the material on rhetoric is placed after that on the law of evidence, although much of it is older. Very little of the rhetorical tradition fed into law, even though ancient rhetoric was especially designed for use in legal contexts. While it might seem natural to have discussed Greek rhetoric before Roman law, the truth is that the rhetorical and juristic attitudes to evidence are so different that the traditions are almost independent. The purpose of rhetoric, in the view of the ancient masters, is the construction of persuasive arguments in, especially, courts of law. The impression given by the Sophists and Aristotle that procedure in Athenian courts was essentially a rhetorical free-for-all is apparently correct.[1] In a situation in which logic was on a par with appeals to honor and exhibitions of the wife of the accused in tears, persuasiveness rather than logic was the saleable commodity. A Roman jurist, ancient or medieval, on the other hand, regarded himself as a professional evaluating evidence according to objective rules and principles handed down by authoritative tradition. The Roman judge preferred interrogation to the cut and thrust of parties and their lawyers, and, as we saw, took a view of evidence shaped by that preference.

The Greek Vocabulary of Probability

The speeches produced by Greek rhetoricians were intended to be heard by assemblies of ordinary citizens and were usually designed to be delivered by ordinary citizens in the course of defending themselves in lawsuits. So what was said about probability in rhetoric is continuous with ordinary language usage, and it is convenient to survey that first.

The resources of the Greek language included two words, *pithanon* and *eikos*, which had some probabilistic content, at least potentially. *Pithanon* means persuasive, or plausible; it can be used of both speakers and what they say. It can have the suggestion of mere persuasiveness.[2] But sometimes it simply means what is in fact convincing: Herodotus speaks of "the most plausible of the many stories of Cyrus' death"; Plato says that "such has been the constant tradition, and is plausibly true"; Aristotle argues that "it is equally plausible that every compound is a composition or that none is."[3] The word *eikos*, literally "like," is often used in exactly the sense of its English descendant "likely."[4] From Sophocles:

> *Electra:* These counsels mean that thou wilt not share my deed.
> *Chrysothemis:* No, for the venture is like to bring disaster.
> *Electra:* I admire thy prudence, thy cowardice I hate.[5]

A speech in Thucydides has the reasoning: "For many reasons we are likely (*eikos*) to prevail: firstly, because we are superior in numbers and military experience . . . so if we win a single victory at sea, they are most likely (*kata to eikos*) defeated."[6] Plato says "Chance will bring them, as is likely, just the sorts of things that usually happen to orphans."[7] He posits the following chain of arguments:

> Any excess often brings about a reaction to the opposite, in the seasons,
>   in plants, in animal bodies, and above all in political societies.
>   It is likely, he said.

> And so the likely outcome of too much freedom is only too much slavery, in the individual and the state.
>
> Yes, that is likely.
>
> It is likely, then, that tyranny develops out of no other constitution than democracy.[8]

Readers of Plato's dialogues recall the many occasions in which Socrates' victim, his ignorance exposed, is reduced to agreeing with a series of Socrates' propositions. The admissions conventionally rendered as something like "It seems so, O Socrates," often in fact say simply, "It is likely (*eikos*)."[9] Plato claims that the *like* in *eikos* means "like truth" but gives this idea no further content.[10] It is not clear that anyone else understood the word in this way except for later authors following Plato, so it may just be a piece of amateur etymology.

What is difficult to say is how much these usages represent the innate resources of the Greek language and how much they are an effect of the rationalist spirit of the fifth century. Certainly there is nothing like them in Homer, even at places where there would seem to be natural opportunities[11] and even though Homer shows a characteristically Greek preoccupation with evidence and the reasons for belief.[12]

A second essential piece of the Greek cultural background is the explicit idea of the conflict of opinion and its resolution. The saying, "Hear both sides," is quoted by choruses in Greek plays as if it is part of ancient wisdom; Aeschylus mentions it at the same time as doubting whether oaths are reliable, thus marking the transition to deliberately rational thought, which the Greeks made in this area as in so many others.[13] Those who later found Greek thought exasperating often felt there was far too much hearing of both sides, to the exclusion of other desirable goals, like reaching a conclusion.

The Sophists Sell the Art of Persuasion

A conscious awareness of probability or of anything like the grading of evidence apparently begins with the treatment of plausibility by the Sophists, who were paid precisely for teaching how to make conclusions plausible. The invention of rhetoric was traditionally ascribed to Corax and Tisias, Greeks of Syracuse about 460 B.C. The story told of them (for a certain sum, Corax promises to train Tisias to win any case of law; Tisias receives the training and refuses to pay; who wins the case?), if not true, certainly gives the flavor of the Sophist movement.[14] From the beginning there was an emphasis on likely arguments, in the sense of what could be made to seem convincing.

Aristotle explains the nature of such arguments and expresses the general disapproval later felt for the whole immoral business. "It is of this line of argument that Corax's Art of Rhetoric is composed," he says. "If the accused is not open to the charge, for instance if a weakling is being tried for assault, the defence is that he was not likely (*eikos*) to do such a thing. But if he is open to the charge—if he is a strong man—the defence is still that he was not likely to do such a thing, since he could be sure that people would think he *was* likely to do it. And so with any other charge: the accused must be either open or not open to it: there is in either case an appearance of likely innocence, but whereas in the latter case the likelihood is genuine, in the former it is only in the special sense mentioned. This sort of argument illustrates what is meant by making the worse argument seem the better. Hence people were right in objecting to the training Protagoras undertook to give them. It was a fraud; its likelihood was not genuine but an appearance, and has a place in no art except rhetoric and eristic."[15]

Sicily and Athens in the fifth century B.C. shared, in addition to a rationalist spirit and a love of disputation—entities intangible in the small but undeniable in the mass[16]—a long period of democracy. Legal as well as political decisions were generally entrusted to popular assemblies, leading to a demand for skill in public speaking. The immediate stimulus for the success of Corax and Tisias was the succession of democracy in Syracuse upon the overthrow of a tyranny that had forcibly moved large populations between cities over considerable periods of time. There resulted disputes about property and citizenship for which there was little evidence available to the parties. It is in such conditions of lack of evidence that arguments from likelihoods must be called on to fill the gaps.[17]

The use made of the notion of likelihood can be followed in the only Sophists whose work has survived in more than small fragments, Gorgias and Antiphon. Gorgias of Leontini was the archetypal Sophist, renouncing the pursuit of truth, wisdom, and virtue in favor of pure rhetorical skill. Plato's dialogue *Gorgias* represents him as claiming to be able to answer any question put to him.[18] His earnings from the teaching of rhetoric were large. Plato elsewhere mentions Gorgias and Tisias, "who saw that likelihoods (*eikota*) are more to be esteemed than truths, who make small things seem great and great things small by the power of their words." (Gorgias is said to have remarked, "How well Plato knows how to satirise.")[19] One of his surviving works is the *Defence of Palamedes*, a hero in Homer. The point of choosing a well-known fictional character seems to have been to produce a model speech exhibiting various defensive strategies, which could be adapted to real cases as appropriate. The case is one in which there is little evidence, so some of the arguments illustrate the use of likelihoods. Palamedes, a hero said to have invented the alphabet, lighthouses, and dice, offended Odysseus, who had Palamedes convicted and executed for treason by hiding a forged letter from the Trojans in his tent. Gorgias writes the speech Palamedes is to give at his "trial." Only one of the arguments explicitly contains the word "likely" ("Someone will say that we made the contract for money, he giving it, I taking it. Was it for little? But it is not likely that a man would take a little money for a great service."). But it is clear that Gorgias regards nearly all his arguments as similar in principle to this one and as merely expressed differently for literary reasons ("For much money? Who was the go-between? How could one person bring it? Perhaps there were many? If many brought it, there would have been many witnesses to the plot.").[20]

The Athenian sophist Antiphon, "an object of suspicion to the multitude because of his cleverness," was executed in 411 after the fall of the Four Hundred. Although the story that he conducted a pain and grief clinic at Corinth, offering therapy by *logoi*, is probably an invention of the comic stage, his dedication to the art of words and debate is unquestionable. His book *On the Interpretation of Dreams* is said by Cicero to interpret dreams intelligently but without force or natural plausibility.[21] His surviving speeches comprise both exercises on fictional cases and speeches actually delivered in court. The remarkable complexity of his arguments from likelihoods can be judged best from the speech for the defense in an imaginary case in which the facts are disputed.

A man's enemy has been killed in the street. A slave attacked with him has, before dying, named the accused as one of the attackers. The defendant addresses the jury: "If on grounds of likelihood you suspect me because of the intensity of my hostility, it is still more likely that before I did the deed I should foresee the present suspicion falling on me. . . . It is not, as the prosecution alleges, unlikely, but likely, that a man wandering about in the middle of the night should be killed for his clothes."

"But is it not likely," he continues, "that those who hated him not much less than I did, and there were many of these, killed him rather than I? They could see plainly that suspicion would fall on me. . . . Further, what is the value of the attendant's testimony? It was not likely that he would recognise the killer in the panic of the moment. What is likely is that he would assent to the persuasion of his masters. When slaves are not as a rule held trustworthy as witnesses—otherwise we should not subject them to the examination—how is it fair to destroy me on the evidence of this one? But if anyone thinks that likelihood has the weight of proven truth against me, let him on the other hand consider by the same token that it was more plausible that I should wait for a safe opportunity for the plot and not be present when the deed was done."[22]

There ensues a second speech for the prosecution, followed by another for the defense, involving further duels of likelihoods. The use here of a fictitious case is doubtless responsible for some of the author's self-indulgence in following various lines of reasoning, but Antiphon's speech in a real case, in which the defendant is accused of murdering someone who vanished without trace, is hardly less complicated and certainly no less explicit in exploiting likelihoods.[23] The idea that a jury of common citizens could be expected to follow the balancing of various contrary-to-fact suppositions and comparisons of likelihoods would be incredible were it not for the fact that the Athenian dramatists routinely expected similar feats of concentration from their audiences.

Aristotle reports also an interesting saying of Antiphon the Dramatist, which no doubt represents the influence of the Sophists: "One might perhaps say this was likely / That many things not likely do by chance happen."[24] Socrates, Plato, and Aristotle were concerned to distance themselves from all this. Plato in particular hoped to raise the intellectual and moral tone of discussion by contrasting his true logic with the Sophists' immoral use of mere persuasiveness; the English meaning of *sophistry* is a sign of his success. (The consequences for Plato's philosophy, and hence all later philosophy, are considered in chapter 8.)

The question arises as to whether the picture of the Sophists in Plato and Aristotle is fair. No more than fragments of the Sophists have survived, and men of genius are not known in general for giving fair treatments of their opponents' views. The distinction between the Socratic school with its high-minded devotion to the truth and the word-chopping Sophists, clear enough no doubt to Plato, was less so in the popular mind, as is apparent from the satire in Aristophanes' play *The Clouds*. There was, then, every motive to exaggerate the differences. The arguments of Gorgias and Antiphon above are not unreasonable, though they do leave the impression that the authors are fundamentally not interested in the truth. The idea of dismissing arguments as merely plausible is a trick of the Sophists themselves, parallel with the rhetorical trick of dismissing one's opponent's arguments as mere rhetoric. Gorgias's *Defence of Palamedes* includes the argument: "That you have no knowledge of your accusations is clear. Hence they must be opinion, and you are the most villainous of men, to bring a capital charge relying on opinion—which is a most unreliable thing—and not knowing the truth. Conjecture is open to all in everything, and you are no wiser than anyone else in this. One must believe not conjecture, but truth."[25] Indeed, the most perfect example of this argument occurs in the writings of Plato himself. At his trial, Socrates opens his defense with the claim that he can only speak the truth in plain language, as he is making his first appearance in court, does not know the language of the place, and is too old to learn new tricks. This is of the same form as Corax's argument about the strong and weak men charged with assault: "How could someone so old and with so little experience in verbal argument (!) be able to use skilful speech to deceive?"[26]

The later Attic orators, of the fourth century B.C., also provide some evidence that in rhetoric there is more than the shallow appeals to plausibility and emotion that Plato suggests. They often argue well on what is the more likely explanation of the facts.[27] There is a rather high standard of argument in the orators' speeches that deal with cases in which probabilities are relied on, as in Antiphon and Gorgias, because hard evidence is not available. One of the most accomplished is Lysias's speech *On the Sacred Olive*. Lysias, a pupil of both Tisias and Socrates, was an orator considered strong in reasoning though weak in pathetic appeal. In this case, an Athenian landowner is accused of digging up and removing the stump of a sacred olive on his property. There appears to be no actual evidence that he did so, nor any trace of the alleged stump. The defendant, in the speech written for him by Lysias, argues that witnesses have said there was no olive on the spot, that there was no profit from removing the stump, that doing so would have put him in his slaves' power, that the accuser did not act at the time, that "there are many sacred olives and burnt stumps on my other plots which, had I so desired, it would have been much safer to clear away,"[28] that the site was visible to a road and neighbors, that he has been zealous in his service of the state, and that the accuser refused his offer to have his slaves tortured. "You should remember, gentlemen, which side you ought rather to credit, those for whom many have borne witness, or one for whom nobody has ventured to do so; whether it is more likely (*eikos*) that this man is lying, as he can without danger, or that in face of so grave a danger I committed such an act."[29]

The mention of torture indicates that a perhaps inevitable reaction to the frustration arising from the juggling of likelihoods is to demand "not conjecture but truth." Torture was not confined to the absolutist legal system of imperial Rome and its heirs. Demosthenes assures a jury that torture is the most certain of all methods of proof and that it is right to torture witnesses, especially slaves.[30] Many legal systems have used torture but usually only on suspects. Torturing innocent witnesses is another level of horror entirely.

Aristotle's *Rhetoric* and Logic

Until well into the nineteenth century it was almost possible to regard logic as a subject that had sprung fully formed from the head of Aristotle. By this was meant, of course, deductive logic, and it would be too much to expect him to have given birth to nondeductive logic as well. If anything, one would expect the founder of deductive logic to have spent his time emphasizing the distinction between deductive arguments and all "mere persuasions." The case is almost the exact opposite. Aristotle's *Rhetoric* is the most intrinsically important work on probability before Pascal.

While it is true that Aristotle presents rhetoric as primarily an art of persuasion—an argument "is persuasive (*pithanon*) because there is someone whom it persuades"[31]—the whole tenor of the book is to distinguish sharply between arguments that ought to persuade (studied by dialectic) and sophisms (studied by rhetoric strictly so called).[32] His objections to merely sophistical arguments are primarily moral ones, but he has in addition a faith that, in general, the truth will out: "Both these arts [dialectic and rhetoric] draw opposite conclusions impartially. Nevertheless the underlying facts do not lend themselves equally well to the contrary views. No, things that are true and things that are better are, by their nature, practically always easier to prove and easier to believe in."[33] Deductive arguments, he says, are rarely useful, since deliberation about actions concerns the contingent; hardly anything is determined with necessity. One must use likelihoods (*eikoton*) and signs, the likely (*eikos*) being what usually happens.[34] The arguments Aristotle actually gives are also clearly rational and are meant to be taken as such. The ones relevant to the topic of probability are induction (*epagoge*), or argument from example, argument from signs, and analogy.

Induction covers arguments from the particular to the general, the argument from observing that a number of As are Bs to the conclusion that all As are Bs. "When we base a proof on a number of similar cases, this is induction in dialectic, example in rhetoric."[35] The actual cases given of argument by example are of argument from a number of As having been Bs to the conclusion that a new A is also a B. "The example has already been described as one kind of induction. . . . The argument may be, for instance, that Dionysius, in asking as he does for a bodyguard, is scheming to make himself a tyrant. For in the past Peisistratus kept asking for a bodyguard in order to carry out such a scheme, and did make himself a tyrant as soon as he got it; and so did Theagenes at Megara; and in the same way all other instances known to the speaker are made into examples, in order to show what is not yet known, that Dionysius has the same purpose in making the same request: all these being instances of the one general principle, that a man who asks for a bodyguard is scheming to make himself a tyrant."[36] Cases of induction are also given where the argument is from examples to an explicitly general conclusion, such as that a country becomes prosperous when its leaders become philosophers.[37]

Signs, though at first said to correspond to what is necessarily true, are then divided into fallible and infallible signs. "'The fact that he breathes fast is a sign that he has a fever.' This argument also is refutable, even if the statement about fast breathing be true, since a man may breathe hard without having a fever." Using one instance as a certain proof is classified, of course, as a fallacy.[38]

Arguments from the past to the future also rely on examples: "Examples are most suitable to deliberative speeches [ones recommending future actions], for we judge of future events by divination from past events."[39] While it is easier to supply examples by inventing fables, it is more valuable for the political speaker to supply them by quoting what actually happened, since in most respects the future will be like the past.[40]

Similarly for speeches in court there can be arguments from likelihood. An example of an argument that involves grades of likelihood is, "Again, the argument that a man who strikes his father also strikes his neighbours follows from the principle that, if the less likely thing is true, the more likely thing is true also; for a man is less likely to strike his father than to strike his neighbours."[41] Other modes of argument to past fact use common knowledge of what things succeed others: "If a man was 'going to do something,' he has done it, for it is likely that the intention was carried out. . . . Of all these sequences some are inevitable and some merely usual."[42] Aristotle is not entirely explicit at first about whether he understands these arguments to be rationally persuasive or merely ones that happen to persuade. That *rationally* persuasive is what he means becomes clear in his discussion of how such arguments may be refuted:

> Now as the likely (*eikos*) is what happens mostly but not always, enthymemes based on them can, it is clear, always be refuted by raising some objection. The refutation is not always genuine; it may be appar-

ent, for it consists in showing not that your opponent's premiss is not likely, but only in showing that it is not inevitably true. Hence it is always in defence rather than in accusation that it is possible to gain an advantage by using this fallacy. For the accuser uses likelihoods to prove his case, and to refute a conclusion as unlikely is not the same as to refute it as not inevitable. Any argument based on what mostly happens is always open to objection: otherwise it would not be a likelihood but an invariable and necessary truth. But the judges think, if the refutation takes this form, either that the accuser's case is not likely or that they must not decide it; which, as we said, was a false piece of reasoning. For they ought to decide by considering not merely what must be true but also what is likely to be true: this is, indeed, the meaning of "giving a verdict in accordance with one's honest opinion." Therefore it is not enough for the defendant to refute the accusation by proving that the charge is not bound to be true; he must do so by showing that it is not likely to be true. For this purpose his objection must state what is more usually true than the statement attacked. It may do so in two ways: either in respect of frequency or in respect of exactness. It will be most convincing if it does so in both respects; for if the thing in question both happens oftener as we represent it and happens more as we represent it, the likelihood is more.[43]

Aristotle further advises on how to deploy likelihoods in connection with external evidence, such as witnesses, and in defending implausible statements. The treatment lacks something of the high moral tone approved of elsewhere. [44] "In dealing with the evidence of witnesses, the following are useful arguments. If you have no witnesses on your side, you will argue that the judges must decide from what is likely; that this is meant by 'giving a verdict in accordance with one's honest opinion,' that likelihoods cannot be bribed to mislead the court, and that likelihoods are never convicted of perjury. If you have witnesses, and the other man has not, you will argue that likelihoods cannot be put on trial, and that we could do without the evidence of witnesses altogether if we needed to do no more than balance the pleas advanced on either side."[45] Torture likewise is easily argued either to be reliable or unreliable.[46] An even more cunning, or confusing, line of argument "refers to things which are supposed to happen and yet seem incredible. We may argue that people could not have believed them, if they had not been true or nearly true: even that they are more so because incredible. For things believed are either facts or likelihoods: if, therefore, a thing that is believed is incredible and unlikely, it must be true, since it is certainly not believed because it is likely or plausible (*eikos kai pithanon*)."⁴⁷

If later ages had been familiar with Aristotle's *Rhetoric*, they would have had a sound basis for developing a theory of probable argument. That was not the case. While in most centuries Aristotle's eminence was agreed on and his authority respected, he was not read from beginning to end. For one thing, his total work is enormous and is not understood without great effort. Second, some of the works considered less important were more or less lost for long periods of time. A few of the details of what was known at what times are mentioned later as appropriate, but in general it can be said that while something of Aristotle's logical works, especially the *Prior Analytics* and the *Topics*, were known to the learned of almost all times, the *Rhetoric* was hardly known in Western Europe between the time of Quintilian and the thirteenth or fourteenth century. It is important, then, to mention briefly the view of probability found in the logical works. The account of induction, in particular, is much less probabilistic than that in the *Rhetoric*; it never really becomes clear whether these arguments are supposed to be deductive or not. Sometimes such inductions appear to be inconclusive, and it is said that if someone refuses to admit the conclusion, one is justified in asking what objection he has.⁴⁸ On the other hand the *Prior Analytics* suggests that induction must be complete, that is, all the instances must be surveyed. The example given is the argument to the conclusion, "All animals that have no bile are long-lived" from "Horses, men and mules are long-lived." Such an argument from species to genus needs only to survey the small number of species in the genus to make the induction complete.⁴⁹ Argument from example also appears in the logical works, but its status is most obscure.⁵⁰

More generally, Aristotle's ideal of science requires that the universal propositions used in demonstration should be established by a process of understanding or insight, as happens in establishing the axioms of geometry. Ideally, one will know that all As are Bs with certainty because one will understand *why* any A must be a B—even in very empirical sciences like medicine.⁵¹ Plainly this reduces the need for considering inductive arguments in the modern sense.

These passages on induction give substance to the Enlightenment picture of Aristotle as the father of the old order and the prime obstacle in the way of the rise of experimental science. In life Aristotle was a genuine experimenter and observer, especially in biology. In fact he notices that the ideas of rhetorical probability can apply to argument in biology: he remarks on the reasons for thinking that semen comes from all parts of the body: "These opinions are made probable by the witness of such facts as that children are born with a likeness to their parents, not only in congenital but also in acquired characteristics."[52] But the Aristotle of the logical works, as taught in the schools and universities of medieval Europe, was indeed a supporter of the view that pure thought is the method of science, not "argument from examples."

On the other hand, the logical works do give a certain prominence to arguments from "what happens for the most part." The *Prior Analytics* says that "a likelihood (*eikos*) is a generally accepted premiss (*endoxos*); for that which people know to happen or not to happen, or to be or not to be, mostly in a particular way, is likely, for example, that the envious are malevolent or that those who are loved are affectionate."[53] There follow some examples of arguments using premises of this kind, which produce some truth but that "can always be refuted."[54] Their status is not as clear as it might be.

The *Topics* is said to be about dialectic; it opens with a division of arguments into three kinds: the demonstrative, the dialectical, and the eristic (fallacious). In demonstration, necessary truths follow from necessarily true premises, as in geometry. Then "reasoning is dialectical which reasons from generally accepted opinions (*endoxon*). . . . *Endoxa* are those which are agreed to by all or most or the wise, that is, to all of the wise or most or the most notable and distinguished of them. Reasoning is eristic if it is based on opinions which appear to be *endoxa* but are really not so."[55] This passage produced an important result when translated. In the standard Latin translation of Boethius, *endoxa* appeared as *probabilia*,[56] the same word standardly used to translate *pithana*. As a result, the formula "Agreed to by all or most or the wise, that is, to all the wise or most or the most notable and distinguished of them" became the standard medieval definition of *probable*. Such a definition is far from making probability an objective logical relation, though it is also far from making it a matter of subjective plausibility. It turned out to fit in all too well with the already excessive medieval habit of evaluating propositions in accordance with the weight of the authorities who approved them. The wide knowledge of this dictum was a factor in causing the plague of probabilism that swept the courts and confessionals of Europe in the seventeenth century (described in the previous chapter).

Probability, in this sense, belongs to premises, not arguments, and there is not much in the body of the *Topics* about probable arguments. Almost the only arguments discussed are from analogy, or similarities: "It is an accepted principle that what holds good of one of several similars, holds good for the rest." It is difficult, however, to decide which things are really similar.[57] Aristotle's *Poetics* makes use of a notion of the "likely" in literary criticism. "The unlikely one has to justify either by showing it to be in accordance with opinion, or by urging that at times it is not unlikely. For there is a likelihood of things happening also against likelihood. . . . There is no possible apology for unlikelihood of plot or depravity of character, when they are not necessary and no use is made of them."[58] Probability as applied to a plot or a character is not obviously the same as probability in general, but Aristotle clearly does not mean to make any such distinction here.

Besides the *Poetics*, Aristotle's works on literature include one on the reconciliation of apparent contradictions in Homer. From the surviving references to it, it sounds rather like fundamentalist treatments of the Bible. It seems a particularly inappropriate exercise of his genius, and it is perhaps fortunate for his reputation that it is lost. At least it seems that he recognized the uncertainty of that kind of argument: "Agamemnon is disparaged by Thersites as a womaniser. . . . But it is hardly likely (*eikos*), Aristotle says, that this number of women was for use—it was rather a mark of status; after all, it was not for getting drunk that he had a large supply of wine."[59]

The *Rhetoric to Alexander*

Among the body of works attributed to Aristotle, a number are definitely spurious. Of these the most interesting is the *Rhetoric to Alexander*; it purports to be a treatise written by Aristotle for his pupil Alexander the Great but is generally agreed to be "a work proceeding from an entirely different and inferior order of mind and character."[60] No one has a good word to say for it. It is much more a practical guide to oratory than Aristotle's *Rhetoric*, in which, as in all his works, the practicalities are for the most part excuses for introducing theoretical principles of extreme generality. In fact, it fulfills its purpose well. The author has a sound grasp of the subject and a clear view of what is and what ought to be convincing. One of the divisions of proof is the "likely":

> Something is likely (*eikos*) when the audience have examples in their own minds of what is said. For instance, if someone said that he desired his country to be great, his friends prosperous and his foes unfortunate, and the like, his statements would seem likely, because each of his hearers is aware that he has such wishes on these and similar matters. . . .
>
> With these definitions made, when we are seeking to persuade or dissuade, we must show with regard to the matter in question that the action we are urging or opposing does have the effect that we assert, or if

not, that actions similar to this action either generally or invariably turn out as we say they do. This is how one must apply likelihoods in regard to actions. With regard to persons, when accusing you should prove, if you can, that the party has often committed the same act before, or if not, acts like it. Also try to show that it was to his advantage to do it, for most men, themselves preferring their own advantage, think that others too look to their own advantage. This is how to use argument from likelihood if you can derive it from your opponents personally. Failing that, infer what is normally the case for people resembling them, for example, if the man you are accusing is young, argue that he has acted as those of that age usually do act, for your allegations against him will be believed on the ground of similarity. In the same way, you will be believed if you can show that his companions are the same sort of persons that you say he is, since it will be supposed that he has the same pursuits as his friends, because of his association with them.[61]

The argument about the young man is a clearer example than anything in Aristotle of the proportional syllogism, that is, the argument from a proportion in a reference class to a particular case: Young men generally act thus, so this particular young man probably acts thus. The author helpfully adds something on how to reply to such arguments. "If you have never done anything of the kind, but some of your friends do happen to have done such things, you must say that it is not just that you should be discredited because of them, and must show that others of your associates are honest men; you will thus throw doubt on the accusation. If they show that other people who resemble you have done the same things, declare it to be absurd if the fact that certain others can be shown to have done wrong is to be a proof that you too have committed the offences alleged against you. That is how you must make your defence from likelihood (*eikoton*) if you are denying that you have done the thing alleged, because this will make the charge improbable (*apithanon*)."[62] There follows a section on arguments from examples, which are said to be weaker than arguments from likelihoods, and it is explained how to choose examples for each side of a case. To reply to such arguments, "when your opponents use this device, you must show that their instances were just good luck, and say that such things happen rarely, while your examples have often occurred."[63]

Then there is argument from signs. "A sign of a thing is what usually precedes or accompanies or follows it. . . . A sign may produce either belief or full knowledge; the latter is the best kind of sign, while one that produces a very probable (*pithanotaten*) opinion is second best."[64]

There is later some discussion of how to deal with the evidence of witnesses, whether given freely or under torture and whether probable or not.[65] The treatment is similar to Aristotle's but is more detailed. The advice is shrewd. In general, the author has a very sound grasp of what he is doing. Perhaps he does not have Aristotle's theoretical range and philosophical viewpoint. Perhaps he does not want it.

Roman Rhetoric: Cicero and Quintilian

Cicero was probably the individual most responsible for the translation of Greek ideas into Latin. As we see in chapter 8, he wrote Latin versions of some Greek philosophical debates involving probability but without much success in explaining the notion of probability involved. Cicero's expertise was principally as a legal orator, expert in persuasion on matters of fact. Here he is on home ground and is prepared to classify the reasons that make probable arguments persuasive:

> That is probable which for the most part usually happens (*quod fere solet fieri*) or which is the general opinion or which has in itself some likeness to these, whether it be true or false. In the class of things which usually happen are the probable of this kind, "If she is a mother, she loves her son," "If he is avaricious, he disregards his oath." Under the head of what is the general opinion are probables such as "Punishment awaits the impious in the next world," "Those who write philosophy do not believe the gods exist." Likeness is found mostly in contraries, in analogies and in things that fall under the same principle. In contraries, such as "For if it is right to pardon those who harm one unintentionally, there is no need to be grateful to those who assist one because they have no choice." In analogies, thus, "For as a place without a harbour cannot be safe for ships, so a mind without integrity cannot be relied upon by friends." The probable in things that fall under the same principle is considered, "For if it is not base for the Rhodians to farm out the customs, nor is it base for Hermocreon to take the contract." Arguments of this kind are sometimes true, such as "Since there is a scar, there was a wound," sometimes like true (*veri similia*), such as "If there was much dust on his shoes, he must have just taken a journey."
>
> Everything probable (to make definite subdivision) that is used in argument is either a sign, or credible, or judged or a comparison. A sign is something that falls under the senses and signifies something or seems to follow from that; it may have occurred either before the event or in it or following after it, but still needs further testimony and weightier confirmation, such as blood, flight, pallor, dust and the like.[66]

Cicero keeps Aristotle's connection between probabilistic reasoning from what happens for the most part and persuasion using the opinions of the audience. But he introduces a confusion by conflating these with the kind of argument now found mostly under the name of jurisprudential analogy. A good deal of legal and moral argument is about whether a new case is relevantly analogous to past ones in which decisions have been reached; the reasoning is typically like that in Cicero's argument above: "If it is not base for the Rhodians to farm out the customs, nor is it base for Hermocreon to take the contract." It has never been clear whether this kind of legal argument is supposed to reach its conclusions with certainty or not. In any case it seems quite different from genuinely probabilistic arguments about matters of fact.[67]

As with Aristotle, the works attributed to Cicero include the *Rhetoric to Herennius*, a book that is contemporary but that is not by him. Anonymous works on rhetoric seem particularly sad, since there can be little point in studying rhetoric, let alone writing books on it, other than to achieve fame. The work is close to the *Rhetoric to Alexander* in tone, explaining how to make the most of one's own side: "For things after the crime we consider what signs usually accompany the guilty and the innocent. The prosecutor will say, if he can, that his adversary, when come upon, blushed, was pale, faltered, spoke uncertainly, collapsed, or made some offer,—which are signs of a guilty conscience. If the accused has done none of these things, the prosecutor will say his adversary had so premeditated what would happen to him that he stood his ground and replied most confidently—which are signs of confidence, not innocence."[68] The author argues that by putting different pieces of evidence together, one may produce "knowledge, not suspicion" and justifies this by a "not by chance" argument, a form of argument Aristotle uses in astronomy (discussed in chapter 6): "To be sure, some one or two things can by chance have happened in such a way as to throw suspicion on him, but for all the things to agree among themselves from first to last, it is necessary that he committed the crime; this cannot have happened by chance."[69]

By the time of the late Republic and early Empire, skill in speaking was the path to public office—at first, no doubt, because it persuaded but later perhaps more because it provided a system of formal credentials. Qualifications in rhetoric were regarded as quite the thing for a young man entering public life, proving his readiness to imbibe technicalities on the instructions of his seniors, to regurgitate them on demand, and to put together a few thoughts on themes not too intellectually demanding. The abstractness of Aristotle's treatment disappeared. Quintilian, for example, has a section on arguments from "conjecture," by which he means arguments for doubtful matters of fact or intention. He has little advice to give, even in cases of conflicting evidence, for example when considering this "controversial case: 'A tyrant, suspecting his doctor had given him poison, tortured him, and when he still denied he had done so, sent for a second doctor; the latter asserted that poison had been given, but that he would give an antidote; he gave him a potion, drinking which the tyrant died. The doctors both claim the reward for killing a tyrant.' Now just as in cases of mutual accusation where each party shifts the guilt to the opposite side, so here the characters, motives, means, opportunities, documents and testimonies of the claimants are compared."[70]

But Quintilian does at least transmit some of the basics of Greek thinking on the combination of evidence. "There are other non-necessary signs, called *eikota* in Greek. Even if these are not sufficient by themselves to remove doubt, they may be of the greatest value when taken together with others. . . . But bloodstains on a garment may be the result of the slaying of a victim at a sacrifice or of a bleeding nose. . . . Hermagoras would include among such non-necessary signs an argument such as, 'Atalanta is not a virgin, since she has been roaming in the woods with young men.' If we accepted this view, I fear we should make all inferences from a fact into signs."[71]

He has an important passage that explains the grading of evidence. "One who is to handle arguments correctly must also know the power and nature of everything, and what usually happens. It is from this that arise what are called *eikota*. There are three kinds of credibility: First, the firmest, based on what mostly happens, such as that children are loved by their parents. Secondly, there is the very likely (*propensius*), such as that a man in good health will survive to the morrow. The third is the not unlikely, as that a theft in a house was done by one of the household. So Aristotle in the second book of the *Rhetoric* has examined very carefully all that commonly happens to things and persons."[72]

Where Quintilian gives a weakened version of arguments with conflicting evidence, his contemporary Seneca does the same for argument from what happens "for the most part." Of one who died in boyhood, Seneca philosophizes that "he has lost nothing except a hazard where loss was more assured than gain (*nisi aleam in damnum certiorem*). He might have turned out temperate and prudent; he might, under your care, have been moulded to better things, but—as one would more justly

FIGURE 5.1

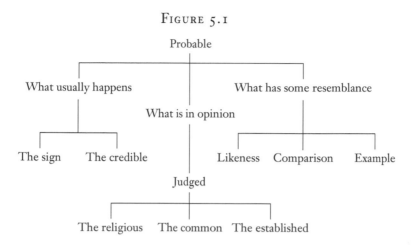

fear—he might have become like most. See those youths, whose extravagance flung them from the most noble homes into the arena. . . ."[73]

The diversion of serious thinking into these tedious moralizings, so much admired by the Elizabethans, to us shows the decay of the Silver Age. Let us draw a veil.

There was a continuing Latin tradition of commentary in rhetoric during the later Empire. These works in general lack much interest, being merely summaries and codifications of Cicero and Quintilian. An example is Marius Victorinus's commentary on Cicero's *De Inventione*, which includes a diagram of Cicero's classification (see figure 5.1). The explanation follows Cicero closely.[74]

In considering the transition from antiquity to the Middle Ages, it is important to remember how few of the ancient texts were known to most medieval writers. From the purely historical point of view, the works of Boethius assume a significance quite out of proportion to their intrinsic merits. For many medieval authors, Boethius was the first, and often the only, point of contact with the more abstract levels of ancient thought. For them, the classical statement about probability was the definition from Boethius's *On Different Topics;* which is based on the definition of *endoxa* in Aristotle's *Topics* and thus has a distinctly subjectivist tone. "Something is readily believable if it seems true to everyone or to most people or the wise. . . . The truth or falsity of the argument makes no difference, if only it has the appearance of truth."[75] But Boethius's definitions of *induction* ("discourse by means of which there is a progression from particulars to universals") and *example* were likewise standard for the next thousand years. There is nothing subjectivist about these definitions or the accompanying examples. He is also admirably clear of the fallibility of induction (an issue raised earlier in philosophy, as we see in chapter 8, but not in logic).[76]

Islamic Logic

While the West underwent the simplification of its culture to a semi-tribal level, the torch of knowledge passed instead to the countries of Islam, which proved sympathetic to abstract thought during its early centuries. The important works of Aristotle were translated into Arabic, and the chief thinkers were more or less close followers of him. Unusually, the Islamic writers regarded Aristotle's *Rhetoric* and *Poetics* as part of his logic.[77] His *Rhetoric* was thus taken rather more seriously than it later was in the European Middle Ages; al-Farabi, the early leader of the Aristotelian school in the ninth century, is said to have read the *Rhetoric* two hundred times and written seventy books on it.[78] (The cavalier attitude to figures evident here is arguably one of the obstacles to the development of a mathematical theory of probability.) His division of logical arguments into the demonstrative, tentative, sophistic, rhetorical, and poetic allows rhetorical arguments to approach certainty.[79]

The most influential of this school is Ibn Sina, known in the West as Avicenna. His *Treatise on Logic* is heavily dependent on Aristotle but has something to add on the distinction between induction and analogy. "Argument by analogy is weaker than induction. It is a proposition about the particulars of a class based on observing things similar to those particulars. For example, 'The soul is a power which ceases to exist after the death of the body, just like sight which ceases to exist after the destruction of the eye.' This argument is often used in politics and theology. But the proposition is never certain, because it is possible that a proposition about one class of particulars will be contrary to a proposition about a class of similar particulars. For there are many things which are like each other in some respects and unlike each other in other respects. But even though analogy can never give us certainty it can provide us with some kind of satisfaction."[80] He wisely remarks that one of the difficulties with arguments such as analogy that use the properties of things is that it is difficult to enumerate all the properties a thing has. (Avicenna's opinions on the problem of induction and on experimentation in medicine are mentioned in later chapters.)

The last noted Islamic thinker is Ibn Rushd, called Averroes in the West. Commenting on Aristotle's *Rhetoric*, he considers testimony. "Assent to testimonies and reports of sense-perceived matters which have not been witnessed is strengthened and weakened in accordance with the number of the reporters and other considerations relating to them. Thus, the most powerful assent resulting from reports is what a group which cannot be enumerated reports it has perceived or what a group reports on the authority of another group which cannot be enumerated but which has perceived it."[81] The reason for his concern to place a high value on the testimony of groups soon becomes clear—he is talking about religious tradition: "Certainty with regard to diverse matters— like the sending of the Prophet, the existence of Mecca and Medina, and other things—may result from this." In general, Islamic treatments of *endoxa*, the "widely accepted" premises of Aristotle's *Topics*, strongly emphasize the element of community assent. Islamic analyses of the certainty of traditions dating from the time of the Prophet contain substantial discussions of the independence of witnesses, the confirmation of reports by circumstantial evidence, and the improbability of everyone conspiring in a lie.[82]

The Scholastic Dialectical Syllogism

The revival of learning in the Gregorian reform was as much logic led as law led. At the same time as the rediscovered *Digest* was having its profound impact, Anselm offered his ontological argument, according to which God could be proved to exist purely by the analysis of concepts. It must represent a valuation of logic than which no greater can be conceived.[83] Abelard's rational questioning of theology is not far behind: "If Adam's sin was so grievous that it could only be expiated by the death of Christ, what atonement shall be offered for the crime of them who put Christ to death?"[84]

As in the case of legal reasoning, one can find before the discovery of the ancient texts certain discussions that indicate minds prepared to receive the gifts of the ancients. Basic discussions of the distinction between probable and necessary and of Boethius's definitions are found as far back as 1000.[85] Cicero's *De Inventione* and the *Rhetoric to Herennius* were known from about 1080 and stimulated a wide interest in rhetoric.[86] Chartres was the center of synthesis of this first version of the ancient heritage with Christian thought. Symbolizing the harmony still perceived, at least by some, between the ancients and Christianity, a statue of Aristotle, representing dialectic, appears on the west front of Chartres Cathedral.[87]

Being restricted at first to texts in Latin, the European mind had to concentrate on what would now be called the "soft" sciences; perhaps this allowed it a necessary period of grace before being exposed to the hard discipline of Aristotle's scientific works and the mathematics of Euclid and Ptolemy. It also gave probability time to become established as an object of study.

Original thought in the area dates at least from Abelard's *Dialectic*, of about 1120. The issue is the still live one of whether probabilistic inference from premises to conclusions, inference that falls short of deductive implication, should be admitted as part of logic. Abelard thinks it should not but usefully summarizes the view of his opponents thus: "They ascribe probability, according as the consequent is easily acquiesced in following the proof of the antecedent. Often one who judges the consequence false, still acquires belief in the consequent from the admission of the antecedent, although he does not concede the consequence and the necessity of the inference. For although I know that there is not necessarily any love-making when a girl is often caught at night speaking with a young man in secret, nevertheless I easily suspect and concede love-making from such conversations, from the fact that we never see conversations of this kind happen except between lovers. So from admitting an inference, I acquire belief, and whatever does not seem to suffice for inference, still seems very much worthy to produce belief, and whatever probability does not seem to suffice for inference, does support the belief."[88]

Abelard refuses to admit such reasoning into logic, arguing that "probability is casual and often adjoined to falsity" and that to admit such reasoning would require the abandonment of established rules of logic such as "whatever implies the antecedent implies the consequent." On the other hand, not all of Abelard's opinions are hostile to probability. When commenting on Boethius, he is clear, as Boethius is not, on the distinction between arguments from probable premises and arguments that are probable as inferences, that is, where the premises, whether certain or not, provide only partial support for the conclusion; he asserts that both are possible.[89] The brief treatments of probable, dialectical, and inductive reasoning in later medieval works on logic do little more than repeat a few formulas from Aristotle and Cicero.[90]

Rhetoric as a study in itself appealed little to the Scholastic mind, and its place in the curriculum shrank after the twelfth century.[91] At the end of the thirteenth century, Aristotle's *Rhetoric* finally began to make some impact in the West.[92] Despite Aristotle's authority by that time, it never became a widely studied work. The first Latin commentary on it explains that the difference between belief (caused by rhetoric) and opinion (caused by dialectic) is not that opinion is accepted more firmly but rather that rhetoric moves the mind to assent by appetite (that is, appeals to emotion), while in dialectic the intellect is moved by its proper object (that is, what is reasonable).[93] Some translations were made also of Arabic commentaries on the *Rhetoric* and the *Rhetoric to Alexander* but had little influence.[94] The *Rhetoric* and the *Rhetoric to Herennius* were discussed in the next century but not widely.[95]

Meanwhile, the Scholastics had found something of interest in Aristotle's much better known logical works. Albert the Great was one of the most enthusiastic followers of Aristotle, whose texts were then, in the first half of the thirteenth century, becoming widely read and causing controversy. Like his pupil Thomas Aquinas, Albert was convinced of the fundamental harmony between Aristotle and the Christian revelation. His distinctions on the grades of belief, discussed in the previous chapter, were plainly suggested by his reading of Aristotle, and his commentary on Aristotle's *Topics* shows more of a genuine meeting of minds than many commentaries. In discussing Aristotle's definition of "probable" in the *Topics*, Albert distinguishes two senses of the word. "The probable (on which the dialectical syllogism is based) is the likely. Likely is taken in two ways. Either things are likely in themselves, in that the predicate's being in the subject is itself likely, because the predicate is not in the subject *per se* nor the subject in the predicate, nor both in both, nor does the predicate necessarily and essentially inhere in the subject, but it is taken as likely not from necessary causes but from signs. Or the inherence is necessary, but is only understood through signs; and this is the probable commonly understood, even though it is necessary in itself: for example that the sun is larger than the earth (because it looks the same size everywhere)."[96]

He goes on to describe dialectical, legal, and medical reasoning as all probabilistic. "I call arts of potency those that do not establish one side of a contradiction with solid argument, but fluctuate between the two sides with infirmity of argument: this is to use the suitable and probable to persuade of what we wish to persuade. For the rhetorical advocate does not persuade completely and every time, being prevented in three ways: by the evil of his cause, the perversity of the judge and the weakness of his suit. Nor does the doctor always heal, for three reasons: the fallacy of experience, the evil of the disease and the disobedience of the patient, so the desired end does not always follow. Likewise the dialectician is prevented by the weakness of the probable he adduces, by the power of his opponent and by the difficulty of the problem in dispute."[97] Albert was also prepared to take seriously argument from what happens for the most part and to accept it as a part of logic parallel to the syllogism. He says definitely, unlike Aristotle, that induction is a probable argument, since it cannot survey all the instances.[98]

The legal mind will be happy that proof by two, or three, or twenty witnesses fulfills the prescription, No condemnation on suspicion. The philosophical mind will not. To an abstract thinker, it will be obvious that no amount of multiplying the number of witnesses or other signs of guilt, even confessions, can convert an uncertain argument into a certain one, one beyond suspicion. This was clear to Thomas Aquinas. Aquinas is not one to say that any piece of received wisdom is rubbish, but he is certainly prepared to reinterpret the terms in it where necessary. He agrees in theory with the jurists in holding that passing judgment on suspicion is unlawful, but what he means is something different: "According to Cicero a suspicion involves an ill opinion founded on light indications."[99] Suspicion is thus defined as essentially a vice, perhaps better translated as suspiciousness, and is therefore something to be avoided always. So whereas for the jurists no condemnation on suspicion meant that the highest standard of proof was required, for Aquinas it means only that a very low standard is excluded.

Aquinas is insistent on the need for merely probable reasoning in law and other practical affairs. Considering the objection to the biblical two-witness rule from the possibility that two or more people might conspire in lying, he answers that "in human affairs it is impossible to have demonstrative and infallible proof; it suffices to have a certain conjectural probability (*conjecturalis probabilitas*) such as the rhetorician uses to persuade. So, although it is possible for two or three witnesses to agree in a lie, it is not easy or probable that they do, and so their testimony is taken as true, especially if they do not hesitate in it and are not otherwise suspect."[100] Considering directly the question whether two or three witnesses suffice, he cites the dictum of Aristotle's *Ethics*, that "the same certainty is not to be sought in every matter" and adds that "it is not possible to have demonstrative certainty in what concerns the contingent and variable. And so a probable certainty suffices, which attains the truth in most cases (*ut in pluribus*), though in a few cases it fails to. And it is probable that what many say more contains the truth than what one says."[101]

Because of the variability of things, laws cannot cover all cases; Aquinas adduces the *Digest* as well as Aristotle in support of this proposition.[102] On the issue of conflict of testimony, Aquinas reflects the understanding of the lawyers, saying that conflict on matters of substance takes away from the efficacy of testimony and that in general the judge must use his mind to decide, taking into account the number and standing of the witnesses. He mentions the lawyers' half-proof without any special comment.[103]

In theology, there are also probable arguments, namely those from authorities less than the best. Argument from scripture is necessary, but that from doctors of the church or pagan authorities is only probable.[104] The significance for the later citation of authorities in moral theology is clear.

These statements constitute most of what Aquinas has to say explicitly about probable arguments. They involve a full acceptance of probabilistic reasoning, and there could be no more prominent or influential a place to showcase such an acceptance than the *Summa Theologiae*. On the other hand, there is no serious examination of probabilistic reasoning in itself and not much explanation of how to perform it. Aquinas does certainly approve of reasoning from what happens "for the most part." Besides the connection noted with the evidence of witnesses, the subject is treated under the heading, Is memory a part of prudence?: "Prudence deals with contingent matters of action. . . . Now to know what is true for the most part one must consider experience. . . . It is from the past that one ought to take a kind of argument to the future; and so memory of the past is necessary for being well advised as to the future."[105] And he does make some worthwhile general remarks about Aristotle's dialectic, the science of probable reasoning. "Opinion, caused by a dialectical syllogism, is on the way to [full] knowledge, which is acquired through demonstration. Once this is acquired, the cognition that came from the dialectical syllogism can remain, as a kind of consequence of the demonstrative knowledge, which is based on knowledge of causes. For one who knows the cause can better know the probable signs from which the dialectical syllogism proceeds."[106]

Interestingly, Aquinas does not take the view that because dialectic leads to uncertain conclusions it is itself variable, contingent, and unworthy of study. On the contrary, he asserts that the study of dialectic as such, showing the ways in which it reaches probable conclusions, is a demonstrative science.[107] He is here relying on a distinction, common in the logic of the time and perhaps going back to Abelard, between *dialectica utens*, the art of constructing probable arguments, and *dialectica docens*, the science dealing with the theory of probable argument.[108] Conversely, one can know probably that a proposition is demonstrable without knowing it demonstratively, that is, by having understood the demonstration.[109] He further compares the fallibility of dialectical argument with the operations of nature, which act correctly for the most part but fail sometimes, "as when from seed something monstrous is generated." Just so, the dialectical syllogism can cause belief or opinion—but not reliably.[110]

Probability in Ordinary Language

In view of Aquinas's influence, it is significant that he uses the word *probable* in an unusually free way in ordinary language: "It is probable that parents [in ancient times] addressed certain prayers to God on behalf of their newly-born children"; "It is more probable that [the star that appeared to the Magi] was a newly-created star, not in the heavens, but in the air near the earth."[111] This sort of language is part of everyday speech now, but that has not always been so. Even so developed a language as classical Greek, as we saw, had words for "probability" only in moderately learned contexts, and there is no sign of such concepts in languages like Old English. The significance is that the century or two after Aquinas was the period when the vernacular languages of Western Europe became capable of supporting abstract discussion and adopted the vocabulary of the Scholastics to do so.[112] Particularly in the hard intellectual times after the Black Death, universal learning in the international language was a luxury that had often to be dispensed with,[113] though demand for books by moneyed but less educated customers like merchants also played a part. Popularizations and translations of serious works in the vernacular languages were one of the vehicles by which the thought and vocabulary of the learned reached the audience that created the written languages of modern Europe.

Brunetto Latini's encyclopedia in French, *Li livres dou tresor* of about 1265, for example, helped introduce the distinction of the necessary and the likely into French. Mostly following Cicero, Latini's chapter on probable arguments defines *voirsemblables argumens* as "from things that are accustomed to happen often, or from things that are believed to be, or from things that have some appearance of being true or likely." Signs, examples, and presumption are mentioned.[114] The French language was particularly fortunate to secure the services of Nicole Oresme, among whose many talents was an aptitude for mathematical probability (discussed in the next chapter). His translations of some of Aristotle's works into French show that the vulgar tongue, suitably supplemented, could support the highest level of abstract discourse.[115] His translation of the passage in the *Nicomachean Ethics* about the certainty of mathematics being unattainable in other sciences contains apparently the first occurrence of the word *probable* in French; *probabilité* first occurs in his translation of Aristotle's *Politics*.[116]

English was a somewhat less developed language, but again the first uses of *probable* and *likely* in English reflect the context of Latin learning. The first known occurrence of the word *probable* in English is in Trevisa's translation of Higden's *Polychronicon*, of about 1385. It says of snakes in Ireland: "It is more probable [Latin *probabilius*] and more skilful that this lond was from the bygynnynge alwey without such wormes."[117]

It might be thought that English already had a native word, *likely*, for probability concepts, but that is not the case. *Likely* is the English version of the Latin *verisimilis*, and the first uses of *likely* and its derivatives are in the *Canterbury Tales*, of about the same date as Trevisa.[118] The early uses of *probable* and *likely* in English are nothing like "It will probably rain tomorrow" but more like "What if ever either of the said premyssis concludyng for feith, y have not oonli likli euydencis, whiche ben clepid in scolis of logik 'probable euydencis' or 'probabilitees,' but y have sure certeynte bi kunnyng or bi experience."[119]

Most modern European languages have acquired a pair of synonyms corresponding to English *probable* and *likely*, French *probable* and *vraisemblable*, and so on.[120] They are not native to any of these languages but are borrowings from the Scholastic *probabilis* and *versimilis*, themselves descended from the Greek *pithanon* and *eikos* via Cicero. Although the two Greek terms are not quite synonyms, the two in Latin and its derivatives are in practice perfectly synonymous, despite occasional individual attempts to find small differences between them.

Humanist Rhetoric

Later humanism did not produce developments important for probability. Where Scholastic logic kept almost wholly to deductive logic, the humanists saw rhetoric as an art of discovering arguments and of embellishing them so as to speak well. There is no serious discussion of the middle ground as there is in Aristotle's *Rhetoric*. The frequent mention of probability in humanist rhetoric is to be seen in this context.[121]

In line with this, Aristotle's *Rhetoric* itself, though occasionally read and lectured on, had a low profile compared with the works of Cicero and Quintilian.[122] On the other hand, Renaissance literary theory was dominated by Aristotle's *Poetics*. It was common to connect what Aristotle said about the probable in the *Poetics* with his definition in the *Topics*, and there were some attempts to distinguish between *probabilis* (what seems so the wise) and *verisimilis* (what seems so to the vulgar).[123] Savonarola went so far as to classify poetry under logic: the poetic art is "knowledge of the syllogism called Example."[124]

The most influential humanist work in the field was Rudolph Agricola's *De Inventione Dialectica*. Probability, when it occasionally appears, is a purely rhetorical property. "I am aware that really probable things can be said in matters which are abhorrent, not only to our convictions about the world, but even to our mental powers themselves: Lucian on the matter of men being changed into birds or Macrobius' puzzle as to the priority of the chicken or the egg: either side you take on them appears improbable, yet both sides may be spoken of credibly. And the view of Heraclitus and his followers that good and evil are the same, as well as the position of the later Academy that nothing can be known, and very many other things of the same stamp can not only be argued credibly but can claim authors who have so believed. Hence, dialectical probability will be defined by us in this manner: as the art of discoursing with probability about any matter being considered, according as the nature of each subject can render this possible."[125] Plainly this material is not on or adjacent to any route toward a genuine understanding of probability. Henry VIII's order to the students of Cambridge in 1515 to study Agricola instead of the "frivolous questions" of Scotus is as useful as most government edicts that academic knowledge should be more "relevant."[126]

One humanist whose work is a partial exception to this unhelpfulness is Thomas Wilson, whose *Logique* of 1551 and *Rhetorique* of 1553 were the first complete works on logic and rhetoric in English.[127] The *Rhetorique* is brought down to earth by illustrating its techniques with arguments in legal cases; Wilson was familiar with both English and Continental law. He provides a model speech for the prosecution to use when there is no direct evidence:

> And whereas God revealeth to the sight of men the knowledge of such offences by divers likelihoods, and probable conjectures: . . . This soldier being desperate in his doings, and living by spoil all his life time, came newly from the wars, whose hands hath been lately bathed in blood, and now he keepeth this country (where this farmer was slain) and hath been here for the space of one whole month together, and by all likelihoods he hath slain this honest farmer . . . his soil also (where he was born) giveth him to be an evil man: considering he was bred and brought up among a den of thieves, among the men of Tindale and Ryddesdale, where pillage is good purchase, and murdering is counted manhood. . . . No greater gamester in a whole country, no such rioter, a notable whoremonger, a lewd roisterer among ruffians, an unreasonable waster, today full of money, within a sevenight after not worth a

groat. . . . And who will not say that this caitiff had little cause to fear, but rather power enough to do his wicked feat, seeing he is so sturdy and so strong, and the other so weak and unwieldy: yea, seeing this villain was armed, and the other man naked. Doubt you not (worthy judges) seeing such notes of his former life to declare his inward nature, and perceiving such conjectures lawfully gathered upon just suspicion: but that this wretched soldier hath slain this worthy farmer.[128]

This is plainly a collection of the standard examples of Continental law put together implausibly in a single case. Most of the evidence would be inadmissible in modern English law.

Wilson's expertise in the massaging of evidence was put to use in 1571, when he examined on the rack two of the Duke of Norfolk's servants, in connection with the Duke's alleged treason, and wrote a preface to the Casket Letters, a set of apparently edited, if not exactly forged, documents implicating Mary Queen of Scots in the murder of her husband.[129] (Wilson's book on usury is mentioned in chapter 10.)

Late Scholastic Logics

Through the late sixteenth and seventeenth centuries and beyond, the Scholastics, who still had charge of university syllabuses, kept up a flood of books summarizing the standard course in logic and philosophy. They usually included a very small section on the "topical syllogism," discussing the difference between knowledge and opinion.[130] It was usual to explain also how opinion ("a probable assent from an intrinsic medium, or some natural connection in things") differed from faith ("from an extrinsic medium, namely the authority of a speaker").[131] The work on which Galileo based his early lectures on logic laments that the probable syllogism is often neglected in the schools and spends several pages remedying the defect.[132] Another popular logic, in a long section on the dialectical syllogism, warns against arguments from silence, that is, from lack of testimony, like "This phrase is not in Cicero, therefore it is not Latin," "All the ancients were unaware that the magnet points to the pole, therefore this is not true."[133] Another text gives as maxims of logic rules of the kind found earlier as legal presumptions: "If what seems more likely is not true, neither is what seems less likely."[134] The 1638 *Logic* of the Jesuit Smiglecki, much used as a textbook at Oxford, said that multiplying probable reasons can make an opinion "infinitely more probable" but cannot make it certain, any more than making a body "infinitely more corporeally perfect" can make it into a spirit.[135]

On the same question of the gradual emergence of belief as the result of the accumulation of evidence, John of St. Thomas says, in 1631, "In faith and opinion there can be moral evidence and certainty, not with regard to the truth attained, which always remains obscure and incertain—this incertitude is called intrinsic or metaphysical, for it concerns truth itself—but with regard to credibility or probability. When there is such an accumulation of motives that credibility becomes evident and no room remains for disbelief, all extrinsic fear is removed. . . . Everything that pertains to wavering and doubting in belief and opinion is suppressed or diminished in proportion to the growth of motives and of the reasons leading to credibility or probability; these reasons may reach evidence, and then they entirely rule out this kind of fear." He notices also that the regress of probable reasons must end in something probable, since the contingent does not follow from the necessary.[136]

The most popular of the Scholastic logics was that of Burgersdijck, of 1626. It contains what may be one of the rare incursions of reality into pure logic; his example of a universal proposition is "All crows are black," his countrymen's recent extraordinary discovery in New Holland having rendered inoperative the previous example, "All swans are white." Burgersdijck displays no tendency to inductive skepticism as a result of this contretemps. He believes that an induction that does not survey all the instances is adequate provided it is clear that the unsurveyed instances follow the "same reason." So one cannot argue, "The foot, hand and ear should be amputated if they are so badly infected that there is no hope of recovery otherwise, so the head should be too," because there is a reason why not. His chapter on the dialectical syllogism mentions probability but avoids discussing the concept.[137]

And so on. None of this is especially interesting in itself, but the intellectual climate of the time cannot be appreciated without knowing what the Scholastics said, as the curriculum of the universities was of their making. Burgersdijck's *Logic*, for example, was a staple as far away as Harvard, and its vogue lasted well into the eighteenth century, when it disgusted such students as Swift and Burke and was burned "according to annual custom" in Dublin.[138] The Scholastics were thus studied perforce by those later most critical of them. We will see such makers of modern thought as Galileo and Descartes using the Scholastics' vocabulary to explain the novelty of their own thought.

The most famous of seventeenth-century logics, the *Logic of Port-Royal*, contains an amalgam of ideas from older theory and Pascalian discoveries on probability. (It is considered briefly in the epilogue.)

CHAPTER SIX

Hard Science

The big bang theory of the universe is much more probable, on present evidence, than the steady-state theory. But it is a rare scientist who can be found to say exactly how much more probable—or even approximately how much. Physics and astronomy, both ancient and modern, have few explicit probability calculations. Nevertheless, probabilistic reasoning is essential to both sciences, and it is visible at both ends of the theoretical spectrum. At the most theoretical end are overarching theories intended to explain a large and disparate body of observational data. The more complicated the relation of a theory to all relevant data, the more difficult it becomes to isolate and calculate the relation between the theory and any particular piece of data. Hence discussion of the probability of large theories is generally in the qualitative and "anything goes" style of legal arguments, with individual experts giving different weight to particular measurements or alleged measurements, to background theories, and to alternative hypotheses. The widely accepted Duhem thesis, asserting that observation confronts theory only on a holistic basis, was originally asserted only for physics ("an experiment in physics can never condemn an isolated hypothesis but only a whole theoretical group").[1] It is much more plausible there than in, say, anatomy.

At the most detailed level, on the other hand, where a formula or curve is fitted to a series of measurements, there is a place for formalizable probabilistic methods. In astronomy, especially, measurements are subject to random errors, from the limited accuracy of the instruments and from atmospheric fluctuations. Modern mathematical statistics descends from the application of least squares methods to predict the orbits of comets from a few data points.[2] Modern methods are essentially a refinement of the older idea that a more accurate measurement of a doubtful quantity can be gained by averaging several inaccurate measurements.

Observation and Theory

If a few of the factual claims in the vast corpus of Aristotle's works are false, who can be surprised? He and his contemporaries lacked most of what now seem to be the minimal prerequisites of serious research on matters of fact—institutions of learning endowed with secure money, large libraries, training in research techniques like observation, ways of communicating with experts in distant places. It is astonishing that a largely correct and coherent body of science appeared at all. When it did, however, support followed. An adequate research infrastructure was made available to science in later antiquity, especially in Alexandria, the intellectual center of the world for a record term of eight hundred years, and it is there that one finds the development of techniques for combining observations.

The difference a tradition of research makes to science is obvious in the treatment of the roundness of the earth in early and late ancient science. The basic discovery was made by Parmenides or some other pre-Socratic around 500 B.C. It needed a single flash of inspiration, based largely on its ability to give a coherent account of eclipses. It is a prime example of the kind of probabilistic reasoning now called inference to the best explanation. Its initial appeal lay more in its cleverness than in any detailed observational support. Almost the only directly observable facts it explained were the roundness of the dark patch on the moon during an eclipse, caused by the roundness of the earth's shadow, and the fact that eclipses occur at full moon.[3] But by the time of Ptolemy, about 150 A.D., there was much more convincing evidence, resulting from the combination of observations from widely spaced locations. The roundness of the earth in the north-south direction is proved by the gradual appearance of new stars over the southern horizon as one travels south. The roundness of the earth in the east-west direction is shown by the fact that the reported time of an eclipse is earlier in the day as one travels westward.[4] Facts like these are not collected and collated without a community to nourish a tradition of research.

The accusation that the pioneers of science had overindulged in "mere plausibility" is found in Diodorus Siculus's discussion of the cause of the floods of the Nile. The early theory of Democritus, he says, can be admired for its ingenuity, but now "the precise knowledge derived from experience prevails over the plausibility (*pithanoteta*) of mere argument. . . . [Democritus's theory] is not only advanced without proof, but it does not possess, either, the credibility that is accorded to facts established by observation."[5] (The basic thought that experience is to be preferred to theory, and theory accepted only when it agrees with observation, is of course found in that old empiricist, Aristotle.)[6] Diodorus expresses the converse also: that if there is observation then skepticism should be restrained. The example argument is one that would be hard to formalize in modern terms, though it is obviously sound: the claim that snakes a hundred cubits long are to be found in Ethiopia is not to be dismissed out of hand, since a snake thirty cubits long was caught and included in the zoological collection of Ptolemy II.[7]

A high level of reasoning can be expected from the acknowledged masters of logic, the Greek mathematicians. Of course, their concern was mostly for deductive logic, since their achievement was to organize mathematics as a system of theorems deductively derived from axioms. But one cannot pursue mathematics with deductive logic alone; before a theorem has been proved, and even before the proof has been seriously attempted, the mathematician has reason to think the theorem might be true, so that effort expended on the attempted proof may be worthwhile.[8] The ancient Greeks began the tradition, which modern mathematicians have continued, of tidying up the deductive structure of their mathematics before publication, so that it is impossible to see how they arrived at their results. The effect is elegant but almost unreadable in all but the simplest cases. This gives their mathematics an air of high-powered mystery, no doubt as intended. The single instance in which we are permitted a glimpse at the workings inside is in a letter of Archimedes. In it Archimedes describes a method for finding that certain areas of differently shaped figures are equal, without strictly proving it, by cutting the areas into infinitesimal slices. "Certain things," he says, "first became clear to me by a mechanical method, although they had to be demonstrated by geometry afterwards because their investigation by the said method did not furnish an actual demonstration."[9] The method helps find a proof but is not itself a proof.

Aristotle's Not-by-Chance Argument

Some of Aristotle's discussions of chance reveal a rather acute sense of the subject. While not exactly quantitative, they constitute the closest approach in the ancient world to a mathematical theory of probability.

There is a clear case of a qualitative statistical argument in *On the Heavens*. Aristotle is discussing whether the stars in their daily revolutions around the heavens move independently or whether they are all fixed to some sphere. It is observed that those stars that move in large circles (near the celestial equator) take the same time to rotate as those near the polestar, which rotate in small circles. "If, on the other hand, the arrangement was a chance combination, the coincidence in every case of a greater circle with a swifter movement of the star contained in it is too much to believe. In one or two cases it might not inconceivably fall out so, but to imagine it in every case alike is a mere fiction. Besides, chance has no place in that which is natural, and what happens everywhere and in every case is no matter of chance."[10]

There is a suggestion that the same kind of argument will give something similar to what was later called the argument from design for the existence of God: the heavens cannot exist by chance, since they are highly ordered. The conclusion, however, is only to some kind of necessity in the heavens rather than to a divine orderer.[11] For the weaker conclusion that nature, especially biological nature, acts for a purpose, Aristotle is firmer and again uses a not-by-chance argument, among other arguments for design in nature. "Why should it not be a coincidence that the front teeth come up with an edge, suited to dividing the food, and the back ones flat and good for grinding it, without there being any design in the matter? And so with all other organs that seem to embody a purpose. In some cases where a coincidence brought about such a combination as might have been arranged on purpose, the creatures, it is urged, having been suitably formed by the operation of chance, survived; otherwise they perished, and still perish, as Empedocles says of his 'man-faced oxen.' Such and suchlike are the arguments which may be urged in raising this problem; but it is impossible that this should really be the way of it. For all these phenomena and all natural things happen always or mostly, and this is contrary to the meaning of luck or chance."[12]

An essentially similar argument, ruling out the hypothesis of chance, occurs briefly at the beginning of Ptolemy's *Almagest*, though no doubt the argument is older. The stars that rose tonight must be the same as rose last night, for how could the order in their size and number, intervals and positions be restored by a random or chance process?[13]

Averaging of Observations in Greek Astronomy

Aristotle's science has plenty of observations and plenty of theory. The observations are relevant to the theory, but the theory is often not exactly based on the observations, in the direct sense of a modern (or Galilean, or Copernican) mathematical model, in which numerical data agree with the numerical predictions of the theory. The reasons for this are that Aristotle's interest in physics was mainly in its qualitative aspects, while his purely scientific work was mostly in biology, where classification, structure, and function are more important than numbers.

Aristotle is interested in almost everything except numerical measurements.

The Aristotelian attitude is not appropriate in astronomy, where numerical observations are the input, and the desired output is numerical predictions of future eclipses and positions of the planets. In Aristotle's time, Eudoxus attempted a model of the universe as a system of nested spheres, but it was agreed that, like Democritus's theory on the Nile floods, it was a brilliant idea that did not fit the facts. In the following two centuries, the amount of data increased, as observations from different locations were collected, and the huge store of Babylonian astronomical data became accessible to the Greeks. Babylonian astronomy of the last three centuries B.C. had developed an elaborate series of arithmetical schemes for predicting celestial phenomena.[14] They were purely numerical: as far as is known, they did not fit a geometrical model of any kind to the phenomena, and it is even hard to tell whether the Babylonians thought of the sky as spherical. There is nothing probabilistic in them either; although there were some inevitable inaccuracies in the predictions, the phenomena used were ones like the times of eclipses and of risings and settings, which are not subject to substantial observational error. Predictions were made purely by tables of numbers with recurrent patterns, which predicted numbers from numbers.

The attempt to fit a geometrical model to the real data was made by Hipparchus about 150 B.C. with great subtlety and insight and an enormous appetite for calculation. He begins by rejecting theories that are ingenious but implausible and unable to fit the data, like heliocentrism. Then he takes the Babylonian and earlier Greek data and finds a suitable combination of circular motions for the stars and planets that will, as far as possible, produce the observations. In contrast to both the Babylonians' pure arithmetic and the older almost purely geometrical schemes of Eudoxus and others, Hipparchus is concerned with adjusting the geometry to fit the arithmetic, that is, the Babylonian data.[15]

What Hipparchus is doing is similar to modern regression analysis, in which a straight line or other simple curve is made to give as good a fit as possible to a series of points representing observations. The line of best fit does not go through all the points exactly; the deviations of observed points from the true line are regarded as error (due either to observational error or to the intrinsic variability of the subject matter; the latter cause is not relevant in astronomy). It is an exercise in good probabilistic judgment to decide how much to complicate the model to provide a better fit to the points; that is, how far simplicity should be traded off against exactness of fit. A more complicated model can fit the data better, but fitting to noise degrades the predictive performance of the model.

Hipparchus does not fit straight lines, since the characteristic feature of astronomical data is that it is periodic (or almost so). Circular motions are the obvious choice for periodic data—Hipparchus does not discourse on the philosophical perfection of circles. As Kepler says, "It is an immediate presumption of reason, reflected in experience, that their gyrations are perfect circles."[16] Hipparchus's solution is to postulate a not unreasonably large system of eccentric orbits with epicycles. The sun, moon, and planets move on eccentric circles, that is, circles whose centers themselves rotate around the earth. In addition, there are epicycles, smaller circular orbits attached to the larger ones. His picture, as improved by Ptolemy, was not bettered as a fit to the data until Kepler.

Not much is known about how Hipparchus chose the number and speeds of the eccentrics and epicycles to fit the Babylonian data.[17] But Ptolemy's *Almagest* preserves a good deal of Hipparchus's reasoning on some of the more subtle aspects of the theory, which show that he had a clear understanding of the problems of combining different error-prone observations. The main difficulty is that certain phenomena, especially those with long-term cycles, require observations of the *angle* between heavenly bodies, taken at widely separated times. Measurement of angles in the sky, especially large ones, is subject to serious errors with the instruments the Greeks had available, and Hipparchus had very few measurements available from earlier periods. Hipparchus was forced to "conjecture rather than predict" with the uncertain data, as Ptolemy says.[18]

Nevertheless, Hipparchus's most famous discovery was of the precession of the equinoxes, a phenomenon that is barely visible above observational error even in a quite long series of observations. The problem arises because there are two natural ways of defining the length of a year. The first is the time from one spring equinox to the next (or one summer solstice to the next, and so on). But the sun also appears to move against the background of the fixed stars, moving through the signs of the zodiac in a year. The length of this year differs from the one defined by equinoxes by a small fraction of a day, so that the position of the sun against the star background at one spring equinox is slightly different from its position at the spring equinox a year later. This slight movement—not much more than one degree per century—is called the precession of the equinoxes; plainly it can only become evident from accurate observations over a long period of time. Hipparchus was initially afraid that the first definition of the year did not make sense, since he was not sure that the solstice-to-solstice or equinox-to-equinox times were the same every year. Ptolemy describes him as perhaps over-scrupulous in recording the observations that might suggest the length of the solstice-to-solstice year is not constant but concluding that the available accuracy of the observation is not enough to suggest the discrepancies are genuine. Ptolemy, with the benefit of another three hundred years of observations, reaches the conclusion that "there is nothing to be disturbed about here. We become convinced that these intervals do not vary, from the successive solstices and equinoxes which we ourselves have observed by means of our instruments. For we find that [the times] do not differ by a significant amount from those derivable from the [365] ¼ day year. . . . But we also guess from Hipparchus' own calculations that his suspicion concerning the irregularity [in the length of the tropical year] is an error due mainly to the observations he used. . . . And in general, we consider it a good principle to explain the phenomena by the simplest hypotheses possible."[19]

For all Hipparchus's deep insight into numerical issues in calculations, and how errors can propagate, his estimate of the distance from the earth to the sun was defeated by the numerical fact that dividing by a small number magnifies any error in that number. His estimate is too small by a factor of twenty.[20]

It is an important fact of probability that a way to get a more accurate value of a quantity that one can only observe inaccurately is to take the average, or mean, of a number of observations. Errors in the individual observations tend to cancel out. Ptolemy does not, strictly speaking, take a mean of observations to get a more accurate figure. But he does do something essentially the same in determining the length of the year, and explains clearly why it works:

> Now it is already clear to us from Hipparchus' demonstrations that the length of the year, defined with respect to the solstices and equinoxes, is less than ¼-day in excess of 365 days. The amount by which it falls short [of ¼-day] cannot be determined with absolute certainty, since the difference is so small that for many years in succession the increment [over 365 days] remains sensibly the same as a constant ¼-day increment. Hence it is possible, when comparing observations taken over quite a long period, that the surplus days [over 365] which have to be obtained by distributing [the total surplus] over the years of the interval [between the observations], may appear to be the same whether one takes [observations over] a greater or lesser number of years. However, the longer the time between the observations compared, the greater the

accuracy of the determination of the period of revolution. This rule holds good not only in this case, but for all periodic revolutions. For the error due to the inaccuracy inherent in even carefully performed observations is, to the senses of the observer, small and approximately the same at any [two] observations, whether these are taken at a large or a small interval. However, this same error, when distributed over a smaller number of years, makes the inaccuracy in the yearly motion [comparatively] greater (and [hence increases] the error accumulated over a long period of time), but when distributed over a larger number of years makes the inaccuracy [comparatively] less.[21]

Ptolemy then compares Hipparchus's observations with some of his own. According to modern astronomers, Ptolemy's are a day wrong, giving rise to a suspicion that he faked some of his results.[22] On the other hand, the problem may arise from the way he conceives the relation of observation to theory. Rather than take means of individual observations as a normal practice, or try to make the theory fit all the observations, as Hipparchus does, he allows theory to confront the set of all observations in a more holistic fashion. This means that the observations to be exhibited are chosen, to some extent, in accordance with the theory. So older observations may be selected, or adjusted, in accordance with what the theory predicts.[23] It is a dangerous practice, but arguably necessary within reasonable limits. It is allowed in modern statistics under the name rejection of outliers, to deal with the awkward variability of real data.

The Simplicity of Theories

Ptolemy makes a beginning on discussion of the vexed subject of the simplicity of theories and why, and to what extent, one should prefer simple theories. As we saw, he regarded it as "a good principle to explain the phenomena by the simplest hypotheses possible," but his own theory of epicycles has been widely regarded as the paradigm of a hypothesis that does not satisfy this principle. His excuse is that the heavens can be expected to operate on different principles:

Now let no one, considering the complicated nature of our devices, judge such hypotheses to be over-elaborated. For it is not appropriate to compare human [constructions] with divine, nor to form one's beliefs about such great things on the basis of very dissimilar analogies. For what [could one compare] more dissimilar than the eternal and unchanging with the ever-changing, or that which can be hindered by anything with that which cannot be hindered even by itself? Rather, one should try, as far as possible, to fit the simpler hypotheses to the heavenly motions, but if this does not succeed, [one should apply hypotheses] which do fit. For provided that each of the phenomena is duly saved by the hypotheses, why should anyone think it strange that complications can characterise the motions of the heavens when their nature is such as to afford no hindrance. . . . Rather, we should not judge "simplicity" in heavenly things from what appears to be simple on earth, especially when the same thing is not equally simple for all even here.[24]

This is all very well, but if one truly believes that God, or nature, favors simplicity, then Ptolemy's system is probably a little beyond what one will find credible. Proclus already complains that Ptolemy's hypotheses "do not have any probability, but some are far removed from the simplicity of divine things, and others, fabricated by more recent astronomers, suppose the motion of the heavenly bodies to be as if driven by a machine."[25]

Some of the best thought of late antiquity comes from the Greek commentators on Aristotle.[26] The art of commenting on ancient texts, so central a part of medieval intellectual life, has a tendency to direct attention to something like logical probability. It is in the nature of the exercise that arguments for and against the opinions of the ancients should be collected, sifted, compared, weighed—anything but definitively decided for or against. Alexander of Aphrodisias asserts that the most general principles of Aristotle's physics cannot be demonstrated, since there is nothing prior that they can be demonstrated from. The principle that things act according to their natures, which they have according to what is appropriate to the order of the universe, is defended with an argument like the modern inference to the best explanation, with an emphasis on its nondeductive character. The principle is "not only more than any other in accord with the divine government; it is also the one that more than any other is appropriate for speculation and that is verified because of its agreement with, and close relation to, the things visible. . . . It alone among the various opinions preserves the coherence and the order of the things that are produced because of this coherence and order. If it should occur to anyone that some [point] among those that we have set forth requires further and more subtle investigation, he should not, because of a slight difficulty that might become manifest, bring about a slackening of our vigilance and effort toward defending this opinion in its totality."[27] The last point is the one emphasized by modern philosophers of science like Duhem and Quine: that one will not easily abandon a good and wide-ranging theory for the sake of a few minor anomalies.

The story of the decline of science in the West and its survival in the East is a familiar one. The twelfth century saw the reappearance in Western Europe of the scientific point of view on the world[28] and the recovery, translation, and assimilation of all the main ancient scientific texts. What the later Middle Ages made of their scientific legacy is, by common consent, less impressive than their achievements in such fields as philosophy, theology, and law. The long debate on why the scientific revolution did not take place before it did is not subject to conclusive resolution, but there is wide support for the view that the Scholastic method, relying too much on conceptual and textual analysis, failed to devote enough attention to experiment and measurement and their relation to theory. Nevertheless, there are areas of science in which purely conceptual work is entirely appropriate, namely the more mathematical sciences, and it is there that medieval science is strongest. Optics and astronomy, in particular, were regarded as actually part of mathematics and were central to both teaching and research.

While a concentration on the purely mathematical aspects of those or any other sciences does tend to obscure any role of probability in them, by emphasizing deduction, there was, in astronomy at least, one countervailing influence. This was the knowledge that there were two mutually incompatible theories of the heavens, those of Aristotle and Ptolemy. While Ptolemy's was the later, more developed, and more observationally accurate theory, it lacked the physical naturalness of Aristotle's. One response was to maintain that Ptolemy's epicycles were merely fictions or hypotheses.[29] Another, pursued more by Arabic and Jewish astronomers than by Latin ones, was to try to develop alternative theories better supported by observations.[30] One very occasionally finds, in both Arabic and Latin writings, the idea that the probability of different theories may be compared. John Pecham, archbishop of Canterbury in the late thirteenth century, knew of several astronomical theories. There was available a simpler competitor to Ptolemy's, that of al-Bitruji, which contained no epicycles. Pecham says, "I have never heard it said before that one can save all phenomena and explain everything that happens. Nevertheless, the thesis of the mathematicians [that is, of Ptolemy] is more probable; to the reasons they give there has never been given, I think, a reasonable response."[31]

Nicole Oresme on Relative Frequency

By far the most original of the successes of medieval mathematical science were those relating to the conceptual analysis of motion, and of continuous variation more generally, associated with the fourteenth-century Merton school at Oxford and its French counterparts. The most impressive of these thinkers was Nicole Oresme, who was active in Paris in the decades just after the Black Death of 1350; he later became bishop of Lisieux. His most lasting achievement was the invention of the graph, one of the very few mathematical discoveries since ancient times that is familiar to every reader of newspapers. To mathematicians he is known for devising the classic proof of the divergence of the harmonic series. The range of his activities was enormously wide. He was, perhaps, the first to use the long-running metaphor of the universe as a clock.[32] His work on translations of Aristotle into French is mentioned in the previous chapter. As financial adviser to the king of France, he recommended against debasement of the coinage, a primitive Keynesian technique whereby the king spent more than he had.[33] Oresme's book *On Money* contains what looks like a probabilistic concept: "But because the king's power commonly and easily tends to increase (*communiter et leviter tendit in maius*), the greatest care and watchfulness must be used."[34] Oresme is plainly no straightforward monarchist; his commentary on Aristotle's *Politics* makes him one of the last chief theorists of medieval limited monarchy, before political thought was diverted into absolute monarchy during the Renaissance.[35]

It is in Oresme's mathematical works, though, that we must look for a full treatment of his ideas on probability. What is found is a work of sustained genius. A central result is proposition 10 of book 3 of *De Proportionibus Proportionum*: "It is probable (*verisimile*) that two proposed unknown ratios are incommensurable, because if many unknown ratios are proposed it is most probable that any would be incommensurable with any other."[36] ("Commensurable" here means that one is a rational power of the other; for example, the ratios 4/1 and 32/1 are commensurable since $32/1 = (4/1)^{5/2}$, but 4/1 and 5/1 are not commensurable. The idea of fractional powers was itself a discovery of Oresme's.)

To show this proposition, he considers first an easier case; here appears, among other things, the clearest statement up to that time of the connection between probability and relative frequency, that is, between what one should believe about an individual case and the frequency in a class of which the case is a member. "With regard to numbers, we see that however many numbers are taken in series, the number of perfect or cube numbers is much less than other numbers and as more numbers are taken in the series the greater is the ratio of non-cube to cube numbers, or non-perfect to perfect numbers. Thus if there were some number as to which it were completely unknown what it is or how great it is, and whether it is large or small—as perhaps the number of all the hours that will pass before Antichrist—it will be likely that such a number would not be a cube number. It is similar in games where, if one should inquire whether a hidden number is a cube or not, it is safer to reply that it is not, since that seems more probable and likely (*probabilius et verisimilius*)."[37]

The argument for proposition 10 then proceeds by arguing that commensurable ratios thin out like cubes among numbers: taking first just those ratios that are whole numbers, of the 4,950 pairs of whole numbers in the range 2–101, only 25 are commensurable, and if one took the first 200, 300 numbers and so on, the proportion of commensurable pairs would be even less. Further, this is the best possible case; if one took pairs of arbitrary ratios instead of pairs of whole numbers, the proportion of commensurable pairs would be even less. The argument lacks something in rigor, but all except possibly the last step would be acceptable today in an informal mathematical lecture. The last step involves a transition from relative frequencies in finite sets to comparisons in infinite sets, and Oresme's intuition, though sound, would need explication in the mathematical machinery of 1900, not that of 1350. The important thing is that the *probability* argument is perfectly clear, and perfectly correct; it is an instance of the probable argument schema: "The vast majority of As are Bs. This is an A (about which there is no more relevant information). Therefore, this is a B." Oresme is equally clear that what possesses probability is a proposition in an argument: "The consequent is evident, . . . and the antecedent is probable because the first part is true by assumption and, as already proved, the second is probable, so that the proposition is probable."[38]

A related passage connects these relative frequency ideas with symmetry notions of insufficient reason. He divides the possible into three. "Either it is equally possible, or it is improbable, or it is probable. An example of the first way: The number of the stars is even; the number of the stars is odd. One is necessary, the other impossible. However, we have doubts as to which is necessary, so that we say of each that it is possible . . . sometimes in such cases we have no reason for one part; and sometimes we do have a reason, and then it is called a 'problem.' . . . An example of the second way: The number of stars is a cube. Now, indeed, we say that it is possible but not, however, probable or credible or likely (*probabile aut opinabile aut verisimile*), since such numbers are much fewer than others."[39]

Oresme's interest in these questions was partly for their own sake as pure mathematics, but he did have an application in mind. The next chapter of *De Proportionibus Proportionum* applies the results to ratios of continuous physical quantities, such as distances and velocities. His conclusion is that it is probable that a day and the solar year are incommensurable, as are any other two celestial motions. The practical point now becomes clear. Celestial motions, their conjunctions and repetitions, are the basis of astrology. Oresme attacked astrology in a number of his works, including some in French, and always referred to his result that motions of the heavens were probably incommensurable. This would imply that a conjunction of planets would never be exactly repeated, and hence, according to Oresme, astrology could make no predictions.[40]

The criticism Oresme is advancing is one that would now be described by saying that astrology is not structurally stable. If the genius of ideas is measured by the time taken for them to become common currency, then this one is a winner. It is almost totally original; although a few thirteenth-century writers had realized that there would be some problem if the ratios of the orbits were irrational,[41] only Oresme understands what the problem really is. On the way to his result, Oresme, in 1350, has come up with the ideas of fractional powers and probability based on relative frequency, ideas that took three hundred years to reappear; structural stability did not appear again until Poincaré's work of the 1890s.[42] The idea is that a theory dealing with physical things must make predictions based on measurements, which do not have infinite accuracy. Therefore the theory must, at least usually, make similar predictions for sufficiently close initial conditions. For example, the Newtonian theory makes predictions about the orbits of planets; if one puts into the equations a slightly wrong position, the theory's prediction of the orbit will be wrong, but only slightly. Similarly, Euclid's results on perfect spheres apply to real spheres, which are imperfect, not because of some magical idealization but because of the purely geometrical fact that the volume of, for example, an approximate sphere is close to the volume of a perfect sphere that it approximates. Physical systems that are not like this are unpredictable, even though deterministic; they are now much studied under the name of "chaos"[43] (another word possibly introduced into French by Oresme).[44] Oresme is saying that astrology is not structurally stable and so cannot make any predictions.

Oresme has one further conclusion to draw. He holds that the probable existence of so many incommensurable motions is not only true but is a good thing. The general tenor of his argument is that geometry, with its continuous quantities, contains many noble and exciting things that arithmetic lacks; diversity and variety are good things in themselves and contribute to a harmonious whole.[45] The idea that diversity and variety are necessary to beauty has had a long history since then, although as modern architecture and town planning show, it is far from being completely accepted.

The relevance of diversity and variety to matters of explanation is one of the themes of another of Oresme's works, *On the Marvels of Nature*. The general thesis is that marvels can be explained naturally: "The natural causes there assigned and the manner of finding them are possible and much more likely than that demons or unknown influences are the causes of the aforementioned effects." There is simply too much credulity among people, Oresme thinks. "It seems to me that to believe easily is and has been the cause of the destruction of natural philosophy; and also in faith it makes and will make great dangers, and it will be the cause of receiving Antichrist and the introduction of a new law."[46]

One of his lines of argument is that certain things commonly thought to require explanation do not need any at all. Variety as such needs no cause; if some men are sodomites and others only attracted to black women, it can be put down simply to the variety of things. Similarly there is no need to posit demonic or astral influences, or direct intervention by God, to explain a run of coincidences. Oresme gives the same kind of naturalistic explanation of prophecies and their apparent success: "I also say that, if someone says many things, it is difficult for him not to speak the truth sometimes. I also say that if by chance it happens that he predicts some marvel among a thousand lies, that marvel will be much noted."[47]

Perhaps the most famous passage in Oresme's works is his discussion of the possibility that the earth rotates, not the heavens. After explaining why everything would look the same whichever theory were true, and disposing of Scriptural passages that would imply, if taken literally, the immobility of the earth, Oresme presents the positive reasons for preferring a rotating earth. His reasoning looks backward to the Scholastic discussions of Ockham's Razor (discussed in chapter 8) and forward to the ideas of Copernicus, Kepler, and Galileo on the choice of scientific theories according to criteria of simplicity:

> If neither experience nor reason indicates the contrary, it is much more reasonable, as stated above, that all the principal movements of the simple bodies in the world should go or proceed in one direction or manner. Now, according to the philosophers and astronomers, it cannot be that all bodies move from east to west; but, if the earth moves as we have indicated, then all proceed alike from west to east—that is, the earth by rotating once around the poles from west to east in one natu-

ral day and the heavenly bodies around the zodiacal poles: the moon in
one month, the sun in one year, Mars in approximately two years, and
so on . . . [whereas to have the outer sphere of the fixed stars spinning
incredibly fast once a day would disrupt this orderly progression]. All
philosophers say that an action accomplished by several or by large-scale
operations which can be accomplished by fewer or small-scale operations is done for naught. And Aristotle says in Chapter Eight of Book I
[of *On the Heavens*] that God and nature do nothing without some purpose. Now, if it is true that the heavens move with diurnal motion, we
must assume an excessively great speed; for, if we consider thoughtfully
the height or distance of the heavens, their magnitude, and the immen-
sity of their circuit, mindful that this circuit is traveled in but one day's
time, no man could imagine or conceive how marvelously swift and excessively great, how far beyond belief and estimation their speed must
be. Since, then, all the effects we see could be produced and all appearances saved by substituting for the diurnal movement of the heavens a
smaller operation, namely, the diurnal motion of the earth, a very small
body as compared with the heavens, and by so doing avoid the multiplication of operations so diverse and so outrageously great, then it follows that God and nature must have created and arranged them for
naught; and this is an inadmissible conclusion, as we have often said.[48]

But in the end, he agrees, unenthusiastically, with the common opinion that it is the heavens that move, "For 'God hath established the
world which shall not be moved,' in spite of the contrary reasons, because they are persuasions which do not conclude evidently."[49]

The detectable influence of Oresme's opinions concerning probability and incommensurability is almost zero. One of the few to know
anything about them is Gerson (noted in chapter 4 as the inventor of
the phrase "moral certainty"), who mentions that Oresme shows that
questions of the commensurability of the heavenly motions and the influence of the stars are completely uncertain and can be discussed only
in terms of "rhetorical probability."[50]

Copernicus

The Renaissance took place along an axis stretching from Rome to London. Actual additions to knowledge in the period are associated with regions far from that axis: Salamanca, for example, and the ports from
which the Indies were discovered. At the other end of Europe it was still
possible to receive a solid medieval education in cities such as Cracow.[51]
After leaving the University of Cracow, Copernicus traveled to Italy for

postgraduate study in 1497. It is common to assert that he was influenced there by the new humanist learning, a proposition for which there is no evidence. What he was actually studying, apart from astronomy, was law and medicine, the two subjects in which the evaluation of evidence had reached its most developed state. Copernicus studied canon law for some years at Bologna, eventually taking a doctorate in it, and studied medicine to professional standard, though without taking a degree.

One of the books he took with him to Italy was the classic of medieval observational astronomy, the *Alfonsine Tables*. While in Italy, he familiarized himself with the full range of ancient and Arabic observations. His task was then to do for these ancient texts what Gratian had done for the canons of the popes—to find general principles by which they could be harmonized. In his dedication of *On the Revolutions* to the pope, Copernicus wrote, "Nothing except my knowledge that mathematicians have not agreed with one another in their researches moved me to think out a different scheme of drawing up the movement of the spheres of the world."[52] In presenting his theory that the earth revolves around the sun, he keeps referring to it as "demonstrated," but what he actually does demonstrate is that the effects follow from the theory. Then he argues, "Since then so much and such great testimony on the part of the planets is consonant with (*consentiant testimonia*) the mobility of the earth. . . ."[53] A doctor of canon law does not use such a phrase lightly.

On more subtle points where Copernicus could not reach certainty, there is argument in a style that recalls both the Scholastics and Hipparchus's difficulties with unreliable data, while his belief in the regular progression of data resembles Oresme:

> There now arises a greater difficulty over the inconstancy of the solar apsis since, though Ptolemy thought that it was fixed, some have thought . . . Arzachel opined that this movement also was irregular. . . . Wherefore it is believed by many that some error had crept into the observations. But each mathematician is alike in his care and industry, so that it is doubtful which one we should follow in preference to the other. At all events, I confess that nowhere is there greater difficulty than in determining the apogee of the sun. . . . Hence in placing the apogee at 6 ⅔ degrees of Cancer, we were not content to rely upon the instruments of the horoscope, unless the eclipses of the sun and moon gave us more certainty, since if any error lay concealed in our observations, the eclipses would uncover it without fail. What therefore was most likely

(*vero simillimum*) we can apply our intelligence to conceiving the movement as a whole: it is eastward, but irregular, since after that standstill between Hipparchus and Ptolemy the apogee has appeared to be in continuous, orderly and increased progression down to our time, with the exception of the movement which occurred erroneously, it is believed, between al-Battani and Arzachel, as all the rest seems to be in harmony.[54]

Copernicus obviously believed that his theory was, in the main, certainly true, though perhaps in a few parts only highly likely. But a preface to the book was inserted anonymously by the Lutheran theologian Andreas Osiander (one of the few Reformed theologians to show any sympathy with the Catholic doctrine of probabilism).[55] In an attempt, temporarily successful as it turned out, to save the Copernican theory from theological attack, he took it upon himself to deny Copernicus's realism and to assert the hypothetical nature of the theory. "It is not necessary that these hypotheses should be true, or even likely (*verisimiles*); but it is enough if they provide a calculus which fits the observations. . . . Maybe the philosopher demands probability (*veri similitudinem*) instead; but neither of them will grasp anything certain or hand it on, unless it has been divinely revealed to him. Therefore let us permit these new hypotheses to make public appearance among old ones which are themselves no more likely (*verisimiliores*)."[56]

Kepler Harmonizes Observations

Much has been written on the status of scientific theories in the century of the scientific revolution. The most prominent fact is that the leading writers—Bacon, Galileo, Descartes, and Newton—stated emphatically that their results were certain. Their statements were undoubtedly sincere, but one should not be too quick to conclude that the science of the time was universally thought to be a search for certainty. It is proverbial that very successful people tend to possess, besides unlimited energy, an utter conviction that their own opinions are right. There is less talk about certainty in the lesser, but still exceptionally brilliant, scientists of the seventeenth century. There is also less insistence on certainty in some of the works of the great men themselves, in places where they stopped advertising their projects and dealt with the detailed business of explaining their results.[57]

Kepler and Galileo share the same project, the mathematization of science, support the same unorthodox thesis, heliocentrism, and often propose the same arguments. They both interpret astronomy realistically—Kepler was the first to call attention to the fact that the preface to *De Revolutionibus* was not written by Copernicus, and Galileo rejected the suggestion that he regard Copernicanism as fictitious to reconcile it with theology. They distinguish between probable and demonstrative arguments for that thesis in the same way. Their genius is equally exceptional. It has therefore often been wondered why they took so little notice of each other; Galileo, for example, never took seriously Kepler's principal discovery, that the planetary orbits are elliptical. Of course, geniuses are very busy people and often have no time to read one another's works. But in this case, their styles of thinking and writing show a fundamental incompatibility sufficient to make them almost incomprehensible to each other.

Kepler devoted his life to fitting the observed phenomena of the heavens to an overarching geometrical theory that would explain everything. The theory has three levels. The bottom one is heliocentrism. Next is the theory that the orbits of the planets around the sun are ellipses, not circles with epicycles, as Copernicus had thought. The third is the theory that the orbits fit neatly into a figure formed by nesting the five Platonic solids. The first and second of these are true, the third false.

In defending heliocentrism, Kepler claims that Copernicus's reasoning is geometrical and therefore demonstrative, in contrast to the "merely probable" reasoning of the ancients.[58] But he adds a number of probable arguments for the theory as well. The first is the one that most impresses the modern mind, the fact that (with Kepler's modifications) the theory can do without the complicated systems of epicycles: with the "superfluous multitude of spheres and movements" removed, "it is much more probable that there should be some one system of spheres." Other arguments involve physical plausibility: "For it is more believable that the body around which the smaller bodies revolve should be great. For just as Saturn, Jupiter, Mars, Venus, and Mercury are all smaller bodies than the solar body around which they revolve; so the moon is smaller than the Earth around which the moon revolves; so the four satellites of Jupiter are smaller than the body of Jupiter itself, around which they revolve. But if the sun moves, the sun which is the greatest, and the three higher planets which are all greater than the Earth, will revolve around the Earth which is smaller. Therefore it is more believable that the Earth, a small body, should revolve around the great body of the sun."[59]

There is also no need for intelligences to steer the planets in complicated orbits but only a simple magnetic power emanating from the sun. Such a power is also an economical hypothesis for the movement of the earth. "And so it is probable that the Earth is moved, for a probable cause of its movement appears; and it is probable that the sun remains fixed." And there is a probable cause provided for the precession of the equinoxes and other things unexplained on the Ptolemaic theory. The last argument replies to the objection that man's dwelling place deserves to be in the center of the universe: only by being transported around the sun could "man" (that is, Kepler) view and measure the heavens and work out their true system.[60]

It has been common to see, in consequence of the (true) ellipse theory and the (false) Platonic solids theory, two Keplers, or at least two opposing aspects of Kepler's thought—one, rationalist and looking forward to modern scientific method, with its curve fitting and hypothesis testing; the other backward looking and "mystical," misled by all manner of unlikely Pythagorean speculations. Kepler's works reveal no such split. The difference between the two theories is simply that the ellipse theory turned out to be right, the solids theory wrong. Kepler's manner of thinking, and of argument, is the same in both cases and is the one that lies at the bottom of the success of modern mathematical science. It involves finding the harmonies of the world, to use Kepler's words, or in the language of a more pedestrian century, argument from concomitant variation.[61] Some reasonably simple and intuitively plausible theory is proposed to explain the facts. It will make some predictions about observations, except that typically there will be some parameters not determined (such as the diameters of the ellipses and of the Platonic solids or some constant of proportionality). These parameters having been determined by a sufficient number of observations, exact predictions can be deduced for future observations, and one can see how closely observations agree with the predictions. Some lack of agreement is inevitable; the mathematical precision of the original theory has the advantage that close agreement can be taken as confirming the theory strongly and that the need for any excuses to explain away discrepancies can be evaluated. Kepler differed from Galileo, and indeed from everyone else up to his time, in demanding an extremely high standard of agreement.

Kepler was prepared to change completely his theory of the orbit of Mars to account for a disagreement of only eight minutes in longitude between the observed value and that predicted by his earlier theory.[62] This is an error below the accuracy of Ptolemy's observations. It is not the kind of thing found in Galileo at all.[63] Nor is the phenomenal amount of calculation that Kepler required to decide whether hypothesis and observation did agree, an amount found to be nontrivial even by modern computer.[64] To achieve results with so much calculation requires skill not only in the actual calculating but also in dealing with the errors in the original data and those introduced by the approximations made in the calculations. Kepler himself says this is an "unmathematical game-of-chance method," but of course that is not true.[65] Kepler's use of methods like the finding of the most plausible values of unknowns by averaging the solutions of redundant systems of equations makes a substantial beginning on the treatment of random errors.

The elliptical orbit theory at last fell into place when Kepler realized that it could be made to work by supposing that a planet moved faster when closer to the sun. This is an unlikely hypothesis from a purely geometrical point of view, but it makes sense if the motion of the planets is due to some force emanating from the sun, which could reasonably be supposed to become weaker as the distance from the sun increases. Such a supposition is not susceptible of proof in the way that purely geometrical facts may be. Kepler sees the divide between necessary and probable reasons for his theory as a result of introducing physics into astronomy. "As is customary in the physical sciences, I mingle the probable and the necessary and draw a plausible (*probabilem*) conclusion from the mixture. For since I have mingled celestial physics with astronomy in this work, no one should be surprised at a certain amount of conjecture. This is the nature of physics, of medicine, and of all the sciences which make use of other axioms besides the most certain evidence of the eyes. . . . Thus the physical difference is now immediately apparent [among the three opinions of Ptolemy, Copernicus, and Brahe]: by way of conjecture indeed, but yielding nothing in certainty to conjectures of doctors on physiology, or to any other natural science."[66] Since things work so nicely, probability is added to the original heliocentric theory, on which all is based.[67]

Kepler never makes very explicit his notion of probability, but it is clear from what has already been said that it is closer to an objective logical relation than to, for example, arguments from authority. He does distinguish what he is doing from arguments from common opinion, as in Aristotle's *Topics*. He also distinguishes *eikota* that support his theory—including considerations of theological harmony, numerology, and the like—from geometrical reasoning from observations.[68]

The elliptical orbit theory explains the observed movements of the planets. The Platonic solids theory is an altogether more ambitious undertaking. The full theory, heliocentrism plus elliptical orbits plus the solids, aims to explain *everything* about the system of the heavens, including facts of kinds that modern astronomy makes no attempt to explain, such as why there are six planets, why the angle of the sun's disc seen from the earth is exactly a half a degree, and why the orbits of Mars and Mercury are especially eccentric. Kepler remarks that Aristotle asked this kind of question and "judged that in these questions you should not suppress or be silent about probabilities any more than about fully explored certainties." He adds, "I took my solution from the Archetype of the harmonic cosmos: whence it is established that this cosmos cannot be better than it is."[69] In that sort of cosmos, it is true, one expects a lot to be explained.

The reasoning from principles of harmony, which reaches its full flowering in Kepler's *Harmony of the World,* can be seen on a smaller scale in an argument that resembles Oresme's from the gradually increasing periods of revolution of the planets. But more quantitative information is now to hand as well as observations of a whole new "solar system," that of the moons of Jupiter, recently discovered through the telescope. Kepler argues that "belief in this thing is confirmed by the comparison of the four satellites of Jupiter and Jupiter with the six planets and the sun." He extrapolates from this "sense-perception," saying that "exactly as it is with the six planets around the sun, so too is the case with the four satellites of Jupiter: in such fashion that the farther any satellite can digress from Jupiter, the more slowly does it make its return around the body of Jupiter. And that indeed does not occur in the same ratio but in a greater, that is, in the ratio of the 3/2 power of the distance of each planet from Jupiter: and that is exactly the same as the ratio which we found above among the six planets. . . . That is so much the more probable (*tanto probabilius*) because we see that, just as the sun is greater than all the planets which it moves, so too the Earth is much greater than its moon, and Jupiter than its satellites, and for that reason, like the sun, fit for moving them. The remaining probabilities (*verisimilitudines*) have to do with the moon again."[70]

The theory that the orbits of the planets somehow fitted into the five regular solids was the subject of Kepler's first book, the *Mysterium Cosmographicum* of 1596. The idea of the theory is to find some correspondence between the nested five solids, with their inscribed and circumscribed spheres, and the spheres determined by the planetary orbits. (Since the orbits are elliptical, each planet is actually associated with three spheres: one for its closest distance from the sun, one for its greatest distance, and one for its mean distance.) Since it is demonstrable that there are only five regular solids—the tetrahedron, the cube, the octahedron, the dodecahedron, and the icosahedron—the correspondence would explain why there were as many planets as there were. It is remarkable that any reasonably close correspondence is possible, and Kepler was perfectly rational in pursuing the theory. Still, it simply could not be made to fit. So he concluded that looking at the spheres was wrong and that of course he should have realized this all along: "It will be apparent that it is not very likely that the most wise Creator should have taken thought most of all for harmonies between the actual planetary paths . . . who will benefit from harmonies between the paths, or who will perceive these harmonies?"[71]

So he looks instead at the apparent daily arcs of the planets as seen from the sun. He considers the ratios between these arcs at closest and furthest distances (that is, the ratio for each planet and the ratios for adjoining planets) and finds these are close to small integer ratios. Thus Mars moves 26'14" per day when farthest from the sun, 38'1" when closest, the ratio of these being nearly 2:3, which is the fifth in music. These correspondences, Kepler says, cannot have happened by chance.[72] This suggests to him that the reason things do not work out neatly for the solids theory is the same reason they do not work neatly for the musical scale, namely, that it is mathematically impossible that everything should harmonize. One cannot divide the octave so that all the harmonies are correct—if one places the notes so that the fifths are correct, with the ratio of their frequencies at exactly 3:2, one cannot also have the major thirds and major sixths at their correct ratios of 5:4 and 5:3. The usual equal-tempered scale is a reasonable compromise; in it only the octave itself is exactly correct, but none of the others are too far wrong. Kepler did not accept proposals, such as those made by Galileo's father, for a tempered scale, but favored "just intonation," in which fifths and minor thirds are wrong but everything else is exactly right.[73]

Naturally, Kepler is still faced with a great deal of calculation to explain the discrepancies. In the middle of a particularly convoluted mass of them, he breaks off to ask whether God is as fond of this activity as he is himself. He answers that "it is not absurd for God to have followed even these [reasons], however thin they may appear, since he has ordered nothing without reason. For it would be far more absurd to declare that God has snatched these quantities, which are in fact below the limit of a minor tone prescribed for them, accidentally. Nor is it sufficient to say that He adopted that size because that size pleased Him. For in matters of geometry which are subject to freedom of choice it has not pleased God to do anything without some geometrical reason or other, as is apparent in the bordering of leaves, in the scales of fishes, in the hides of wild beasts, and in their spots and the ordering of their spots, and the like."[74] Kepler's reasoning was criticized by Kircher in 1650 for permitting large errors; Kircher says that it would be easy to find musical proportions in anything, given the latitude Kepler allows himself.[75] Kepler at least knew the danger of playing with symbols and analogies and claimed that he accepted an apparent fit of phenomena to a pattern only when he understood the causes of the connection.[76] Analogies with the human body, and speculations about life on the sun, appear too, but even Kepler has to stop somewhere.[77]

Kepler's writings on subjects outside his grand scheme of the universe are meager. He did revolutionize the optics of the eye by discovering focusing on the retina, but this was for him a sideline, appearing in his work on the "optical part of astronomy." Even arguments for the existence of God became attracted into the orbit of astronomy: writing on the nova of 1604, he argued against the notion that such a thing might be a chance concourse of atoms, and presented "not my own opinion, but my wife's. . . . Yesterday, when weary with writing, I was called to supper, and a salad I had asked for was set before me. 'It seems then,' I said, 'if pewter dishes, leaves of lettuce, grains of salt, drops of water, vinegar, oil and slices of eggs had been flying about in the air from all eternity, it might at last happen by chance that there would come a salad.' 'Yes,' responded my lovely, 'but not so nice as this one of mine.'"[78]

A rare occasion when Kepler was forced to think about more mundane affairs found him becoming an expert in the other kind of probability, the legal. In 1615 he received news that his mother had been accused of witchcraft. The charges involved the usual cows, pigs, dead children, assorted aches and pains. With a magistrate present, Frau Kepler was urged to restore someone's health by countersorcery, but fortunately she refused. Matters dragged on for years, with 280 pages of testimony coming from more than thirty witnesses. In 1620 she was imprisoned and confronted with witnesses; she admitted nothing but suspiciously failed to weep. The final document for the defense was written partly by Kepler, with legal assistance. It is adept at the manipulation of presumptions and indications and references to Menochio, Mascardi, and Bartolus. The legal faculty at Tübingen sent its advice that in law there was sufficient evidence for torture but that mercy should prevail. Frau Kepler was taken to the torture chamber and threatened with the instruments if she did not confess. On still protesting her innocence, she was released. She died the next year.[79]

Galileo on the Probability of the Copernican Hypothesis

Understanding Galileo's opinions on scientific theories is much complicated by the fact that his main works were written with a view to avoiding problems with the Inquisition.

In the *Dialogue Concerning the Two Chief World Systems*, of 1632, he makes fun of the suggestion, perhaps made by the Pope, that the Copernican theory can be regarded as just a "hypothesis" to be studied, without inquiring whether it is true or false. He goes on to claim that Copernicus's theory is true with certainty, but probability enters the story in two ways. First, Galileo does speak often about the "probability" of different opinions and arguments. Second, the sentence against Galileo of 1633 found him "vehemently suspect of heresy" for holding two things: that the sun is the center of the world and that "any opinion may be held and defended as probable after it has been declared and defined as contrary to Holy Scripture."[80]

The reasoning behind the condemnation takes us back to the end of the previous century, when the Jesuits, by then leaders in the world of academic science, applied the Scholastic method to astronomical questions, treating them as if they were problems of canon law or moral theology. The following passage, written about 1590, is typical of the style of argument; it is the same style as found in, for example, the debates on probabilism, with its mixture of considerations of plausibility and appeals to authority. It is worth quoting at some length, to convey the exact consistency of the intellectual bog from which modern science had to extricate itself.

> Fourth question: Are the heavens incorruptible?
>
> The first opinion is that of Philoponus . . . since the heavens are a finite body, if they were eternal they would have an infinite power . . . it seems that Holy Scripture teaches generally that the heavens are corruptible, especially Isaiah ch 51, "the heavens shall vanish like smoke," and 34, "the heavens shall fold up like a book." . . .
>
> The second opinion is that of Aristotle, who was the first, as Averroes notes, to teach in this book that the heavens are ungenerated and therefore incorruptible. . . . Finally, the intelligences achieve their perfection in moving the heavens, and so the heavens must be incorruptible.
>
> For the solution of the difficulty, note that in truth we can speak of the heavens, just as we can of anything created, in two ways: first, from their very nature, whether, namely, they have by their nature some in-

trinsic principle through which they can be corrupted; second, whether only through the absolute power of God, whereby God can make everything return to nothing, they are corruptible . . . the Council of Constance defines that angels and human souls are immortal by divine grace. . . .

I now say, first: if we speak of the heavens according to their nature, and if corruptible be taken to signify anything that has in itself a passive potency whereby it can be corrupted by an active power proportioned to it, it is probable that the heavens are corruptible. . . .

Second proof: because the heavens were made especially for man; therefore they ought not to be incorruptible, for otherwise they would be more noble than man. . . .

I say, second: it is more probable that the heavens are incorruptible by nature. Proof of this, first: because it is more conformable to natural reason, as is apparent from the arguments of Aristotle. . . . The second argument of Aristotle is drawn from experience: for it has been found over all preceding centuries that no change whatever has taken place in the heavens. And this argument has the greatest force. . . . A third argument is drawn from the consensus of all peoples. . . . Fourth, from the etymology of the word.[81]

As evidence of the forms of thought that modern science had to overcome, this is particularly convincing in that the author of this rubbish is the young Galileo himself. It is from unpublished notes, which have been found to be a collage of the lecture notes of a number of Jesuit professors at the Collegio Romano, dating from the late 1570s and the 1580s. These professors made considerable play of probabilities and their comparisons, in the same way as the passage just quoted.[82] Galileo later wrote, "If what we are discussing were a point of law or of the humanities, in which neither true nor false exists, one might trust in subtlety of mind and readiness of tongue and in the greater experience of the writers, and expect him who excelled in those things to make his reasoning more plausible, and one might judge it to be the best. But in natural sciences whose conclusions are true and necessary and have nothing to do with human will, one must take care not to place oneself in the defense of error; for here a thousand Demostheneses and a thousand Aristotles would be left in the lurch by every mediocre wit who happened to hit upon the truth for himself."[83]

Galileo's reference to law and the humanities is exactly right. Men like his opponents, and his earlier self, do not believe anything is established in science, or ever will be. They expect science to continue like law, the disputes of the learned stretching out in commentaries and glosses till Judgment Day. In their favor it must be admitted that very little *had* been established in science for centuries. They were not to know that the telescope was about to be invented and Galileo about to exist. What is most blameworthy in their activities is the same tendency as in moral theology, to exaggerate the worth of arguments from tradition or "extrinsic probability."

Though Copernicus himself was no opponent of authorities and indeed had founded his system mainly on observations found in books, his theory became the main locus of the conflict between the two ways of thinking. The Jesuit Cardinal Bellarmine, in his letter of 1615 to Foscarini, who had attributed a high probability to the Copernican theory, advised that Copernicus's theory should be regarded as hypothetical and that, since there was no demonstration of it and it was not likely that there would be, "in a case of doubt, one may not depart from the Scriptures as explained by the Holy Fathers."[84] Such caution had served the cardinal well enough in his long controversies over grace and the Anglican Church, but history has judged harshly his conception of "safety."[85]

Galileo's main statement in favor of heliocentrism was his *Dialogue Concerning the Two Chief World Systems,* of 1632. At certain points in the argument of the *Dialogue,* Galileo does treat of probability rather extensively. It is clear that he means the force of arguments. Some probable arguments he ranks "little lower than mathematical proof," and real probability is distinguished from mere rhetorical force: "It cannot be denied that your argument is ingenious and carries something of probability, but I say that this is a probability in appearance only and not in reality."[86]

The first crucial question on the way toward showing the truth of the Copernican hypothesis is whether the earth rotates once a day or is still while the heavens rotate around it. It is natural to decide this question before considering the Copernican opinion that the earth revolves around the sun; the Copernican theory attributes two motions to the earth, the daily rotation and the annual revolution around the sun, while the Ptolemaic theory denies both motions. The biblical texts that created problems tell against any motion of the earth; they have nothing to do with whether the earth or sun is the center of the universe. Galileo is prepared to grant, as Oresme had been, that since motion is relative, there may be no observational difference between the theory that the earth rotates once a day and its denial. (He is a little worried about centrifugal force but leaves the problem aside.) It follows that there would be no demonstration of either theory. Consequently, he admits that at this point his argument is probable; both theories can explain the effects, but one is more reasonable. Discussing why the diurnal motion must more probably belong to the earth, he argues: "If, throughout the whole variety of effects that could exist in nature as dependent on these motions, all the same consequences followed indifferently to a hairsbreadth from both positions, still my first general impression of them would be this: I should think that anyone who considered it more reasonable for the whole universe to move in order to let the earth remain fixed would be more irrational than one who should climb to the top of your cupola just to get a view of the city and its environs, and then demand that the whole countryside should revolve around him so that he would not have to take the trouble to turn his head."[87] Two further arguments for the same conclusion are given, both described as probable. One concerns the opposite directions that the motions of the fixed stars and the spheres of the planets would have; the other is an interesting argument about the proportionality in the rotations of the spheres. The argument is similar to Kepler's, but Galileo does not bring to it Kepler's background theories of divine harmony:

> I do not assume the introduction of two [circular motions] to be impossible, nor do I pretend to draw a necessary proof from this; merely a greater probability (*una maggior probabilità*). The improbability (*l'inverisimile*) is shown for a third time in the relative disruption of the order which we surely see existing among those heavenly bodies whose circulation is not doubtful, but most certain. The order is such that the greater orbits complete their revolutions in longer times, and the lesser in shorter: thus, Saturn, describing a greater circle than the other planets, completes it in 30 years; Jupiter revolves in its smaller one in 12 years, Mars in 2; the moon covers its much smaller circle in a single month. And we see no less sensibly that of the satellites of Jupiter the closest one to that planet makes its revolution in a very short time, that is in about 42 hours; the next, in three and a half days; the third in 7 days and the most distant in 16. And this very harmonious trend will not be a bit altered if the earth is made to move on itself in twenty-four hours. But if the earth is desired to remain motionless, it is necessary, after passing from the brief period of the moon to the consecutively larger ones, and ultimately to that of Mars in 2 years, and the greater one of Jupiter in 12, and from this to the still larger one of Saturn whose period is 30 years—it is necessary, I say, to pass on beyond to another incomparably larger sphere, and make this one finish an entire revolution in twenty-four hours.[88]

The debate between the participants in the dialogue then takes a theological turn. In Kepler, the theology supported the science, but in Galileo it threatens to undermine it. Salviati, the character in the dialogue who speaks for Galileo, remarks that, "up to this point, only the first and most general reasons have been mentioned which render it not entirely improbable that the daily rotation belongs to the earth rather than to the rest of the universe. Nor do I set these forth to you as inviolable laws, but merely as plausible reasons. For I understand very well that one single experiment or conclusive proof to the contrary would suffice to overthrow both these and a great many probable arguments. So there is no need to stop here; rather let us proceed ahead and hear what Simplicio answers, and what greater probabilities or firmer arguments he adduces on the other side." Simplicio answers: "It seems to me that you base your argument throughout upon the greater ease or simplicity of producing the same effects. . . . To this I answer that it seems that way to me also when I consider my own powers, which are not finite merely, but very feeble. But with respect to the power of the Mover, which is infinite, it is just as easy to move the universe as the earth, or for that matter a straw. And when the power is infinite, why should not a great part of it be exercised rather than a small?"[89]

The logical affiliation of the argument is with older debates on the absolute power of God (discussed in chapter 9). A later age would perhaps have replied that of course the Mover would prefer to accomplish his tasks deftly, with a minimum of effort. Salviati replies not too convincingly with some considerations on the nature of infinity but is at least able to call the simplicity of his theory in its support. "Paying attention to the many other simplifications and conveniences that follow from merely this one, it is much more probable that the diurnal motion belongs to the earth alone than to the rest of the universe excepting the earth. This is supported by a very true maxim of Aristotle's which teaches that, 'It is pointless to use many to accomplish what can be done with fewer.'"[90]

The matter recurs at the end of the last dialogue, when Galileo puts into Simplicio's mouth the argument advanced by the Pope: "He [God] would have known how to do this [arrange the tides] in many ways which are unthinkable to our minds. From this I forthwith conclude that, this being so, it would be excessive boldness for anyone to limit and restrict the Divine power and wisdom to some particular fancy of his own."[91] The reader is expected to find this argument ridiculous, a point not lost on the Pope, but it is unclear whether Galileo has actually explained what is wrong with it. Perhaps his most definite attempt at justifying his method—in a case in which a nondemonstrative argument is intended to defeat a theological argument—occurs in some unpublished notes on Bellarmine's letter to Foscarini. Galileo writes, "Not to believe that a proof of the earth's motion exists until one has been shown is very prudent, nor do we demand that anyone believe such a thing without proof. Indeed, we seek, for the good of the holy Church, that everything the followers of this doctrine can set forth be examined with greatest rigor, and that nothing be admitted unless it far outweighs the rival arguments. If these men [Copernicans] are only ninety percent right, then they are defeated; but when nearly everything the philosophers and astronomers say on the other side is proved to be quite false, and all of it inconsequential, then this side should not be deprecated or called paradoxical because it cannot be completely proved."[92] It is certainly true that it is useful to consider the comparison of hypotheses here. As with modern creationism, the defense of improbable hypotheses discredits religion.

There was much debate on the probability of the Copernican theory in the years around Galileo's condemnation. The Scholastic opponents of the Copernican theory were not entirely unequipped with probable arguments themselves. The Jesuit Amicus suggested that though the earth was smaller than the heavens, it was heavy, and as water was more mobile than earth, air than water, and fire than air, so the celestial bodies were more suited to motion in their place than the earth in its.[93] Mersenne discussed the Copernican theory in rather favorable terms but never attributed either truth or probability to it. His being a Catholic priest is no doubt sufficient explanation. He does however explain why it is not rational to either decide to believe one side by an act of will or simply to suspend judgment: "It is necessary," he says, "that the understanding have some probable reason in order to judge something false or true; otherwise the will would command the understanding in vain; as, for example, if the will commanded it to judge and conclude that the heavens were solid, and it had no reason however little probable for that conclusion, it could not do it, for it would have no grounds for truth."[94]

Mersenne's discussion of the simplicity of the Copernican hypothesis casts some light on why the new theory was accepted so slowly. He considers the fact that Copernicus can explain the phenomena more simply to be the strongest argument in his favor. But he answers with an argument like the one Galileo had put in the mouth of Simplicio: God did not choose the shortest possible route to assuring our salvation, and need not have chosen a simple arrangement for the planets either.[95]

Discussion of the simplicity of hypotheses has advanced little since Mersenne. Everyone agrees that the simplicity of a scientific theory is a reason to believe it (or at least to "accept" it, in the jargon of the century of noncommitment). But why should it be? Does a theory's simplicity make it more probable? If so, why? If not, why does it constitute a reason for preferring the theory? Newton's "Nature is pleased with simplicity, and affects not the pomp of superfluous causes" is magisterial but unsupported by reasons.[96] God may have the potential to be "pleased" with simplicity, but why Nature?

In the meantime, a comparatively minor technical dispute on evidence for the Copernican hypothesis occasioned a different contribution by Galileo on the relation of observation to theory, one that is the ancestor of all modern statistical techniques of fitting theories to data. It concerns methods for combining discordant and error-prone observations. This matter had been treated to some extent by Ptolemy and Copernicus, but Galileo's version is considerably more precise. He is refuting the theory of Scipio Chiaramonti, published in 1621. The subject of the dispute is the place of the nova of 1572, a particularly bright star which had been visible in broad daylight for several weeks. It was debated whether it was closer to the earth than the moon; if it proved to be farther away, the doctrine of the Aristotelians that there could be no change beyond the sphere of the moon would be refuted. In theory the question could be decided by measuring the elevation of the star above the horizon at different points on the earth's surface and comparing the results.

The difficulty was that such observations were hard to make accurately, and small errors in the observations could lead to large errors in the estimated position of the star. Observations made by thirteen astronomers were available; Chiaramonti's book compares twelve pairs of these observations and shows that the estimated distance of the nova derived from each pair was, though always different, in each case less than the distance of the moon. He concludes that the nova was actually closer than the moon. Galileo points out what is wrong with this procedure, namely, that of the sixty-five possible pairs of observations, Chiaramonti has obviously chosen just the twelve that support his conclusion. Galileo undertakes to explain the correct method of reconciling discordant observations.

He puts forward the principles that observations are "equally prone to err in one direction and the other" and that careful observations are "more likely to err little than much." Hence one should choose the position that requires the least correction in the whole body of observations. He then calculates that a position of the nova in the sphere of the fixed stars (that is, at an effectively infinite distance) requires much less correction in the observations than any position closer than the moon. He concludes, "How much more clearly and with how much greater probability it is implied that the distance of the star placed it in the remote heavens."[97] Galileo's solution of choosing a position that makes the sum of the corrections least is not quite the same as that commonly used in modern statistics, which is to make the sum of the squares of the corrections least. But it is close enough and is completely adequate for his conclusion.

CHAPTER SEVEN

Soft Science and History

A stronomy has many difficulties with errors in measurement but very few with variability in the subject matter itself. The more mundane sciences may be able to measure more directly and accurately, but the variability in what is being studied makes the extraction of laws and the prediction of new cases very hard. Superficially, it is easy to know whether an herb cures a certain disease, by testing it on many cases and seeing if a cure results. Just occasionally, it is as easy as that, but almost always there are so many variables, spontaneous cures, extenuating circumstances, and possible excuses that it is almost impossible to extract a truth from any reasonable amount of data. It was only from the late nineteenth century that the sophisticated statistical techniques of biometry (modern statistics) were developed in agriculture and genetics and were applied in psychology, sociology, and drug trials.[1] The mathematics involved is difficult—much more so than the combinatorics of dice. Without those methods, hardly anything quantitative can be said with confidence in the biological or social sciences. Even with them, there is little to be done with questions like the authenticity of texts and other historical matters except rely on the commonsense evaluation of evidence by well-informed humans. It will still be some time before a historical thesis is proved at the 95 percent level of significance.

Lacking any idea of such techniques, ancient study of the "soft" sciences could not make much sense of the mass of experience and observation, however assiduously collected. The situation was made worse by the fact that some of the "sciences" in which experience was most attended to were ones like physiognomics and divination, in which the data are, on a modern understanding, all noise and no laws.

The *Physiognomics*

One of the various works falsely attributed to Aristotle is the *Physiognomics*, on the art of inferring character from external appearances. It is interesting for its comments on a science that could not reasonably be forced into a deductive mold (though the real Aristotle was certainly willing to try).[2] The general idea is to put together arguments from various analogies and to classify men by the signs they have in common ("The Pathic is weak-eyed and knock-kneed, his head hangs on his right shoulder . . . and he casts a furtive gaze around, like Dionysius the Sophist"), by comparison of the features of men with those of animals ("Small eyes mean a small soul, by congruity and on the evidence of monkeys"), by inference from the characteristic differences of male and female ("Breasts too devoid of hair indicate impudence, as in women"), and by causal speculations ("Men of abnormally small stature are hasty, for the flow of their blood having but a small area to cover, its movements are too rapidly propagated to the organ of intelligence").[3] The allowable kinds of sign are therefore various and must be put together: "In general it is foolish to rely on a single sign: you will more likely have confidence in your conclusions when you find several signs pointing the same way."[4] This is indeed the main reason that a nondeductive science will look so different from a deductive one. Euclid does not collect different proofs for the same theorem, since the theorems are given the highest possible degree of belief by a single proof. But when conclusions are not susceptible of deductive proof, it will always be worthwhile to collect more "signs pointing the same way." This means that wide experience is an advantage in sciences like physiognomy. One will need long experience to discern the difference between pallor due to terror and that due to fatigue and, generally, the congruity between different subtle shades of expression and conditions of mind.[5] In such a science there must also arise problems when the signs do not point the same way, that is, when evidence conflicts: "It is safest, however, to refrain from all positive assertion when you find that your signs are inconsistent and contrary to one another in detail, unless they belong to classes, some of which you have determined to be more trustworthy than others."[6]

Physiognomy remained of interest from time to time in later centuries and was supposed to be useful in medicine, astrology, rhetoric, and so on.[7] But it can hardly be said to have made any advance, then or since. This is remarkable, since everyone surely regularly takes action on the basis of physiognomic conjectures, and feedback on their truth seems perfectly possible. Yet Central Casting deploys no greater body of knowledge than the ancients.

Contemporary with the *Physiognomics* was Theophrastus's treatment of signs concerning the weather, for which some predictions may be made but for which the inability to be certain is legendary. "It is a sign of rain when a toad takes a bath, and still more so when frogs are vocal," says Theophrastus, and recognizes not only the grading of predictive value in signs but also the possibility of having many pointing the same way: a red sky at sunset suggests rain but less certainly than a red sky at dawn.[8]

Divination and Astrology

Many ancient authors extolled the virtues of experiment and observation.[9] But the Alexandrians, for all their research grants, found it a good deal harder to learn from experience than to praise doing so. The fields in which the careful recording of observational data perhaps went furthest in antiquity were divination and astrology. Cicero reports the Stoic arguments for divination and their replies to the Skeptics. It is an important passage both for its suggestion that in sciences without certainty a reasonable level of mistakes can be tolerated without that making the science worthless, and also for its use of random phenomena similar to dice throwing as a model of uncertainty:

> You ask, Carneades, do you, why these things so happen, or by what rules they may be understood? I confess that I do not know, but that they do so fall out I assert that you yourself see. "Chance," you say. But can that be true? Can anything be by chance which has all the marks of truth? Four knucklebones are thrown and a Venus throw results by chance; but would you think it chance if four hundred bones gave a hundred Venus throws? It is possible for paints flung on a surface to form the outlines of a face; but surely you do not think a chance throwing could produce the beauty of the Venus of Cos? If a pig should form the letter A in the ground with its snout, you would not therefore suspect it could write Ennius' *Andromache?*
>
> Carneades had a story that in the Chian quarries a stone was split open and there appeared a head of Pan; I believe there may have been some resemblance in shape, but certainly not such as would make you ascribe it to Scopas. For it is undeniably true that chance never imitates truth perfectly.
>
> But, it is objected, some things predicted do not happen. But what art—and I mean an art that involves conjecture and opinion—is not like this? Is not medicine considered an art? And how many mistakes it makes. Do not pilots make mistakes? . . . Surely the fact that so many illustrious captains and kings suffered shipwreck did not make piloting not an art? And is military science nothing, because a very great general recently lost his army and fled? Again, is statecraft not reasonable and prudent because many times Gn. Pompey, sometimes M. Cato, and sometimes even you [Cicero] were mistaken? So it is with the replies

of augurs and all divination involving opinion; for they produce conjecture and cannot go beyond it. Perhaps it misleads sometimes, but
most often it leads to the truth; for it has been repeated from all eternity, during which almost countless instances of the same things following the same antecedent signs have been observed and noted, resulting in the art.[10]

Astrology, too, enthusiastically collected data. It was claimed that the
Babylonians had taken the horoscope of every child, and tested it, for
470,000 years.[11] Until modern times, Ptolemy's *Almagest* was no more
famous than his companion work on astrology. His defense of astrology
is moderate, perfectly reasonable, very willing to admit the limitations
of the science. His list of excuses for the frequency of wrong predictions
is a long one: unlike astronomy, astrology must deal with the "weakness and unpredictability of material qualities found in individual
things"; individual fortunes naturally need much more study than the
gross effects of the heavenly bodies like seasons and tides; the unin-
structed—"and there are many, as one would expect in an important and
many-sided art"—make many mistakes. But his main excuse is the one
that still serves so well for sciences with poor predictive power, like economics and meteorology, the excuse from complexity: "The ancient
configurations of the planets, upon the basis of which we attach to similar aspects of our own day the effects observed by the ancients in theirs,
can be more or less similar to the modern aspects, and that, too, at long
intervals, but not identical, since the exact return of all the heavenly bodies and the earth to the same positions, unless one holds vain opinions
of his ability to comprehend and know the incomprehensible, either
takes place not at all or at least not at all in the period of time that falls
within the experience of man; so that for this reason predictions sometimes fail, because of the disparity of the examples on which they are
based."[12] Oresme, as we saw in the last chapter, harassed astrology with
vigor over its predictive failures, to little effect. Kepler, who supported
himself with his astrological work, resuscitated Ptolemy's defense that
astrology can only predict tendencies.[13]

The Empiric School of Medicine on Drug Testing

It is in medicine that there is most pressure to find correct methods of
learning from experience. And there is always money available for it.
The subject matter, unfortunately, is recalcitrant; though there are true
generalizations to be found, in contrast to divination or astrology, it is
much harder than in astronomy to discover simple hypotheses true uni-

versally. That is what gives credibility to Cicero's comparison of medicine with divination. As in divination, so in medicine, and especially in pharmacology, one must generalize from experience, without knowledge of underlying causes: "The power of scammony root for purging . . . I see, which is enough; why they have the power I do not know."[14]

Not everyone agreed. Aristotle argued for a rational search for certain causes in medicine as everywhere else. He provided a geometrical explanation for the fact that circular wounds heal more slowly than long narrow ones.[15] He presumably hoped other medical facts would eventually be understood in some similar way. One of the Hippocratic writings, *The Art*, believes that true doctors know the genuine science of healing, and that if patients think otherwise, they are mistaken, and any failures to recover are the patients' fault: "It is far more likely (*polu ge eulogoteron*) that the sick are unable to carry out the instructions than that the doctors prescribe the wrong remedies. . . . There can be no doubt that the patients are likely to be unable to obey and, by their disobedience, bring about their own deaths."[16]

A rationalist approach, combined with experiment, was pursued in early Alexandria by doctors like Erasistratus, whose neurophysiological experiments on criminals show a zeal for finding causes equaled only by the worst of the Nazi doctors.[17] In opposition to these and other Dogmatist schools seeking the causes of medical phenomena, there arose the Empiric school, which denied there was any knowledge of causes to be found and restricted itself to describing the succession of phenomena. Such an attitude also had some basis in the works of Hippocrates. The *Aphorisms* of Hippocrates open with the famous saying, "Life is short, art long, opportunity fleeting, experiment deceptive, judgement difficult."[18] The genuine writings of Hippocrates do not contain much about causes or the certain knowledge of them. There is at most some distinction between signs that indicate death or recovery with certainty and those that do not.[19] Book 3 of his *Epidemics* contains a number of case histories; some ancient manuscripts of this work had at the end of each case certain annotations, apparently shorthand. The first letter of each of these was *pi*, which Galen says stands for *pithanon*, introducing a note on what probably caused the disease to develop as it did.[20] Generalizations true "for the most part" are common in Hippocrates: "Persons who are naturally very fat are more apt to die quickly than those who are thin"; "Epilepsy in young persons is mostly removed by changes of air, of country and modes of life."[21] The Hippocratic *Prognostic* contains a number of these and concludes with some acute remarks on how much to take account of the variability of conditions: despite the variability

in the subject matter of medicine, most signs, it is suggested, can be interpreted independently of local conditions. "Anyone who is to make a
correct forecast of a patient's recovery or death, or of the length of his
illness, must be thoroughly acquainted with the signs and form his
judgement by estimating their influence one on another. . . . The physician must be quick to think of the trend of any diseases that are epidemic
from time to time, and the climatic conditions must not escape him. It
should, however, be observed that the indications and signs have invariably the same force, the bad being always bad and the good good, in
every year and under all climatic conditions. The truth of those described in this treatise has been proved in Libya, in Delos and in Scythia.
It should therefore be realised that there is nothing remarkable in being
right in the great majority of cases in the same district."[22]

The significance of the passage is its choice of a middle path in finding the correct reference class for a generalization, still one of the hardest problems to solve in probabilistic reasoning.[23] One can easily give
vacuous advice like, Take account of all circumstances (which would result in no generalization, as nothing agrees with a given case in *all* circumstances) or Generalize as much as possible, which sets no practical
limits. The *Prognostic* says what characteristics are relevant: one may
generalize over time and space but not over differences of weather or
the presence or absence of an epidemic.

The Empiric school itself did not arise until there was a sufficient
number of Dogmatist schools to combat. Like any skepticism, it throve
best in the presence of dogmatisms in collision. In general, its lines of
argument were taken from the Skeptics of the Academy and were a
straightforward application of the skeptical method to medicine.[24] The
school existed from about 200 B.C., but its main theoretician seems to
have been Menodotus of Nicomedia in the second century A.D.[25]
Though his view was that little can be known, his manner was said to be
dogmatic and loquacious in the extreme; his opinion that the aim of a
doctor is to acquire fame and a lot of money is doubtless to be attributed to the *enfant terriblisme* endemic among Skeptics.[26] Though Sextus
Empiricus was a member of the school shortly after Menodotus, he tells
us little about any aspects of the Empirics' views other than the purely
negative. Our knowledge of the positive side of their doctrines, on how
to learn from experience, comes from two obscure books of Galen, *An
Outline of Empiricism*, which exists only in a Latin translation of 1341,
and *On Medical Experience*, most of which survives only in Arabic.[27] Both
works were almost unknown until modern times.

The Empirics sought to justify certain inductive arguments without appealing to hidden causes. According to the account in *On Medical Experience*, they began with some polemic on the possibility and necessity of learning purely from experience in the ordinary course of life. "And what think you, moreover, of the peasant? Is he, until he has learned from one of the philosophers something of the nature and substance of the soil, and what is the nature and substance of rain and wind, and how they come about, unable to know by experience what seeds to sow at certain times and on what soil, if they are to spring and flourish and attain completion and perfection? . . . You know that men taken as a whole, of whatever type they may be, do not feel bound to examine into the nature of wine, but that they know perfectly well that too great indulgence in drinking wine is harmful. And so it is with mushrooms. . . ."[28] Even Erasistratus, it appears, argues from experience, at least when replying to objections.[29]

The Empirics rejected inference to unseen causes for various reasons, among others that a cause could in principle be inferred from an effect that had been observed only once. Dogmatists were not, therefore, in a position to object to the Empirics' use of inference from what has been observed "very many times." To admit that in doubtful cases a second case is useful, as the Dogmatists do, opens the door to considering the whole question of inductive skepticism and the worth of large numbers of cases.

> Experience has shown that what has produced a like result in three cases can produce the reverse in three others. . . . A thing seen may be seen exactly as before, and yet belong to those things which are of both kinds (*amphidoxos*), or to those things which happen often, or to those things which take place but rarely. It . . . may be seen only thus and yet belong to those things which happen *amphidoxos*. But . . . it is not impossible that it should belong to the things which are frequent or those which are rare, and yet can be seen in this manner. . . . What is to prevent the medicine . . . from having a given effect on two hundred people and the reverse effect on twenty others, and that of the first six people who were seen . . . and on whom the remedy took effect, three belong to the three hundred (*sic*) and three to the twenty without your being able to know which three belong to the three hundred, and which to the twenty . . . ? surely do not construct your *logos* on this. . . . You must needs wait until you see the seventh and the eighth, or to put it shortly, very many people in succession.[30]

The Dogmatists naturally demanded to know how many times counted as "very many" and pointed to the absurdity of taking, for example, forty-nine times as not enough to support an inference but fifty times as enough. They asked again how experience worthy of inference could be composed of experiences each of which was not worthy of inference.[31] A natural reply would have been that the probability of the induction increases gradually with the number of observed instances, but the Empirics were diverted from making this discovery by having on hand another intellectual structure into which the case would fit. They simply answered that the situation is the same as a heap: it is impossible to say exactly how many grains of wheat are needed to make a heap; nevertheless it remains true that very many grains form a heap.[32]

To the Dogmatist argument that any experience has simply too many features to remember, such as, "If your patient should have happened to sleep, or to speak, or what garments he was wearing," the Empirics replied that memory, here as elsewhere, manages to hold relevant things of frequent occurrence. As to the certainty or otherwise of their methods, the Empirics, though apparently not admitting the fallibility in principle of induction from what has been seen "very many times," did admit to making mistakes because their experience was limited. They hastened to add, of course, that the Dogmatists did too, without admitting it. Dogmatism, they said, could at best only reach the level of plausibility and likelihood.[33]

The *Outline of Empiricism* goes further into the positive aspects of the Empirical method. To Dogmatic "reason" is opposed "the experience of those things which have been found to happen for the most part and in a similar way."[34] The most interesting aspect is a grading of relative frequencies on a scale: never, rarely, as often as not, mostly, always. "These are the four differentiations of theorems. Hence we will also say that a theorem is the knowledge of something which has been seen often but a knowledge which involves at the same time a distinct knowledge of results to the contrary. This will be a distinction between what happens always (as something whose contrary never makes its appearance), what happens for the most part (as something whose contrary does appear, but rarely), what happens either way, as it may chance to be (as something whose contrary appears equally often), and finally what happens rarely (because its contrary does appear, not just sometimes, but for the most part)."[35] Examples are "Always, as death in the case of a heart wound; for the most part, as purgation from the use of scammony resin; half the time, as death in the case of a lesion of the dura mater; rarely, as health in the case of a cerebral wound."[36]

Besides direct experience, the Empiric will acquire knowledge through "history"—what has been written down by others. Here the

Empirics make a start on discussion of the credibility of testimony, or what was called by later historians the "signs of true histories." Agreement among authorities is important, but not everything repeated by many writers is to be believed (for example, things written under the influence of dogmatic misconceptions):

> The first and foremost criterion of true history, the empiricists have said, is what the person who makes the judgement has perceived for himself. For, if we find one of those things written down in a book by somebody which we have perceived for ourselves, we will say that the history is true. But this criterion is of no use if we want to learn something new. For we do not need to learn from a book any of those things which we already know on the basis of our own perception. Most useful and at the same time more truly a criterion of history is agreement. For it is possible that I have never used mace (this is a drug brought from Arabia, the bark of a tree). But all who write on *materia medica* say about it that it constipates. Shall we, then, believe or disbelieve them? I, for my part, say that one should believe those who are in such agreement. But I say this, since we talk about matters which can be perceived. For agreements concerning what is not manifest [that is, observable] may be widespread; yet they are not supported by everybody who writes on the subject. And, even if one granted that it is not ruled out that such an agreement could ever occur, at least the Empiricist will have no share in such an agreement. But, whatever agreements come about among all men concerning matters which can be perceived, such agreements can be trusted in practical life. For, though we ourselves never have sailed around Crete or Sicily or Sardinia, we have come to believe that they are islands, because those who have perceived the matter for themselves are all in agreement with each other on this. Now, we have acquired in everyday life the experience that those who report on matters which can be perceived agree with each other. . . . Another criterion for history is whether what is said resembles those things we have come to know through our own observation.[37]

These considerations are in principle equally applicable to works of history in the usual sense, but historians in ancient times do not seem to have developed any theory of evaluation of written evidence as self-conscious as that of the medical writers.

It is significant that the Empirics ask *why* agreement should be taken to be the sign of truth and answer simply, "It is known from experience." They give the same reason for believing in "transition to the similar," that is, argument from analogy. "Empirical transition . . . relies on what is known naturally, not because it is plausible (*suasibile*, probably translating *pithanon*) that something similar should have similar effects, lack similar things, or be similarly affected; it is not because of this or because of anything else of this sort that one insists on transition, but only because we know from experience that similar things are like this."[38] So there is no concession to Dogmatic a priori reasoning; all conclusions are to depend on experience. Because of this, one can learn things about analogy that would not be suggested by pure reason. As an example, one finds that one should sometimes proceed not to the similar but to the contrary, which could only be learned from experience. The discussion of continuous gradation in probability and similarity is remarkable:

> The different degrees of expectation and trust [literally, "the more and the less of hope and faith"] do not directly correspond to the degree of similarity in each of the cases mentioned. For similarity in one case amounts to more than similarity in another case. For one learns about similarity in one case not by accident and going about the matter haphazardly. For what is similar in shape and color and softness and hardness has been least observed to produce similar effects, whereas what is similar in odor or in taste has been found usually to lead to the same result, and, among these, more so in the case of what is similar in taste and even more so in the case of what is similar in both respects, that is, in odor and taste. But if in addition shape, color and consistency enter, too, one will see that things are similar to the highest degree and produce the same effects. And among the things themselves which are similar in taste, one should not judge similarity just by some single quality, such as sharpness, astringency, bitterness, sweetness, harshness, sourness or saltiness but should put one's mind to the peculiar character of the taste as a whole . . . to the extent that the things to which we make the transition differ in similarity, there is a different degree of expectation of the possible outcome [defect in text] if it is recommended by a trustworthy man against diarrhea and if it seems very similar to things one knows from experience. But it is clear that this kind of case will raise the highest expectation as to the possible outcome, and perhaps someone will dare to trust it even before having gained a practical experience of it. But, for what neither is confirmed by history nor is similar, there is little expectation. But thus it is reasonable to have a higher or lower expectation [literally, "it is appropriate to hope more or less"] also in the case of the transition from one disease to a similar disease, depending on the similarity between the diseases, an expectation which is diminished or increased, depending on whether it is confirmed by history or

> not. And, in similar fashion, with the transition from one part of the body to another, to the degree that there is the more or less difference, to that degree there will be differences of expectation [literally, "as much as there is more or less, so much will there be of hope"].[39]

Not only is this the clearest statement in antiquity that reasonable belief should be proportioned to evidence but it also shows some understanding of the very difficult problems in integrating different kinds of evidence to produce a single degree of belief in a proposition.

Galen proudly describes some of the knowledge he has gained by means of these methods of learning from experience. He has discovered a medicine made from snakes, for example, by observing accidental cures from a patient drinking wine into which a snake had fallen. Perhaps this shows that the purely qualitative methods of evaluating evidence, which Galen describes so well, are not enough to give good results if the material is resistant. Galen himself says elsewhere, "If ever, when you are dissecting a limb, you see something that contradicts what I have written, recognise that this happens infrequently. Do not prejudge my work until you yourself have seen, as I have, the phenomenon in many examples."[40]

The Talmud and Maimonides on Majorities

Jewish law discusses a number of cases that involve proportions in populations. The style of reasoning resembles that of the Greek medical writers, but the examples are quite different, though a few of them are also medical.

Jewish law deals with far more than what other societies regard as legal. It lays down an order for all activities of life. There is immense pressure on the interpreters of the law to produce answers for all doubtful cases and methods that can be thought to solve in principle all future doubts. While a decision procedure for doubtful cases need not be one that aims at producing the answer that is most probably true, it can be expected that there will be some connection between what the law lays down and what is most likely true.

Among the most frequently discussed cases are doubts as to whether various objects are permitted or not, when, for example, they are found in the street and their origin is unknown. The rule normally applying in such cases was "Follow the majority" or "What is separated derives from the majority." "If nine shops sell ritually slaughtered meat and one sells meat that is not ritually slaughtered and he bought in one of them and does not know which one—it is prohibited because of the doubt; but if meat was found one goes after the majority."[41] It would be too much to read the rule itself as probabilistic, but it is certainly a procedure that sometimes agrees with the proportional syllogism, the inference from the proportion in a population to a particular case.

There was some dispute about what to do in cases in which the majority is not actually counted, as in Most children will grow up fertile. However when the majority was overwhelming, such as "The scribes of the court know the law," following the majority was allowed by all authorities.[42] As in the Greek medical writers, proportions of half and half were envisaged, but that was the limit of quantification. "If an abandoned child was found there, if the majority [of inhabitants] were gentiles, it may be deemed a gentile; if the majority were Jews, it must be deemed a Jew; if they were half and half, it must be deemed a Jew. R. Judah says: we must consider who form the majority of those who abandon their children [that is, Gentiles]."[43]

A difficulty in applying "Follow the majority," as with any inference from population to sample, is that of specifying the reference class: two different proportional syllogisms may apply to a case and give conflicting answers. The case just mentioned is an example, since the case belongs both to the class of all inhabitants and to the narrower class of inhabitants who abandon their children, and the proportions of Jews in these two classes, it is said, are different. A case even more explicit about the reference class problem is the following. "One sold grain to his fellow who planted it and it failed to grow . . . in the case of flaxseed, the seller is not liable. Rabbi Yose says: He must give him the value of the seed. But they [his colleagues] said to him: Many buy it for other purposes. . . . Both of them follow the majority—one of them goes after the majority of people [who buy flaxseed for many purposes, for example, food and medicine, and not just for planting] and one goes after the majority of seed [which is mostly used for sowing]."[44] Some follow-the-majority arguments may be weakened by further evidence, and doubts may be added to each other to become "double doubts."[45] Thus to decide how long a pregnancy has lasted, in order to determine paternity, "When a woman bears at nine months, her embryo is recognisable in most cases after a third of her pregnancy, and with this woman, since her embryo was not recognised after a third of her pregnancy, the majority is weakened."[46]

There are some decision rules corresponding roughly to inference from a sample to a population. *Tefillin* are ritual objects, one worn on the arm and one on the head; they must be prepared according to exacting specifications. Checking whether the specifications have been met is very laborious. So "If a man buys a supply of *tefillin* from one who is not certified, he checks two for the arm and one for the head, or two for the head and one for the arm," by way of establishing the credentials of the maker. But if they are bought from several people, how many need checking? Do the reputations need checking for the arm and head separately? The rule that a presumption is established by two or three cases becomes very difficult to apply to such a case.[47]

The inferring of a pattern from two or three cases can be regarded as too risky to trust. Even modern authors who approve of inductive inference in general commonly require that a considerable number of cases should be surveyed before any generalization or action is ventured. But one's skepticism on induction might change if one were on the receiving end, as in the following cases from the Talmud:

> It was taught: If she circumcised her first child, and he died, and a second one also died, she must not circumcise her third child; thus Rabbi. R. Shimon ben Gamaliel, however, said: She circumcises the third, but must not circumcise the fourth child. . . . It once happened with four sisters at Sepphoris that when the first had circumcised her child he died; when the second [circumcised her child] he also died, and when the third, he also died. The fourth came before R. Shimon ben Gamaliel who told her, "You must not circumcise." But is it not possible that if the third sister had come he would also have told her the same? . . . It is possible that he meant to teach us the following: That sisters also establish a presumption.
>
> Raba said: Now that it has been stated that sisters also establish a presumption, a man should not take a wife either from a family of epileptics, or from a family of lepers. This applies, however, only when the fact had been established by three cases.[48]

There follows a discussion of whether it is safe to marry a woman who has had two husbands die. It is felt that if the deaths were obviously due to some chance event, such as falling out of a palm tree, there is no need to worry, but if not, there may be some hidden cause in the woman, which the prospective husband would do well to take into account.

Some recognition of how the reliability of the inference varies with the size of the sample appears in the rules about when to declare a plague. A plague is declared in a town that supports 500 fighting men if there are three deaths in three days (but not in one or four days). For a town that supports 1,500 fighting men, it is nine deaths in three days.[49]

Maimonides comes close to evaluating "doubts" by counting cases that could be regarded (though he does not say so) as equiprobable. If a female, whether human or animal, has a firstborn who is a male, he belongs "to the Lord," and something must be given to the priests to "redeem" him. Maimonides considers the case where two she-asses give birth for the first time, producing a total of two females and one male, but it is unknown which belong to which mother and in which order they were born. Is there an obligation in respect of either she-ass? "If they bore two females and a male . . . the priest gets nothing and even the ritual of exchange for a lamb is not required, because there are here many doubts—perhaps the one bore a male and the second bore two females, or perhaps she bore a female and the other a male followed by a female or a female followed by a male."[50] (To make sense of this passage, it must be supposed that the aim is not to evaluate, in modern language, the probability that at least one mother had a firstborn male but to ask, for each mother, whether she probably had a first-born male.)

Maimonides treats a similar case involving human children by counting cases (rather unclearly) and concluding that redemption must be paid because the husband is "free in only two cases," and the chance of there being no obligation is "remote."[51] Also of interest is a fourteenth-century passage that explains that one cannot derive expected relative frequencies simply by counting logical possibilities. "Half of all [children] are male and half female, it is certain and necessary, for thus did the King of the Universe establish it for the preservation of the species. Therefore of necessity, of all pregnant women those who bear males are a minority, for some abort, and this is inescapable. But, here we cannot say that of all those who have illicit relations, half do so after betrothal and half before . . . for whence do we know that it is half and half? It is only that we say the matter is in doubt, for the one or the other is possible; and even if we add all the cases of rape together with those not under him [the eventual husband], the willing adulteress after betrothal is not certainly in the minority, and we can still say the one or the other is possible . . . and the doubt still exists."[52] A number of Jewish writers show an explicit grasp of the idea that what is possible in one instance is unlikely to keep happening time after time.[53]

Vernacular Averaging and Quality Control

High theory is interesting and easy to learn about from texts. One would be equally pleased to learn how the ancients and medievals coped with the more day-to-day matters that need probability as "the guide of life." Unfortunately, technology in general has left few written records, while archaeological evidence is not much help in revealing how people conceived of what they were doing. Nevertheless, a little shows through here and there in the written record. Almost the only evidence in ancient writing of something of this kind is the story in Thucydides of the siege of Plataea. The city was surrounded at some distance by a brick wall built by the enemy, and the defenders proposed to break out by scaling the wall with ladders. Their problem was to arrive at the wall with ladders of the right length, since the height of the wall was not easy to estimate from the distance of the city. "These [layers of bricks] were counted by many persons at once; and though some might miss the right calculation, most would hit upon it, particularly as they counted over and over again, and were no great way from the wall, but could see it easily enough for their purpose. The length required for the ladders was thus obtained, being calculated from the breadth of the brick."[54]

Thucydides seems unaware of the statistical nature of the argument. He remains so even when he himself uses arguments that would later be called statistical, in that they draw conclusions about a population from facts about a sample. For example, he infers that the ancient inhabitants of the Aegean islands were mostly Carians from the fact that most of the graves on Delos were found to have Carian remains and argues that the size of the Greek expedition against Troy was smaller than commonly thought (by taking a mean of Homer's—incomplete—figures for the complements of ships).[55]

An idea logically the same as that of estimating the length of the ladders is found in sixteenth-century Germany, where averaging was used to reduce the variation in the length of units that depended on body parts, like feet and inches. A rood should be determined, the surveyor Köbel advises, by taking sixteen men as they come out of church, "tall ones and short ones, as they happen to come out"; the rood is the sum of the lengths of their left feet.[56]

According to the logic of induction, the price of a thing that might or might not work should be based on its performance to date. Medieval Welsh law says, of the price of an important item of food preservation technology, imported from the East: "The value of a cat, fourpence. The value of a kitten, from the night it is born until it opens its eyes, a legal penny; and from then until it kills mice, two legal pence; and after it kills mice, four legal pence, and at that it remains for ever."[57]

Quality control in engineering was plainly a serious problem, particularly in the more ambitious medieval projects like cathedral building. It was most discussed, however, in the context of minting coins. The point of coins is that they should be exactly the same, at least up to the limits of perception. Otherwise, there is money to be made by spending the bad ones and melting down the good ones. Knowing whether coins were the same or not was difficult, but it was recognized that some people were far more expert than others, though "hardly anyone can be found so fully expert as not often to be mistaken in this."[58] It was incumbent on kings to impose strict requirements on the mints, without demanding the impossible. The outcome, in the English mints of the thirteenth century, was a process called the trial of the *pyx*, which incorporates an implicit understanding of averaging and random sampling. From each day's minting, one coin was saved in a secure box, the *pyx*. At irregular times, once every few years, the king would declare a trial of the *pyx*, supervised by an independent jury of goldsmiths. The *pyx* was opened and the contents counted and weighed. Then a sample from them would be assayed. Assaying involved melting down the coins and was the only accurate method of determining their composition. A tolerance, or "remedy," was allowed, the master of the mint being threatened with various punishments if the result was not within the remedy. Since assaying was difficult, it was laid down that the best of three careful assays should be used; if their results were different, "judgement should always be given for the assay which weighs heaviest . . . because silver can easily be lost and can never be gained in the fire."[59]

The most interesting feature is a random mixing process for deciding which of the store of coins taken from the *pyx* should be counted and weighed before being assayed. "When the Master of the Mint has brought the pence, coined, blanched and made ready, to the place of trial, for example, the Mint, he must put them all at once on the counter which is covered with canvas. Then, when the pence have been well turned over and thoroughly mixed by the hands of the Master of the Mint and the Changer, let the Changer take a handful from the middle of the heap, moving round nine or ten times in one direction or the other, until he has taken six pounds. He must then distribute these two or three times into four heaps, so that they are well mixed. Then he must weigh out, from these well mixed pence, three pounds, well and exactly, by a standard pound of 20 shillings which is correct to a grain."[60]

Some understanding along the same lines was available to Kepler, who advised the Senate of Ulm that the total weight of a large number of coins could be accurate even if the weights of the individual coins were not.[61]

Experimentation in Biology

Avicenna was aware of one of the major problems in experimentation in the low sciences: that the variability of the subject matter makes it difficult to sort out the causes of any given effect. He proposes seven rules for medical experiments, which go some way toward resolving the problem. His main concern is to be able to distinguish the natural, or *per se*, effect of a substance from an accidental one—one it has by chance or some acquired characteristic. "The first is, that the medicine be free from any quality acquired from accidental heat or cold, or any quality that attaches to it by alteration of its substance, or from the proximity of something else. For water, though by nature cold, can when heated give heat, as long as it remains hot."

The second rule attempts to deal with composite diseases, which may confuse the effects of remedies, the third with the possible effects of a single remedy on two diseases, one *per se* and one *per accidens*. "The fourth is that the force of the medicine needs to equal the power of the disease it is combatting. There are some medicines whose heat is less than the coldness of a disease, and hence are wholly without effect on it. But when they are administered to a coldness weaker than themselves, they effect heating. One should therefore test first on the weaker [disease], and progress gradually, until the force of the medicine is known with certainty."

The fifth rule is a complicated one concerning the time of onset of the effect of a remedy, and the possibility of a *per se* effect at one time and an accidental one later. "The sixth is, that the operation should be observed to take place constantly or for the most part. For if not, the operation arises from it only *per accidens*, since natural things act according to their natures either always or for the most part. The seventh is, that the experiment should be on a human body. For if the experiment is on a non-human body, it is possible to be mistaken in two ways. One is that it is possible the medicine is hot relative to the human body, but cold relative to the body of a lion or horse."[62]

A modern statistician would say that Avicenna has a sound appreciation of the confounding effects of several variables. Even now, it is difficult to infer causes from observations in complex medical situations because of the problems Avicenna describes.

Some such rules of experimentation were commonplace in Latin Scholastic medicine, though it is very hard to tell if experiments were conducted in accordance with them.[63] Certain refinements were made to the rules. One authority suggests the need to take a sample of an herb that is average for its kind in color, age, habitat, and so on to get a credible result. He advises that one should test drugs successively on birds, then on dumb animals, then in hospitals, then on Franciscans, in case they prove fatal.[64] There was some desultory argument as to whether medicine really had the certainty Aristotle demanded of a science or whether it might be excused on the grounds that it was an art. Arnald of Villanova, about 1300, adds an excuse reminiscent of the ancient Empirics but with theory taken from astronomy: any method of calculating the intensity of a drug might be acceptable if it gives the right results; just as astronomers calculate with epicycles and eccentrics that may not really exist, "so too here . . . it is enough for the physician to use such measurements as will provide certainty (insofar as possible) in calculation, even though they do not actually correspond to reality."[65]

Biology was especially subject to domination by unhelpful arguments from analogy, whereby superficial aspects of plants and animals caused them to be taken as symbols of something else. One of the things most notably wrong with Renaissance thought was the immense proliferation of signs. Though allegory was certainly popular earlier (medieval bestiaries spend a good deal of space on what virtues animals symbolize), it did not so completely create a veil of signs between man and the world as it did in the Renaissance.[66] The naturalist Aldrovandi arranges his chapter "On the serpent in general" under headings that include the meanings of the word, etymologies, form and description, anatomy, nature and habits, voice, diet, physiognomy, antipathy, sympathy, modes of capture, wounds caused by, remedies, epithets, prodigies, mythologies, gods to which dedicated, emblems of, proverbs, coinage, miracles, riddles, heraldic signs, dreams, statues, use in diet, use in medicine, miscellaneous uses.[67] Paracelsus says, "Behold the *Satyrion* root, is it not formed like the male privy parts? No one can deny this. Accordingly magic discovered it and revealed that it can restore a man's virility and passion. . . . *Siegwurz* root is wrapped in an envelope like armour, and this is a magic sign showing that like armour it gives protection against weapons. And the *Syderica* bears the image and form of a snake on each of its leaves, and thus, according to magic, it gives protection against any kind of poisoning."[68] It is argument from analogy gone mad.

It has been argued that probability around 1600 was "a child of the low sciences," and as evidence there has been adduced the book of Fracastoro on the signs of plague. "Each contagion has its own signs (*indicia*). Some are warnings of the future, some reveal the present. The ones called premonitory are taken variously from the sky, the air, and terrestrial and aqueous things. Some of them are true for the most part, others often, so that one ought not to prognosticate from them universally, but only probably."[69] It will be obvious that there is nothing in these concepts beyond what is common in Galen. Nor is there anything in Fracastoro's collection of signs from comets, planetary conjunctions, and swarms of locusts to suggest any development of rationality in practice.

Though mostly too busy writing voluminously in praise of experiments to actually perform any, Francis Bacon does report an interesting experiment on the germination of seeds that involves the use of a control. Seeds steeped in claret, urine, water, and a number of other liquids are compared for time of germination with unsteeped seeds. Urine is good, claret bad. Bacon adds that it would be "a great matter of gain, if the goodness of the crop answer the earliness of the coming up; as it is like it will." "Likelihood" is thus used to fill in with pure conjecture where experiment is lacking.[70]

The Authority of Histories

With the low sciences one at least has the opportunity of confirming or disconfirming a hypothesis by collecting further evidence. The peculiar difficulty of historiography is that this is rarely possible, as the sum of evidence on a given question about the past is usually fixed. The problem was made worse in antiquity by the circumstance that the most prominent "histories" concerned the doings of gods as well as of men.[71] Sextus Empiricus's skepticism about a "criterion of history" is perhaps more justified than some of his other attempts to cast doubt on all subjects. "They deserve to be laughed at," he says, "who assert that even if the subject-matter of history lacks method, yet the judging of it will be a matter of art, by means of which we ascertain what is falsely related and what truly." He continues with this reasoning: "Firstly, the Grammarians have not furnished us with a criterion of true history, so that we might determine when it is true and when false. In the next place, as the Grammarians have no history that is true, the criterion of truth is also non-existent; for when one man says that Odysseus was killed in ignorance by his son Telegonus, and another that he breathed his last when a sea-gull dropped on his head the spike of a roach, and yet another that he was transformed into a horse, surely it is a hard task to try to discover the truth in such incoherent accounts. For we must establish first which of these dissentient narrators is telling the truth, and then inquire as to the facts; but when all relate what is improbable (*apithana*) and false no opening is given for a technical criterion."[72]

The social sciences in general were in a poor state of development in the Middle Ages and Renaissance. While legal science made some progress on the problem of evaluating the testimony of witnesses, history did not. As is well known, medieval histories, chronicles, travelers' tales, bestiaries, and so on generally combine the perfectly true with the totally impossible, with little attempt to sort out the two. In the absence of any generally recognized concept of probability, readers' attitudes to the contents of such works could only oscillate unstably between credulity and skepticism. *Mandeville's Travels* of the mid-fourteenth century, for example, achieved immense popularity by stringing together the colorful portions of everyone else's tales of the wonders of the East. It retails without skeptical comment stories of paper money in China and rivers of jewels in the lands beyond Java, while being perfectly well aware that the spearhead that pierced Christ's side cannot be in both Paris and Constantinople and that there are problems with the total content of the alleged parts of John the Baptist's head.[73] The frequently repeated tale that the fragments of the True Cross would have added up to a whole forest appears to be a modern myth.[74] Indeed, it is only one of a cluster of widely credited myths about the Middle Ages (in the Middle Ages it was believed the earth was flat; lords enjoyed *droit de seigneur* over peasant women; the Scholastics debated how many angels could dance on the head of a pin),[75] raising questions about which age is really the more credulous. Is the level of common "knowledge" about the Middle Ages actually negative?

One has of course to allow for the higher medieval valuation of the a priori likelihood of a miracle. This is more a difference in philosophy than a matter of rationality, being dependent on belief in the absolute power of God. One medieval historian writes, "Marvels are not to be entirely disbelieved. As Jerome says, 'You will find many things that are incredible and not likely, which are nevertheless true. For nothing in nature can prevail against the Lord of nature.'"[76]

Even so, one can often see in medieval historians that a critical skill is not absent; it is just that it is not their practice to remold the stories they are preserving in the image of their own critical opinion. They will, for example, retail contradictory stories and perhaps note that one is "scarcely credible," rather than excising one of them. Compared to modern historians, they prefer to present evidence rather than conclusions and leave evaluation to the reader.[77] Their practice is perhaps more comparable to that of a present-day archivist than a historian.

As might be expected, at least a beginning on the critical evaluation of evidence was made in the twelfth century, when Guibert of Nogent saw that excessive demands for certainty in matters of faith could simply produce belief about obvious falsities. His *On the Relics of the Saints* observes that the two monasteries claiming to possess the head of John the Baptist cannot both be right and demolishes the evidence in support of the authenticity of various relics. He contrasts the credulity in these claims with the caution of the church on belief in the Assumption of the Blessed Virgin. "Restraint in the common language of the entire holy

Church is so great that the Church does not dare to say that the body of even the mother of the Lord has been glorified by resurrection, on the grounds that it cannot be proved by necessary arguments . . . we dare not at all say that her body was resurrected, and this simply because we cannot assert it except with probable indications (*probabilibus indiciis*)."[78] Guibert is not, however, a good candidate for any kind of proto-Skeptic or proto-Rationalist. He allows that belief in the Assumption is actually more pious, and elsewhere he asserts that Jewish magicians are agents of the devil and that his mother was assaulted by an incubus.[79]

Retailing an unlikely story does not necessarily indicate an inability to tell the likely from the unlikely. William of Newburgh, in the 1190s, relates a prodigy in the reign of King Stephen. During harvest time in East Anglia, two green children appeared, wearing clothes of an unknown material. At first they would eat only beans, but after becoming accustomed to English food they gradually lost their green color. On learning English, they claimed to have been suddenly transported from St. Martin's Land, a Christian country where it was always twilight. William understands perfectly well how grossly improbable this story is but regards its improbability as outweighed by the weight of testimony in its favor. He introduces the story with these remarks: "I myself had protracted doubts over this, though it was reported by many, and it seemed to me absurd to accept as genuine an event whose rational basis was non-existent or most obscure. But finally I was so overwhelmed by the weighty testimony (*pondere testium*) of so many reliable people that I was compelled to believe and marvel at what I cannot grasp or investigate by any powers of the mind."[80]

In similar vein, the most popular medieval hagiography, Jacobus de Voragine's *Golden Legend*, in the course of relating uncritically the implausible story of the martyrdom of the eleven thousand virgins, suddenly remarks that the claimed date of A.D. 238 cannot be right, "for at that time neither Sicily nor Constantinople was a kingdom, whereas we have said here that the queens of these kingdoms were with the virgins. It seems more likely that the martyrdom took place long after the reign of the Emperor Constantine, when the Huns and the Goths were raging."[81]

William of Newburgh shows also how in history, as in canon law, the conflict of authorities could stimulate the development of a critical spirit. His prologue is an extended attack on Geoffrey of Monmouth's "history" of King Arthur and his forebears. He is particularly incensed at Geoffrey's exaggeration of the qualities and achievements of the Britons, contrary to the truth recorded in sober historians like Gildas and Bede.

Gildas writes, for example, that the Britons were neither brave in war nor trustworthy in peace, and "it is no slight proof of integrity that he does not spare his own nation in revealing the truth." And Geoffrey could not be right in having three archbishops present at Arthur's celebrations, since there were at that time no archbishops in Britain. Arthur's alleged conquest of the Romans, Libyans, Syrians, and so on, with thirty other kingdoms, is false, as there did not exist thirty kingdoms beyond those named, and "how could the historians of old, who took immense pains to omit from their writings nothing worthy of mention, and who are known to have recorded even modest events, have passed over in silence this man beyond compare?" And Geoffrey's confidence in the prophecies of Merlin, on the grounds that his father was an incubus, is misplaced, since it is well known that demons cannot see the future but only conjecture it.[82]

One of the more acute medieval historians is Bernard Gui, whose skills in evidence evaluation in history are close to those in his work as inquisitor. Occasionally even the actual evidence is the same; he has inquisitorial records sent to him under seal to assist his writing on history.[83] His extensive researches in a wide variety of oral and written sources leave no doubt of the sincerity of his statement, "I have inquired (*inquisivi*) into the truth of what has gone before."[84] He is of course concerned with the lack of harmony in the sources, which he attributes to the errors of copyists and (being one of the first to do so) to the "diversity of authors' opinions and positions." In cases in which the authorities are equally trustworthy and the issue is minor, he is prepared to simply "present the contrary arguments and leave it to the reader to judge and choose."[85] But otherwise he argues for his view as to which authority should be preferred. One of the more extended examples concerns two accounts of the martyrdom of Saints Ciricus and Julita, in the later Roman Empire. The disagreements concern both chronology—perhaps the main source of contradictions that medieval historians had to face—and what the martyrs said to their persecutors.

> First, in the account of Theodore it is said that a certain Alexander had been recently appointed by Diocletian as ruler of the region of Seleucia, which is a city of the Hysaurian nation. But in the other account it is maintained that the ruler Alexander, under whom the martyrs Ciricus and Julita suffered, lived under the Emperor Alexander, his like in both name and wickedness. The latter seems to be correct, since the Emperor Alexander, under whom Saint Ciricus and his mother Julita suffered, preceded Diocletian as emperor by some years, as can be gathered from

the chronicles of the emperors. So their suffering was not under Diocletian, but some years earlier.

Secondly, in Theodore it is not written that Saint Ciricus said much before the ruler, beyond voicing his agreement with his mother: "I too am a Christian." But in the other writing there is a description of his constantly discoursing on many things of the faith and its content, marvellously beyond his years and the law of nature, and confuting the ruler. But this problem is quickly dissolved, since in writing about events, one can include what another omits, or conversely omit what another describes. Even sacred scripture is occasionally found to be silent.[86]

On the other hand silence does worry Gui when he considers the question that is perhaps the touchstone for critical thinking in medieval historians, the legend of King Arthur. The early editions of his world history include a summary of Geoffrey of Monmouth, without any suggestion of skepticism. But by the revision of 1320, he has considered, like William of Newburgh, the silence of other authors. He adds, "Truly a certain lack of belief is engendered in our mind by the fact that there is no mention of Arthur in other historians, even though he is said to excel in nobility and outshine all his predecessors as King of Britons."[87]

The Authenticity of Documents

Some progress was made on evaluating the authenticity of documents, which were often as deserving of skepticism as relics. Men of law possessed the relevant skills, being trained both in the evaluation of evidence and in close attention to documents.[88] Since Roman law gave a high evidentiary value to documents, it had encountered the problems of their forgery and loss. Both the *Digest* and the *Code* have sections "on faith in documents" that raise the problem of determining authenticity but say little about how to resolve it beyond a vague mention of "comparison of handwriting," and some doubts about the value of "private documents."[89] As usual, the Glossators and their followers built a substantial conceptual structure on these vague hints. As we saw in chapter 2, "private documents" were one of the first examples used to illustrate the notion of half-proof. While there was concern that the unscrupulous were thereby given the opportunity of forging a half-proof in their favor, it was recognized that there were often cases in which such evidence was the only kind available. "One does not have recourse to comparison [of handwriting] when other proofs are available," Baldus says.[90] But very light proofs can suffice, in the absence of others, especially when dealing with past events for which there are no surviving witnesses. Two cases of recurrent concern were merchants' account books and holographic wills. It was accepted that account books were the only evidence normally available on merchants' debts and could be taken as half-proof (and so could be supplemented by oath to make sufficient proof).[91] Holographic wills, that is, those apparently in the handwriting of the deceased but lacking the normally required two or three signatures of witnesses, were normally accepted in the absence of evidence of forgery, under the "comparison of handwriting" rule, despite the recognized weakness of the proof.[92]

Forgery was especially rife in the twelfth century, as the spread of literacy left such institutions as abbeys without documents confirming their privileges—or at least without documents as explicit as the granters of the privileges no doubt would have intended. A quite extraordinary proportion of surviving charters purporting to date before 1150 are forged, and it appears that a number of the early historians now most respected were involved directly in forgery.[93] Innocent III was concerned with this, as with so many other matters concerning evidence. In this case, however, it is possible to see that he bases his opinions on those of his canon law master at Bologna, Huguccio. Huguccio says, "If there is doubt about a decretal, one must refer to the register; if it is not found there, one must resort to presumption, namely what is regarded by men as being a decretal because it is generally in the style of the curia, and does not disagree with the equity of the canons . . . anything added by a forger, or subtracted or altered, whether in the bull or the parchment, the thread or the letters, suggests forgery."[94] It is interesting that such diverse kinds of evidence as style, physical alterations, and agreement of the text with "what such a text would be expected to say" are combined on a more or less equal footing.

Within a month or two of Innocent III's election, a nest of forgers of papal letters was uncovered, necessitating a number of instructions from the pope on methods for being sure of the authenticity of papal documents.[95] A long decretal of 1199 represents probably the best of medieval thinking on the critical evaluation of documentary evidence. Innocent's decretal decides a conflict between the archbishop of Milan and a monastery over the ownership of certain lands. The monastery has tendered documents, notably one granting the lands to it, allegedly in the name of Luitardus, bishop of Luca. The archbishop's representative finds an extraordinary number of different pieces of evidence that the document is a forgery. There is erasure, "especially where falsification was most easily to be apprehended, in the statement of the tax imposed." The writing seems more recent than the charter itself. There is a seal that appears to be a king instead of, as it should be, a bishop. There is new wax where there should be old. Then there is some fine detail about the spacing of letters. "Furthermore, on the seal itself, no letters appeared except a proper name with the addition of 'Dei gratia.' But, although the name of the bishop was Luitardus, in the name shown on the seal, two letters had been erased: the second, between L and T, and the sixth, between R and U, in such a way that, if the second had been O and the sixth I, without doubt it would have read not LUITARDUS but LOTARIUS. This could also be proved from the fact that, from the spacing of the other letters, there was space between the L and the T for no more than one letter, although to form the name LUITARDUS two letters are needed between L and T. Moreover between R and U the space was so small that it seems to have contained not the letter D, which occupies greater space, but the letter I, which occupies the least space."

All this is about the physical aspects of the document, but there is also a mention of a problem in what the alleged grants say. The reasoning here is an argument from silence, of a kind that would be given a certain weight by a modern historian. One of the monastery's documents supposedly gave a verdict in the monastery's favor by a former archbishop of Milan. The decretal says: "In all the imperial privileges which were more recent, no mention at all was made of that verdict, although in some of them mention was made of Luitardus, who was said to have made the grant."[96]

The theory is good. Nevertheless, there is a clear case of Innocent being deceived and personally declaring genuine a forged document after examining it carefully.[97] However, there is some possible confusion, in these cases in which the pope "authenticates" a document, between the factual question as to whether the document is forged and the legal one as to whether the claim made in the document is sustainable. The use of *authentic* or *authorized* for both notions, or a conflation of the two, is a symptom and possibly also a partial cause of the excessive medieval reliance on "authorities" generally.[98]

A law case of 1410 that turns on the authenticity of a historical document allows us to see the medieval mind forced to be explicit about principles of historical evidence and reaching for legal principles as the nearest things relevant. The case is that of the head of St. Denis. The monks of the Abbey of St. Denis believe themselves to possess the complete skeleton of the saint, but for some two centuries the canons of Notre Dame have maintained that they possess the saint's skull. The monks rely on a document written by a certain Rigord two hundred years earlier. Rigord was employed by the Abbey of St. Denis, but his history was then inserted into the authoritative *Great Chronicle of France* by order of the king. Counsel for the canons suggests that the monks have substituted for the correct title for Rigord, "chronographer of the church of St Denis" the unsupported "chronographer of the Kings of France" and denies the "authenticity" of the document on the grounds that it is "only a private document or private attestation by those who have an interest in the case." To the contrary, counsel for the monks relies on the royal "authentication" and argues that even if Rigord were not the author, "on account of the form and publication before the Prince, and its location conserved in a public archive, it is an approved and authorised history." The argument of the opposing side about the danger of private documents is cunningly turned to the opposite end: "in public custody resides faith, therefore the converse also holds. Otherwise there would be too great danger, for one could write what one liked in one's own favour, creating a title for one's interest."[99]

Valla and the *Donation of Constantine*

The humanists, for all their criticism of science, logic, law, and so on, were not incompetent at everything. Two arts in which they particularly excelled were philology and invective. Petrarch was strong on both, as in a letter to the emperor in which he mocks a document allegedly written by Julius Caesar for anachronistically using the royal plural.[100] But these skills, and their combination, reached a high point in the writings of Lorenzo Valla. Counterintuitive theses maintained by him include: that all Scholastic disputes on metaphysics are caused by misuse of words; that the Latin of medieval law is barbarous ("to the highest degree gothic"[101]—the persistence of this word to describe things medieval is a sign of Valla's success); that historians and rhetoricians are superior to philosophers; that logic is a subdivision of rhetoric, not vice versa;[102] that the lay state is better than the religious; and that Quintilian is a better stylist than Cicero. "There would be no reason for later generations to write anything," he says, "if there were nothing to disagree with their predecessors about."[103] Central to his ideas is his admiration for Cicero and the "last of the erudite," Boethius, which led him to prefer dialectic and rhetoric to deductive argument.[104] "Faith, in Latin," he says, "is properly speaking proof, as when I produce faith by documents, arguments, witnesses. The Christian religion does not rest on proof, but on persuasion, which is superior to proof."[105]

His project was to replace Scholastic logic and metaphysics with dialectic and rhetoric, as understood by Cicero, Quintilian, and Boethius.

This involved a revival of the arguing on both sides of a question, the urbane skepticism and topical argument of classical rhetoric, in opposition to Scholastic "dogmatism."[106] As in Cicero, an emphasis on the likely followed. After dividing premises of arguments into the necessary and the likely, Valla draws distinctions among the latter. "But as long as the reason is not plainly true but half-true and half-certain, then the conclusion is not necessary but half-necessary, which when it has much force is called likely and credible, that is, exceedingly possible, and when it has scant force is called possible, that is, scarcely likely or credible."[107]

The most influential of Renaissance works on history is Valla's *Treatise on the Donation of Constantine* of 1440. The *Treatise* begins, "I have published many books, a great many, in almost every branch of learning. Inasmuch as there are those who are shocked that in these I disagree with certain great writers already approved by long usage, and charge me with rashness and sacrilege, what must we suppose some of them will do now!"[108] The notion that one may congratulate oneself for attacking tradition is entirely modern. The work proceeds with much railing against the greed and lust for power of popes and so on, but among it all there is some serious argument to show the inauthenticity of the alleged *Donation of Constantine*. The *Donation* was actually written about 750 and is included in the *False Decretals* and Gratian's *Decretum*. It purported to be a document in which the Emperor Constantine, following his baptism and cure from leprosy at the hands of Pope Sylvester, recognized the superior dignity of the pope by holding the bridle of his horse and assigning to him the entire Western Empire, to remain forever under the control of the Holy See. The document was accepted as genuine almost universally from the ninth to the fifteenth centuries and quoted by many popes in their disputes with emperors.[109] One of the very few doubters was Ockham. He offered to prove that the *Donation* must be a forgery by comparing it with respectable writings, but the text breaks off before he does so.[110] There was interest in the question even in such an obscure language as English, where the *Donation* was called "ascryued vnlikeli to Constantyn."[111]

The most convincing attack appeared a few years before Valla's, in 1433, and was known to him. It occupies a few pages in Nicholas of Cusa's *Catholic Concordance*. Where Valla's argument concentrates on the text itself, Cusa's involves research he has conducted on other texts. These include many original documents, not an easily available resource in the fifteenth century. The form of his argument is the argument from silence as found briefly earlier in Innocent III's decretal. "[The *Donation*] does not appear in authentic books or approved histories. I have collected all the histories I could find, the acts of the emperors and Roman pontiffs, the histories by St Jerome who was very careful to include everything, those of Augustine, Ambrose, and the works of other learned men; I have reviewed the acts of the holy councils which took place after Nicaea and I find no confirmation (*concordantiam*) of what is said about that donation. . . . Hence we always read that the emperors up to that time and earlier had full legal rights over Rome. . . . And we read that the Roman pontiffs acknowledged the emperors as their overlords."[112]

There is more along the same lines, with appropriate details and references to the original documents. It is customary to say that arguments from silence are weak, but enough of them together, as in this case, can be strong enough. Cusa also mentions that he has found a more complete version of the alleged donation, and "examining it carefully, I have found a clear argument in the story itself for its being invented and false, which it would be too long and unnecessary to insert here now."

Length and lack of necessity are no object to Valla, and it is to this internal evidence he turns. Among his sound arguments are his observations that many claims in the *Donation* are simply reflections of old stories, such as the Old Testament story of Naaman's gift when the prophet Elisha cured him of leprosy.[113] Then, there is no evidence, as one would expect, of Sylvester's accepting the donation nor any evidence of the supposed rule of Rome by Sylvester:

> To make us believe in this "donation" which your document recites, something ought still to be extant concerning Sylvester's acceptance of it. There is nothing concerning it extant. But it is credible, you say, that he recognised this "donation." I believe so, too; that he not only recognized it, but sought it, asked for it, extorted it with his prayers; that is credible. But why do you call credible what is contrary to the opinion of men? For the fact that there is mention of the donation in the document of the deed is no reason for thinking it was accepted; but on the contrary, because there is no mention of an acceptance, it should be said that there was no donation. So it appears more that [Sylvester] refused the gift than that [Constantine] wished to give it. . . . Where is any taking possession, any delivery. . . . It is likely, you say, that any one who makes a grant, gives possession of it also. See what you are saying; for it is certain that possession was not given, and the question is whether the title was given! It is likely that one who did not give possession did not want to give the title either. . . . What an infinite number of coins of the supreme pontiffs would be found if you ever had ruled Rome. But none

such are found, neither gold nor silver, nor are they mentioned as having been seen by any one.[114]

Valla goes on to examine the language of the document. Certain words, such as "satrap," are used in the document in ways not found in other texts of the period; phrases like "the clemency of our imperial serenity" are barbarous; Constantinople is mentioned before the city had that name; certain phrases imitate the Apocalypse, which Constantine had not read. "And so it should be apparent that he who spoke thus lied, and did not know how to imitate what Constantine would be likely to have said and done." As to the threat contained in the document that those who do not uphold its provisions shall be burned in the lowest hell, "This terrible threat is the usual one, not of a secular ruler, but of the early priests and flamens, and nowadays, of ecclesiastics."[115]

Not all these arguments are fully convincing. Opponents pointed out that arguments about what Constantine was likely to have done, a style of argument Valla has taken from Quintilian, are weak, and that a number of his criticisms about barbaric language are a result of his using a corrupt medieval text instead of available older ones.[116] But Valla's case was clear enough, and he was successful in convincing everybody of his thesis. Erasmus rightly remarked that Valla "forced even the learned to express themselves henceforth with more circumspection."[117]

What Valla meant by the "likely" in matters of history can be gathered from a dispute over his *History of King Ferdinand of Aragon*, in which he became embroiled with another humanist, Facio. Facio is enraged that Valla represents a king as pretending to fall asleep during an ambassador's audience and the scholars at the court as envying the wealth of the court jester. His objection is that such things are not "fitting to the royal majesty," "for your narrative must be not only true, but also likely, if you want anyone to believe it . . . my good expert, you must write in such a way that the dignity of such persons is maintained, otherwise your narrative will not be probable and trust will not be placed in it."[118] (Has the moral sense of "probable" as something like "approvable" affected his opinion?) In Valla's reply there is an element of the argument that such things do happen, fitting or not. From anyone but Valla the defense that history is written for the "memory of posterity," not for engaging in present-day battles, would be reasonable. On the other hand, he argues that, as a form of rhetoric, history must be allowed invention, for example, of speeches.[119]

Further good judgment on what someone is likely to have said is apparent in Valla's emendations of the corrupt texts of the classical authors,

Quintilian, Thucydides, and Livy. His principles of emending are not usually explicit, but he does sometimes argue that a manuscript reading is in error by showing what mistake the scribe could easily have made.[120]

A more serious attempt to explain the principles of emendation was undertaken shortly after by Poliziano. He argues that manuscripts that could be shown to be descended from others should be regarded as useless as evidence. He expresses the reason in the old legal maxim, "The testimonies of the ancients should be not so much counted, as weighed."[121]

Erasmus was one of the acutest of the emenders of texts, especially of the text of the Bible, for which he collected an unprecedented wealth of variant readings in support of his conjectures about the original.[122] Experience with the variants made it clear enough that a number of them had been written in the interests of various theological positions, and it is in this context that Erasmus formulated what is still one of the strangest principles of probabilistic reasoning, the principle *lectio difficilior potior*, literally, "the harder reading is the stronger." It states, in simple terms, that the less likely reading is more likely. The reasoning behind the principle is that a scribe is more likely to substitute an "easier" reading for a difficult one (where "easier" may mean either what is natural for linguistic reasons or what is considered less offensive to pious ears). Therefore, in case of doubt, one should regard the apparently less likely reading as more likely to be the original. Erasmus explains: "Whenever the Fathers report that there is a variant reading, that one always appears to me to be more esteemed which at first glance seems the more absurd—since it is reasonable that a reader who is either not very learned or not very attentive was offended by the spectre of absurdity and changed the text."[123]

A different kind of humanistic project in history is that of Machiavelli's *Discourses*. He attempts to make history a science by extracting maxims from history: generalizations that are summaries of and lessons taught by the many individual events of the past. A typical example is "Dictatorship is advantageous in times of emergency." Naturally, there are continual problems with the wealth of counterexamples to the maxims; Machiavelli attempts to explain them away individually, rather than saying that the maxims only hold true for the most part.[124] His advice on what action to take in case of doubt is unhelpful: "In all discussions one should consider which alternative involves fewer inconveniences and should adopt this as the better course; for one never finds any issue that is clear cut and not open to question."[125] To consider only the pay-

off, and not the probability of occurrence, will not lead to a satisfactory
decision theory.

Cano on the Signs of True Histories

The pose of skepticism affected by a section of the humanist movement
sometimes included skepticism about historical evidence.[126] The rhetorical temperament of certain humanists was impressed by simplistic arguments such as the dilemma posed by one of them: an observer is either privy to the counsels of the Prince or he is not. If not, he cannot
tell the truth. If so, he is a partisan of the prince and will not tell the
truth.[127]

A call for an answer to these frivolities came from the needs of the
Reformation and more especially the Counter-Reformation. When the
Protestants insisted on "Scripture alone" as a basis for proving theological propositions, the Catholic Church was left with a large body of
propositions needing justification, many of them historical. The Protestant call for a return to the purity of the early church led to competition
in the "more traditional than thou" stakes, since the obvious Catholic
reply was to assert that, according to Scripture itself, Jesus had left not
a book but a church, so that the tradition of the church was a reliable
source of theological knowledge.

Melchior Cano was a leading figure in the Counter-Reformation as
the emperor's nominee to the Council of Trent. (His additions to Vitoria's ideas on probability in moral theology are noted in chapter 4.) His
book *De Locis Theologicis* came to be the main Catholic text on the
sources of theology. He recognizes that in history there is a middle
ground between credulity and doubt and tries to lay down conditions
for it. While admitting that "apart from the sacred authors . . . there is
no historian who has not lied about something," he distinguishes between good and bad historians: "Serious and trustworthy historians, of
which sort some undoubtedly exist both on ecclesiastical and secular affairs, provide the theologian with a probable argument either for corroborating propositions of his own, or for refuting false opinions of his
adversaries."[128] One may however disbelieve what even serious historians like Pliny say if it is incredible. In reading serious historians, "both
are justly reprehended: one who believes too quickly, and one who believes too slowly."

Cano refers to the two-witness rule, which theologians especially
should follow; even in canon law human histories are regarded as having sufficient probability, if other evidence is lacking. As to the idea of
being skeptical in principle, "There are a number of people in this age

of ours who perversely, not to say imprudently, cast doubt on things which have been attested by very serious historians. If they were to give relevant and probable reasons they should perhaps be listened to. But since they give none they are to be despised." Cano argues that "if all tested and serious historians concur about the same event the argument drawn from their authority is certain." He is clearly thinking of a situation like that of Hume's *On Miracles*, where the testimony is about something antecedently unlikely: "It would be as if the Mediterranean peoples were to deny the existence of the ocean, or if those who were born on an island in which they saw nothing but hares and foxes should not believe in the lion or the panther, or if, indeed, we should mock at him who speaks of elephants." He further considers "whether there are any rules or signs of history, by which the faith and truth of a historian may be investigated," but he can suggest only the integrity and probity of the historian, his prudence in selecting and judging his informants, and any pronouncements the church may have made on him.[129]

Later works on the same subject are not in any theoretical way in advance of Cano, but Jean Bodin's *Method for the Easy Comprehension of Histories*, of 1566, contains some sound advice on matters of detail. A serious historian may be biased on some particular points by patriotism or fear, so a writer who praises his country's enemies is to be believed. A history containing eulogies or entertainment is suspect, since the author has purposes other than the simple narration of the truth. He improves on Cano on the difficulties where there is agreement among authors: "One author may be misled by the error of another." He suggests that certainty is greatest when authors are in agreement on a point when they conflict on related matters.[130]

If the theory of evaluation of historical evidence remained rather primitive, there was a project that in large measure succeeded in integrating well a huge mass of information of different levels of certainty. Even now, one of the most difficult problems in the history of the ancient world is the establishment of absolute chronology. Ancient historians used their own internal systems of dating, of varying accuracy, but how is one to make them match up and determine their dates in our system? Biblical events are particularly difficult, and the dating of the Crucifixion is still a matter of controversy. Historians of the twelfth century and later made a good deal of progress. They were able, for example, to assign reasonable absolute dates to lists of abbots stretching over several centuries, despite the problems caused by rounding (when whole years of reign only were listed), interregnums left out, and scribal errors. On such matters as these and the tangled affairs of Merovingian and

Carolingian rulers, modern investigators in general have only the advantage of more data rather than a better method.[131]

It was also the medievals who suggested the idea of using astronomical data to harmonize ancient dates.[132] The harmonizing of both astronomical and historical data is plainly a task of the same logical kind as Copernicus's and Kepler's fitting of observations to geometrical models. But it is much larger and requires humanistic skills as well as mathematical. The most successful attempt, in large measure confirmed by later work, was Scaliger's *De Emendatione Temporum* of 1583.[133] It is no surprise to find Kepler dabbling in the same areas—and not entirely without success. He sees some connections with astrology; while he accepts some influence of the stars in shaping history, he avoids, unlike in his work on the heavens, finding numerological patterns in historical dates.[134]

CHAPTER EIGHT

Philosophy: Action and Induction

Philosophy and religion are old enemies of probability. Philosophers from the earliest times have wished to distinguish themselves from the spinners of mere rhetoric by offering certainty. Parmenides distinguished sharply between truth, associated with Being, and the opinion of men, called "likely" and associated with non-Being.[1] Logical reasoning is intended, by Parmenides, Plato, Aristotle, and their followers, to establish the foundations of knowledge beyond all doubt, and correspondingly likelihoods are banished as other people's business. Plato has Socrates say to Theaetetus, "You are not offering any argument or proof, but relying on likelihood (*eikoti*). If Theodorus, or any other geometer, were prepared to rely on likelihood when doing geometry, he would be worth nothing. So you and Theodorus must consider whether, in matters as important as these, you are going to accept arguments from plausibility and likelihood (*pithanologia te kai eikosi*)."[2]

It is an unfortunate start to discussion of probability. It is true that Plato's *Timaeus* gives a speculative account of the universe and its origins that is described as likely (which it certainly is not).[3] But it is utterly mysterious what this means; it cannot mean probably true, since the point is to contrast the theory with the real and unknown truth. Nor does it seem to mean like the truth in any way. The word "likely" is clearly intended to situate the theory somewhere between the completely worthless and the certainly true, but beyond that it seems to have no meaning. It confirms rather than undermines the contrast between likelihood and truth.

Those religions, too, that offer doctrines (as opposed to practices and wisdom) usually believe those doctrines are certain, doubt being often regarded as an insult to the God who has deigned to reveal the truth. But as history has progressed, it has been found impossible to ignore the fact that the more certainties that are offered by philosophy and religion, the more those supposed certainties conflict. In philosophy, one reaction has been moderate skepticism or fallibilism, which emphasizes the merely probable character of all theory and the sufficiency of probability as a guide to life. That reaction is rare before the twentieth century.

Carneades' Mitigated Skepticism

The Stoic school inherited most of the opinions of Aristotle. The Stoics did, however, concentrate strongly on the certainty of knowledge, to the neglect of the beginnings Aristotle had made on nondeductive inference.[4] The Stoics had a place for the plausible (*pithanon*) and the reasonable (*eulogon*), but their use of these words made little of their probabilistic aspects. They are reported as offering these definitions: "A plausible (*pithanon*) judgement is one that induces to assent, for example, 'Whoever gave birth to anything, is that thing's mother.' This, however, is not necessarily true, since the hen is not mother of the egg. . . . A reasonable (*eulogon*) proposition is one that has many things in its favour, for example, 'I shall be alive tomorrow.'"[5] But these ideas are not expanded. The Stoics' exaltation of the reasonable as the highest good in action led to a number of jokes at their expense, to the effect that the reasonable did not seem to be reasonable *for* anything.[6]

Stoic understanding of *pithanon* can be further clarified by their division of impressions into those that are *pithanon* ("It is day"; "I am conversing"), those that are *apithanon*, that is, implausible ("If it is dark, it is day"), those that are both (such as puzzle arguments, which seem now convincing, now not), and those that are neither ("The number of stars is even").[7] It is clear that the Stoic meaning of *pithanon* is not close to any of the more objective senses of *probable* but is more like what actually convinces, or appears credible.

The reason for the Stoics' avoidance of anything like a genuine probability is their demand for certainty: "The Stoic doctrine to which he [Ariston] attached most importance was the wise man's refusal to hold mere opinions. And against this doctrine Perseus was contending when he induced one of a pair of twins to deposit a certain sum with Ariston and afterwards got the other to reclaim it."[8] Many saw in such arguments the main challenge to the Stoic doctrine of certainty. Since the time of Democritus, one of the chief irritants producing the pearl of philosophy has been the challenge of skepticism, the worry that false sense impressions are sometimes indistinguishable from true ones. The experience of illusions of perception is familiar—the oar that appears bent in water, the tower that appears round from a distance but square nearby. As Aristotle explains, "Which, then, of these impressions are true and which are false are not obvious, for the one set is no more true than the other, but both are alike. And this is why Democritus, at any rate, says that either there is no truth or to us at least it is not evident."[9]

This symmetry argument is the driving force behind skepticism: "There is no mark to distinguish the true and the false," as Cicero puts it.[10]

The challenge to break the symmetry was taken up by the Stoics with a theory of the comprehensive presentation, a sense impression so strong that it could not possibly be false. It is no easy task. The Stoics' difficulty on the point is illustrated by an event in Alexandria around 220 B.C., when the Stoic Sphaerus met King Ptolemy Philopator. "One day when a discussion had arisen on the question whether the wise man could stoop to hold opinion, and Sphaerus had maintained that this was impossible, the king, wishing to refute him, ordered some waxen pomegranates to be put on the table. Sphaerus was taken in and the king cried out, 'You have given your assent to a presentation which is false.' But Sphaerus was ready with a neat answer: 'I assented not to the proposition that they are pomegranates, but to another, that there are good grounds for thinking them to be pomegranates. Certainty of presentation and the reasonable (*eulogon*) are different.' Mnesistratus having accused him of denying that Ptolemy was a king, his reply was, 'Being of such quality as he is, Ptolemy is indeed a king.'"[11]

The skeptical position is an easy one to propose but has always been dogged by the objection that genuine skepticism would result in paralysis.[12] Carneades' mitigated skepticism is an answer to this objection. A native of Cyrene, leader of the New Academy in Athens for many years around 150 B.C., Carneades was a champion of the power of argument even among the school of Skeptics that, in the nature of its doctrine, depends more on argument than any other philosophical school. He wrote no works himself, but his ideas are known in reasonable detail from accounts in the much later authors Cicero and Sextus Empiricus. Of the two, Cicero is much the more sympathetic to Carneades' position. But though Cicero continually writes about probability, his account of what probability is supposed to be is curiously content free. Sextus Empiricus is a follower of a more extreme skeptical school and does not believe in probability, but his account of what Carneades said is much clearer. The difference is significant for the later history of probability, because Cicero was one of the best-known ancient authors, while Sextus Empiricus was almost unknown until nearly 1600.[13]

The key phrase in Carneades' theory, according to Sextus, is "probable presentation" (*pithane phantasia*). The discussion is in the context of perceptual errors, such as a coiled rope that one may at first take to be a snake. The presentation, or what appears to the observer, is either true or false, but which it is, according to earlier skeptical arguments he has given, cannot be known. Distinctions can be drawn, however, in the way the presentation appears. A first division is between presentations that appear true and those that appear false. The latter are not of interest, since the observer will not tend to base action on them—even if, by chance, they are true.[14] More important are the divisions among apparently true presentations. The relevant passage deserves to be quoted at some length, because it is a comprehensive theory of belief and action on the basis of probable reasons. A presentation, or sensory appearance, is probable simply if it appears true with some vividness. More reliable is the presentation that is probable and also "irreversible," in which the judgment is strengthened by the concurrence of probable presentations, for example when a person is recognized by appearing with all his usual qualities:

> So whenever none of these presentations disturbs our faith by appearing false, but all with one accord appear true, our belief is the greater. For we believe that this man is Socrates from the fact that he possesses all his customary qualities—colour, size, shape, converse, coat, and his position in a place where there is no one exactly like him. And just as some doctors do not deduce that it is a true case of fever from one symptom only—such as too quick a pulse or a very high temperature—but from a concurrence, such as that of a high temperature with a rapid pulse and soreness to the touch and flushing and thirst and analogous symptoms; so also the Academic forms his judgement of truth by the concurrence of presentations, and when none of the presentations in the concurrence provokes in him a suspicion of its falsity he asserts that the impression is true.
>
> And that the "irreversible" presentation is a concurrence capable of implanting belief is plain from the case of Menelaus; for when he had left behind him on the ship the wraith of Helen and had landed on the island of Pharos, he beheld the true Helen, but yet though he received from her a true presentation, yet he did not believe that presentation owing to his mind being warped by that other impression from which he derived the knowledge that he had left Helen behind in the ship. Such then is the "irreversible" presentation; and it too seems to possess extension inasmuch as one is found to be more irreversible than another.

Carneades then discusses an even higher grade of presentation, that which is irreversible and also tested; that is, the possible sources of error in the organ of perception and in the environment have been checked and found unproblematic. How much inquiry is necessary in any particular case depends on the importance of getting the right answer. "Just as in ordinary life when we are investigating a small matter we question a single witness, but in a greater matter several, and when the matter investigated is still more important we cross-question each of the witnesses on the testimony of the others—so likewise, says Carneades, in trivial matters we employ as criterion only the probable presentation, but in greater matters the irreversible, and in matters which contribute to happiness the tested presentation."[15] Whether Carneades actually asserted this theory is a matter of controversy. Carneades' skill was essentially in dialectic, and even in ancient times it had become impossible to tell what, if anything, he accepted and what he put forward as a *reductio ad absurdum* of his opponents' theories. It has been maintained that Carneades in fact asserted none of the above probabilist theory but only discussed it as part of a strategy to exclude the possibility of any criterion of truth; the matter remains unresolved.[16] Nevertheless, for the history of probability, the most significant fact is that the theory appears as a positive one in the two main sources, Cicero and Sextus.

There is much here that accords with later theories of probability: the various grades of probability, the high degree of probability that provides a good criterion for action, the increase in probability through the coherence of different pieces of evidence. What is unique about Carneades' theory is that his probability is a property of *perception* rather than of propositions or beliefs and is genuinely equated with appearance of truth or likeness to truth.

This idea owes something to Plato's remark on the supposed etymology of *eikos* as connected with likeness (discussed in chapter 5) and something to the most ancient Greek poems, which speak of false things that are "like true."[17] But the main reason for it is obviously the Skeptics' concentration on the argument from perceptual illusion, which is naturally expressed by calling appearances true or false, rather than statements. The difference between Carneades' theory and other theories should not be exaggerated, since the Skeptics were capable of distinguishing between a presentation and the assent to it and saying that their skepticism was really about assent.[18] Furthermore, if assent in general depends ultimately on sensory evidence, it is fair to consider the status of sensory evidence first; Carneades certainly makes essentially this point.[19] Still, the difference is important in its effects. Making probability a property of appearances is a natural idea, but it can hardly be called a good one. It directs attention away from the areas in which the theory of probability later developed—arguments, dice, relative frequencies—and reinforces Plato's unfortunate suggestion that probability is deceptive, a mere appearance or illusion of truth.

Cicero's phrase "deceitful probability" (*captiosa probabilitate*) sums up the worst of ancient thinking on the subject. Unless Sextus is missing something, Carneades has given no adequate reason why those appearances that *are* like the truth are in fact reliable guides for action. His only reported attempt at explaining this is an appeal to reasoning "for the most part": "But the rare occurrences of this kind—the kind I mean which imitates the truth—should not make us distrust the kind which for the most part reports truly; for the fact is that both our judgements and our actions are regulated by the standard of 'for the most part' (*epi to poly*)."[20] This fails to explain why things that look like the truth should be for the most part true. The inadequacy of Carneades' reply was not unnoticed in ancient times.[21]

Cicero's account of these matters in his *Academica* consists of Carneades' theory minus the best parts. Cicero merely reports Carneades as holding that "the wise man withholds assent" in theory but follows probability in practice, where it is sufficient ground for action.[22] As to what probability is, Cicero is silent, and as to how to discover what is probable, the following is all he offers. While not in itself negligible, it hardly constitutes a sufficient exposition of probability: "When a wise man is going on board a ship, surely he has got the knowledge already grasped in his mind and perceived that he will make the voyage as he intends? How can he have it? But if for instance he were setting out from here to Puteoli, a distance of thirty *stadia*, with a reliable crew and a good helmsman and in the present calm weather, it would appear probable that he would get there safe."[23] In expounding his own philosophy in other works, Cicero again approves of advancing doctrines as only probable, but one is left none the wiser as to what this means. In addition to this he inserts into his long works on argument a few remarks of more or less Aristotelian tenor on probability as a property of arguments. Dialectic, he says, "conveys a method that guards us from giving assent to any falsehood or ever being misled by deceitful probability."[24] Among the things one should consider in constructing arguments is "the nature of the event, whether it is wont to happen commonly or unusually and rarely."[25]

The Epicureans on Inference from Signs

One of the pleasures of studying the ancient world is that a large proportion of what one has to read is of good quality, since the better works have enjoyed a higher survival rate over the centuries than the rubbish. Scholars of ancient philosophy have been less than overjoyed, therefore, at the recovery of part of a large library of Epicurean philosophy in the ruins of Herculaneum, the city destroyed with Pompeii in the eruption of Vesuvius in 79 A.D. The library is a small sample of the vast production in late antiquity of books about words. Among the least uninteresting of the papyri so far deciphered is the end of a book *On Signs* by the Epicurean Philodemus. It is at least genuinely about argument, not just tricks of persuasion or speculations on etymology. Philodemus led a school of Epicurean philosophy in the area of Naples in the mid-first century B.C., with which Cicero had some connection. The work describes the debate between the Stoics and Epicureans over, essentially, the problem of induction, or the inference to general facts from observations. To infer "All men are mortal" from "All observed men are mortal" requires, according to the Stoics, following Aristotle, rational insight into the nature of man. The Epicureans maintain that there are no such rational insights into natures and that one can only make the inference from suitably repeated and suitably checked experiences. The debate is in principle the same as the modern one about whether inductive argument needs laws of nature and a uniformity of nature principle, or whether it is a purely statistical procedure like arguing from sample to population.[26] In the ancient debate, the logical details on both sides never become completely clear. The surviving section of the papyrus roll of *On Signs* opens with the Stoic arguments for the fallibility of induction or of analogy that proceeds merely from repeated observations:

> Furthermore, the method of analogy seems not to be cogent in the case of the unique objects observed in the places that are within our experience, if there is among the great variety of stones one kind that draws iron, which is called the magnet or the stone of Heraclea, if only amber is capable of attracting chaff, and if only one square, the square of four, has its perimeter equal to its area. How then can we say that there is not some race of men who alone do not die when their heart is cut open— so that it is not in accord with necessity to infer from the fact that men among us die when the heart has been cut open, that all men do?
>
> There are also in our experience some infrequent occurrences, as for example the man in Alexandria half a cubit high, with a colossal head that could be beaten with a hammer, who used to be exhibited by the embalmers; the person in Epidaurus who was married as a young woman and then become a man; and the person in Crete who was forty-eight cubits tall, by inference from the bones that were found; and further the pygmies that they show in Acoris, who are quite similar to those which Antony recently brought from Syria. If these things go beyond

all that we are familiar with and are not similar, we may ask whether any of those things about which we make inferences may also be exceptional, since it is not [unintelligible text], and some things in the universe are extraordinary.[27]

The Stoics go on to say that argument from analogy "assumes that things in unperceived places are like those in our experience" and that in general one does not know which kinds of similarity will allow inferences—for instance, how general the similarities may be: "We shall not, therefore, use the [inference] that since the men among us are mortal the men in Libya would also be mortal, much less the inference that since the living beings among us are mortal, if there are any living beings in Britain, they would be mortal." Any attempt to decide which similarities could be used in inference to the unobserved would result in an infinite regress.[28]

The Epicurean reply includes a component of admitting that some of their arguments were only probable: "But for us it will be sufficient to have been persuaded by what is according to probability in these matters and in matters discovered by trial, for example, that we shall be safe when sailing in summer."[29] (The word in the text here translated "probability" is *eulogian*, which, despite the Stoics' distinctions mentioned earlier, came to be used in later Greek virtually as a synonym for *eikos*.[30] The example is the same as Cicero's example of sailing to Puteoli—Puteoli is in the Bay of Naples, site of Philodemus's school.)

The main part, however, of the Epicurean defense is to argue that inferences from experience can produce certainty if the proper precautions are taken. Thus they say that one can make observations to see what properties are constant among things and what variable, so that one will know which to project, and that "it is ridiculous that anything inferred from clear perception about things unperceived should contradict the clear perception." Sometimes careful observation will lead to inferences, with conclusions only true for the most part, other times to conclusions true universally, the difference between the two being itself learned from experience.[31]

Inductive Skepticism and Avicenna's Reply

Sextus Empiricus shows the fallibility of induction continuing to be a live issue in philosophy. It would be a mistake, he says, to suppose that because most animals eat by moving the lower jaw, all do, since the crocodile moves the upper jaw.[32] There was some discussion among the ancient medical writers and rhetoricians of the uncertainty of inductive reasoning (see chapters 5 and 7). Sextus goes further, and poses the problem of induction as a general skeptical philosophical challenge: "For, when they [Stoic dogmatists] propose to establish the universal from particulars by means of induction, they will effect this by a review either of all or of some of the particular instances. But if they review some, the induction will be insecure, since some of the particulars omitted in the induction may contravene the universal; while if they are to review all, they will be toiling at the impossible, since the particulars are infinite and indefinite."[33]

Avicenna became known in the West chiefly as a medical writer; he and Galen were the most respected authorities until well into the sixteenth century. In this capacity he transmitted something of the ancient medical writers' emphasis on reasoning from what happens "for the most part." He notes the fallibility of induction, referring to Sextus's example of the crocodile.[34] But he suggests an answer that has considerable appeal: he expands a little the connection briefly found in Aristotle between reasoning from what happens "for the most part" and the exclusion of the hypothesis of chance: "Indeed, when we have the repeated sensation of [connection] between one thing and another, such as between purging and scammony, and between the observed movements and things in the heavens, this sensation repeats itself in our memory; following which, we acquire experience of it, the cause of it being a concomitant syllogism or memory: If this thing, for example purging relative to scammony, were due to chance, accidental, with no necessity in nature, it would not be possible in the majority of cases for the same to happen—so that, when it is not thus, the mind finds that fact extraordinary and looks for a reason . . . we have experienced this often and then reasoned that if it were not owing to the nature of scammony but only by chance, this would happen only on certain occasions."[35] That is, Avicenna has tried to use the Aristotelian not-by-chance argument to show that inductive reasoning is justified. His reasoning here seems to have had little influence.

As discussed in the next chapter, inductive skepticism in Islam later became caught up with considerations on the absolute power of God, which changed the complexion of the problem entirely.

Aquinas on Tendencies

Aquinas admits that a good deal of common knowledge concerns things about which one can make conclusions true conjecturally or true only in most cases. This includes knowledge even of such a crucial and inti-

mate matter as whether one is in a state of grace: "something can be
known conjecturally by signs; and in this way one knows that one has
grace, in that he sees he delights in God, and despises worldly things."[36]
Inferring the intentions of others from their actions is of course also
probabilistic.[37] Most truths of the lower or practical natural sciences are
likewise conjectural: "Hence indeed when consideration is of things that
are movable and do not behave uniformly, cognition of them is less firm,
because the demonstrations about them, for the most part, are subject
to the fact that things sometimes happen otherwise; and so when some
science approaches closer to particular things, as in the practical sciences
such as medicine, alchemy and morals, it can have less certainty on account of the multitude of things considered in such sciences, which if
any are omitted, cause error, and on account of their variability."[38]
Aquinas is not content with a simple fallibilism about these sciences of
the variable and understands that some account needs to be given of how
exactly a mixture of causes can lead to outcomes of differing degrees of
universality. His solution is a theory of dispositions, tendencies, or inclinations. These are properties inherent in things that incline them to
produce determinate outcomes "for the most part" and that therefore
allow prediction of those outcomes but without certainty:

> Other future effects, however, do not exist so determinately in their
> causes that something else might not happen; their causes are merely
> disposed more to one effect than another; and these effects are contin-
> gent events, which happen more or less often as the case may be. As a
> consequence, effects of this type cannot be known in their causes with
> infallibility, but only with conjectural certitude. Moreover, other future
> effects come from causes that are indifferently related to opposite ef-
> fects; and these effects, especially those that depend upon free choice,
> are called "contingent to opposites."
>
> Now the Commentator [Averroes] proves, that an effect cannot come
> from a cause indifferent to opposites and in a certain respect in potency,
> unless this cause is determined to one effect more than to another by
> means of another cause. Consequently, an effect of this sort cannot be
> known in any way through causes indifferent to opposites if these causes
> be taken merely by themselves. Yet, if we consider these causes, which
> are indifferent to opposites, together with those things that incline them
> more to one effect than to another, we can get some conjectural certi-
> tude about their effects. For example, we can conjecture about future
> effects depending on free choice by considering men's habits and tem-
> peraments, which incline them to one course of action.[39]

As Aquinas recognizes, this is a substantial qualification of the Aristotelian thesis that "there is no science of the contingent": "There can be no science of future contingents considered *per se*. But there can be science of them considered in their causes, in that some sciences can know there are certain inclinations to such and such effects."[40] One could, in principle, use this reasoning to justify a science of dice throwing, which certainly deals with inclinations to future contingents.[41] But Aquinas himself uses it to reconcile astrology and free will. Because the human will is free, the stars cannot determine it. But they can "dispositively incline" it, so that, since most men follow their passions, astrologers often predict the truth, "especially in public events, which depend on the multitude."[42]

This last remark, with its distinction between the individual and the multitude, introduces Aquinas's most important contribution to the theory of probability. He understands that in cases of variable outcomes, one can predict what a group will do more confidently than what the individuals that make it up will do. It is the idea behind the modern "law of large numbers," the assertion that random effects in individuals even out as one takes a large sample. His first example, relevant to life in the mendicant orders such as Aquinas's own Dominicans, concerns how one can rely on alms when one is unable to know whether an individual man will give alms to beggars. The answer is that donors are not predictable individually but are in the mass: "For [the mendicant] does not depend on the will of one, but the will of many. And it is not probable that in the multitude of the faithful there would not be many who with a quick heart would provide the necessities to those they respect for their perfection in virtue."[43] Likewise with angels. Did the number of angels who sinned equal the number who did not? Most did not sin, since "sin is against their natural inclination, and what is against nature happens in the minority of cases."[44]

But the multitude can have its less benign side, for the same reasons. "A multitude for the most part (*ut in pluribus*) follows its natural inclinations, inasmuch as the men in the multitude give in to their passions; but the wise overcome these passions and inclinations by reason. So it is more probable of a multitude that it acts as a celestial body inclines it to do, than that an individual man does, who by reason is able to overcome the inclination. It is like the case of a multitude of choleric men, which does not easily avoid anger, while it can happen more with one."[45]

If Aquinas had had a somewhat more favorable estimate of the natural inclinations of the majority, he might have been tempted to favor democracy. This course was taken two centuries later by Nicholas of

Cusa, who held that the majority of citizens could be expected to make the right decision. "Otherwise it would happen that a natural appetite would be frustrated in many cases, which is considered most unfitting by the philosophers." So in church councils, "the opinion of the majority has a great presumption in its favor so that if that opinion remains fixed and unchanged among them, the presumptive judgement is very great."[46] Whether it is relevant that Aquinas was a scion of the nobility and Nicholas the son of a boatman is impossible to know at this distance in time.

Scotus and Ockham on Induction

Aquinas does not give the impression of being genuinely concerned by the problem of induction or by skepticism in general. His successors in the fourteenth century did reply to the newly revived ancient skeptical arguments. Duns Scotus advances a not-by-chance argument combined with a realism about causes, similar to Avicenna's: "Even though a person does not experience every single individual, but only a great many, nor does he experience them at all times, but only frequently, still he knows infallibly that it is always this way and holds for all instances." Scotus considers that this person "knows this in virtue of this proposition reposing in the soul: 'Whatever occurs in a great many instances by a cause that is not free, is the natural effect of that cause.' This proposition is known to the intellect even if the terms are derived from erring senses, because a cause that does not act freely cannot in most instances produce an effect that is the very opposite of what it is ordained by its form to produce. The chance cause, however, is ordained either to produce or not produce the opposite of the chance effect. Consequently, if the effect occurs frequently it is not produced by chance and its cause therefore will be a natural cause if it is not a free agent. But this effect occurs through such a cause. Therefore, since the latter is not a chance cause, it is the natural cause of the effect it frequently produces."[47]

The "principle of the unfree cause" is required to carry a heavy philosophical load, that of supporting the reliability of all sense knowledge. It has to solve the problem of perceptual skepticism as well as the problem of induction. "But how can a person be certain of those things which fall under the acts of the senses, for instance, that something outside is white or hot in the way that it appears to be?" Scotus's answer is that, "regarding such an object, either the same things appear opposite to different senses or they do not appear so but rather all the senses knowing such an object judge the same about it. If the latter be the case, then we have certitude of this thing perceived by the senses in virtue of the aforementioned principle, viz. 'What occurs in most instances by means of something that is not a free cause is the natural effect of this thing.' Therefore, if the same change repeatedly occurs in the majority of cases when such an object is presented, it follows that the transformation or image produced is the natural effect of such a cause, and thus the external thing will be white or hot or such as it naturally appears to be according to the image so frequently produced."[48]

It might seem to follow from this reasoning that an unfree cause so understood must act always in the same way and thus that induction is infallible. That is not Scotus's opinion. The principle, "Whatever occurs in a great many instances by a cause that is not free, is the natural effect of that cause," results only in the "lowest grade of scientific knowledge" and "perhaps only knowledge of what is apt to be the case" in instances in which the observation shows only what happens "in most cases," as with the effect of herbs. As in Aquinas, only tendencies are being dealt with; "commonly secondary causes can be impeded, and a cause that can be impeded, however much it is not impeded, is not necessary."[49] Even among inductions from conjunctions without exceptions, it appears that Scotus means to distinguish between ones that lead to certainty, of very commonly observed truths such as "all fire is hot," and those that result only in "probable opinion, faith, or persuasion." As to how many cases one needs to observe, he maintains that it varies "depending on the diversity of the matter and of the understanding of the intellect." So in principle, "experimental knowledge, however frequent, does not imply necessarily that it is so in all cases, but only probably."[50]

William of Ockham presents the same combination as Scotus, asserting that causal connections are in some way necessary, while undermining the supposed absoluteness of the necessity. Causal connection is such that it can in principle be inferred from a single observation, "since there is no reason why one heat should be more calefactive than another." But in practice, if we observe that administering a certain kind of herb results in a reduction of fever, we cannot be sure what it is in the herb that effects the cure.[51]

Ockham confused discussion of probability by asserting that probability is a kind of necessity: "Probables are those that, though true and necessary, are not known *per se* nor derivable from what is known *per se*, nor evidently known by experience nor following from it, but which because of their truth seem true to all or most, etc. . . . It follows also that not every topical syllogism always involves doubt and fear, but frequently results in firm belief, free from doubt."[52] But this idiosyncratic usage of *probable* was not followed, and Ockham himself says that *prob-*

able is understood "more widely" simply as what appears to be true to the wise, whether it is true or false. He also retains categories of "contingent to the most part, contingent to the lesser part, or contingent to either side or equally."[53] So the ground of probability is covered adequately, even if in different words.

In the early fourteenth century, philosophy came to be pursued by an established community of academics. There is much pushing of intellectual positions to their "logical" conclusions, irrespective of whether those conclusions contradicted a balanced and reasonable view of the world.[54] More and more, writing was devoted to critiques of one another's works, finer logical distinctions were made, and a wave of modesty swept through philosophical style. Such a culture naturally supported an attention to finer classifications of propositions than to the gross categories of true and false, and the grades of probability that could attach to propositions became something of an obsession. The beginnings are found in Scotus: "But [God's] omnipotence, taken in this way, though not sufficiently demonstrated, can nevertheless be shown probably to be true and necessary—and more probable than certain other beliefs, for there is nothing inappropriate in some beliefs being more evident than others."[55] The standard use of *probable* later involved a balancing of reasons for and against: "The reasons of John [Duns Scotus] which he advances on this side are more probable than the replies to them."[56]

A typical phrase of the period, in theology, philosophy, and elsewhere, was "more probable than their opposites from the point of view of reason." In these discussions, probability came to be a kind of honorific attachable to false and scandalous propositions, in much the same way as it did later in moral theology:

> Although it is erroneous to say that God is not three and one and that the world had no beginning, it is nevertheless not erroneous to assert that, faith aside, it is more probable that God is not three and one or that the world never began than to assert their opposites. For nothing prevents some false propositions from being more probable than true ones. For I understand "probable" so widely (however erroneous this may seem to many) that, faith aside, it is in a way probable or even provable that everything happens of necessity.[57]
>
> I say, that it does not seem to me demonstrated, that there is only one God, but it seems that this comes from faith, as Rabbi Moses [Maimonides] says. . . . But I say, that the reasons adduced are strongly probable, and much more so than those which can be adduced to the opposite.[58]

But speaking in the third mode of necessity, Aristotle perhaps and many other philosophers have held that God produces the world of necessity. . . . But concerning that way of assertion it can be argued with reasons equally probable as or more probable than the reasons in favour.[59]

If it is said, Faith is in opposition, therefore this is not probable, I say: the consequence is not good. For although it follows, "Faith is in opposition, therefore it is not true," it does not however follow that the opposite is not probable; indeed the opposites of some articles are more probable to us than the articles themselves.[60]

(The same author writes that his meaning of *probable* is not that of Aristotle but that it means "anything that is neither known nor believed from the Catholic faith nor determined by the Church nor stated by one whose statement none ought to deny or to assert its opposite. 'Probable' is commonly so described."[61] The change of tone may reflect the author's difficulties over a condemnation of his opinions.)

One theoretical issue about the notion of probability was taken up by these writers. If probability can increase, and one can speak of "equally probable," "more probable," and "much more probable," can probability if sufficiently increased become certainty? Or are probability and certainty of their nature different? Opinion was divided. Gregory of Rimini writes that, "if one is in doubt over some conclusion, the more probable reasons he has for it, the less he will fear concerning it, as is clear; indeed he may so multiply probable reasons, that he will adhere to it completely without fear." Against which John Baconis says that "on opinion it is said that fear is so much part of its concept that it [fear] does not vanish when opinion is strengthened. [Augustine] says that never, when many probable reasons are adduced, is opinion so strengthened that fear disappears."[62]

The language used here suggests that probability, if not exactly a quantity to which numbers can be attached, is at least capable of continuous variation. The early fourteenth century is chiefly known in the history of science for its analysis of the notion of continuous variation, under the name of "the intension and remission of forms." Buridan and Oresme in Paris and the Merton school in Oxford studied such continuously varying quantities as speed and acceleration (which they were the first to distinguish), heat and the lightness and darkness of surfaces.[63] Their discoveries were the necessary preliminaries to the calculus of the seventeenth century. Potentially, probability too is capable of being seen as a continuously varying quantity. In the century before, Boethius of

Dacia had written that "this property (probability) admits of more and less, so that a proposition is called more or less probable. So one proposition is called probable, another more probable, and another very highly probable."[64] John of Dumbleton, of the Merton school, does in fact consider the intension and remission of belief and knowledge with respect to evidence, but the idea was never developed.[65]

These ideas date to the 1330s and 1340s. Time for the development of ideas was running out.

Nicholas of Autrecourt

Before the disaster of the Black Death in 1348–49, there appeared two men of immense genius, Nicholas of Autrecourt and Nicole Oresme. The plague deprived them of their audience, and they were not influential figures in the history of thought. Their writings were simply too profound to be understood by the intellectual pygmies by whom they had the misfortune to be succeeded.

Nicholas of Autrecourt has been called "the medieval Hume" for his criticisms of the received views on sense knowledge and causality. The parallel with Hume is indeed very close. His opinions appear first in his *Letters to Bernard of Arezzo*, written in the 1330s. Bernard is an orthodox Ockhamist Aristotelian and so believes that, for example, although there could be the appearance of a star without there being a star, if God created a vision, that does not happen in the normal course of nature. He believes also that philosophy consists of demonstrated certainties. Nicholas argues, in his *First Letter to Bernard*, that no propositions of this sort be known with "evidence reducible to that of the law of contradiction." His criticism, like Hume's, keeps to a purely logical plane:

> But you will perhaps say, as I think you wished to suggest in a certain disputation over at the Preaching Friars', that although from the fact of seeing it cannot be inferred, when that seeing is produced or conserved by a supernatural cause, that the seen object exists, nevertheless when it is produced precisely by natural causes—with only the general concurrence of the First Agent—then it can be inferred.
>
> But to the contrary . . . since from that antecedent it cannot be inferred evidently by way of intuitive cognition, "therefore whiteness exists," we must then add something to that antecedent—namely, what you suggested above, that the [vision of] whiteness is not produced or conserved in existence supernaturally. But from this my contention is clearly established. For when a person is not certain of some consequent,

unless in virtue of some antecedent of which he is not evidently certain whether or not the case is as it states it to be—because it is not known by the meaning of its terms, nor by experience, nor is it inferred from such knowledge, but is only believed—such a person is not evidently certain of the consequent.[66]

Nicholas goes on to show without difficulty that if certainty means the certainty of logical demonstration, almost nothing is certain: not the existence of the objects of the senses, nor whether any proposition exists.

The *Second Letter to Bernard* is a more elaborate treatise on "evident certainty." Nicholas argues that every certitude is reducible to the "first principle" that contradictories cannot be simultaneously true. It follows that "the certitude of evidence has no degrees" and that "in every consequence reducible to the first principle by as many intermediates as you please, the consequent is really identical with the antecedent or with part of what is signified by the antecedent." The domain of what can really be demonstrated is thus very narrow:

> From the fact that some thing is known to exist, it cannot be evidently inferred, by evidence reduced to the first principle or to the certitude of the first principle, that some other thing exists. . . . But Bernard objects . . . it follows. . . . 'Fire is brought into contact with the fuel, and there is no impediment, therefore there will be heat' . . . they conclude nothing against my position. Because in these consequences which he states, if the consequent is really identical in its signification with the antecedent as a whole or with a part of the antecedent, then the argument is not to the point, because in that case I would concede the consequences to be evident, and nothing against my position would be adduced. . . .
>
> From this rule, so explained to anyone having the grasp of it, I infer that Aristotle never possessed an evident cognition concerning any substance other than his own soul. . . .
>
> From which it follows, whether you like it or not, and not because I make it so but because reason determines it, that Aristotle in his whole natural philosophy and metaphysics had such certitude of scarcely two conclusions, and perhaps not even of one. And Father Bernard, who is no greater than Aristotle, has an equal amount of certitudes, or much less.[67]

At that point, the letter is almost finished, so the next paragraph is unexpected. Not only is certainty unattainable, so is probability: "I have an argument which I cannot refute, to prove that [Aristotle] did not have

[even] probable knowledge. For a person does not have probable knowledge of any consequent, in virtue of some antecedent, when he is not evidently certain whether the consequent will at some time be true together with the antecedent: For let anyone really consider well the nature of probable knowledge—as for example that because it was at one time evident to me that when I put my hand in the fire it was hot, therefore it is probable to me that if I should put it there now it would be hot. But from the rule stated above, it follows that it was never evident to anyone that, given these things which are apparent without inference, there would exist certain other things—namely those others which are called substances. It therefore follows that of their existence we do not have probable knowledge. I do not assert this conclusion; but let this argument be resolved."[68] The passage is interesting for the suggestion that if inductive arguments are the only probable arguments, there is a problem with the knowledge of their premises: if the premises are not absolutely certain (as Nicholas has argued), whence do they derive their probability? It is hard to point to any progress on this question since Nicholas's time.

The *Letters* are purely negative in intent, and one would not expect from them to find that Nicholas had any special skill in devising positive views. As with Carneades, it is surprising to find that the opposite is true. There is another treatise by Nicholas, the *Exigit*, which expounds an anti-Aristotelian, and apparently totally original, picture of the world. The details of Nicholas's worldview are remarkable. His idea is to oppose to the Aristotelian world picture, whose necessity he has denied, his own view, which he asserts is "probable." Of the infinitely many things that could possibly exist, or ways things could be, he needs some probable principle to determine which ones do in fact exist. His central probable principle is that a thing exists if it is *good* that it should exist. His notion of good seems to be as much aesthetic as moral. "It may be taken that all entities of the universe are rightly disposed, that is, it may be taken that those things exist if it is good that they exist, and those things do not exist if it is bad that they exist. This proposition comes to the understanding from inspecting what happens in things of nature and things of art. In things of art the artificer is the measure of the good . . . the fore teeth of man are sharper for cutting food and the back teeth flatter for chewing, an example Aristotle uses."[69] It is to be noted that he does not derive this principle from anything about God, whose existence he regards as not demonstrable.[70]

There is a superficial innocuousness to this principle. It can seem hardly more exciting than the lawyers' maxim, "Everything is presumed rightly done."[71] The corollaries Nicholas draws from it become progressively more extraordinary, until the theory resembles Leibniz's "best of all possible worlds."

> The second principle is that the beings of the universe are connected among themselves, so that in some way one thing seems to be on account of another . . . take away the continuance of heat and rain, and you seem to take away the usefulness and desirability of houses. . . .
>
> The third principle seems to follow from the preceding: the universe is so connected, that nothing exists unless it is good for the whole multitude of things that it exist; whence this entity is for that and that for another and so on.
>
> The fourth principle is, that the universe is always equally perfect. . . .
>
> Philosophers have used these principles which, now they are stated, I wish to use as probables to show that this particular thing that now exists always exists; I argue thus: everything that now exists and is for the good and ornament of the whole multitude of a whole always equally perfect, always exists, but this thing is such a thing. . . .
>
> Again from the principles posited above it seems that nothing in the universe, particular or universal, can be in vain since, if it were, it would be better for it not to exist. Therefore since whatever exists is good, it seems impossible to remove anything without producing a deformity in the whole.[72]

Of course Nicholas admits that some things *seem* to come into and go out of existence, but he asserts that this cannot really be so. They must consist of invisible parts that combine and disperse; the parts are eternal, while combination and dispersion are only changes of relation, and relations do not exist. So the next corollary is a thoroughgoing atomism. "Thus I have here sufficiently probable ways for concluding that the conclusion of the eternity of things is probable, but because I cannot show those tiny whitenesses coming and going like grains, some will perhaps not believe in them; but it is not to be denied on this account. Perhaps they will interrupt in anguish to say that I am denying things that are known *per se*, like a blind man without knowledge talking to someone who knows. Let them consider that there are many things which are not born to come into the evidence of the senses. Whence as I will say below in treating of indivisibles, there is a certain wheel of a clock that moves, but that, however closely one stares at it, does not appear to move; similarly an arrow in the air, which, the faster it moves, the less it is seen to move."[73]

Atomism in turn implies the doctrine of the eternal return; the atoms eventually reconstitute themselves just as the planets eventually return to the same positions. This includes mental entities: "We can however say that it can be sustained probably, and it is probable, that the acts of our mind are eternal by being recapitulated."[74] Further conclusions are that all wishes are fulfilled (if not now, then later); that qualities do not exist; and that everything that appears to be really is. The meaning of this last conclusion is that, for example, if something tastes sweet to one person and bitter to another, both tastes exist; a Protagorean relativism ensues: "I say therefore that this conclusion is probable, that everything that appears properly and fully to be is, and what appears to be true is true. I put forward this conclusion as more probable than the opposite, not as more true."[75] The obvious objections are replied to: as for appearances in dreams, they are said to be not clear and so not to really have a proper appearance of truth. As to the natural suggestion that sense appearances are only veridical when the sense organ is healthy, the medium between the organ and the object well disposed, and so on, Nicholas observes that one cannot have knowledge that these conditions hold, which is more evident than sense knowledge itself, so that making this qualification would make all sense knowledge uncertain.

The conclusions, stated baldly (as they are by Nicholas), are unusual. A little meditation on them will show that principles very like them are assumed time and time again, without being explicitly stated. Of course, it is true that no modern philosopher (or almost none)[76] would state anything so optimistic as "everything is well-disposed." But at a lower level of generality, things are otherwise. There are Ockham's Razor, for instance, the currently fashionable "inference to the best explanation," and various older "principles of least action." Certain modern philosophers have been happy, in unguarded moments, to admit to a preference for desert landscapes, as if this were an intellectual virtue that all would commend. It appears to be, as William James says in another context, "a sort of philosophic faith, bred like most faiths from an aesthetic demand."[77] And the popularity of the theory of evolution in philosophical circles is surely partly attributable to its licensing of speculations about the supposed advantages of anything whatsoever in the living world. What are all these but particular versions of Nicholas of Autrecourt's "all entities of the universe are rightly disposed"? At least he states this explicitly as a probable principle, and at least his theory contains a God, whose tastes may be conjectured to lean toward a classical simplicity. The rational support for the modern versions of his principle is not so clear.

It will have been observed that whenever Nicholas puts forward a new outrageous conclusion, he is careful to say it is only probable. Scattered remarks explain his understanding of probability, the largest being this: "If these reasons are not found absolutely conclusive, nevertheless positing them is probable and more probable than the reasons for the opposite conclusion. To anyone not predisposed more to one side than the other, the degree of probability will appear to predominate (*gradus probabilitatis excedens*) in these reasons. I speak thus because in the books of others I have seen few reasons for their obscure conclusions that I cannot give probable answers to, and if they say that I deny principles known *per se*, it is amazing how they express notorious falsehoods. . . . Was it not sufficiently proved above that to say that [the proposition] when blackness succeeds, whiteness no longer exists, is not a principle known by the meaning of its terms, nor through our experience? It is remarkable too how they consider as known *per se* principles the opposites of which were agreed to by almost everyone before Aristotle, at least the most serious."[78]

Nicholas states explicitly that the existence of God must be held merely as probable *because* merely explicating the meaning of terms will not lead to any proposition about what exists.[79] The real point of all this argument never becomes entirely clear. The anti-Aristotelian polemic is obviously sincere, but beyond that it is impossible to discover what Nicholas really thought. He never asserts his worldview as being true but develops it only to show that it is a probable view (according to reason) and, hence, that the Aristotelian view cannot be proved true, as the Aristotelians claim. He even says of some of his own assertions that, "although they appear far more probable, as it seems to me, than those posited by Aristotle, nevertheless just as the assertions of Aristotle whose probability is now perhaps diminished did long seem probable, so there will come one who will take away the probability of these."[80] He insists that a position cannot be true if faith is in opposition but avoids saying whether his position is consistent with faith or not.

The professionals in that field decided that faith indeed was in opposition. Nicholas was summoned to the Papal Curia in Avignon around 1340, and after some years of investigation thirty-two propositions from his works were chosen for condemnation. They involve the skeptical theses of the *Letters to Bernard*. The condemned propositions include:

(2) From the fact that one thing exists, it cannot by any evidence deduced from the first principle be inferred that another thing exists.

(6) Evident certainty has no degrees.

(8) Of the existence of material substance distinct from our own soul we have no evident certainty.

(10) This consequence is not admitted with any evidence deduced from the first principle: "Fire is brought near to tow and there is no impediment, therefore the tow will burn."

(12) We do not know evidently that any cause which is not God exercises efficient causality.

(21) Whatever thing is known to exist, no one knows evidently that it is not God; if by God we understand the most noble being.[81]

It is one of the most philosophical condemnations made by the church. In fact the propositions condemned are not strictly assertions of theology or even philosophy but of logic, a subject as inapt for theological intervention, in its way, as astronomy.

Nicholas insisted to the commission of inquiry that all his assertions were made disputatively rather than definitively and that he always meant that one should adhere strictly to the faith. The commission called this a "foxy excuse" and sentenced him to burn his own writings in Paris. He did so in 1347 and is last heard of a few years later as dean of Metz. A few contemporary references show he was known and understood; one of his readers was Nicole Oresme, certainly a man capable of appreciating Nicholas's achievement, though Oresme's own genius was more mathematical than philosophical.[82] Nicholas's arguments were still known some fifty years later, but he was thereafter unknown until modern times.

The Decline of the West

Nicholas of Autrecourt and Nicole Oresme belong to the last generation that grew up before the Black Death. Of those mentioned in this story, Nicholas and William of Ockham were probably plague victims, as were those, unknown to us, who would have been their successors. The plague's impact on population and economic life, though catastrophic, was not quite an isolated event—there had been decline before, and recurrences of the disease prevented recovery later. But in matters of the mind, there was a distinct break in the intellectual tradition. The urban monasteries, the centers of intellectual training, were hit by deaths much more than the general population, and elementary education suffered severely. "Few could be found in the good towns and castles who knew how or were willing to instruct children in the rudiments of grammar." Oxford, become "like a worthless fig tree without fruit," was estimated to have lost 80 percent of its students.[83]

What passes for philosophy around 1400 combines either originality with stupidity, as with Wyclif, or learning with complete unorigi-
nality, as with Pierre d'Ailly. Ailly repeats the phrases of the Ockhamists,
among them, "The probable taken more widely is what is true, either
necessarily or contingently, but is not evident" and "More probable in
natural light than its opposite."[84] But the life has gone out of the exercise. The spirit of the age is represented better by Petrarch. His work of
1370, *On His Own Ignorance and that of Many Others*, was composed after
he left Venice, stung by the remark of some young Averroists that he was
a "good man, but a scholar of little merit." Its tone may be gathered
from its remarks on skepticism: "Socrates says: 'This one thing I know,
that I know nothing.' This most humble confession of ignorance Arcesilas blames as still too bold. . . . The Academy is disproved and rebutted
long since, and it is established that something can be known when God
reveals it. Therefore, it may be sufficient to know as much as is necessary for salvation."[85]

Bacon and Descartes: Certainty? or Moral Certainty?

It was seen in chapter 6 that in Galileo, for example, there was an insistence on certainty in science, combined with some use of probability
in practice. The demand for certainty is not confined to science, of
course. It is characteristic of the period 1630–60, especially, also in philosophy, moral theory, and political theory.[86]

In Bacon there are a few hints that there are degrees of certainty in
science,[87] but on the whole Bacon is the most insistent on certainty of
all the founders of the scientific revolution. He is against the simple inductive enumeration of the logicians, simply on the grounds of its fallibility: "For to conclude upon a bare enumeration of particulars (as the
logicians do) without instance contradictory, is a vicious conclusion; nor
does this kind of induction produce more than a probable conjecture."[88]
At first sight, this is rather surprising, in view of Bacon's championing
of experience and his professional knowledge of law. But it appears that
his conception of induction was formed in an area of law other than the
law of evidence. His rivalry with Coke concerned matters of principle
as well as personality and included a fundamentally different view of the
relation of legal principles to cases: while Coke kept very closely to
precedents, Bacon favored the more axiomatic, rule-based approach of
Continental law.[89] To decide what law is applicable, one collects as many
relevant cases as possible, both positive and negative, and infers the rules
that will cover the positive cases, while not being contradicted by the
negative ones. The cases do not "support" the rules probabilistically but

instead impose deductive constraints on what the correct rule could be.[90]
In the scientific context, this raises the problem of the reliability of the
reports of the positive and negative cases themselves, since Bacon proposes to have investigators using many reports, not only those of their
own experiments. Here Bacon is reasonably subtle and allows some reports to be corrected by theory, in the way one would correct misprints
by taking account of the context.[91] But the passive processing of reports,
like the passive process of induction by enumeration, is not his ideal of
science. Bacon's actual legal experience was not so much in pleading factual cases in court as in conducting examinations on behalf of the government. His most original remarks on natural philosophy are those that
urge the examination of nature "under constraint and vexed; that is to
say, when by art and the hand of man she is forced out of her natural
state, and squeezed and moulded."[92] So certainty is attained because nature will be made to yield up her secrets by torture. This is not an innocent metaphor. Bacon was involved in a number of murky investigations into Catholic plots in the 1590s, which required his presence at
torture. Later, as lord chancellor, he advised its use in treason cases. "In
the highest cases of treason," Bacon says, "torture is used for discovery,
and not for evidence."[93] The legal point of torture, it will be recalled, was
to give rise to certainty by revealing facts that could, ideally, be checked.

Descartes too is on the side of certainty as the only worthwhile ideal.
Rule 2 of his *Rules for the Direction of the Mind* is, "We should attend only
to those objects of which our minds seem capable of having certain and
indubitable cognition."[94] That is clear enough, and there are many other
antiprobabilist passages. He dismisses arguments from authority with a
sophism: "It would be of no use to count heads, so as to follow the view
which many authorities hold. For if the question at issue is a difficult
one, it is more likely that few, rather than many, should have been able
to discover the truth about it."[95] He attributes various evils to the failure to separate probabilities and certainties. "We would be well-advised
not to mix any conjectures into the judgements we make about the truth
of things. It is most important to bear this point in mind. The main reason why we can find nothing in ordinary philosophy which is so evident
and certain as to be beyond dispute is that students of the subject first
of all are not content to acknowledge what is clear and certain, but on
the basis of merely probable conjectures venture also to make assertions
on obscure matters about which nothing is known."[96] He recommends
the study of mathematics to philosophers as a way of learning how to
argue demonstratively instead of probably.[97]

When Fermat suggested to him that in calling a proposition in optics "easy to believe," he must have meant "probable," Descartes angrily denied it: "In this he is very far from my intention. For I consider as almost false whatever is only likely; and when I say that a thing is easy to believe, I do not consider that it is only probable, but that it is so clear and evident, that there is no need for me to stop and demonstrate it."[98] But what certainty is available outside mathematics and pure philosophy? Can Descartes maintain that the truths of empirical science, including his own discoveries, have the certainty of demonstrations?

In the exuberance of youth, his answer was yes. But even an intellectual ego as robust as the Cartesian proved subject to the erosion of certainties from long experience of life. The source of trouble was argument "from effects to causes"; very frequently in Descartes' science he infers the truth of a theory from its ability to explain phenomena. But surely this cannot be a demonstrative argument, since different arrays of causes might produce the same observable effects? As he himself says, "I have never noticed anything in them which I could not explain quite easily by the principles which I had discovered. But I must also admit that the power of nature is so ample and vast, and these principles so simple and so general, that I notice hardly any particular effect of which I do not know at once that it can be deduced from the principles in many different ways."[99] The passage continues with an idea like Bacon's: crucial experiments can be devised to decide between competing hypotheses with certainty.

But in his last works, when he comes to present his deductive science in full, Descartes finds he cannot quite do it.[100] One thought he toys with is that God would be deceiving us if he made things as if they followed from a hypothesis, when the truth was otherwise.[101] But his principal defense is to fall back on the Scholastic standby in such cases: moral certainty. In an important passage he argues that "it would be disingenuous, however, not to point out that some things are considered as morally certain, that is, as having sufficient certainty for application to ordinary life, even though they may be uncertain in relation to the absolute power of God." [The French edition adds: "Thus those who have never been in Rome have no doubt that it is a town in Italy, even though it could be the case that everyone who has told them this has been deceiving them."] Descartes continues, "Suppose for example that some one wants to read a letter written in Latin but encoded so that the letters of the alphabet do not have their proper value, and he guesses that the letter B should be read whenever A appears, and C when B appears, that is, that each letter should be replaced by the one immediately following it. If, by using this key, he can make up Latin words from the letters, he will be in no doubt that the true meaning of the letter is contained in these words. It is true that his knowledge is based merely on a conjecture, and it is conceivable that the writer did not replace the original letters with their immediate successors in the alphabet, but with others, thus encoding quite a different message; but this possibility is so unlikely [the French edition adds: especially if the message contains many words] that it does not seem credible."[102]

This passage is interesting for several reasons. The first sentence and the first French addition are strictly Scholastic; they are almost word-for-word identical with the Scholastic Arriaga's *Cursus philosophicus* of 1632.[103] But the example of the encoded message is new. It suggests that moral certainty is only a high degree of probability, rather than being different in kind. (For what if the cipher contains only a few words? Would there not be only a small probability that it meant something?) The other Scholastic theme, of the absolute power of God, acts to distract attention from the point by suggesting that the falsity of the proposition is a mere logical possibility, which need not be given any weight in practice.

Still, Descartes lacks the usual Scholastic smugness about moral certainty as the cure for all doubts. The paragraph before the one just quoted has a weaker theory on the permissibility of regarding scientific theories as true simply because true observable consequences follow from them:

> Just as the same craftsman could make two clocks which tell the time equally well and look completely alike from the outside but have completely different assemblies of wheels inside, so the supreme craftsman of the real world could have produced all that we see in several different ways. I am very happy to admit this; and I shall think I have achieved enough provided only that what I have written is such as to correspond accurately with all the phenomena of nature. This will indeed be sufficient for application in ordinary life, since medicine and mechanics, and all the other arts which can be fully developed with the help of physics, are directed simply towards applying certain observable bodies to each other in such a way that certain observable effects are produced as a result of natural causes. And by imagining what the various causes are, and considering their results, we shall achieve our aim irrespective of whether these imagined causes are true or false, since the result is taken to be no different, as far as the observable effects are concerned.[104]

The idea is similar to the "saving of the appearances" of astronomical hypotheses, as Descartes notes elsewhere.[105] Earlier Scholastic thought made some attempt to grapple with the problem of why an astronomical theory's agreeing with the observations should be a reason for believing it and whether the belief should amount to certainty or not.[106] Kepler tried to find something in common between the Copernican and Ptolemaic theories, which would account for the appearances and thus explain why they made the same predictions, although one was right and one wrong about the underlying causes.[107] But Descartes is asserting something over and above a simple recommendation of inference to the best explanation. He asserts that there is no need to find the true underlying model to make a correct prediction. It is the first clear statement of the dream of modern statistical inference: to make true predictions independently of difficult inquiry into inner causes. The modern economic modeling that attempts to forecast unemployment, interest rates, and so on without any commitment to grand economic theories is a continuation of Descartes' project.[108] More relevant to Descartes' own time is the problem of the truth of the atomic hypothesis: it could, more or less, explain appearances, but is that a good reason for believing it true, in the absence of any more direct evidence?[109]

Further blurring of the distinction between probability and certainty in Descartes is not hard to find. In a passage in a letter on a point like the cipher, he says that pieces of evidence that are separately only probable can together have the force of demonstration.[110] And in a late letter, he is prepared to balance probabilities for and against an uncertain opinion, the very thing his younger self had so forcefully condemned. "However, while I regard it as demonstrated that it cannot be proved that there is any thinking in animals, I also do not think it can be demonstrated there is none, because the human mind cannot penetrate their heart. But, on examining what is the most probable, I see no reason in favour of thinking in animals but one, that since they have eyes, ears, tongue and other sense organs like us, it is likely they feel like us, and since thinking is included in our way of feeling, a like thinking should be attributed to them. This reason, being very obvious, is part of the thinking of all men from a very young age. But there are many more stronger reasons, not so obvious to all, which are quite persuasive to the contrary. Notable among these is, that it is not as probable to give all worms, gnats, caterpillars and other animals immortal souls, as to have them move like machines."[111]

The argument then continues by adducing analogies between animal bodies and machines.

Another probabilistic argument Descartes suggests is that "an example of composition by way of conjecture would be our surmising that above the air there is nothing but a very pure ether, much thinner than air, on the grounds that water, being further from the centre of the globe than earth, and air, which rises to greater heights than water, is thinner still. Nothing that we put together in this way really deceives us, so long as we judge it to be merely probable, and never assert it to be true; nor for that matter does it make us any the wiser."[112] (It is a disputed point of Cartesian scholarship as to whether the negative in the last clause should be there.)[113]

Descartes also admits that, "since in everyday life we must often act without delay, it is a most certain truth that when it is not in our power to discern the truest opinions, we must follow the most probable. Even when no opinions appear more probable than any others, we must still adopt some; and having done so we must then regard them not as doubtful, from a practical point of view, but as most true and certain, on the grounds that the reason which made us adopt them is itself true and certain."[114] The obvious nonsequitur is another outcome of trying to regard as certain, in some sense, what is not. Descartes tries to confine the operation of probability to subjects in which he is fundamentally not interested.

Among the many objectors to Descartes, Caramuel inquired most searchingly into whether the results that Descartes claimed were certain really were so. Caramuel's observations were not published.[115] However, the general structure of Gassendi's well-known replies to the *Meditations* is the same, although there is nothing explicit about certainty and probability. Gassendi's strategy is the same as the one Nicholas of Autrecourt deployed against the Aristotelians: to put forward hypotheses, or possibilities contrary to what Descartes claims to have proved, "suggestions which do not concern the actual results which you have undertaken to prove, but merely the method and validity of the proof."[116]

The Jesuits and Hobbes on Induction

Certain Jesuit Scholastics of the seventeenth century pursued some interesting ideas on the problem of induction. In theory, they take up the problem where Scotus and Ockham left it and take no notice of purveyors of novelties like Descartes, but in fact their problem is very like Descartes' one of predicting the behavior of clocks whose internal mechanisms one does not know but whose past behavior one has observed. In place of the clock they have the will of God. Given that, as Ockham had emphasized, the course of nature can be changed arbitrarily by the will of God, what reason has one to believe that He will not actually do so and, for example, annihilate the world in the next second?

The Jesuit cardinal Lugo argues that it is a matter of presumption based on long observation. "As long as the contrary be not proved, we always presume that God without miracles or violence allows secondary causes to act naturally, although indeed God could morally do the opposite. . . . Because in case of doubt there suffices a presumption founded on long induction of effects, a bare moral possibility is not sufficient for a prudent doubt or judgement of the opposite. For many things are morally possible, which as long as they are not proved, are not posited, but rather are presumed not to be so; for example, it is possible, morally speaking, that a baptising priest has not a right intention or does not use the right form [of ceremony], but as long as that is not apparent, we always presume and judge that children are validly baptised, and so in other cases."[117]

That may be what we do presume, but it hardly explains why the presumption should be so and not otherwise. A pupil of Lugo's argues that it is a matter of probability, founded on relative frequencies: "Every reasonable presumption must be founded on the probability of the thing presumed being greater than that of its opposite . . . we presume the natures of things are not disturbed from the fact that through long induction we can see that God ordinarily keeps to this mode of governing the visible world."[118] Another Jesuit argues for the moral certainty of the axioms that like causes produce like effects, and that in case of doubt about the future, one should conjecture what occurred most frequently in the past.[119] This still fails to explain why exactly God should be on the side of the presumption founded on a long observation of effects.

There is a genuinely new thought on this matter in the writings of the Jesuit Esparza in the 1650s. He argues that it is a necessary fact that induction can only fail in rare cases. The reason for believing that the world will not end immediately, he says, does not lie in the fact that God prefers the world to continue, since he could, by compensating souls suitably, produce just as good an outcome by destroying the world. Instead, the reason for expecting the world to continue lies in the "metaphysical necessity" that the end of the world is the kind of event that *could* only happen very rarely:

> The metaphysical necessity of not perturbing the natural order of things except rarely (and even with compensation being made) supplies a sufficient foundation for the certainty of the common persuasion of the continuance of the world. For an evil that is of such a kind as to happen only rarely is hardly ever to be feared, and in individual cases one can hope that the future will be quite safe. Although sometimes it does

happen that a house collapses suddenly without cracks appearing beforehand, or being preceded by any other sign of ruin, and similarly it
does happen that a tile falls from a roof and breaks the skull of a passer-
by without any wind dislodging it or anyone moving the roof, nevertheless people can live in houses and walk around cities safely and quite
without worrying, as long as there is no actual sign of danger, as if such
accidents never happened. Likewise it is possible for men to be entirely
untouched by any concern for the annihilation of the world, even if God
were quite indifferent to that event; because that event belongs to the
class of events that can happen but rarely.[120]

While one can still ask whether a particular case might not be one of the
rare ones, and also ask what exactly a metaphysical necessity is, the idea
that induction can only fail rarely has found some support among the
better modern attempts to justify inductive reasoning.[121]

Hobbes, though inclined to follow Bacon and Descartes in exalting
deductive reasoning, does insist on the fallibility of induction, while allowing that it is a good bet. He even assigns a number to the degree of
reasonableness of an inductive argument, "for the signs are but conjectural, and according as they have often or seldom failed, so their assurance is more or less; but never full and evident: for though a man have
always seen the day and night to follow one another hitherto; yet can he
not thence conclude they shall do so, or that they have done so eternally:
experience concludeth nothing universally. If the signs hit twenty times for
one missing, a man may lay a wager of twenty to one on the event; but
may not conclude it for a truth."[122] The "twenty to one" may be assigned
only by way of example, rather than as an assertion that this is the correct degree of belief in the inference. If it is intended as the right answer,
it must mean that the correct odds for an induction are equal to the proportion in past cases. This of course contradicts what Hobbes has said
in the previous sentence, because such a rule prescribes infinite odds,
that is, a judgment of certainty, when the previous cases have been all of
one kind. That it has been unable to discover what the correct odds are
has been a notorious source of embarrassment for the theory of probability ever since. The problem has stayed so comprehensively unsolved
that it has been more or less agreed to forget about it.[123]

Pascal's Deductivist Philosophy of Science

Pascal's most successful experimental work in science was his investigation of pressure in fluids. His debate on his results with the Scholastic
Père Noël provided the occasion also for considerations on the philos-

ophy of science, which raised many of the issues still current in that subject. They include the role of deduction versus probability in the evaluation of alternative hypotheses, the significance of crucial experiments, the place of metaphysics in science, and the degree to which hypotheses can survive adjustments to fit the experimental facts. At issue was the explanation of experiments in which inverted glass tubes of water or mercury of sufficient height were observed to have spaces under the glass at the top of the tube. Were these essentially vacuums, as Pascal thought, or had some subtle matter passed through the glass to fill the space? And was the correct explanation of the phenomenon that of Pascal, according to which the pressure of the outside air could support a column of a certain height but no higher, or the Scholastics' maxim that "nature abhors a vacuum"?

Pascal first admits to a principle of conservatism: the onus is on him to show that the existing hypothesis is wrong. He says that, "seeing that all the effects of this last experiment with the two tubes, which is so naturally explained by the mere pressure and weight of the air, can also be explained, probably enough (*assez probablement*), by the abhorrence of a vacuum, I still hold to that ancient principle, although I am determined to seek a thorough elucidation of this difficulty by means of a decisive experiment."[124] In discussing the "decisiveness" of his experiment, he subscribes to a fully deductivist ideal of science: Only what is clear and evident to the senses or reason, or logically deducible from such principles, is certain; "and all that is based on neither of them passes for doubtful and uncertain. We pass decisive judgment on things of the first kind, and leave the others in a state of indecision, so that, according to their merit, we call them now *vision*, now *caprice*, at times *fancy*, sometimes *idea* and at most a *fine thought*; and because we cannot affirm them without temerity we incline rather to the negative."[125]

Referring to the usual example of competing astronomical hypotheses, Pascal is clear on the difference between deriving a hypothesis from established truths and deriving established truths from the hypothesis: "to have a hypothesis seem obvious, it is not sufficient that all phenomena follow from it. . . . It is thus that, when we discourse humanly about motion and about the stability of the Earth, all the phenomena of the movements and the retrogradation of the planets follow perfectly from the hypotheses of *Ptolemy*, of *Tycho*, of *Copernicus* and of many others that one may name . . . and who can, without courting error, uphold one at the expense of the others?" He suggests that since the Scholastics have an Aristotelian ideal of science that is likewise based on deduction, they should also believe there is a vacuum at the top of the tube, if they cannot demonstrate that there is any matter there. "According to your maxims would that not suffice to assure that that space is empty?"[126]

Père Noël, in his answer, is well aware that Pascal has not achieved the demonstration he claims. He quotes this last question and says:

> I reply simply no. If, for lack of a mathematical demonstration, that is, an evident and convincing one, that some matter enters through the glass upon the descent of the quicksilver, I were to say there were empty space inside, then by the same reasoning one would deny that there were any matter between our atmosphere and the firmament, by concluding as follows: all men together could not demonstrate mathematically that those great spaces are full of any body, and therefore I would say that those great spaces are only an immobile and penetrable vacuum, sufficient to sustain and transmit the light of the stars, and affect their movements. If that were my assertion and opinion, what could you say? Now, in just the same way as naturalists think they have adequate proofs and physical reasons for having assurance that those great spaces are filled with an impenetrable and movable body, though they have no mathematical demonstration for that; so, though I have not the same conviction, I think I have nevertheless enough natural proofs to say that there passes through the pores of the glass and enters inside the glass some matter, which I call subtle air.[127]

He relies, therefore, on the usual claim that there are "physical" proofs with a force lower than mathematical demonstration. He goes on to argue that, having regard to such physical reasons, the explanation of either side is equally likely, but that the "vacuum" that Pascal posits is metaphysically very strange. It is "a space that is neither God, nor creature, nor body, nor spirit, nor substance, nor accident; which transmits light without being transparent, resists without resistance, which is immovable and is transported with the tube, which is everywhere and nowhere, does everything and does nothing."[128]

Pascal bizarrely replies, apparently with a straight face, that "since Father Noël concludes from the appearance of these effects that this space is filled by a matter which bears light and causes this retardation, we may with just as much reason conclude from these same effects that light exists in a vacuum and that motion takes place there in time. In addition, so many other things favor this latter opinion that in the judgement of scholars it was incomparably more plausible (*sans comparaison plus vraisemblable*) than the other, even before it had the support which these experiments lent to it."[129] Somehow, one does not expect to see the author of the *Provincial Letters* relying on arguments from the authority of the "judgement of scholars."

After accusing Père Noël of misunderstanding some of the experiments, and thus deriving an experimental conclusion different from the one observed, Pascal suggests there is a logical problem with theories that can accommodate any result. Père Noël's hypothesis "is so accommodating that he has proved by a necessary succession of his principles why the finger no longer feels any pull, although that is absolutely false. I believe that he could just as easily make the contrary seem reasonable by the same principles. But I do not know in what esteem judicious persons would hold his fashion of proving the affirmative and the negative of the same propositions with the same forcefulness."[130]

CHAPTER NINE

Religion: Laws of God, Laws of Nature

Religion too is an old enemy of probability—and for much the same reason as philosophy. It has certainty to offer, and acceptance of that certainty is not only a privilege but a duty. A strong though not always dominant strand in religion has been the condemnation of half measures, epitomized in the Book of Revelation's "you are neither hot nor cold. I wish you were one or the other, but since you are neither, but only lukewarm, I will spit you out of my mouth."[1] When St. Paul mentions the distinction between probable and necessary reasons, it is to claim that he has the latter: "My speech and my message were not in the plausible words (*pithanologia*, Vulgate: *persuasibilibus verbis*) of wisdom, but in demonstration of the Spirit and of power."[2]

Still, when two faiths, or a faith and a nonfaith, exist in the same place, there is a motive to find some rational way, common to the disputants, of resolving the deadlock. There is a need to defend faith in general, as well, through arguments for the basic propositions that religions share. In deciding on the reasonableness of faiths that assert some historical facts, like Christianity, Islam, and Judaism, there is room to debate the reliability of historical knowledge. And one reaction to conflicting dogmatisms has been a moderate position that denies the possibility, and necessity, of knowing some of the matters in dispute between the warring sects.

The Argument from Design

Of all arguments for the existence of God, the most durable is the design argument. It comes in two flavors, deductive and nondeductive. The deductive type, of which the most famous example is Thomas Aquinas's Fifth Way, argues that teleology, or directedness, in nature necessarily implies an Orderer, in much the same way as a street sign's having a meaning implies that someone wrote it to have that meaning. This is offered as a philosophical intuition into the nature of things. Take it or leave it. The other, more popular, kind of design argument is quite different. Exemplified by such modern versions as "What is the chance of a tornado blowing through a junkyard and leaving a 747?" it is a nondeductive argument and hence subject to infinite discussions of its plausibility. Is there an alternative explanation of apparent order? Can nature self-organize? Can order come to be by chance? Can the analogy between the world and an artifact be defeated by some parallel argument, like an argument from disorder?

These moves and countermoves have all been known since ancient times. Sextus Empiricus reports ancient design arguments. "Just as the man who is familiar with ships, as soon as he sees in the distance a ship with a favouring wind behind it and with all its sails well set, concludes that there is somebody who directs its course and brings it into its appointed havens, so too those who first looked up to heaven and beheld the sun running its courses from east to west and the orderly processions of the stars sought for the Artificer of this most beautiful array, conjecturing that it had not come about spontaneously but by the agency of some superior and imperishable nature, which is God."[3] The argument is explicitly probabilistic and almost explicit in being a rejection of the hypothesis of chance as unlikely.

The opposition "chance or design" is explicit in a version of the argument attributed to Socrates. He continues, "Do you not think that he who created man from the beginning had some useful end in view when he endowed him with his several senses? . . . The eyeballs, being weak, are set behind eyelids, that open like doors . . . with such signs of forethought in these arrangements, can you doubt whether they are the works of chance or design? . . . do you suppose that wisdom is nowhere else to be found, although you know that you have a mere speck of all the earth in your body and a mere drop of all the water?"[4] (To which Sextus mentions the reply: "You possess also a small portion of the great quantity of gall existing in the Universe, and phlegm and blood. It will follow, therefore, that the Universe is gall-making, and productive of blood." Analogies are always in danger of being outflanked by parallel arguments.)

If a design argument is not presented as deductive, there is always a need to consider the probability of the alternative hypothesis, that the order in the world arises from natural causes, through some kind of self-organization of matter. Sextus says that "the Stoics and their supporters try to demonstrate the existence of the Gods from the motion of the Universe. For that the Universe is in motion everyone will admit, being driven thereto by many things. It is moved, then, either by nature or by will or by vortex and of necessity. But that [it is moved] by vortex and of necessity is not probable (*eulogon*). For the vortex is either disorderly or orderly. And if it is disorderly, it will not be able to move anything in an orderly way."[5] Analogies between a house or a ship and the universe were standard among the Stoics. Chrysippus says simply, "If we saw a handsome mansion, we should infer that it was built for its masters and not for mice; so therefore we must deem the world to be the mansion of the gods."[6]

These arguments were taken up by the revealed religions, when their apologists saw the need to take seriously the problems of the relation of faith and reason. Christians were encouraged by St. Paul's saying that "ever since the creation of the world his invisible nature, namely his eternal power and deity, have been clearly perceived in the things that have been made."[7] Most of those themes, developed with such effort from the Middle Ages on, are found in the Jewish contemporary of Jesus, Philo of Alexandria, and the design argument is among them. Philo's conception of the argument seems to be probabilistic. "Entering the world as into a well-ordered city they have beheld the earth standing fast, highland and lowland full of sown crops. . . . They have as I said advanced from down to up by a sort of heavenly ladder and by a probable calculation (*eikoti logismoi stochasamenoi*) happily inferred the Creator from his works."[8]

The Church Fathers

According to the church fathers, the relation of theology to probability consists mainly in accusing other disciplines of being restricted to "mere probability," the implication being that theology could of course do better. Thus Lactantius says that, "rightly, therefore, did Socrates and the Academics, who followed him, remove knowledge. . . . It remains that there can only be opinion in philosophy. . . . Therefore they do not know the truth, for knowledge is of the certain, opinion of the uncertain." Clement of Alexandria writes similarly, adding that "conjecture, which is only a feeble supposition, counterfeits faith; as a flatterer counterfeits a friend, and the wolf the dog." In any case, he says, the assent needed for ordinary life is already a kind of faith.[9] Lactantius and Clement believe that the certainty offered by faith will appeal to those who wish to transcend the uncertainty and endless arguing of the philosophers. No doubt they are right.

A less expected intervention of theology occurs in one of the many late Latin commentaries on rhetorical theory, Marius Victorinus's commentary on Cicero's *De Inventione*. It is a very brief remark, but in it antiquity has one last pearl to throw. Commenting on Cicero's examples of necessary argument, "if he was born, he will die; if she has given birth, she has lain with a man," Victorinus says that, "moreover, according to the opinion of the Christians, the argument, 'if she has given birth, she has lain with a man,' is not necessary, nor is, 'if he was born, he will die.' For, according to them, there has appeared one born without a man, and not dead."[10] (Of God's power to overcome things otherwise necessary, much more later.)

An argument "for the most part" is invoked by St. Cyprian in dealing with the perennial problem of disagreement among experts on the content of faith. The church may be infallible, but which sect represents the opinion of the true church? He writes, in words that recall the definition of the probable in Aristotle's *Topics* as "what is agreed to by all or most or the wise": "'For God is true but every man a liar.' The majority of the confessors, and the better ones (*pars maior et melior*), stand strong in their faith and true to the Lord's law and discipline." This text was later taken as a support for the principle of majority rule in assemblies, especially but not exclusively in church councils on matters of faith.[11]

Finally there appeared a professional teacher of rhetoric who, in middle life, stopped talking about persuasion and began to persuade in earnest, converting a whole civilization to his vision. Augustine's early work against Skepticism, the *Contra Academicos*, contains a brief summary of Cicero's account of Carneades' theory of probability. Augustine makes a number of jokes at the expense of Carneades, but it is not completely clear what he thinks is wrong with the theory. Characteristically, he is worried most about the moral consequences; young men, he thinks, will use probability as an excuse to seduce other men's wives.[12] Despite Augustine's dominant position in medieval thought, the *Contra Academicos* attracted little attention.

Little of the thought of antiquity was available in the Dark Ages, but a rudimentary vocabulary of probability did survive. Probabilistic argument of a basic kind can be found in the texts of the period. In his *History of the Franks* of about 590, Gregory of Tours recounts an incident in which twenty Frankish raiders sack a monastery, murder the monks, and steal their possessions. Escaping across a river, their boat sinks, killing all but one, "a man who had rebuked the others for what they were doing. If anyone thinks that this happened by chance (*fortuitu*) let him consider the fact that one innocent man was saved among so many who were doing evil."[13] This is an Aristotelian not-by-chance argument. It is so simple that it might have been thought of spontaneously, but the word *fortuitu* is a learned term in what is, for the author, a technical language —one that, according to Latinists, he could barely write at all. It is most likely that Gregory's ability to use the word is a survival of antiquity.

Inductive Skepticism by Revelation

By the time of Averroes, in the twelfth century, European scholars were
journeying to Islamic countries to translate into Latin the major Islamic
authors and the Greek classics that could be found in Arabic translation.
They were only just in time. Averroes, for example, though later well
known in the West, was virtually unknown to later Eastern scholars. In
his time Islam had become in most places the anti-intellectual religion
it remained everywhere till the nineteenth century, and remains in many
countries today. What happened is not entirely clear. Undoubtedly wars
and unsettled times played a part in disrupting the tradition of scholarship. Civilization was almost totally destroyed in Central Asia, Avicenna's homeland, by Genghis Khan, while Cordoba, the home of Averroes and not long before his time the largest city in Europe, fell to the
Spanish some forty years after his death. But at least as important were
purely intellectual reasons. Islam adopted a fundamentalism that had no
use for reason; it was left in the position Christianity would have been
in if the Reformation had resulted in total victory for the Calvinists.

The classic of this school, al-Ghazali's *Destruction of Philosophy*, attacked reason with great success. In this book, and not for the last time,
the question of the absolute power of God became entangled with nondemonstrative reasoning. The general idea of this is natural enough: the
monotheistic religions generally attribute to God the power to accomplish anything except, perhaps, a contradiction. Consequently, the conclusion of any reasoning less than strict demonstration may be made true
or false according to the will of God; and who will be so rash as to
fathom His inscrutable will? Conversely, a little meditation on the power
of God can be useful in revealing how restricted the domain of deductive logic is: if God could arrange it so that A is true but B false, there
is no point in trying to demonstrate that B follows logically from A.

Al-Ghazali, "the renewer of Islam," applies this line of thought to
the connection between cause and effect, which he argues can be broken by the will of God and hence cannot be truly necessary. Causal connection cannot be logical connection. He says, in opposition to the
philosophers:

> According to us the connexion between what is usually believed to be a
> cause and what is believed to be an effect is not a necessary connexion;
> each of two things has its own individuality and is not the other, and nei-
> ther the affirmation nor the negation, neither the existence nor the non-
> existence of the one is implied in the affirmation, negation, existence,

and non-existence of the other—for example, the satisfaction of thirst does not imply drinking, nor satiety eating, nor burning contact with fire, nor light sunrise, nor decapitation death, nor recovery the drinking of medicine, nor evacuation the taking of a purgative, and so on for all the empirical connexions existing in medicine, astronomy, the sciences, and the crafts. . . . The agent of the burning is God, through His creating the black in the cotton and the disconnexion of its parts, and it is God who made the cotton burn and made it ashes either through the intermediation of angels or without intermediation. For fire is a dead body which has no action, and what is the proof that it is the agent? Indeed, the philosophers have no other proof than the observation of the occurrence of the burning, when there is contact with the fire, but observation proves only a simultaneity, not a causation, and, in reality, there is no other cause but God.

An obvious objection is that one would be left without knowledge of the course of events in the world and would be forever wondering about far-fetched possibilities ("I only know that I left a book in my house, but perhaps by now it is a horse which has soiled the library with its excrement"). Al-Ghazali replies simply that "God has created in us the knowledge that he will not do these possible things . . . protracted habit time after time fixes their occurrence in our minds."[14]

Al-Ghazali's *Destruction of Philosophy* of about 1090 destroyed philosophy. Averroes' reply, *The Destruction of the Destruction*, went unread.[15]

John of Salisbury

The man of the European twelfth century who made most of the classical heritage was John of Salisbury, student of Abelard, later bishop of Chartres. He has attracted the label *humanist* and indeed shares many of the characteristics of the later Renaissance humanists—a style full of classical allusions, an admiration for Cicero and an imitation of his language, a skepticism about the higher intellectual pursuits such as science and philosophy, and a tendency to write at length in praise of virtue and the delights of philology and rhetoric. His reading included Aristotle's logical works (in Latin translation), but not the *Rhetoric*, and much of Cicero. His books *Policraticus* and *Metalogicon* of 1159 contain a more or less Ciceronian account of probability. In an age when Scholasticism and its claims to demonstrate so much were gaining ground, he argued strongly that probability is needed because very little of knowledge is really demonstrable. His argument for this truth is impressive:

Only that which cannot possibly be otherwise is necessary. Since no-one, or hardly anyone, ever fully comprehends natural forces, and since God alone knows the limits of possibility, it is frequently both dubious and presumptuous to judge that a thing is necessary. For who has ever been absolutely sure about where to draw the line between possibility and impossibility? Many ages took the following principle, "If a woman gives birth to a child, she must have had previous sexual intercourse, whether voluntary or involuntary, with someone," to be a necessary axiom. But finally, in the fulness of time, it has been shown that it is not such, by the fact that a most pure and incorrupt virgin has given birth to a child. Something that is absolutely necessary cannot possibly be otherwise. But something that is conditionally necessary may be modified. Victorinus, in his work on rhetoric, explains this when he discusses necessity. He tells us, that while previous sexual intercourse may be inferred with probability, it cannot be deduced as absolutely necessary from the fact of childbirth. Augustine asserts that necessary reasons are everlasting, and cannot in any way be gainsaid. It is clear, however, that the reasons of probable things are subject to change, since they are not necessary. The great difficulty with demonstration is apparent, as the demonstrator is always engaged in the quest of necessity, and cannot admit of any exceptions to the principles of truth he professes.[16]

John means that if something really is *demonstrated* there cannot be an exception to it, even by miracle. He is right. It remained a thorn in the side of Aquinas and of Scholasticism in general. Plainly, this does not leave much in the sphere of the demonstrable and, correspondingly, transfers much of knowledge into the domain of the probable. John accepts these consequences: mathematics and much ordinary knowledge, he thinks, is indubitable, but a good deal of what is taken for knowledge in the schools, on questions like the status of universals and the source of the Nile, can very well be doubted.[17] His definitions of *probability* repeat Aristotle and Cicero, but John is more insistent than either on the pervasiveness of probability: "The investigation of probabilities, which comprise most of human knowledge, flows, in a way, from the *Topics* as its fountainhead."[18] He also has something to say on the dynamics of probability. Probability is not just a static property of a proposition but something that waxes and wanes with the available reasons, and may approach certainty.

Something that is always or usually so, either is or seems probable, even though it could possibly be otherwise. And its probability is increased in proportion as it is more easily and surely known by one who has judgement. There are some things whose probability is so lucidly apparent that they come to be considered necessary; whereas there are others which are so unfamiliar to us that we would be reluctant to include them in a list of probabilities. If an opinion is weak, it wavers with uncertainty; whereas if an opinion is strong, it may wax to the point of being transformed into faith and aspire to certainty. If its strength grows to the degree that it can admit of no or hardly any further increase, even though it is less than full knowledge, it comes to be equivalent to the latter so far as our certainty of judgement is concerned. This is apparent, as Aristotle observes, in matters which we perceive only by our senses, and which can be otherwise. Thus when the sun has set, we do not know with certainty that it will continue above the earth and return to our hemisphere. For the sensory perception whereby we are apprised of the course of the sun has ceased. Nevertheless our confidence concerning its course and return is so great that it seems, in a way, equivalent to [full] knowledge. . . . A wide knowledge of probabilities prepares expeditious access to all things.[19]

As a eulogy of the importance of probabilities, the *Metalogicon* is excellent. As an aid in actually discovering how probable any proposition is, it is negligible.

John of Salisbury is unusual in his time for emphasizing the importance of probabilities to the extent that he does, but he is not entirely alone.[20] At least the distinction between necessary and probable reasons can be found even in theology, and according to one theory, the need to be content with the merely probable is a consequence of Adam's fall.[21]

Maimonides on Creation

Maimonides shared with Aquinas a major embarrassment that resulted from trying to follow both the Bible and Aristotle. The Bible says that God created the world; Aristotle claims to have proved that the world has always existed. It is a much clearer conflict between science and religion than the post-Darwinian one, and Maimonides and Aquinas both devoted a good deal of effort to explaining how they could accept virtually everything in Aristotle, and accept it as demonstrated, while rejecting just the propositions that happened to conflict with faith.

Maimonides rather unconvincingly argues that Aristotle cannot really have meant that he has *demonstrated* the eternity of the world—if he meant that, why did he expend so much effort collecting authorities in support and vilifying opponents?[22] Maimonides admits the need for a nonarbitrary principle to decide which of Aristotle's opinions are acceptable and which not. He has a suggestion on where to draw the line. It should be drawn at the moon. Aristotle succeeds in demonstrating everything he says about what lies below the sphere of the moon, but most of what he says about angels and other beings beyond the moon is "analogous to guessing and conjecturing." Of propositions about divine things, Maimonides advises instead that one must follow Alexander of Aphrodisias's method of evaluating the total doubts on either side. Then one will see that the hypothesis of the eternity of the world is subject to more doubts than its contrary.

In evaluating doubts, Maimonides, like the canon lawyers, does not approve of simply counting: "Know that when one compares the doubts attaching to a certain opinion with those attaching to the contrary opinion and has to decide which of them raises fewer doubts, one should not take into account the number of the doubts but rather consider how great is their disagreement with what exists. Sometimes a single doubt is more powerful than a thousand other doubts. Furthermore this comparison can be correctly made only by someone for whom the two contraries are equal. But whoever prefers one of the two opinions because of his upbringing or for some advantage, is blind to the truth."[23]

Maimonides revived the probabilistic design argument for the existence of God but with a subtle difference. It has occurred to many that the order in the heavens is a sign of the existence of God, but to Maimonides is due the idea that *disorder* is even better. Order, he thinks, is always in danger of being explained away as necessity, while disorder needs an explanation of why things are in a particular way and not another, so that a potentially more complex explanation, such as the will of God, is needed:

> And it is even stranger that there should exist the numerous stars that are in the eighth sphere, all of which are globes, some of them small and some big, one star being here and another at a cubit's distance according to what seems to the eye, or ten stars being crowded and assembled together while there may be a very great stretch in which nothing is to be found. What is the cause that has particularized one stretch in such a way that ten stars should be found in it and has particularized another stretch in such a way that no star should be found in it? Again the body of the whole sphere is one simple body in which there are no differences. What accordingly can be the cause for the fact that a certain part of the sphere should be more fitted to receive the particular star found in it than another part? All this and everything that is of this sort would be very unlikely or rather would come near to being impossible if it should

be believed that all this proceeded obligatorily and of necessity from the deity, as is the opinion of Aristotle. If, however, it is believed that all this came about in virtue of the purpose of one who purposed who made this thus, (*sic*) that opinion would not be accompanied by a feeling of astonishment and would not be at all unlikely.[24]

On the other hand, Maimonides understands that there are similar cases in which one should *not* look for an explanation. Many of the particular requirements of the law have no explanation, he holds. There is no point in asking why a law prescribes seven lambs and not eight, since one could ask the same question whatever the number was, and some particular choice had to be made.[25]

Are Laws of Nature Necessary?

Aquinas took the conventional view that at least philosophical and mathematical arguments achieve the ideal of demonstration, even if one has to be content with less in sciences like physics, which deal with variable matter. This view was denied, indeed reversed, by the *enfant terrible* of thirteenth-century philosophy, Siger of Brabant. Where Aquinas was concerned to show that the church had nothing to fear from the new Aristotelian science, his flank was continually being turned by Siger, his contemporary at the University of Paris. Siger represented the Averroist view that Aristotle's opinions, being the best science to be had, should be adopted in full, regardless of any conflict with faith. The proposition that the world is eternal was again the focus of concern, in view of Aristotle's claim to have demonstrated it. The contradiction between science—that is, Aristotle—and faith seemed clear. Aquinas argued at length that neither the proposition nor its opposite could be proved, so that one was free to accept the Christian revelation that the world was created. Siger asserted the opinion of Aristotle nevertheless. He evaded the full force of his conclusion, or the full force of the penalties likely to ensue from it, by asserting that the conclusion was only probable.[26] He went so far as to say that *all* metaphysical truths were only probable, in contrast to physical arguments, which were demonstrative. He was indeed the first in a line of philosophers to speak of probability as a kind of detachable feature of propositions, almost as something opposed to truth: "Although indeed this opinion is probable, nevertheless it cannot stand."[27]

Siger discusses what happens when one has equally probable reasons for both sides of a contradiction. Does one then hold both opinions? No, that is impossible. Reasons are like causes: some, such as the demonstrative syllogism, cannot be impeded in their action, but the dialectical syllogism is like an ordinary physical cause; it can be impeded by another that tends to the opposite conclusion.[28]

Siger's need to find a mode of expression acceptable to the inquisitors in his audience prevents us from understanding his opinions correctly. His precautions were not adequate. Many of his propositions were condemned by the bishop of Paris. He was cited by the Inquisition, and on appealing to Rome he was committed to the care of a minder, who eventually went mad and fatally stabbed him. Yet Dante placed him in paradise, as one true to his mission, and the idea of considering the probability of controversial philosophical propositions was widespread by the first half of the fourteenth century.

It seems even that the idea of defending propositions as merely probable was not entirely ineffective, politically speaking. John Pecham, later archbishop of Canterbury, says, when criticizing a heretical opinion, "No one but a pernicious heretic will assent to it; he will defend it as probable unless he is insane and totally ignorant of sacred and human learning." Since Pecham issued an important condemnation of dangerous philosophical opinions current at Oxford, his opinion on the matter is significant.[29]

But the most influential condemnation of Averroist opinions was that issued by the bishop of Paris in 1277. The disquiet of the ecclesiastical authorities was well founded. The extraordinary list of 219 condemned propositions reveals a tendency to an extreme Enlightenment rationalism. Some of the propositions are directly irreligious ("That happiness is had in this life and not in another"; "That one should not confess except for the sake of appearance"; "That the only wise men in the world are the philosophers"; "That there are fables and falsehoods in the Christian law just as in others"). Some assert a logical conflict between faith and reason ("That creation is not possible, even though the contrary must be held according to the faith"; "That the separated soul is not alterable, according to philosophy, although according to the faith it is altered"). Others assert that human fate is strictly determined by biology and the stars. On principles of knowledge, there are both antiauthoritarianism ("That man should not be content with authority to have certitude about any question") and a strict deductivism ("That one should not hold anything unless it is self-evident or can be manifested from self-evident principles").[30] The worldview implicit in these propositions is of a consistent and strict necessity, with no room for divine intervention or free will or for consideration of authorities. Nor is there any need for reasoning less logically rigid than strict demonstration.

The condemnation succeeded in suppressing extreme Averroism, but the logical issues remained problems for more orthodox thinkers. Though a few of Aquinas's views were implicated in the condemnation of 1277, they were specially excepted and declared permissible again in 1325.[31] His difficulties in effecting a synthesis of faith and reason lay rather more on the side of reason. Though Christianity did reach an accommodation with Aristotle, the obstacles to doing so were prima facie more serious than for the other religions of the book. The Jewish and Islamic God, "the Desired, the Existent, the Sole, the Supreme," was already a somewhat remote and almost philosophic figure and tended to create philosophical problems for his followers only in such marginal areas as the creation of the world. The Christian God did much more extraordinary things, such as walking around a Roman province at a not too distant time in history.

It was difficult for an Aristotelian like Aquinas to admit the creation of the world, or indeed any miraculous interventions in the course of nature at all, since Aristotle certainly leaves the impression that he thought that the causal actions of things follow from their natures and could not be otherwise. It is bad enough for God to give things, at least sometimes, actions that do not follow from their natures—for example, healing powers to relics.[32] But there are much more essential Aristotelian doctrines than that. Surely there can be no exceptions to the basic metaphysical structure of the world? For accidents to exist without a subject (for example, color without a thing that is colored) is surely impossible— a kind of category mistake, like saying, "Today is crystalline." Aquinas quotes Aristotle as saying that "the being of an accident is by inherence."[33] Or as one of the propositions condemned in 1277 has it, "to make an accident exist without a subject has the nature of an impossibility implying contradiction." But that is exactly what Aquinas admits does happen in the case of the Eucharist. Rejecting the suggestion that the whiteness seen in the host really attaches to the body of Christ, and Abelard's theory that it inheres in the surrounding air, Aquinas holds that the color itself is miraculously preserved in existence without it being in any substance, that is, without its being the color of anything. Further, "a more probable opinion" on why the host, if left, generates worms is that by divine power the accidents continue to act as they normally do.[34] His excuse for this denial of a fundamental Aristotelian tenet makes use of a phrase that remains part of the vocabulary of science today, "law of nature": "Nothing prevents something being ordained according to the ordinary law of nature (communem legem naturae) the contrary of which being however ordained according to a special privilege of grace, as is clear with the resurrection of the dead and the giving of sight to the blind. In human affairs too certain concessions are granted to some beyond the ordinary law, by special privilege."[35]

The idea of Marius Victorinus and John of Salisbury, that what can have exceptions by miracle cannot be really necessary, has flowered here into a new conception of nature. Though it has some antecedents in ancient Jewish emphasis on God's power,[36] it is a fundamentally medieval idea that the necessities of nature are just God's contingent laws, to which he may easily make exceptions. (The plural is found once in an obscure corner of Aquinas's works: "Those mutual inclinations of things to their proper ends, which we call natural laws." It is in Augustine and is common in twelfth-century writers and Roger Bacon.)[37] This conception of natural laws hardly changed from the thirteenth century to the time of the early Royal Society, when the phrase was again used to suggest a harmonious relationship between religion and science.[38]

It is clear that Aquinas's carefully woven synthesis of faith and reason was in danger of unraveling if anyone should care to pull the thread at this point. The most energetic in doing so was William of Ockham, in the early fourteenth century. His characteristic saying that God could create the sight of a star without creating the star itself, since "anything is to be attributed to the divine power, when it does not contain a manifest contradiction," is no more than Aquinas was constrained to admit.[39] But his insistence on it sets the tone of the philosophy of his century. If the truths of philosophy are subject to more or less random exceptions by the will of God, one would be better advised to study theology instead, to find out what God's will is.

Another nexus Ockham was concerned to break was that between ecclesiastical and temporal power. His arguments make use of the lawyers' concepts of probability and presumptions. The antipapal *Dialogue on the Authority and Power of the Popes* has much discussion of heresy, of popes and others. Ockham distinguishes between inner, or mental, and outer pertinacity, that is, persistence in heresy: "We can only be convinced of the first [inner] pertinacity through a probable or violent presumption. For in such matters outer and inner do not necessarily correspond; often one thing is held inwardly and another outwardly."[40] The same language of presumptions is used of doubts about the standing of church councils. "If something is erroneously defined by a general council or by an assembly that is regarded by the multitude of the faithful as a general council, anyone who does not know that the general council or such assembly errs ought to presume in favor of such council or assembly—not, however, with a presumption so violent that no proof to the contrary may be admitted: just as a presumption must be made in favour of a judge's sentence, even if in truth it was unjust and wicked, until the contrary is proved or becomes certain."[41]

Probability is sufficient to justify drastic action, "and so where the Catholic faith would be saved by the mere chaining and captivity of a heretical pope the laity should not proceed to further punishment. But if danger to the faith were feared with probability from a heretical pope merely detained in captivity, and the faith would be saved by his death, the laity could proceed to the bodily death of the heretical pope, with zeal for the orthodox faith."[42] (By laity Ockham means emperor. His employer, the emperor Louis IV, seized Rome in 1328, declared the pope deposed for heresy and induced the "Roman people" to elect a new one.)[43]

Ockham is now best known for the principle of economy in reasoning known as Ockham's Razor. This is a misnomer for two reasons. First, he did not originate it; there is the inevitable origin in Aristotle. The phrase "We consider it a good principle to explain the phenomena by the simplest hypotheses possible" is in Ptolemy. Formulations like "It is vain to do with more what can be done with fewer" and "A plurality is not to be posited without necessity" are Scholastic commonplaces from the early thirteenth century.[44] Second, though he does repeat these ideas frequently, Ockham's own contribution is more to restrict the operation of the principle, in the interests of God's absolute power: "God does many things by means of more which He could have done by means of fewer, simply because He wishes it, and no other cause is to be sought. From the very fact that He wishes it, it is done suitably, and not in vain." In the Eucharist, especially, Ockham holds that a plurality of miracles is to be postulated, simply because that pleases God.[45]

In Ockham's lifetime, criticism of views on causal connection by the Inquisition became more intense. The determination of fate by biology and the stars, as condemned in 1277, plainly contradicts miraculous intervention in nature by God and, indeed, micromiraculous intervention, so to speak, by human free will. The first university scholar burned by the Inquisition was Cecco d'Ascoli, in 1327, found guilty of relapse into heresy. The details of his opinions are far from clear but appear to include asserting that the events of Christ's life follow necessarily from his horoscope—though at one point he agrees with Aquinas's opinion that the zodiac "is not the cause of our will or intellect except as a tendency (*nisi dispositive*)."[46]

Cecco, as a believer in necessity, no doubt expected the next flame to be hot. The peasants of Montaillou, on the other hand, were familiar with the promise of the heresiarchs that God would take upon himself the pain of the Inquisition's fire.[47]

The Reasonableness of Christianity

Religious controversy in the post-Reformation period was the setting for many of the applications of probability in the two countries, England and Holland, that attempted a moderate course in matters of religion.

The Church of England, as constituted by the Elizabethan Settlement of 1559, was hardly at first in a position to trade arguments with anyone. Only in England was it possible to decide theology by majority vote in Parliament: the abolition of the mass passed the House of Lords by just three votes; it was voted against by all the bishops and spoken against eloquently by the only English archbishop, who then ended his days under house arrest. The compromise was imposed by the new queen and supported by the exhausted English people, but there is some doubt whether it was at first actually believed in by anybody—not by all, or most, or the wise, certainly; not by the old leaders of the church, and not by the new ones either. The Calvinist divines, returning from the Continent to enter into their kingdom, always remained a minority. "Walk with the few godly, in the Scriptures narrow path to heaven but crowd not with the godlesse multitude, in the broad way to hell."[48] The fewness of the Puritans was sometimes taken by them as evidence of their rightness, considering the advantages enjoyed by their enemies: "Were it not for God's marvellous blessing on our studies, and the infinite odds of truth on our sides, it were impossible, in human probability, that we should hold up the bucklers against them."[49] As a result of their excess of zeal but lack of numbers, they were allowed to have their way with the intransigence of the Thirty-nine Articles but had to accept the queen's compromises elsewhere. The new church's doctrines were the average of the people's beliefs more than the beliefs of the average person.[50]

These sordid beginnings had receded sufficiently from memory by the end of the century. The rising confidence of the Anglican Church is reflected in the measured prose of Richard Hooker, John Donne, and the Authorized Version of the Bible. It was at this period that English thought began to take on its typical and much celebrated cast of reasonableness, compromise, and practicality. Seventeenth- and eighteenth-century English thought takes many of its assumptions from a stream of works on the "reasonableness of Christianity," a topic central to the concerns of the most influential writers. If the causal origins of this temper of mind are to be sought in the Elizabethan Settlement and the longevity of that sovereign, its intellectual origins are to be found in a single book, Hooker's *Laws of Ecclesiastical Polity* of 1594.[51]

What was especially important for later developments was Hooker's concentration on questions of the sources of knowledge. Naturally, this was an especially difficult problem for the Anglican Church, which was searching for a middle way between the authority of Scripture alone and that of authoritative Roman tradition. Hooker's solution was to rely on reasonableness and tradition except when modification was manifestly necessary. "The ground of credit," Hooker writes, "is the credibility of things credited; and things are made credible, either by the known condition and quality of the utterer, or by the manifest likelihood of truth which they have in themselves."[52] The relationship to later English conservative thought is clear. Clear too is the relationship to the past, since reasonableness and tradition are the same as the Scholastics' intrinsic and extrinsic probability. That Hooker was heavily influenced by Thomist thought is well known, though it remains unclear which Thomists he read.[53] Hooker classifies the grades of assurance as the Scholastics did and insists many times that assent should be proportioned to evidence. "Persuasions grounded upon reason are either weaker or stronger according to the force of those reasons whereupon the same are grounded."[54]

He goes on to say that "the greatest assurance generally with all men is that which we have by plain aspect and intuitive beholding. Where we cannot attain unto this, there what appeareth to be true by strong and invincible demonstration, such as wherein it is not by anyway possible to be deceived, thereunto the mind doth necessarily assent, neither is it in the choice thereof to do otherwise. And in case these both do fail, then which way greatest probability leadeth, thither the mind doth evermore incline. . . . Now it is not required or can be exacted at our hands, that we should yield unto any thing other assent, than such as doth answer the evidence which is to be had of that we assent to." At this point Hooker very cunningly suggests that the Calvinists' fervor stems from their mistake on this point and that their zeal is a cover for their lack of evidence. "When bare and unbuilded conclusions are put into their minds, they finding not themselves to have thereof any great certainty, imagine that this proceedeth only from lack of faith, and that the spirit of God doth not work in them, as it doth in true believers; by this means their hearts are much troubled, they fall into anguish and perplexity: whereas the truth is, that how bold and confident soever we may be in words, when it cometh to the point of trial, such as the evidence is which the truth hath either in it self or through proof, such is the heart's assent thereunto, neither can it be stronger, being grounded as it should be."[55] These thoughts on evidence are the basis of his conservative conclusions: "Is it meet that when publicly things are received, and have taken place, general obedience thereunto should cease to be exacted, in case this or that private person led with some probable conceit, should make open protestation?"[56] Like the textual critics and historians discussed in chapter 7, he realized that witnesses could be dependent on each other, so that many witnesses might only be worth the testimony of one.[57] His point is again a conservative one: the fact that many reformed churches have adopted the practices of Geneva is not a good reason for the Church of England to do likewise.

One of the most interesting of the late Scholastic writers is the Jesuit Leonard de Leys, or Lessius. He was best known for his activities in the controversies on grace, in which he defended the comparatively lax position of predestination consecutive to the divine prevision of merits. His books were condemned at Louvain in 1587, but the condemnation was quashed, and later Jansenist efforts to have his views condemned came to nothing. Among other works of religious controversy, Lessius wrote *On the Providence of the Deity, and the Immortality of the Soul, Against Atheists and Politicians*, which, together with a similar book by Grotius, is the origin of the long English tradition of natural theology. Lessius relies on the Stoic arguments for the existence of God reported by Cicero, especially the argument from the consensus of peoples and that from the regularity of the heavenly motions.

Lessius's version of the design argument actually depends on the division of the world into sublunary and superlunary parts (and so would be wrong if Copernicus were right). But in contrast to the reaction of some Scholastics to the telescope, he was excited by the discoveries made by the "dioptric fistula, recently invented by someone in Holland," which show that the mechanism of the heavens is even more complex than previously thought. His argument is expressed, in the manner of Aristotle and Maimonides, as the ruling out of the hypothesis of chance. "Who of sound mind would think, that those things established with the highest reason, and exceeding the devices of all the arts, are produced without reason, without art, but only by the motion, and chance concourse of atoms? It would be as if you said that a most beautiful palace were not constructed by the high art of some artificer, but came to be by chance, when, an earthquake disturbing a cliff, the fragments of rock fell by chance arranged in this form; or the annals of Ennius or the commentaries of Livy were not devised by those writers, but were achieved by a chance concourse of letters. For if the parts of the world, and the disposition of the parts, and the bodies of animals and plants. . . ."[58]

Lessius's work was translated into English in 1631 under the odd title *Rawleigh his Ghost*. In an added preface Sir Walter Raleigh's ghost appears and begs that the book be translated. The choice of ghost is due to a suspicion of atheism that attached to Raleigh in life but from which he would of course have been converted after death.[59] (Lessius's opinions on commercial risks are discussed in the next chapter.)

Grotius's work *On the Law of War and Peace* is examined in chapter 4. In his own time, Grotius was equally well known for his *Truth of the Christian Religion*.[60] This book enjoyed enormous success; it was the subject of a number of books in England and was one of the very few works of apologetic used by Pascal in his *Pensées*.[61] Grotius presents Christianity as a historical thesis, for which it is appropriate to consider arguments and testimonies. With a reference to Aristotle's *Ethics*, he says that one ought to be satisfied in matters of religious truth with "testimonies void of all suspicion," though these cannot attain the certainty of mathematical proof. It is indeed God's will that revelation should not be as clearly true as sense knowledge and demonstration, "so that the Gospel is, as it were, a touchstone to try men's honest dispositions by."[62]

Grotius's chapter on the design argument for God's governance takes up the ideas of Lessius on natural theology. Besides containing brief accounts of arguments from purposiveness in biology, originally put forward by Galen, it develops the argument from the order of the heavens expressed as the ruling out of the hypothesis of chance. "Moreover, the most *perfect forme* and *figure* of this world, to wit its *roundnesse*, as also the parts thereof, shut up as it were in the bosome of the heavens, and disposed with a marvellous order, doe all expressly declare, that they were not tumbled together, or conjoyned as they are, *by chance*, but wisely ordained by *such an understanding* as is endued with super-eminent excellencie. For what Ninny is there so sottish, what man so foolish, as to imagine, that so compleat and exact a worke as this, came to passe *by hap-hazard?* He might as wel think that the stones and trees of any building were united into the forme of some house *by chance*, or scattered syllables and words became a Poem by *meere fortune:* A thing so unlikely, that even a few *Geometricall figures* espied on the *Sea-shore*, gave the beholder just ground to argue, that some *man* had beene there; it being evident enough that such things could not proceed from *meere chance.*"[63]

Grotius's ideas found their most influential support in England, at least in the longer term. A circle of Anglicans, meeting at Great Tew in the 1630s, maintained the traditions of Hooker and Grotius of moderate skepticism, toleration, and political conservatism. Though dispersed during the troubled times of the Civil War, their surviving members formed the thought of the Restoration church.[64] The central writer of the circle was William Chillingworth, whose *The Religion of Protestants a Safe Way to Salvation*, of 1638, is one of the most acute works of Catholic-Anglican controversy. Chillingworth had been converted briefly to Catholicism and had studied at the Jesuit College at Douai before returning to Oxford and preferment in the Anglican Church. He had been led to Rome in the first instance because he was convinced of the need for an infallible judge in matters of faith.[65] His book, though of course largely taken up with particular points of doctrine, is at bottom an attack on the notion of infallibility, based on considerations about the logic of certainty and probability. The direction of his thought is indicated by his objections to probabilism in moral theology; his treatment is brief and standard, except to suggest the Jesuits will use it to justify killing the king and blowing up parliament. But the example he gives of the laxity to which it leads is unusual in that it is itself about belief: "For upon this ground I knew a young scholar in Doway, licensed by a great casuist to swear a thing as upon his certain knowledge, whereof he had yet no knowledge, but only a great presumption, because (forsooth) it was the opinion of one doctor that he might do so."[66] Chillingworth argues, like Hooker, that the correct degree of belief in a proposition is that proportional to the reasons for it and that God cannot require more:

> God desires only that we believe the conclusion, as much as the premises deserve, that the strength of our faith be equal or proportionable to the credibility of the motives to it. Now, though I have and ought to have, an absolute certainty of this thesis—all which God reveals for truth, is true—being a proposition, that may be demonstrated, or rather so evident to one that understands it, that it needs it not; yet of this hypothesis—that all the articles of our faith were revealed by God—we cannot ordinarily have any rational and acquired certainty, more than moral, founded upon these considerations: first, that the goodness of the precepts of christianity, and the greatness of the promises of it, show it, of all other religions, most likely to come from the Fountain of goodness. And then, that a constant, famous, and very general tradition, so credible, that no wise man doubts of any other which hath but the fortieth part of the credibility of this; such and so credible a tradition tells us, that God himself hath set his hand and seal to the truth of this doctrine, by doing great, and glorious, and frequent miracles in confirmation of it. Now our faith is an assent to this conclusion, that the doctrine

of christianity is true; which being deduced from the former thesis, which is metaphysically certain, and from the former hypothesis, whereof we can have but a moral certainty, we cannot possibly by natural means be more certain of it than of the weaker of the premises; as a river will not rise higher than the fountain from which it flows. For the conclusion always follows the worser part, if there be any worse; and must be negative, particular, contingent, or but morally certain, if any of the propositions, from whence it is derived, be so.[67]

God is not even morally permitted to require certainty of belief where it is not to be had. "If I should send a servant to Paris, or Rome, or Jerusalem, and he using his utmost diligence not to mistake his way, yet notwithstanding, meeting often with such places where the road is divided into several ways, whereof every one is as likely to be true, and as likely to be false, as any other, should at length mistake, and go out of the way, would not any man say, that I were an impotent, foolish, and unjust master, if I should be offended with him for so doing? and shall we not tremble to impute that to God, which we would take in foul scorn if it were imputed to ourselves? Certainly, I, for my part, fear I should not love God, if I should think so strangely of him."[68]

He argues too that the Catholic doctrine of infallibility is logically incoherent, since, being not fully evident from Scripture, it needs to be supported by reasoning drawn from Scripture and tradition, which is not itself certain. Hence to demand certain faith in the infallibility of the Catholic Church is to "build an infallible faith upon motives that are only highly credible, and not infallible, as it were a great and heavy building upon a foundation that hath not strength proportionable."[69] The Jesuits argued that uncertainty could not be admitted into religion anywhere without everything becoming uncertain. This is a mistake in logic too, according to Chillingworth: "A traveller is not always certain of his way, but often mistaken; and doth it therefore follow that he can have no assurance that Charing-cross is his right way from the Temple to Whitehall?"[70] He argues that Catholics in any case make salvation rest on fallible grounds, since they require that one receive the sacraments from priests, who are real priests only if descended by a chain of valid ordinations from the apostles. That this is so is uncertain; Chillingworth employs some sophisticated probabilistic argument:

In fine, to know this one thing you must first know ten thousand others, whereof not any one is a thing that can be known, there being no necessity that it should be true, which only can qualify any thing for an object of science, but only, at the best, a high degree of probability that it is so. But then, that of ten thousand probables, no one should be false; that of ten thousand requisites, whereof any one may fail, not one should be wanting, this to me is extremely improbable, and even cousin-german to impossible. So that the assurance hereof is like a machine composed of an innumerable multitude of pieces, of which it is strangely unlikely but some will be out of order; and yet if any one be so, the whole fabric of necessity falls to the ground: and he that shall put them together, and maturely consider all the possible ways of lapsing, and nullifying a priesthood in the church of Rome, I believe will be very inclinable to think, that is a hundred to one, that amongst a hundred seeming priests, there is not one true one: nay, that it is not a thing very improbable, that amongst those many millions, which make up the Romish hierarchy, there are not twenty true.[71]

In the course of exposing the contradiction between Catholic demands for infallibility in one place with its reliance on probabilities when convenient, Chillingworth makes an interesting use of numerical odds:

You . . . ask, "shall I hazard my soul on probabilities, or even wagers?" As if whatsoever is but probable, though in the highest degree of probability, were as likely to be false as true! Or, because it is but morally, not mathematically, certain, that there was such a woman as Queen Elizabeth, such a man as Henry VIII, that is, in the highest degree probable, therefore it were an even wager there were none such! By this reason, seeing the truth of your whole religion depends finally upon prudential motives, which you do but pretend to be very credible, it will be an even wager that your religion is false. And, by the same reason, or rather infinitely greater, seeing it is impossible for any man (according to the grounds of your religion) to know himself, much less another, to be a true pope, or a true priest; nay, to have a moral certainty of it; because these things are obnoxious to innumerable secret and undiscernible nullities, it will be an even wager, nay, (if we proportion things differently), a hundred to one, that every consecration and absolution of yours is void, . . . particularly, it will be at least an even wager, that all the decrees of the council of Trent are void, because it is at most but very probable that the pope which confirmed them was the true pope. If you mislike these inferences, then confess you have injured Dr. Potter in this also, that you have confounded, and made all one, probabilities, and even wagers. Whereas any ordinary gamester can inform you, that though it be a thousand to one that such a thing will happen, yet it is not sure, but very probable.[72]

Chillingworth's skill in argument caused some apprehension even among his own party, and was admired by Locke.[73] (The subsequent great success of these ideas in such authors as Wilkins, Tillotson, and Locke, and their connection with the thinking of the early Royal Society, take us beyond the period covered in this book—and have in any case been well studied.)[74]

On the Catholic side, some of the same ideas appeared in, for example, Kenelm Digby's *A Conference with a Lady about Choice of Religion*, of 1638. He argues that the Catholic faith was "more certaine and more infallible than any naturall science whatsoever," since the Catholic Church could not introduce error without a whole generation conspiring to introduce it, which would not happen.[75] On the whole, however, Catholic writers kept strictly to the doctrine of infallibility: Scripture alone is not a sufficient guide, since one needs an infallible judge to decide which books of Scripture are canonical and to decide on disputed interpretations.[76] This was part of the Catholic Church's "retreat to certainty," itself part of the church's anti-intellectual turn after Galileo. With the condemnation of Galileo, the church lost the control of the intellectual high ground it had held in the West for over a thousand years. There ensued a general inward-looking tendency, a concentration on the study and codification of dogma, and an opposition to intellectual trends in "the world." Until comparatively recent times, in some ways even up to the Second Vatican Council of the 1960s, intellectual innovations were often treated with suspicion rather than interest in official Catholic circles; new books read for their potential for inclusion on the Index of Forbidden Books more than for enlightenment. Thus Cano's beginning in studying the principles of historical evidence as applied to theology was pursued not by those of his own faith but by Anglican thinkers—a fact noted at the time by John Tillotson, archbishop of Canterbury and early member of the Royal Society, who approved Cano for being an exception to the Roman demand for infallibility.[77]

Pascal's Wager

Arguments with a broad similarity to Pascal's wager are traditional in both popular and learned religion.[78] No doubt the "not much to lose" or "insurance policy" aspect to religion has always been part of its appeal as well as a source of the lukewarmness in faith that so exasperates the zealous. The Christian apologist Arnobius, of about 300 A.D., was probably the first to express the argument in a form that recognizably involves the rationality of a decision in a case of doubt. He writes that "there can be no proof of things still in the future. Since, then, the nature of the future is such that it cannot be grasped and comprehended by any anticipation, is it not more rational, of two things uncertain and hanging in doubtful suspense, rather to believe that which carries with it some hopes, than that which brings none at all? For in the one case there is no danger, if that which is said to be at hand should prove vain and groundless; in the other there is the greatest loss, even the loss of salvation, if, when the time has come, it be shown that there was nothing false in what was declared."[79] The true introduction of probability considerations did not come, however, until the seventeenth century. Chillingworth, in the course of arguing that fallible beliefs are sufficient for action, gives the essentials of Pascal's wager in considering the balance of an infinite future happiness and the present inconveniences necessary for attaining it:

> For who sees not that many millions in the world forego many times their present ease and pleasure, undergo great and toilsome labours, encounter great difficulties, adventure upon great dangers, and all this not upon any certain expectation, but upon a probable hope of some future gain and commodity, and that not infinite and eternal, but finite and temporal? Who sees not that many men abstain from many things they exceedingly desire, not upon any certain assurance, but a probable fear, of danger that may come after? What man ever was there so madly in love with a present penny, but that he would willingly spend it upon any little hope, that by doing so he might gain a hundred thousand pounds? And I would fain know, what gay probabilities you could devise to dissuade him from this resolution. And if you can devise none, what reason then or sense is there, but that a probable hope of infinite and eternal happiness, provided for all those that obey Christ Jesus, and much more a firm faith, though not so certain, in some sort, as sense or science, may be able to sway our will to obedience, and encounter with all those temptations which flesh and blood can suggest to avert us from it? Men may talk their pleasure of an absolute and most infallible certainty, but did they generally believe that obedience to Christ were the only way to present and eternal felicity, but as firmly and undoubtedly as that there is such a city as Constantinople, nay, but as much as Caesar's Commentaries, or the History of Sallust; I believe the lives of most men, both papists and protestants, would be better than they are.[80]

In France, the "reasonableness of Christianity" project was sponsored by Cardinal Richelieu as a counterweight to Jansenist enthusiasm. The cardinal well expresses the attitude of officialdom to the conflict of ideas in his remark, "If Luther and Calvin had been locked up when they began to dogmatize, the State would have been spared a great deal of trouble."[81] The man called forth by the cardinal's subsidy was Silhon (whose work justifying the cardinal's notion of "reason of state" is discussed in chapter 4). There is a considerable conceptual affinity, indeed, between reason of state and Pascal's wager in that both involve action on opinions whose low probability is compensated for by a high payoff; deceit for reason of state is justified in terms of the allegedly overriding consideration of the safety of the realm, while acting as if one believed in God is recommended by the wager on account of the infinite gain to be expected, and loss to be avoided, if religion is true.

The work that first called Silhon to Richelieu's attention was his *On the Immortality of the Soul*, of 1634. He draws the usual distinctions between demonstration and moral certainty and treats most questions of controversy as involving the latter.[82] He unusually adds a brief decision-theoretic argument of the kind found in the wager: when the assertions of the existence of God and its opposite are "equally doubtful and equally ambiguous," one should make the pious choice. "In that choice there is no risk to run, nor anything to fear if there is no God and the human soul is mortal, while one hazards much on the opposite choice, and exposes oneself to a final evil and a just punishment if it is true that there is a God."[83]

The full version of the wager, including explicit discussion and balancing of risks and rewards, is first found in the 1637 work *On the Immortality of the Soul* by the Jesuit Antoine Sirmond.

> Without favouring one party or the other, let us suppose there were nothing decided in the matter, and that it were problematic and equally doubtful on either side. . . . There is no man of good sense who would not rather lose a day or an hour of his pleasures than risk an eternity of happiness, or who would not choose to endure in the present a pinprick for a quarter of an hour, rather than put himself in danger of a torment which would have neither moderation of its rigour, nor limit to its duration. Compare the goods of this life with those to be feared or hoped for in the next, if there is a next life, and you will find there is no more proportion between the terms of this comparison than there is between the stakes and the rewards of this choice. What will happen to the man of vice? He prefers to play for the present than to attend to the future. And if the future deals with him otherwise than he thinks, if his soul finds itself taken on leaving the body, and finds itself existing in the midst of the sufferings it will have deserved, what will be his condition? Truly to have inherited an eternal evil for a moment of the pleasures of

which this life has delivered the enjoyment; to have lost eternal happiness that could have been bought at the price of a little pain in the exercises of several virtues contrary to his humour. Will he not reason, and tell himself thereafter: if I die completely when I quit this earthly coil, my lot will be to have avoided the evils of this life, and to have embraced its goods, as far as I could. In addition, I have naught either to fear or to hope, beyond the experience of sixty or a hundred years or so, that will be the most that will roll past me. If, on the contrary, I were to find after death a land where one lives longer than one does here, I would see myself condemned to torments intolerable in their gravity and infinite in duration. I would feel myself excluded from a state happy in proportion, full of all sorts of goods, and assured for eternity. What then is to be done?[84]

Sirmond considers the objection that what is in the present is certain, while the supposed future life is only a prospect. He replies: "It is true that certainty in the present is worth more than uncertainty in the future, as long as there is some proportion between the two. But when it is a matter of eternal life or death, how can there be any comparison with a temporal life or death?" There is a Latin edition of Sirmond's book that generally agrees with the French edition but goes further in comparing the choice between religion and "the world" to a game of chance. "No-one is a man of truly sound mind, who with a not unequal partner, wants to play dice or ball or any kind of game, such that if he wins, he gains a penny, while if not, he loses a most flourishing and opulent and everlasting kingdom. . . . However long and happy the space of this life may be, while ever you place it in the other pan of the balance against a blessed and flourishing eternity, surely it will seem to you, if the weight of things be known, that the pan will rise on high even as if you were to weigh a penny against the weight of gold of the most splendid kingdom."[85]

It is not known for certain whether Pascal read Sirmond's book, though the many close parallels suggest that he did. Pascal must have read another book of Sirmond's, since it is the focus of his attack in the tenth of the *Provincial Letters*. Pascal's indignation with the Jesuits here, over the question of whether salvation is possible without genuine love of God, reaches fever pitch. Pascal, taking the command of Jesus, "Love God," to be about a genuine conversion of the heart, is scandalized by the Jesuits' suggestion that what counts is actions: provided one behaves well, *as if* one loved God, one may be saved. The most extreme example of the Jesuit position is from Sirmond's *Défense de la Vertu*, in which Sirmond maintains the adequacy of "effective" love (that is, action in accordance with charity) in the possible complete absence of "affective" love (that is, genuine feeling).[86] The dispute is close to the issues of Pascal's wager, since much of the indignation directed by later writers at the wager has rested on the premise that Pascal himself advocates a cynical following of rites to attain heaven. His rage at Sirmond will surely suggest that this is a misinterpretation of his intentions; as indeed it is.

After his work on the *Provincial Letters*, Pascal returned briefly to mathematics, discovering some noted results on the cycloid. This was apparently a propaganda exercise, intended to show that some Jansenists were still capable of "hard" thinking and so designed to impress gentlemen of that society with which Jansenism had such an ambiguous relationship, "the world." "People would see that, though Pascal demanded that reason should abase itself before the mysteries of faith, this was because he had an exact knowledge of the scope and limits of reason."[87]

The few remaining years of Pascal's life, until his death in 1662 at the age of thirty-nine, were mostly occupied with prayer, illness, and the composition of a major work on the evidences for Christianity. It was never finished, but the extensive notes for it appeared eight years after his death under the title *Pensées*. Undoubtedly its unfinished state has contributed to its classic status. Just as one admires the artist's sketches for many seventeenth-century paintings more than the paintings themselves, so one feels close to Pascal's thought, before it is smoothed over with unnecessary dressings of rhetoric. One is also at liberty to imagine that some of the sillier pieces would have been excised before publication. In general, the *Pensées* is a work of Augustinian Christianity, with that tradition's insistence on the fallen state of man and the consequent need to trust completely in God, to submit one's will and intellect to him. This is the road to happiness and the only road to certainty. "Submission is the use of reason in which consists true Christianity"; "The heart has its reasons, which reason does not know."[88] Pascal differs from most Augustinian writers, though perhaps not from Augustine himself, in emphasizing not only the fallen state of man but also his greatness.[89]

On the matter of evidence for the Christian religion, Pascal seems to have thought that, uncertainty being a condition of fallen nature, the truth in religion ought to be blindingly obvious but that in all honesty it could not be claimed that it was: "I have a hundred times wished that if a God maintains nature, she would testify to him unequivocally, and that, if the signs she gives are deceptive, she should suppress them altogether; that she would say everything or nothing."[90] A good deal of the strictly intellectual effort of the *Pensées* is devoted to circumventing this problem. On God's part, Pascal argues, "it was therefore not right that he should appear in a manner manifestly divine and absolutely capable of convincing all men, but neither was it right that his coming should be so hidden that he could not be recognised by those who sincerely sought Him."[91]

Going even further on how the balance of the evidence lies, Pascal unpleasantly suggests there is not enough evidence to convince but enough to render culpable anyone not convinced: "But the evidence is such as to exceed, or at least equal, the evidence to the contrary, so that it cannot be reason that decides us against following it, and can therefore only be concupiscence and wickedness of heart. Thus, there is enough evidence to condemn and not enough to convince, so that it should be apparent that those who follow it are prompted to do so by grace and not by reason, and those who evade it are prompted by concupiscence and not by reason."[92] The idea that the will supplies decision where certainty is not available has here got out of hand. Where some would draw the conclusion that one ought therefore to study the probability of the evidence more carefully, Pascal concludes that there is certainty for those who seek sincerely.

Pascal the controversialist comes to the fore in arguing why men in "the world," lacking clear evidence for God, should bother to look for him sincerely. The arguments are many; Pascal is very able at getting under the skin of his opponent: "Atheism shows strength of mind, but only to a certain degree." But most of the arguments take their force from the certainty of death. "The last act is tragic, however happy all the rest of the play." Pascal is masterful in joining this with the folly of not examining religion because its claims are uncertain: "An heir finds the deeds of his house. Will he say, perhaps, that they are false, and not bother to examine them?"[93] It is almost possible to imagine the *Pensées* changing the way someone thinks, a possibility hardly to be envisaged with the worthy but pedestrian approach of Grotius and the defenders of the "reasonableness of Christianity." These thoughts lead up to the famous Pascal's wager:

> "Either God is or he is not." But to which view shall we be inclined? Reason cannot decide this question. Infinite chaos separates us. At the far end of this infinite distance a coin is being spun which will come down heads or tails. How will you wager? . . . You must wager. There is no choice. . . . Let us weigh up the gain and the loss involved in call-

TABLE 9.1. PASCAL'S PAYOFFS

	Decision	
State of Nature	Pray	Don't Pray
God exists	Infinite	Negative infinite
God does not exist	Small loss of pleasure	Small pleasure
Expectation	Infinite	Negative infinite

ing heads that God is. Let us assess the two cases: if you win you win everything, if you lose you lose nothing. Do not hesitate then; wager that he does exist. "That is wonderful. Yes, I must wager, but perhaps I am wagering too much." Let us see: since there is an equal chance of gain and loss, if you stood to win only two lives for one you could still wager, but supposing you stood to win three?

You would have to play (since you must necessarily play) and it would be unwise of you, once you are obliged to play, not to risk your life in order to win three lives at a game in which there is an equal chance of losing and winning. But there is an eternity of life and happiness. That being so, even though there were an infinite number of chances, of which only one were in your favour, you would still be right to wager one in order to win two; and you would be acting wrongly, being obliged to play, in refusing to stake one life against three in a game, where out of an infinite number of chances there is one in your favour, if there were an infinity of infinitely happy life to be won. But here there is an infinity of infinitely happy life to be won, one chance of winning against a finite number of chances of losing, and what you are staking is finite.[94]

This is an argument in what is now called decision theory in which one evaluates the preferability of decisions by looking at their payoffs under various possible "states of nature," the states being weighted by their probabilities of being true. It is clear that Pascal realizes that he has discovered such a science, since he concludes with the phrase: "This is conclusive (*démonstratif*) and if men are capable of any truth this is it."

Generally, decision theory advises one to choose the decision with the highest expectation, that is, the highest product of probability and corresponding payoff. Pascal is right in saying that if the probabilities of the two states of nature, "God exists" and "God does not exist," are nonzero, then the expectations of the decisions "Pray" and "Don't pray" are respectively infinite and negative infinite (see table 9.1).

Pascal recognizes of course that acceptance of the wager does not produce genuine belief, which, according to him, is necessary for salvation. He advises "acting as if they believed, taking the holy water, having masses said, etc." This has led men to belief, he says; he means that God gives belief to genuine seekers.

There are certainly difficulties with this argument: it seems to apply to all religions equally, even ones that no one believes in. As Diderot rightly says, "An imam could reason just as well this way."[95] On the other hand, there is material relevant to ruling out other religions elsewhere in the *Pensées,* and the potential convert whom Pascal is addressing no doubt regards himself as having narrowed the possibilities to the two Pascal gives. As it stands, the argument does depend on the premise, accepted by many at the time besides Pascal (though denied by some lax Jesuits), that there is no salvation without faith in the true religion. The modern world has attached a higher degree of belief to the proposition that, if God exists, "He's a good fellow, and 'twill all be well." But perhaps the truth of this would make no difference to the course of action Pascal recommends; his argument uses a carrot and stick, but the carrot seems sufficient: if the negative infinity in the table is replaced by something finite, the decision "Pray" is still preferable.

It is notable that, even at this late date, and in a context so much to do with belief in opinions (and certainly nothing to do with long-run frequency), Pascal speaks only of chances, not probability. The word *probability* is reserved in the *Pensées* for more thoughts on the Jesuits. Naturally, the remarks about probability are universally disparaging: "But is it *probable* that *probability* brings certainty?" "*Probability.* The ardour of the saints in seeking out the truth was pointless if the probable is safe"; "Ridiculous to say that an eternal reward is offered for morals *à la* Escobar."[96]

Perhaps it is not too much to see in Pascal's tutiorism another application of decision theory. Pascal hints as much by saying, "*Probable.* Let us see if we are sincerely seeking God by making some comparisons with things we care about. It is probable that this meat will not poison me. It is probable that I shall not lose my case by not lobbying."[97] That is, when the outcome is important, we try to guard against small probabilities of harm. This is the line of argument that shows what is wrong with the Scholastic and Cartesian attempt to pretend that moral certainty, a noncertainty close enough to certainty for all practical purposes, is a kind of certainty instead of merely a high probability. However high the probability is, short of certainty, there will be some danger bad enough to make us avoid acting as if the fact were certain.

Likewise in the wager: however small the probability of God's existence might be, there is a sufficiently large reward that will make it rational to act on the assumption that he does exist.

Ronald Knox remarks, "What is least forgivable in the Jansenists is that seeing the world as a *massa damnationis*, rushing on to its ruin, they could find no other remedy for its unhappiness but to make war on the Jesuits."[98] Though a little disfigured in this way, the *Pensées* is the great exception to this generalization, as Knox himself says. It is the classic statement of a certain view of humankind, the view we have seen before in the Islamic thinkers and that can be found in St. Paul and St. Augustine. It is a view skeptical of human nature and the powers of reason but insistent on the need for certitude; seeing an infinite gap between humanity's low state and God; requiring submission, of the reason as well as the will. Pascal's expression of this is the classic one as much as anything because he knows the other side; a great controversialist, one of the founders of probability theory, he is the fittest leader of the return to dogma and certainty.

CHAPTER TEN

Aleatory Contracts: Insurance, Annuities, and Bets

The business world has always been one of the most fertile sources of quantifiable concepts. Money is a powerful incentive for reducing quantities to measurement. And once money is invented as a medium of exchange for merchandise, there is pressure to express all values of things in terms of it. As Aristotle says, prices are really a measure of demand.[1] And the range of things that can be in demand is very wide. It includes far more than physical goods, and explaining what these things are, and how to price them, is often a nontrivial intellectual exercise. The present value of future goods, for example, requires a good deal of thinking in order to decide what to calculate. The pricing of intangibles like the "goodwill" of a business is, if not strictly speaking intellectually difficult, awkward to make sense of and in need of expertise to perform. In general, pricing quantities involving an *if*—such as options or goods that will exist or be transferred only on fulfillment of a condition—is difficult because of the uncertainty in the fulfillment of the condition.

A number of the items that naturally arise in business and have to be priced are probabilities or expectations. If a good is difficult to acquire because of a dangerous sea voyage, the dangers of the voyage will require payment to those undertaking them and need to be factored into the price. Bad debts can be sold, at a discount that reflects the prospect of recovering them. The price at which a life annuity should be sold ought to reflect the life expectancy of the purchaser. The premium of an insurance contract is a price for the risk borne. Obviously, it is possible to have a greater or lesser degree of conscious awareness of what is being priced in cases like these. As always, legal disputes will tend to make understanding explicit, as people argue about what gives them rights to different sums.

One feature of prices, which makes them unusual among quantities, is that they are typically *approximate* in principle. Since what is being measured depends on demand—that is, on the partly disparate wishes of a large number of people—there is typically no exact measure of what the price ought to be—even if there were an exact measure of objective facts like scarcity or risk that affect the price.[2] Hence, although the conceptual requirements of business led to the quantification of probabilities, it did not lead to a mathematics of probability in the exact sense permitted by symmetry considerations with dice.

The Price of Peril

The first known kind of contract that appears to contain some element of payment for risk is the ancient Athenian maritime loan. It was the principal type of loan intended for investment, as opposed to loans to friends or to those in financial difficulty. Loans were advanced for individual voyages at a high rate of interest, with repayment only if the ship arrived safely.[3] The maritime loan thus has an effect like insurance, in that the one who sustains the disaster gains a financial advantage in partial compensation, so that the risk of the venture is spread. This was not necessarily understood or intended at the time; a sufficient explanation for the nonrepayment clause in the loan would be that there was no point in trying to recover money from someone who has just lost all of his goods and possibly his life.

The unusually high interest rates on maritime loans suggest a pricing of risk, and there is a case of different rates applying for what seem to be different levels of risk. A loan of about 340 B.C. for a voyage to the Bosphorus was advanced "on interest of 225 drachmae on the thousand; but if they should sail out from Pontus to Hieron after the rising of Arcturus, at 300 on the thousand."[4] The rising of Arcturus is in mid-September, so that it appears that the higher rate is for the danger of storms at the autumn equinox, which means there is a genuine price given to a risk, though that is not explicitly stated. There is, however, a definite concept of the transferal of risk: it was customary to have witnesses to the moment when the ship left port, at which point the risk moved to the lender.[5] There is a hint, too, that those involved in the business have expertise in evaluating the relative risks of different loans. A man who has exhausted all the usual sources of credit applies to the "men in the Piraeus," but they "are in such a state of mind, that it seems much safer to sail to the [notoriously stormy] Adriatic than to lend money to him."[6]

Roman law adopted the maritime loan, and its treatment of this and other contracts whose fulfillment depends on chance is much more explicit. The treatment of chance in Roman law is not, like Aristotle, concerned with its unknowability. On the contrary, chances are regarded as capable of being known quantitatively, in particular, as capable of being priced. "There can be no sale without a thing to be sold. . . . Sometimes, indeed, there is held to be a sale even without a thing, as where what is bought is, as it were, a chance (*quasi alea*). This is the case with the purchase of a catch of birds or fish or of largesse showered down. The contract is valid even if nothing results, because it is a purchase of an expectancy (*spei*)."[7]

There was talk of "how much the hope of indebtedness is" (apparently meaning the prospects of recovering a debt)[8] and of the sale of the prospect of an inheritance. Ulpian writes that "a sale is acceptable on the terms that 'the inheritance, if there be one, is sold to you,' in effect, the expectation of an inheritance; for the thing sold is uncertain in content as in the sale of the cast of a net."[9] The use of the "cast of a net" as a model for other cases involving chance implies a recognition of expectations as a general concept.

The opposite of a hope or expectancy is a peril or risk. From the time of the Roman Republic, shipwreck was "at the risk of the state" (leading to some predictable frauds).[10] The point in various kinds of sale at which the risk passed from the seller to the buyer was laid down, and the bearing of such a risk could entitle one to benefits.[11] This risk becomes almost a detachable entity in itself when one can specifically agree to retain it when depositing something.[12] The reification goes even further in "maritime interest." Rates of interest were controlled by law; for maritime loans, maritime interest at a higher rate than usual was allowed because of the assumption of the uncertain peril of the voyage (the *Digest* says that "the price is for the peril").[13] It is a major conceptual advance to think of hopes and perils as quantities that can have a price.

As to statistical considerations, there is in the *Digest* what may or may not be a table of life expectancies. The problem being considered is how to decide on a reasonable figure for the present value of a life annuity, that is, an annual payment that will continue for a person's lifetime. "Ulpian says that the following rule should be adopted in making the estimate of maintenance to be furnished. The amount bequeathed to anyone for this purpose from the first to the twentieth year is computed to have lasted for thirty years. . . . From twenty to twenty-five years the amount is calculated for twenty-eight years, from twenty-five to thirty years the amount is calculated for twenty-five years; from thirty to thirty-five years, the amount is calculated for twenty-two years, from thirty-five to forty years, it is computed for twenty years; from forty to fifty years, the computation is made for as many years as the party lacks of the sixtieth year after having omitted one year; from the fiftieth to the fifty-fifth year, the amount is calculated for nine years; from the fifty-fifth to the sixtieth year, it is calculated for seven years; and for any age above sixty, no matter what it may be, the computation is made for five years."[14] It seems unlikely that Ulpian used any data of mortality to arrive at these figures, but the diminishing of the value of an annuity with age does seem designed to reflect at least roughly the diminishing of life expectancy.

Doubtful Claims in Jewish Law

The Talmud also discusses the estimation of prices for quantities that depend on chance. The examples differ from those in Roman law, but the concepts are similar. The *ketubah* is a sum of money specified in the marriage contract, which the husband will pay to the wife if he divorces her or dies before her. (If she dies first, he simply inherits all her possessions.) The Talmud investigates the case in which two witnesses claim that a man has divorced his wife without paying her the *ketubah*, but their testimony is found to be false. The false witnesses are then required to pay compensation to the husband according to the rule, "You shall do unto him, as he intended to do unto his fellow." That is, the false witnesses must pay the husband the present value of the *ketubah*, and the problem is to estimate what that is; it is less than the face value because of the chance that the husband will not have to pay it. The text asks "will he not today or tomorrow eventually pay her her *ketubah?* We estimate how much one would be willing to pay for her *ketubah*, if she became widowed or divorced, while if she died her husband would inherit her possessions. How do we estimate? Rav Hisda said, 'according to the husband.' Rav Nathan bar Ushia said, 'according to the wife.' Rav Papa said, 'according to the wife and her *ketubah*.'" The Talmud is thus envisaging a market price for what would now be called an expectation. The text does not have any word like *expectation*, but some understanding of it is implied by the last comment, in which the value to the husband is correctly distinguished from the value to the wife: what one would be willing to pay the husband for his rights against what one would be willing to pay the wife for hers.[15]

At the same time that the Glossators and canon lawyers were interpreting the ancient Roman legal traditions in the light of "reason," Jewish commentators were doing the same with their inheritance. One of the mysteries of medieval history is how alike the Jewish and Latin commentaries on law are, when there was apparently no interaction between them. Rashi, the most authoritative of the Jewish commentators, lived in Troyes around 1080, just at the time the Latin scholars began treating their texts in almost exactly the same way as Rashi treated his, yet no contact between the two schools has been found.[16] Rashi illustrates as well as any of the Latin writers how the art of commentary can make explicit the abstract concepts implicit in tradition. Commenting on the *ketubah* problem, he introduces a technical term *doubtful claim* for the concept involved, distinguishes explicitly between the value to the husband and that to the wife, and explains the relation between the two: "We estimate *her* doubtful claim, that is, how much one would be willing to pay for her rights. This amount they would not pay, but would pay the remainder, for it is all the remainder that they would make him lose through their testimony."[17]

Maimonides takes the matter further, listing some of the factors that affect the value: "If the wife was sick or old, or if there was peace between her and her husband, the value of her *ketubah* is not as high, should she sell it, as its value if the wife was healthy and young, if there was a quarrel between them, for this last woman is near a divorce and far from death. Also the value of her rights in a large *ketubah* is not the same as the value of her rights in a small *ketubah*. For example, if her *ketubah* was one thousand zuz, it would be sold for a value of one hundred. But if it was one hundred, it would not be sold for ten but for less. And these things are according to what the judges estimate."[18]

In Islamic law, the idea of contracts involving risk survived but only for the purpose of prohibiting them. The Prophet forbade games of chance and, according to authoritative traditions, also any contract that involved *gharar*—any risk, uncertainty, or speculation.[19] Thus one could not sell a bird in the air, even one accustomed to return to its nest, nor a fish in the water, unless easily caught in a shallow pool and owned by the vendor. Even a ripening crop could not be sold; there was debate on the "signs of readiness" permitting a crop still unharvested to be regarded as ripe and salable.[20]

Olivi on Usury and Future Profits

Many of the more abstract entities considered in finance are in reality contracts (of sale on conditions, options, insurance, and the like); that is, they are legal entities. The interface between finance and law is especially apt to generate subtle concepts. Some early developments of the kind are those noted above in Greek and Roman law, such as maritime loans and the pricing of hopes and perils; the ancient word *credit* indicates another. When commerce and Roman law revived in the twelfth century, the evolution resumed. The risk-sharing partnership became widely used. An investing partner gave a seafarer money for a voyage in return for either a share in the profits or a fixed return if the voyage was successful, the loan being written off if the ship sank. The maritime loan was almost the only way of creating large-scale credit, so the return covered the costs of risk, interest, and foreign exchange.[21]

The partnership as such encountered no problems from canon law, but the implicit charge for interest did, because of the prohibition on usury. *Usury* meant any taking of interest on a loan; it was thought of, in effect, as the making of money without doing work, since one took the profits of someone else's work while doing none oneself. The general opinion on the matter resembled the distaste often now felt for those who make fortunes by shuffling money in takeover bids and foreign exchange dealings. On the other hand, there was no objection to partnerships, in which someone with money invested in a voyage undertaken by someone else, with the profits being shared. The difference between the cases is not great to begin with, and then as now, financial practice quickly began to evade the law through complexity, producing hybrid cases, artificial schemes, and technical devices (and usurers, unlike heretics, could afford lawyers to argue their cases, often on the grounds of insufficiency of proof).[22]

What about riskless partnerships, in which the investor does not assume any of the risk of loss in the venture? This was condemned by the canon lawyers as plainly usurious.[23] The incidence of risk was expressly used in the standard commentary on Gratian's *Decretum* to explain the difference between a loan and renting out a house or a horse (on which profit was allowed) and to explain the legitimacy of investment with a businessman (since a partnership is formed, with the risk shared).[24] Aquinas too saw the sharing of peril as the important feature in making partnership licit.[25]

What about sales on credit at a higher price than for cash? The fundamental pronouncement was in the much discussed bull, *In civitate*, of about 1170. One who promises to pay six pounds at a fixed later date for pepper, cinnamon, or other merchandise now worth five pounds commits usury, "unless there is doubt whether the merchandise will be worth more or less at the time of payment."[26]

The phrase "worth more or less" became, with discussion, a kind of technical term indicating a balancing of doubt. Peter the Chanter writes, shortly after 1200, that "a buyer or a seller may be excused from usury if he exposes himself to the risk of receiving more or less. Thus if one buys a crop on the stalk for 20 pounds, and he is in doubt whether it will be worth more or less at the time of harvest, although it may turn out that at that time it is worth more, and he gains, there is no usury. Similarly, if a seller sells his property for 20 pounds to be paid at a fixed later time, and there is doubt whether at that time the thing will be worth more or less, he is not guilty of usury, even though perhaps by then he could only sell it for 15 pounds."[27]

A treatise on usury by an associate of Peter seems to recognize some balance between the size of the risk and the allowed profit: "If he commits capital to the fortuitous risk, hoping to receive something over and above it, there is no usury involved, since the risk turns both ways, provided that the contract is made according to the accustomed course of buying and selling."[28] But one of the documents that made the concession that "a likely doubt" might excuse profit on a loan was the infamous bull *Naviganti* of 1234, which took a hard line on the matter of rewards for peril: "One who lends a certain quantity of money to one sailing or going to a fair, in order to receive something beyond the capital for having taken on himself the peril, is to be considered a usurer."[29] This prohibition goes so much against the trend up to that point that it has been seriously suggested that a "not" was accidentally left out before "to be considered."[30] Nevertheless, the rule stood, though qualified by various distinctions. Thereafter, the proposition that risk itself could provide a title for interest on an actual loan, as distinct from a partnership, was entertained but denied.[31] Public opinion, however, tended to see the presence or absence of risk as morally important in loans.[32]

It occurred to the canon lawyer Hostiensis that, if a "likely doubt" excuses from usury but certainty does not, there is a problem about where to draw the line. His knowledge of Roman law allowed him to see the problem, but at the same time made it difficult for him to make progress on it. He asks, "What therefore if one gives a price at harvest time, with the amount [of grain] to be delivered at Easter? It is clear from the intention of the words that he is to be considered usurious. It is the common and accustomed course in all regions, and common opinion, which are to be taken notice of and expected (*Dig.* 43.12.1.1; 33.7.18.2; *Decretals* 3.28.9). And if you say to me, that sometimes the contrary happens [that is, the price does not rise between harvest and Easter], I reply, that the laws are not adapted to those things which happen rarely (*Dig.* 1.3.5); but rather, to those which happen for the most part, and frequently (*Novels* 94.2)." He adds that the kind of doubt envisaged by the *Decretals*—as excusing an early payment that turns out to be less than the goods are worth at the time of delivery—is to the advantage of neither party. "He is therefore excused by virtue of the doubt, because the buyer and the seller can equally expect advantage and disadvantage from the delay (*Code* 8.43.14)."[33]

Analysis of usury culminated in the work *On Sale, Purchase, Usury and Restitution* of Peter John Olivi, active in the 1280s and 1290s. Olivi's various relations with the topic of uncertainty are distinctly unusual. His *Commentary on the Apocalypse* involves much numerical calculation of the date of the millennium, which he counterinductively places shortly after 1300; the "error of Antichrist," meaning godless Aristotelianism, is already at large.[34] The impression of a fanatical mind is lessened by his remark that some of his predictions are only "probable and conjectural," not infallible.[35] But Olivi is no fallibilist in general. On the contrary, he is the first to put forward in any detail the theory of papal infallibility. He argues that "it is impossible for God to give to anyone the full authority to decide about doubts concerning the faith and divine law with this condition, that He would permit him to err."[36] His own orthodoxy was, however, far from unquestionable, and his opinions were condemned, then rehabilitated, then posthumously condemned again. In life, he was not beyond using the Aristotelian device of claiming his opinions were not asserted but only put forward as "not improbable."[37]

The main reason for Olivi's clashes with authority was his standing as the chief theorist of the Spiritual Franciscans, the party in the order that favored a rule of extreme poverty, on the grounds that Jesus and his disciples had owned nothing.[38] Alarm in high places resulted from the implication that the present body of disciples of Jesus, that is, the church, should own nothing either. It is most surprising, then, that his treatise on merchants' contracts displays a deep understanding of economic matters and very little animus against "filthy lucre."[39] What is especially original in Olivi is his use of the concept of probable presumptions to discuss risk. He distinguishes the perils that excuse profit on a loan from those that do not: the lender must accept his share of both losses in shipwreck and those in trading. He argues that where both the risks are really run by the merchant, not the capitalist, the capitalist is not entitled to any of the profit. He adds that, "in this case there is often another source of usury: for the original lender of money does not take on himself the peril of the sea and the journey, unless he probably presumes that his share in the whole peril is safer and more useful to him than to the merchant, on the grounds that the loss at sea or on the journey is rarer than loss through trade or exchange."[40] The exact meaning is not entirely clear, but it is obvious that the idea of probable presumptions is being tied to the comparison of the sizes of different risks and that this has something to do with the rarity of occurrence.

In evaluating the compensation one is entitled to if the state requisitions one's goods, Olivi actually subtracts one probability (or more accurately, one expectation) from another. Doctors generally agree, he says, that in this case one is entitled to compensation not only for loss actually sustained but also for the probable profit one would have made from the requisitioned goods. On the other hand, "capital forcibly seized could not be lost or put at peril by the owner, as it could have been in trade or business. Therefore there should be subtracted from the probable profit however much the aforesaid certainty outweighs the uncertainty and peril which can happen in trade with respect to capital and profit." The passage continues with remarks on the pricing of future goods. "It is also clear that when someone, by special grace, suffers requisition or sells a crop at a time when it is commonly cheap, but which he firmly intended to keep and sell at a time when it is commonly and probably dearer, he may demand the price which, at the time of requisition, he believed probably would be obtained at the dearer time."[41]

If however he intended only to sell some time in a certain month, the price used to evaluate compensation should be the mean price for that month, not the maximum. Such fine calculations are called for, since it is a matter of making compensation "up to probable equivalence, or to preserve from loss of probable profit." Similar considerations apply in estimating the compensation due for loss of limbs, in regard to the loss of future earnings. It is not the general custom, Olivi says, to require compensation in this case, because the cause is too remote from the effect, but if it is, "if . . . he who lost limbs is now deprived of the mechanisms and occupations from which he could derive profit, then . . . either the depriver is required to restore only as much as the probability of profit weighs (*quantum ponderat probabilitas talis lucris*), or . . . common custom excuses him from any restitution. Perhaps this custom is because often such a probability for the whole life of one who has lost limbs is not at all uncertain, but is of such weight or price as could scarcely be compensated for, at least in popular estimation. I believe though that a rich man who has cut off someone's limbs is bound to support him, if he needs support."[42]

The sixteenth century printed editions of Olivi's work include a further passage, which is not in any of the surviving manuscripts but is believed to be a genuine part of what Olivi wrote. It includes the concept of probability as a tradable commodity. At issue is whether there is usury in the case in which the investing partner risks the loss but takes all the profit by buying it from the trading partner "for the price at which the probability of the profit can be reasonably estimated, before the outcome." There are several reasons for thinking this is not usury:

The second is, that the priceable value of a probability (*appreciabilis valor probabilitatis*), or of a probable hope of profit, from the capital is capable of being traded. For this probability has a value, and a priceable thing can be sold licitly for the price it has at any time.

The third is, that from the fact that the probability is sold for a lower price than it is then believed the future profit from trading with the capital will be worth, it is clear that in the sale, it is always believed probably, that the buyer will in the end make a profit . . . therefore the capital, as much as the principal and final profit from the capital, runs at the peril of the lender. Hence it can hardly be regarded as usury.

They prove this further by an analogy, or equivalence. One who is about to invest his own money in trade sells to another the probability of future profit from it, on the understanding that he will deal with it as diligently as if he had not sold [the profit]; it is agreed that there is no usury in this sale, as it cannot be considered to involve any loan. But this case does not differ from the first, either as to the peril of the capital, or as to the sale of the probable profit, but only as to the mediate and immediate act of trading. Therefore there is no more usury in the first case than in this one.[43]

Olivi agrees with these arguments.

Discussions of these matters continued in the next century. Scotus, not normally thought of in connection with such practical matters, discusses adequately the merchant's entitlement to payment "corresponding to his risks" and various questions about the likely future value of goods. Buridan describes a cunning attempt to avoid the usury prohibition: one lends wine, say, at a time when it is selling cheaply, and asks for repayment of exactly similar goods at a time when they are expected to be dear. Buridan recognizes this as usury.[44] These discussions by philosophers do not, however, use the words *probable* or *presumption* in connection with risk. They confine themselves to the classical Roman law vocabulary of *hopes* and *perils*.

It is fair to ask what the effect on business was of the prohibition on usury in general and of these subtle analyses of morality in cases of risk in particular. It is, indeed, impossible to grasp the context of what writers like Olivi are saying, or its cultural meaning, without some understanding of whether, and how, their ideas could affect commercial practice. If theory can exert an influence on practice, it should be evident here, where the theoreticians were official ideologists to the state and wished to impose an opinion directly contrary to the preferred practice of the capitalist class. Some effect may be presumed. While the moral climate on the bourse naturally differed from that in the cloister, the moral theorists had institutionalized ways of influencing practice. Capitalists had to listen to clerics during sermons, and courts could find a loan usurious and declare that it need not be repaid. The many deathbed "restitutions" of profits show at least a partial acceptance of the content of the sermons.[45] It is a further question whether the usury laws impeded or assisted the economy. A number of the many discussions of this topic have been vitiated by the assumption that any attempts to restrict business for moral reasons must impede the development of capitalism.[46] Plainly this is not so in general: if everything is for sale, then bribery, slavery, the sale of judicial decisions, of industrial secrets and misinformation, and so on, will flourish, none of which is likely to encourage economic growth.

The medieval Law Merchant, developed by merchants for their own dealings, shows that they understood this in principle, and there are good economic reasons to explain the coupling of the growth of justice among merchants with economic growth.[47] For an economy to grow there must be incentives to produce. So those who do the work or incur the risk must receive at least some of the resulting output. That does not happen automatically or easily, as the constant modern litigation over patents and copyright shows. According to Olivi, a principal objection to usury is precisely that it involves taking the profit of someone else's industry: "To sell to another his own industry and his own acts is to sell him what is his, and consequently is unjust; but to sell the utility arising from money solely through the business activity of the recipient of the loan, is to sell him his own industry and acts."[48] Modern underdeveloped economies ossified by debt slavery illustrate the effects of having no such prohibition.

Likewise for risk. Ancient Roman law says that "he who bears the risk should get the benefit." Olivi says that "it is agreed that capital should profit him who runs the risk of it simply" and that "the risk that removes usury should include both ownership and use by the profiter of the thing put in peril. Ownership, because one ought to profit from the thing that is one's own, not a thing now really belonging to another. Use, because the use of a thing from which profit arises ought to be mediately or immediately that of the profiter."[49] If the entrepreneur is to make his contribution to the capitalist economy, then the risk he takes must produce a reward to him, otherwise he will not take the risk. The modern investor in shares in a limited liability company accomplishes this in essentially the same way as a medieval partner.

In evaluating the effect of the discussions of usury, something must also be attributed to the invention of the concepts themselves. The distribution of profit to him who deserves it is a very complicated matter and is not done without a considerable conceptual apparatus. A familiarity with hypothetical and counterfactual reasoning is needed, for example, as when the *Digest* considers liability for profit that would have been made if things had proceeded correctly.[50] East Asian economies were impeded for lack of such an understanding.[51] It is indeed an impediment, since if there is a high chance of being unable to recover the results of one's effort and risk taking, one is not inclined to begin the work. The achievement of Olivi and the other thirteenth-century moralists is to refine these ideas by pricing the probable profit from a venture, not just the profit that might or might not arise.

Pricing Life Annuities

During early capitalism, perhaps more than any other period, ready money was necessary but hard to raise. One tempting source, for states and other powerful bodies, was the sale of something in the future, or the prospect of it, for hard cash in the present. In the thirteenth and fourteenth centuries, a trade grew up in perpetual and life annuities, in which one pays a sum of money and receives a stated annual income, in perpetuity (inheritable by heirs) or for life. There is an element of implicit betting involved in life annuities, in the seller's estimate of the likely date of the purchaser's death. The cities of Germany commonly raised money through the sale of annuities from the thirteenth century. Hamburg standardly offered 10 percent of the purchase price per year on life annuities, as against 6.66 percent for perpetual annuities. Douai in 1324–25 offered 10 percent for life annuities, 5 percent for perpetual ones.[52] The difference represents an implicit quantification of the expected life span. Ghent in the later fourteenth century gave 10 percent for a life annuity, $6\frac{2}{3}$ percent for an annuity on two lives (one payable until both purchasers had died), and 5 percent for perpetual annuities, although these rates varied widely.[53] In the fourteenth century is found the first discrimination on grounds of age, to reflect the different expectancies of life. Nordhausen in 1350 offered one mark a year for a down payment of ten marks for persons between forty and fifty years of age but one mark a year for eight marks to those between fifty and sixty years of age. In Artois in 1399, the rate of interest on a life annuity offered to a man of fifty-eight years old was twice that offered to a child.[54] The problem of record keeping was solved by providing a small reward for whoever brought the happy news of an annuitant's death.[55]

The involvement of the church in the annuity business was substantial. Monasteries were among the principal sellers of annuities, and churchmen common among the buyers.[56] The impression was thereby given that annuities were perfectly licit and had nothing to do with usury. A few of the more rigorous doctors analyzed the nature of annuities in the course of arguing that they were not licit. The first was Goffredo di Trani in the 1240s, who asks, "What of those who give money to churches, and receive from them certain possessions which they hold, and use and derive fruits from, for life; after their death the possessions return to the church, while the money also stays with the church; is this contract licit? It would seem that it is, because of the uncertainty in the conditions (*Code* 2.3.1) and the doubtful event of mortality. For elsewhere usury is excused by doubt, as in the decretals *In civitate* and *Naviganti*. But I think the contrary, because men hope to live, and thus those making this contract believe they will receive more profits from the possessions than the money they gave."[57]

Henry of Ghent argues that annuities are no better than the obviously usurious contract, "I advance you 10 Paris pounds, and at the end of a year you give me back 12; but if I die in the meantime, you give me back nothing, and keep the stake." He suggests that opposition to his views is due to the fact that churchmen everywhere are advising pious women to invest their funds in annuities paid by monasteries.[58]

But much the most remarkable discussion is that in Alexander Lombard's *Treatise on Usury* of 1307. Along with Olivi's treatise, it represents the high point of medieval thought on probability, particularly the quantification of risks. Alexander, who believes life annuities are permitted, first mentions Goffredo's arguments, then adds another against their permissibility. "When in a contract one party has notably the better side, the contract is illicit. But in the case proposed, we see men and women twenty-five years old buying life annuities for a price such that within eight years they will receive their stake back; and although they may live less than those eight years, it is more probable (*probabilius*) that they will live twice that. Thus the buyer has in his favour what happens more frequently and is more probable. The seller has in his favour what is rare, so there is inequality in the contract, and consequently the contract is illicit."[59]

Alexander says on the contrary that it is simply a matter of finding the right price, given the probabilities, and that "such equality can be saved, when life annuities are sold. This is when the price is of such quantity that, after weighing with care and consideration the age of the buyer and his health, and the risks concerning the profits from the possessions, it does not appear that either the buyer or the seller has notably

the better side. If such equity is destroyed, it is certain the contract cannot be made, and is not licit . . . the contract is licit, because the risk and
doubt falls on both sides; for the sale is for a time for which it is doubtful whether he will survive more or less, and the uncertainty of the time
makes equality on both sides." To the argument about the twenty-five
year olds, he replies that, "in the case proposed, [equity] is not saved, because of the age of the buyer, since it is likely (*verisimile*) that he is going
to survive long enough to profit greatly and receive back his stake within
a moderate time. So the condition of the seller is worse, and that of the
buyer stronger, unless perhaps the young woman or man buying is exposed to so many risks and infirmities, that it does not appear which is
more probable: whether the buyer will live so long that the seller suffers
notable loss, or the former will die within eight years."[60]

Besides the clear understanding of the quantification of probabilities, these passages are significant for their regarding a number of the
old concepts as identical. Alexander treats *risk* and *doubt* as synonymous,
uses the terms *probable* and *likely*, and treats *more probable* as meaning
"what happens more frequently." It is important for later developments
that he evaluates the justice of such contracts in terms of the equity or
equality of risks. Like Olivi's work, Alexander's was preserved as a living
part of the Scholastic tradition; his views on life annuities appear, for example, in the standard fifteenth-century work, St. Antonino's *Summa*.[61]

The Roman lawyers came a little later to the problem of life annuities. Their views resemble those of the moralists but are given a slightly
different slant by their need to relate them to the concepts of the *Digest*
and *Code*. The matter was brought to their attention by a case of 1308.
The archbishop of Bremen, on payment of 2,400 livres to the Abbey of
St. Denis, acquired the right to an annuity of 400 livres. There was pro-
vision for a partial refund in the event of his death within two years, and
presumably the Abbey was counting on paying out money for not too
great a multiple of that period. In 1323 the abbot contested the validity
of the contract, unsuccessfully, on the ground that it was usurious and
therefore not binding; the archbishop lived until 1327.[62] It is far from the
only case in the records where churchmen, led by the hope of exchanging present debt for uncertain future expenses, estimated overoptimisti-
cally the likely date at which God would extinguish their obligations.[63]

It will be recalled that the fourteenth-century jurist Baldus developed the theory of evidence in law to its highest point. His writings on
risk in contracts are also impressive. His *Consilia* discuss cases in which
someone buys a life annuity for 300 florins and survives long enough to
receive a total of 600. Is this usury?

The contract is not usurious; for usury is the increase of a returnable stake, but here the stake is not to be returned; therefore nothing accrues to the stake, so it is not a usurious contract. . . . The buyer subjects himself to the peril of fortune, whence the price is of the peril (*Dig.* 22.2.5 on the price of peril). . . . If the buyer had lived only one hour, the seller would not have given the price back to him: certainly not, whence one ought not to demand from someone else what he would not take into account himself, or the contrary of which would not be demanded. . . . Here a hope seems to be sold, so whatever the hope is worth is to be considered, and this is not to be judged from the outcome (as per Bartolus on *Dig.* 39.4.11.5). . . . It seems age is to be considered (*Dig.* 35.2.68), whence is taken conjecture of life and death. . . . An old man is dealt with differently from a young man (*Dig.* 2.15.8.10), on account of age. . . . By reason of the uncertainty the mean is said not to be exceeded (*Code* 4.32.17) . . . because a contract where there is equality of potential loss and gain stands far distant from the evil of usury, or from the attraction of usury, because of the doubtfulness of the peril (*Dig.* 17.2.29.1). . . . Notwithstanding *Naviganti* and *In civitate*, where fraud is presumed: because there even if there was not the certainty of necessity, it was nevertheless very close to certainty, but here it is not, so the case is different.[64]

As to the actual pricing of life annuities, Baldus says: "So it is sold at some suitable and just price, namely, whatever such a hope can commonly be sold for at the time of the contract. And a hope is not worth as much as a thing, since a hope may be interrupted by many chance events: just as a crop is worth less on the stalk than in the barn."[65]

Another jurist considers the interesting question of whether an annuity on two lives sold at 15 percent, irrespective of age, involves deception beyond half the just price (which would in Roman law be sufficient grounds for regarding the contract as void).[66] The problem is correctly regarded as difficult, and no solution is arrived at.[67]

Speculation in Public Debt

With states issuing credit on a large scale, there arose national debts, whose scale naturally tended to become unmanageable. So another quantity that could be the subject of financial speculation was the likelihood of the state's actually paying the debts it had undertaken. As usual, some of the earliest evidence comes from the moralists. Claro of Florence, in his *Cases of Conscience* of around 1260, inquires whether restitution is required from those who invested in town revenues bought cheaply during risks of war and who profited tenfold when peace was restored. No, "they are excused by reason of risk and danger." He adds, indicating how moralists and capitalists might enjoy a symbiotic as well as an adversarial relationship, "but they ought to give very fat charities from their profits."[68] In the fourteenth century, a secondary market in credits in the public debt arose in Italian cities. They were attractive investments in some circumstances; for example, one might purchase credits at a discount, while counting them at face value for purposes of a dowry, and were suitable for long-term speculative investment.[69]

One of the largest such schemes was the Florentine *Monte delle doti*, set up in 1424–25 when the city was particularly hard pressed for cash for war. All fathers could invest in the fund on behalf of their sons and daughters, for periods of from five to fifteen years. On maturity, the daughter received the capital plus interest as a dowry, the son as a fund to begin business. If the claimant had died, the heirs and the state divided the money. The original offer was that 100 florins could be redeemed after five years for 500; the fact that the promised rate of return nearly doubled in the next ten years appears to reflect the city's worsening financial crisis rather than anything about estimates of mortality.[70] *Monte* stock in Florence traded at only 15 percent of par in 1433, when interest payments were suspended after military disasters. The noted preacher San Bernardino of Siena understood the nature of the *monti* well enough, condemning them as instantiating two sins, usury and gambling.[71]

Insurance Rates

The financial practice that most directly involves quantification of a probability is insurance. The premium of an insurance is simply the expectation of loss—the probability of loss multiplied by the size of the loss—plus a margin for the seller's profit.

Ancient Roman law does not have an insurance contract, but it has the conceptual apparatus that, when carefully selected and put together, is sufficient to describe what insurance is. First, as described above, there is the idea of maritime loans, which attract a higher rate of interest for the "uncertain peril" of the voyage, as developed in the twelfth and thirteenth centuries into the risk-sharing partnership. Second, there is the concept of chance disaster (*casus fortuitus*), the kind of rare unfortunate occurrence that "no human counsel can foresee."[72] Third, there is the idea that one can, contrary to usual practice, specifically agree to retain risk when one deposits a thing.[73] The Glossators combine the last two to discuss the transferal of risk by specific agreement in *casus fortuitus*.[74]

Thus they come close to the essential idea of insurance: that the risk of misfortune to a thing is something that can be sold independently of the thing itself. A contract of 1319–20 comes very close to true insurance by this route. A large Florentine firm buys Flemish and French cloth at the Champagne fairs and delivers it to Pisa at the risk of the sellers but at the buyers' expense. There is an extra charge of 8.75 percent, called *rischio*, for assuming this risk.[75]

The premium of an insurance policy is a quantification of the risk involved, and there is money to be made from estimating the risk correctly. Some of the earliest insurance contracts are difficult to interpret, since the premium is hidden by fictitious sales and other such devices. Some authors have thought this was to avoid possible difficulties with the church's prohibition of usury, but it may just as well be explained by the conceptual difficulty of isolating a risk as a thing in itself. Against the theory that the usury prohibition was the problem is the fact that when insurance did appear, it encountered no problems with canon law. No loan, so no usury.[76] If anything, the prohibition on usury would seem more likely to have encouraged the invention of a true sale of risk, since if one was actually doing something other than lending at interest, it paid to explain why.

A Genoese document of 1343 is regarded as the first known true insurance contract. It involves a "loan" from the owner of a cargo of wool bound from Pisa to Sicily to the shipowner who will transport it. It appears that this "loan" is not actually made, but the shipowner accepts an obligation to "repay" it. The obligation is to be cancelled if the goods arrive safely. If they do not arrive, the "lender" will pay a "penalty" of double the value of the loan. Presumably the ship owner has earlier paid a premium for the privilege, but that is not mentioned in the contract. In its roundabout way, the contract has the same effect as a modern insurance contract: if everything goes well, nothing happens, but if disaster strikes, the insurer pays the insured a large sum.[77]

Undoubted marine insurance policies are known from Palermo—not a place at the center of the commercial world—in 1350. In the first, Leonardo Cataneo, merchant of Genoa, insured (*assecuravit*) a shipload of wheat from Sicily to Tunis, assuming all risks, perils, and fortune (*risicum, periculum et fortunatam*) from acts of God, man, or the sea. He received 54 florins and was to pay 300 florins one month after receiving "certain" news of the cargo's loss.[78]

Much lower rates were common later in the century, with variations according to the seasons, war or peace, and reports of pirates in the sea lanes (but not in general according to the type or age of vessel or its cargo). In 1384 rates of 8 percent are found from Cadiz to Sluys or Southampton, 4 percent from the Port of Pisa to Naples or Tunis.[79]

The insurance business soon flourished in the western Mediterranean and Flanders.[80] In Genoa in 1393, a single notary wrote more than eighty insurance deeds in twenty days, and complications like reinsurance were familiar. Maritime insurance was the most common kind, and a number of insurance brokers with specialized knowledge set up business in the main centers; a large number of co-insurers was common, implying some idea of the spreading of risks.[81] Insurance on life is found from 1399, and from the early fifteenth century there are contracts covering the risk of death in pregnancy of both wives and slaves.[82] It is a case in which the risk can be expected to be rather constant and independent of special knowledge, unlike shipping risks.

While the writing of insurance contracts obviously implies an explicit quantification of risks, it is one thing to do this and another to be aware of doing it. There is no theoretical writing on the subject by the merchants, and the conceptualization of risk is not usually apparent in the documentary evidence, which consists of contracts and records of lawsuits. There is however one contract, written in Crete in 1389, which speaks of "certainty" and "doubt" as the determinants of the price and makes these depend on explicitly stated evidence:

> This day 14 October 1389, 13th indiction, as overleaf, I, Niccolino de Fieschi, a Genoese residing at present in Candia, with my heirs, do hereby make known to you Ser Bernardo de Mezzo of Venice, residing in Candia, in your presence and to your satisfaction, and to thine heirs, that as a *griparia*, master Bartolomeo Acardo, entered the port of Candia today and he having said he saw part of a wreck and a barrel on the sea, and suspecting therefore that the ship of Antonio de Barba was wrecked, which sailed a short while ago from here on a voyage to Romania [that is, Constantinople], I insure you for 200 ducats in gold, of good and fair weight, for your part of the cargo existing on said ship; whereby, if the said ship of Antonio de Barba was wrecked, counting from the time she left this port and up to the whole of the present day, understanding that certainty is had of this, I am bound to and must give and pay to you 200 ducats in gold of good and fair weight, here in Candia, safe on land.
>
> And for this insurance, according to our agreement, I have received from you 25 gold ducats.
>
> Nevertheless, should anything new happen in Candia, during the whole of this day, whereby we shall know for certain that the said ship

of Antonio de Barba has sunk, then in that case I shall not be bound to you for the present insurance and this paper must be held of no value and must be null and void, because I issue this insurance to you for the doubt had and that one still has and not for any certainty.[83]

It is clear that this contract is barely an insurance at all but more like a bet. The difference between insurance and betting was not clearly understood, and indeed the distinction is a fine one. In the next century merchants are found making "insurances" on the lives of the pope and the doge of Venice, on horses, and on the duration of a conclave.[84] These practices had to be forbidden, considering the temptation to bring about the event bet on. An intermediate case is insurance against the occurrence of a plague, available at Genoa for 4 percent for a year.[85]

Further difficulties in insurance can arise if the two parties to a contract do not base their action on the same information. In a fifteenth-century case, the Medici representative in London underwrote insurance, at a premium of 50 percent, on a cargo of alum overdue on the short voyage from Zeeland to London. When the loss of the ship was confirmed, he refused to pay, on the grounds that the insured had received secret intelligence of the shipwreck. The insured in fact admitted this but sued the agent in the Venetian courts and seized Medici goods. The outcome of the case is not known.[86]

While the merchants had little to say about the nature of insurance contracts, the jurists soon turned their attention to them. A passage from Baldus was taken by later writers to be a justification of insurance, although it is not clear what contracts exactly Baldus had in mind. He explains that the Roman laws permitting some usury are wrong ("The prohibition of usury is the same in all legal systems. For what is natural is the same in every place, as a stone falls everywhere, and fire rises and heats everywhere, according to Aristotle"). But there are many cases in which no true usury is involved but some genuine quid pro quo. These cases include one in which "there is an accepting of risk, in return for giving something, as in *Code* 4.33. In such relating of one thing to another, there is nothing properly called usury, but a certain recompense owing from this to that."[87]

The matter is explained more fully by the Florentine jurist Lorenzo de Ridolfi in 1403, who says, "Of insurances (*securitatibus*) . . . when you send your merchandise by sea or land to certain parts, and I take upon myself the risk, and agree that for every 100 of value of that merchandise, you pay me a certain quantity of money . . . something is given for something done. It is not given as a loan, since no loan is involved, but something is received for assuring the merchant of his goods, which are at risk by sea or land. Besides, no stake is involved, so it is false to say that something is received over and above the stake; there is no building without a foundation."[88]

The first book on insurance, *On Insurance and Merchants' Bets*, by the Portuguese jurist Santerna, was published in 1552, though written in 1488.[89] Santerna's conception of the contract is expressed in the simple formula, "I undertake the peril, for your giving me money, as is the understanding of Baldus." While admitting that insurance and bets can be a cloak for usury, Santerna argues that generally a good intention in the parties must be presumed. For given the general uncertainty of life, everyone must make bets ("since the Pope does not know the secrets of the heart, and cannot divine them, he judges from likelihoods, and common accompaniments, and so is sometimes mistaken"). There are various cases in which it is important to determine legally whether the price of an insurance is just. Thus if someone charges 16 percent in circumstances for which 8 percent is usual, it may be inferred that the insurance was intended to cover a second voyage, and if a ship's master defers sailing until a more dangerous time of the year, a price previously just may become unjust.[90] Reinsurance is permitted, that is, insuring an insurance against the default of the insurer.[91] There is a complicated discussion of the difference between "normal" though rare events like storms and very rare ones "that occur once in a thousand years."[92]

Can the rule that a contract of sale is invalid if the price exceeds the just price by more than half apply to an insurance? First, it needs to be determined whether an insurance can be said to have a definite value at all, since "it can be said that the insurer sells only the hope of a future outcome, of which there can well exist a sale. . . . From the fact that this hope is uncertain, it might not seem capable of estimation such that in respect of it there could be said to be exceeding of half the just price of its value. But, this is not to be estimated at how much the thing or goods would be worth in case the peril was realised, but at how much the doubtful event should likely (*verisimiliter*) be estimated. In which case the price seems to be constituted with respect to that hope."[93] How is an insurance to be actually valued by the courts? They should use the market price, since "it is to be estimated as worth the percentage that insurers are commonly used to accepting as the price for taking on the peril. Thus it can be decided whether less than the just price was received or not, and whether the insured gave more than half of the just price or not. Suppose for example, if one normally pays 50 aurei for the peril of goods worth one ducat; if I receive only 20, I can be said to have been deceived in that insurance beyond half the just price."[94]

Insurance became standard practice in northern Europe two centuries after its adoption in Italy.[95] English insurance contracts are known only from 1547, written in Lombard Street; the first one is actually in Italian.[96] (Insurance contracts fell under the jurisdiction of the Court of Admiralty, which operated on Continental principles of procedure; it was largely through this route that later English commercial law acquired the concepts developed by the Roman lawyers.)[97] Though seventeenth-century marine insurance became sophisticated in its organization, it did not become more theoretical in its calculation of premiums, which remained intuitive bets very responsive to war and rumors of war.[98]

Renaissance Bets and Speculation

The achievements of the thirteenth and fourteenth centuries in understanding contracts, including contracts under risk, were not lost. They were summarized in Nider's *On the Contracts of Merchants*, written shortly after 1400, and one of the first books on an economic subject to be printed. The language of probability is used throughout, in connection with the uncertainties of a contract's outcome or of its morality: "When a person neither knows nor believes that the community is wrong in the assessing of a price, even if he has a reasonable doubt (*probabiliter dubitat*) that the community may be wrong, then he lawfully can sell his property according to the common estimate. . . . When there is doubt as to the extent to which a thing is worth more or less, a modest probability (*modica probabilitas*) is sufficient for estimating its value."[99] Nider writes acutely of the role of probability in setting the price of a future good. If it is one that is likely to rise in price, "I can properly require more for it on the very ground that I have a reasonable belief (*reputo probabiliter*) that I would be selling at my own loss. In such an instance, however, I cannot properly sell for as much as I think the good will later come to be worth or even is already known to be worth. In setting the price of such a good, however, it is right to deduct an amount representing an estimate of the risk and anxiety (*estimacionem periculi et cure*) which would be involved in keeping the good."[100]

At one point, Nider uses the language of "for the most part" to discuss, in effect, odds. He allows a merchant selling a good in June, with payment delayed until December when the good is usually dearer, to charge the expected December price. But he condemns as usury the case in which the seller retains the choice of the later time for fixing the price, as this virtually guarantees the seller the chance to make a killing merely through the passage of time, because the seller "places himself or his position in line for gain as more often likely to occur (*ut in pluribus*), and him with whom he is contracting is left more often in line for loss. Thus he has the advantage for himself since the odds are for him in more instances and against him only in fewer" [literally: He has for him what happens in most cases and against him what happens in fewer cases].[101] Nider's opinions here, based on the concepts of such earlier authors as Hostiensis and Olivi, are perfectly rational. He looks more to the future in another work, in which he claims that the most sought-after prostitute at the Council of Constance was a succubus.[102] He is one of the authors most quoted by the witch inquisitors in the *Malleus Maleficarum*.

From around 1500 commerce resumed its course toward increasing complexity and toward increasing abstractness of concepts. It also moved from the Mediterranean to northwest Europe. The center of developments was Antwerp, where, by the middle of the century, commerce and finance were concentrated perhaps more than in any one city before or since.[103] By medieval standards, or even modern ones, the bourse of Antwerp was remarkably free of regulation, and the atmosphere of speculation was intense.[104] Little distinction was drawn between entrepreneurial activity and purer speculation like trading in grain futures[105] or outright betting on such events as the next journey of Philip II to the Netherlands and the sex of unborn children. The association of betting with merchants is an old one. The Glossator Azo, for example, mentions betting on whether a child to be born will be male or female, and later authors say that bets on various events such as whether the pope will lose his state are undertaken daily by merchants, "who are accustomed to prognosticate about the future."[106] Even insurance contracts were used as much for speculation as for the reduction of risks.[107]

For those not able to cover their risks, astrological systems were available for predicting the price of pepper and forward exchange rates.[108] Life insurance was conceived in the same spirit, resulting in some macabre frauds.[109] The advancing of huge loans to insolvent sovereigns was a worse gamble than was realized by the lenders, and Antwerp never fully recovered from the round of defaults after 1557 and the sack of the city by the unpaid army of one of the creditors.[110] One theorist remarks that the cheapness of money in Antwerp before the sacking shows the sensitivity of prices to risk.[111]

How widespread the knowledge of the quantifying of risks by merchants was may be estimated from a reference in Shakespeare:

> Or what hath this bold enterprise brought forth,
> More than that being which was like to be?
> We all that are engaged to this loss
> Knew that we ventured on such dangerous seas
> That if we wrought out life 'twas ten to one.
> And yet we ventured, for the gain proposed
> Choked the respect of likely peril feared.[112]

The numerical "ten to one" is significant. The origins of the language of numerical odds are obscure. A work of 1530 in English on the French language translates, "And you sette hym at large nowe, twenty to one he is undone" as *"si vous le mettez au large mayntenant, vingt contre ung il est gasté."*[113] The usage of "to" and "contre" here seems to derive from that in such phrases as "outnumbered twenty to one." Such uses are of interest in themselves, since expressions of relative frequencies in populations are rare enough before the modern period; one does not see numerical ratios like "40 percent of all" or "one in five of all." In its place is only the very occasional use of the language of odds. Thomas More, when asked at his trial why sentence should not be passed against him, discussed at some length the illegality of the Act of Supremacy. When the lord chancellor countered that "seeing all the bishops, Universities and the most learned men in the kingdom had agreed to that Act, it was much wondered that he alone should so stiffly stickle," More suggested a change of reference class; "I doubt not, but of the learned and virtuous men now alive, I do not speak only of this realm, but of all Christendom, there are ten to one of my mind in this matter. . . . I am able to produce against one bishop which you can produce on your side, a hundred Holy and Catholick bishops for my opinion; and against one realm, the consent of Christendom for a thousand years."[114]

In any case, the usage of "to" to express proportions is not unconnected with its use to express probabilities. Shakespeare expects the English troops encamped before Agincourt to convert their estimate of how much they are outnumbered into a probability, and has Henry V say

> . . . take from them now
> The sense of reckoning, if the opposèd numbers
> Pluck their hearts from them.

The numerical "ten to one" appears another five times in Shakespeare's plays, with a meaning related to betting. A different figure occurs also in a mercantile context:

Proteus: But now he parted hence to embark for Milan.
Speed: Twenty to one he is shipped already.

Sir Andrew Aguecheek's "it's four to one she'll none of me" measures a different level of pessimism.[115]

Life annuities remained standard methods for raising public finance. A Florentine document of 1526 includes a note on estimating how long people of various ages can be expected to live. Though it gives the impression that the figures are founded on experience, in fact they are simply taken from Ulpian.[116] Remarkably, they agree reasonably well with modern estimates of mortality for the period. The city authorities of Amsterdam sold a life annuity in the years 1586–90, the records of which were the basis for the calculations of Huygens and Hudde nearly a century later. The records appear to show that the annuities, which were sold for six years' purchase irrespective of age, were underpriced for almost all ages, and that the buyers were concentrated in the ages for which the fixed price was most advantageous.[117]

Some interesting remarks on the pricing of annuities are found in the treatise on commercial law by Charles Dumoulin. A French jurist and the author of (what one would have thought impossible) a Protestant commentary on the corpus of canon law, Dumoulin's lax views on the permissibility of charging interest on productive investments were influential in Calvinism.[118] He notes the difficulty of fixing the "probability" associated with annuities and suggests three factors to be taken into account: the age, health, and "other likelihoods of future life" of the annuitant; the rate of return; and whether the annuity is on one life or several. Quoting Baldus as authority, he suggests a figure of one year for ten as reasonable for someone of middle age (that is, payment of a sum assures the purchaser of a tenth of that sum yearly for life). He discusses why this differs from Ulpian's rules in the *Digest:*

> It could be argued to the contrary, that the rate of one year for ten is not a just price, on the basis of *Digest* 35.2.68, where the maintenance for a man of 35 is estimated not at ten times that of a single year, but at twice this, or twenty times, this being the accumulation of twenty years. The reason is, not that one 35 years old is presumed to live for 20 years, and no more, or that one 15 years old is presumed to have 30 years to live, and no more, and so on (as the Gloss and all doctors wrongly think). Instead the reason is, that in the case of one who is not yet 20, he is presumed commonly to survive to 60 years; one who has lived 35 years but less than 40, is presumed to survive for 40 years. But because this pre-

sumption, on account of extreme uncertainty, cannot be taken for certain, nor estimated for as much as if it were certain, but for less than half as much, a reduction is made in the likely time to around half or thereabouts, to compensate for the said uncertainty.[119]

This reasoning seems completely wrong (though not unknown among modern judges),[120] since any "presumption" of life ought already to have factored in any such uncertainty.

A scheme of compulsory child insurance, similar in principle to the Florentine *monti*, was proposed to the city of Nuremberg about 1565. Estimating that less than half the children born attained to maturity and married, it was proposed to compel every parent to pay a sum of money on the birth of each child. If the child married, three times the deposit would be paid to the child; if the child died, the authorities would receive everything. Nothing came of the scheme.[121]

Commerce continued to expand, although there were no major additions to the concepts used in it in the half century to 1650. Amsterdam took the place of Antwerp as the center of world commerce. That city saw the first of the great modern speculative bubbles, the tulip mania of 1636–37. It involved speculation in the prices of rare bulbs in which, typically, the buyer did not currently possess the cash to be delivered on the settlement date and the seller did not currently possess the bulb. The trade was effectively in risks, not tulips.[122] Speculation in company shares was also sufficiently advanced to permit schemes to rig their prices.[123]

There are signs of one of the most basic forms of statistical reasoning, the projection of a time series into the future. In 1598 Lord Buckhurst, soon to become lord treasurer of England, requested the customs officials to prepare from their records a statement of how much coal had been shipped from Newcastle for the seven years past. What he received from them was a single figure, the total for the seven years. He wrote back: "This is a certificat of confusion more tending to blind than to inform. . . . I require you to set downe every of the said seven yeares in particular, according to the request of my note unto you." The officials did so, but one of them left a note complaining that the treasurer could easily have worked out the seventh part of the total for himself. That is, he did not understand, as the treasurer did, what information the series gave that the total or average did not. The purpose of the extra information was to allow a projection of the trend of the figures—which were in fact rapidly increasing—presumably for use in determining the expected revenue from a proposed tax on coal shipments.[124]

Lots and Lotteries

The financial ingenuity of the Italians extended to the development of lotteries, both private and public. Italian merchants disposed of surplus merchandise by means of lotteries from around 1400.[125] Lotteries to raise public revenue are known in Flanders from 1434 and were common in Germany and Italy in the sixteenth century.[126] Antwerp, appropriately enough, saw perhaps the biggest. Later ones in Holland displayed the Dutch talent for harnessing the desire for wealth to godly ends, by using lotteries with large prizes to pay for lunatic asylums, flood relief, and the like.[127]

Lotteries proved a little harder than insurance to transplant to England. The first English state lottery was organized in 1566. It was planned to sell 400,000 tickets at 10 shillings each, with nearly 30,000 prizes. The public showed a certain skepticism about the government's honesty, and the lottery was finally drawn with reduced prizes in 1569, with only 34,000 tickets sold.[128] Some understanding of the nature of lotteries is evident in Shakespeare's "It is lots to blanks" to mean a near certainty.[129]

Among the riskiest of commercial ventures was colonization. The situation of the Jamestown colony, blamed by the Virginia Company on gambling and other vices, became so desperate that the company was in 1612 granted the exceptional privilege of holding public lotteries to raise funds. Selling the tickets proved difficult, and the lotteries were suppressed in 1621 because of complaints of mismanagement and the disruptive effects on the lower classes. The company soon collapsed, but the lotteries had probably saved the colony.[130]

Long before the invention of lotteries, lots had a history in religious contexts for the purpose of ensuring fair divisions. "The lot causeth disputes to cease, and decideth between the mighty," the Bible says.[131] This is so only if the procedure for drawing the lots is seen to be fair. The various methods given in the Talmud for drawing lots provide for precautions to prevent cheating. Thus the procedure for allocating daily duties to the priests of the temple was as follows. The priests stood in a circle and agreed on a large number. The officer in charge chose a starting place in the circle and counted around the circle until the agreed on number was reached, the priest on whom the count finished coming out first in the lot. This is a fair procedure (unless someone is exceptionally good at mental division of large numbers), and it has been argued that the inventors of the procedure must be granted an implied knowledge of the equiprobability of outcomes.[132] A more explicit understanding is found in the Talmud's commentary on the repetition of favorable outcomes from lots. Each year on the Day of Atonement two goats were selected; one of them was chosen "for the Lord"; the other, the scapegoat, was sent away into the wilderness. The choice was made by lot, the high priest choosing the lots from an urn after shaking, putting in both hands, and drawing out one in each hand. It was regarded as a good omen if the lot "for the Lord" came up in the right hand. It was considered a miracle, therefore, that in the time of the High Priest Shimon the Righteous, "during the forty years that he served, the lot came up in the right hand; thereafter, sometimes it came up in the right, sometimes in the left."[133]

The random processes involved in drawing lots were the subject of substantial debate in moderate Puritanism around 1600. The extreme Puritans had condemned games of pure chance on the grounds that the outcome of a cast is a special determination of God and, though usable for serious matters like choosing magistrates by lot, was not to be taken in vain for casual gaming. Against this, Thomas Gataker in his *Of the Nature and Use of Lots*, of 1619, argues that the outcomes are natural, not providential:

> Again who seeth it not that the lighting of Lots in this or that manner ordinarily cometh immediately from the act of the Creature? For example: In the blending of scrolls or tickets together, the motion of the vessel wherein they are blended (no regard had to the end for which it is done) causeth some to lie this way and some to lie that way, (every new shaking thereof causing a new sorting) and so some to lie higher and nearer at hand, if a man will draw of the next, some lower and further off, not likely to be drawn so soon, unless he dive deeper. Neither can any man say certainly that there is ordinarily any special hand of God, in the shuffling and sorting of them, crossing the course of nature, or the natural motion of the creature, and so causing those to lie higher and so nearer at hand, that would otherwise have lien lower, and those to lie lower and so further from hand that would otherwise have lien higher. So in the shuffling of Cards, the hand of him that shuffleth them is it that disposeth them, and that diversly as he lifteth either to stay or to continue that act of his. In the casting of dice the violence of the Caster causeth the Creature cast to move, till either that force failing, or some opposite hindering it, it cease to move further, and so determine the chance.[134]

Gataker thus understands that a cause is necessary to explain any divergence from the mixing expected from random shuffling but that no cause is needed to explain the random arrangement itself. He also understands that there is no evidence for the divine intervention alleged in choice by lot. For what if the lot were repeated several times? "Were it certain, yea or probable that they should all light upon the same person? Or were it not frivolous, if not impious, therefore to say, that upon every second shaking or drawing God altereth his sentence, and so to accuse him of inconstancy; or that to several Companies he giveth a several sentence, and so to charge him with contradiction and contrariety?"[135] Gataker's use of the word "probable" in this passage seems to be the first time the word was used in connection with random phenomena and, hence, to be the first time a connection was recognized between random phenomena and the strength of belief. His argument seems to have had some effect in the eventual disuse of lots in the British Protestant churches.[136]

Replying to Gataker, the more radical Puritan William Ames held that the lots in which God's providence was to be looked for were only those that were "purely or merely contingent"; he means, in effect, an event with probability a half. "But we do not place a Lot simply in contingency, but in mere contingency: for there are three degrees of things contingent: some often happening, some seldom, and some as far as we can understand, equally having themselves on either part (*equaliter in utramque partem*): for in other Contingents there is some place left to Conjecture by art: but in mere contingency there is none."[137] The ancient division of "happening rarely/as often as not/often" undergoes a subtle change here, in applying to something as easily quantifiable as dice throwing. Gataker replies that actual lots are often not as Ames describes, since it is customary to put thirty or forty blanks in for every lot inscribed with a name, so that it is blanks that usually appear.[138]

Commerce and the Casuists

Debates continued on the morality of the various kinds of contracts. The distinction between usury and insurance threatened to be obscured by the development around 1500 of a complex hybrid, the triple contract. This involved a first contract of partnership between an investor and a venturer, a second contract of insurance on the principal, in which insurance was paid in return for the future probable gain from the partnership, and a third contract in which the uncertain future gain was sold for a lesser certain gain. Each of the contracts appeared to be licit, but the result was a guaranteed fixed interest return, or usury. The debate on the legitimacy of the triple contract dragged on through the usual farrago of scandals, interventions by papal bankers, Jesuit intrigues, appeals to Rome, the setting up of commissions, delays, hopes for an imminent decision raised and dashed, more delays. In the end, very little happened.[139] As one of the first Jesuits observes, the merchants, driven by their lust for gain, have made things so complex that "the least change of circumstances makes it necessary to revise one's judgement."[140]

The basic issue of whether risk excuses interest is considered in the *Discourse upon Usury* of Thomas Wilson (whose *Logique* and *Rhetorique* are considered in chapter 5). He seems to admit that risk could in principle be a title to interest but that, since the merchants have arranged that "it is a hundred to one that by the course and order of the exchange the deliverer of money shall be no loser in the end . . . the gain thereby is indeed very certain to them, let them say what they list."[141]

Domingo Soto, a student of Vitoria and confessor to the emperor Charles V, is noted in the history of science for the first suggestion that falling bodies might be uniformly accelerated and in international law for his strong doubts about the legitimacy of Spanish rule in the Americas. He wrote one of the treatises on doubts of conscience of the kind discussed in chapter 4, and he is known in the history of economics for his accommodating views on commercial practice.[142] Inquiring whether insurance is legitimate, he does not take the question very seriously, thinking that of course it must be legitimate, since it is necessary for business. He notes possible difficulties with the bull *Naviganti* and the view of some that insurance involves paying for something that does not exist, a "peril," but replies with an analysis of the nature of insurance that includes a comparison of it with a fair game of chance. "For anything that can be estimated at a price, one can receive a fee: to render a thing safe [insure it], which is exposed to peril, can be estimated at a price . . . we say of a fair game: whether it will rain tomorrow or not, etc; so in the same way it is permitted to expose a thousand ducats, say, to peril with the hope of making fifty or sixty. There are some who regard it as stupid to allow the peril of someone's ship worth perhaps twenty or thirty thousand, in the hope of making a hundred or a thousand. To this we reply that it is not for us to dispute about prices: these can be just or unjust, but it is for the contracting parties to decide them. But there is no stupidity or folly in accepting this kind of peril at the going price; in fact nothing is more obvious than that insurances can expect to gain. They may lose sometimes, but at other times they accumulate gain."[143]

In the seventeenth century, theoretical discussion of speculation, insurance, betting, and so on was still found mainly in the works of the Scholastic moral theologians. What they say does not make any notable advance on earlier writers on those themes, but they provide a substantial summary of the ideas that were available to the discoverers of mathematical probability. The most complete and authoritative treatment is the section on aleatory contracts in the *De Iustitia et Iure* of the Jesuit Lessius (whose work is considered in two other chapters). Lessius had grown up near Antwerp, at the time of its preeminence in finance, before entering the Jesuits and studying under Suarez. His *De Iustitia et Iure* reflects a considerable understanding of financial practice, and has a claim to be the most significant work of its century in economics.[144] For example, he explains why interest may be taken on loans in compensation for risk; those who are deprived of their money by lending it "value more the lack of their money for five months than the lack of it for four, and the lack of it for four more than three, and this is partly because they lack the opportunity of gaining with that money, partly because their principal is longer in danger."[145]

On life annuities, Lessius says as usual that the uncertainty in how long the buyer will live justifies the price charged, but he remarks also that the price should actually be less than the expected total income in the future, since present money is worth more than future money: An annuity is legitimate even if, he says, "it is much more probable that he [the buyer] will live longer, and hence receive more income; because he is not buying income, but the right of receiving it. Such a right is worth much less than the same amount in the present, both because of the utility of present money, and because there are many ways in which during such a span of time it may turn out that he does not receive it. . . . Note that this price, for a single life of one of good age, is in some places eight years' purchase, in some seven years'; on two lives, ten years'. For what in some country is set down in law or common estimation is presumed to be a just price, unless the contrary is evident."[146]

On the question of estimating the just partition of profits in a partnership, Lessius argues that the share of the investing partner should depend on his risk. "If you contribute 1000 aurei, the risk to which is estimated at 100, and the hope of profit, less expenses, is estimated at 200, then you are regarded as having contributed 300 aurei, whose loss you should bear. Suppose the work of the other [partner] is estimated at 100, so he is regarded as contributing 100, whose loss he would bear. If the profit this work realises is 800, then he will have a quarter of it, or 200." To the objection that it is not 100 but 1,000 gold pieces that the investor stands to lose, Lessius replies that "the uncertain peril of chance should be reduced to a certain price" and imagines a fictitious contract of insurance that the investor could enter into, paying a premium of 100 gold pieces for the safety of the 1,000. Lessius gives commonly accepted figures for some of the relevant chances: for the loss of money invested with an ordinary merchant, 1 percent; for the gain to be hoped for by investing with a merchant, if he is industrious, 10–12 percent, "often much more." But he thinks that normally the "required equality between the price paid to the insurer, and the obligation he accepts" can be left to the judgment of intelligent men.[147]

What this means is expanded by a later writer of a *De Iustitia et Iure*. In the chapter on "gaming, wagers and insurance," the Jesuit cardinal Juan de Lugo (whose views on induction are noted in chapter 8) explains that the equality is between a price and a risk. "The first condition for the justice of an insurance is, that the price be equal to the peril undertaken; certainly that the price paid for the obligation should be as much as that obligation is worth in the judgement of experts. This price is not a definite amount, but has a maximum, mean and minimum, as with buying and selling. As varied circumstances affect the peril, so the just price should be varied. The equality is to be taken from the quantity of the peril at the time of the contract, not after the event."[148]

It was a matter of dispute as to the morality of cases of unequal risk to the two parties, as when one knows the ship is safe, or lost, or knows from astrology that there will be no storms, or from private communications that there are no pirates. Lessius is inclined to think one may still accept the contract at the commonly accepted price, just as a vendor may take advantage of a high price caused by an apparent scarcity which he happens to know is not real.[149]

Lessius's and Lugo's works are full treatments designed for academics, but the same ideas appear at a more applied level. In 1645 the Jesuits in China extracted for the benefit of their local converts a dispensation to charge 30 percent on loans on condition that "there is considered the equality and probability of the danger, and provided there is kept a proportion between the danger and what is received."[150] One of the casuists who wrote manuals for confessors was Bauny, who was, like Lessius, scathingly attacked in Pascal's *Provincial Letters*. Bauny, writing in French, is close to Lessius on the partitioning of profits.[151] (A few further opinions of these authors concerning the particular aleatory contracts known as games of chance are considered in the next chapter.)

The casuists were the last link in the chain joining the pricing of hopes and perils of Roman law with the problem of the just division of stakes in an interrupted game of chance, the problem that led directly to the mathematical theory of probability.

CHAPTER ELEVEN

Dice

S tochastic processes like the throwing of dice and coins have traditionally been the main subject matter of histories of early probability. The present work takes a much wider view of the nature of probability. Nevertheless, this originally peripheral and even sordid byway,
gambling with symmetrical thrown objects, has a special claim to fame.
It proved the first part of probability to be mathematized. The reason
for this is that the mixing or shuffling effect of the many twists as the die
or coin flies through the air "smears out" any asymmetry in the initial
conditions of the throw, so that the falls on each face are equally probable. The result is that the probability of complex outcomes can be determined mathematically, simply by counting the equally probable cases
that comprise them.

That still leaves the problem of finding situations in which it is seen
to be worthwhile to actually determine probabilities involving dice. In
view of all that has gone before on the centrality of legal and moral reasoning, it will be explicable that the problem motivating study of these
problems was a legal-casuistic one: what is the just division of the stake
in an interrupted game of chance?

Games of Chance in Antiquity

Games of chance are found in virtually all societies, even in hunter-gath-
erer societies that lack games of strategy.[1] Perhaps the oldest indications
of some kind of realization of the differing probabilities of outcomes are
to be found in the high scores awarded to unlikely outcomes in games.
A game common among various North American native tribes involved
throwing "hands" of flat objects colored differently on each side—pieces
of elk horn, butter beans, or similar. In the Cayuga version, six smoothed
and flattened peach stones, burned on one side, were put in a bowl,
shaken, and thrown. If all six landed on the same side, the player scored
five points and earned another throw. If five landed on the same side,
the player scored one point and another throw. Otherwise, he scored no
points and passed the bowl to his opponent.[2] The actual probabilities
are $1/32$ for six of a kind and $6/32$ for five of a kind, so there is a reason-

ably close relationship between the scores used and the actual frequencies of the outcomes.

It is not clear if the same is true of the most common game of chance in classical antiquity, the throwing of four astragali. Astragali are the knucklebones of sheep or other animals. They are all much the same and have four sides, with probabilities about 0.4 for the flatter two sides and 0.1 for the narrow two sides. There was much technical terminology concerning the thirty-five possible throws of four but no sign of any understanding of the probabilities of them.[3] The highest scoring throw was the Venus, where all four outcomes were different. It is actually the most likely single throw, so there is no sign of knowledge of the rarity of throws here. But in divination provided by temples in the classical era, using throws of four astragali, it seems that the unfavorable throws were much less probable than the favorable ones. No doubt it was good for business.[4]

In classical antiquity, the closest approach to a mathematics of chance is an isolated remark of Aristotle: "To succeed in many things, or many times, is difficult; for instance, to repeat the Koan throw ten thousand times would be impossible, whereas to make it once or twice is comparatively easy."[5] A similar remark by Cicero was noted in another context in chapter 7, that one could attribute to chance the occasional two or three Venus throws in succession, but one would not think it chance if four hundred Venus throws were observed.[6]

Aristotle's explanation of the apparent success of oracles also reveals an understanding of chance in a game. Those playing games of chance, he says, are "affected in the same way as most people are when they listen to diviners, whose ambiguous utterances are received with nods of acquiescence: 'Croesus by crossing the Halys will ruin a mighty empire.' Diviners use these vague generalities about the matter in hand because their predictions are thus, as a rule, less likely to be falsified. We are more likely to be right, in the game of 'odd and even,' if we simply guess 'even' or 'odd' than if we guess an actual number; and the oracle-monger is more likely to be right if he simply says that a thing will happen than if he says *when* it will happen."[7] Aristotle gives the same explanation for the occasional success of prophecies in dreams. Dreams appear in great number to inferior types of person, so one does not expect them to be reliable as predictions, "their luck in these matters being merely like that of persons who play at 'odd and even.' For the principle which is expressed in the gambler's maxim: 'If you make many throws, your luck must change,' holds good in their case also."[8]

There has been much speculation about why the ancient and medieval world did not go beyond these useful but qualitative remarks to develop a fully quantitative theory of chance phenomena such as dice throwing and coin throwing. There is a great deal of evidence, both archaeological and written, concerning ancient gaming[9] and many surviving symmetrical ancient dice,[10] but the upshot in terms of theory is practically nil. No convincing explanation has appeared of why this should be so. (The next chapter discusses some of the suggestions. Aristotle's remark just quoted and a saying of Xenophon, "nor on the other hand do I commend those dicers who, if they win one success, throw for double stakes, for I see that the majority of such people become utterly impoverished,"[11] at least rule out the theory that intellectuals were not capable of thinking about games of chance at all.)

The Medieval Manuscript on the Interrupted Game

Games of chance were the subject of some debate in legal circles, since Roman law forbade them but without explaining quite why.[12] It was not surprising that canon law forbade gaming by clerics, since clerics are supposed to have higher things on their minds.[13] But this left the impression that gaming was not forbidden for lay people, and it was not at all clear why there was anything unfair in the contract that constitutes a game of chance.[14] Agreement was never entirely reached, but the general consensus among lawyers, as expressed definitively by Baldus, was that there was nothing inherently wrong with gambling but that it was often forbidden by the law of the state because of the many attendant evils, such as blaspheming by the players and the dissipation of patrimonies by feckless youth.[15] The distinction between games of chance and games of skill was drawn by the Glossators.[16] And Baldus says further, in a passage that reveals some thought about the nature of probability in games, that "if a game is one of mixed fortune and skill, as for example the games of *minoretti* or of backgammon, then either the money is lost through the ignorance of the player, who does not know how to work out what the outcomes normally are, and then it is his fault, for it is not outside the skill of the game, as per *Dig.* 11.5 (l. *alea*). Or he loses by the chance of the dice, because the victor throws an unusual point, which is not likely to happen except rarely."[17] Backgammon is particularly interesting from a probabilistic point of view because of its mix of chance and skill. Much of the skill involves estimating chances, but despite the popularity of the game, there seems to be no explicit mention of this.[18] Also from the legal world comes a small example of the use of dice games to model another probabilistic phenomenon. A jurist around 1300 mentions the opinion that the ancient Roman "maritime interest" is "not prohibited, just as games of dice are not, because of the risk and by reason of the doubtful outcome."[19]

"Prognostication about the future" was one of the main "uses" of dice in the Middle Ages; fortune-telling using dice probably produced most of the writing on the subject. There seem to be no ideas of interest for probability in this material.[20] The rest of what was written on dicing consists mostly of condemnations by clerics—describing its invention by the devil, urging the faithful to include dice and backgammon boards in the "bonfires of the vanities," and so on.[21]

Writings on dice games simply as games are rare in the Middle Ages but not nonexistent. There is, for example, a short treatise on dice in the compilations of the thirteenth-century written under the name of Alfonso X, king of Castile. It at least recognizes that dice should be symmetrical and that it is possible to cheat by biasing the die.[22] Various ordinances providing for punishments for using false dice[23] argue for a wide distribution of some kind of implicit knowledge of probabilities. Arguably, biased dice are a stronger sign of knowledge of probabilities than symmetrical ones.

At the end of the Middle Ages, a strange connection between probability, in the sense of the judgment of uncertain matters, and dice occurs in Rabelais' story of Judge Bridlegoose, who decides doubtful cases by the throw of dice. Rabelais was a former law student and knew what he was parodying. "How come you to know, understand and resolve the obscurity of these various and seeming contrary passages, in law, which are laid claim to by the suitors and pleading parties? 'Even just,' quoth Bridlegoose, 'after the fashion of your other worships: to wit, when there are many bags [of documents] on the one side, and on the other, I then use my little small dice, after the customary manner of your other worships, in obedience to the law, *Semper in stipulationibus ff. de reg. jur.* and the law *versale* verifieth that in the same title, always in obscure matters we follow the smallest things.'" The "continuation of Bridlegoose for so many years, still hitting the nail on the head, never missing the mark, and always judging aright, by the mere throwing of the dice" is a sign of divine help.[24]

These various qualitative remarks on dice are interesting. Serious mathematical thought on the subject, however, needs a sufficiently numerical culture. Such a culture became established in fourteenth-century Italy, which was responsible for a number of the crucial steps in creating a tradition of skill in applied numerical calculation. It was the scene of many of the first mechanical public clocks, for example, of the invention of double-entry bookkeeping and, as we saw, of insurance.[25] Especially notable is the discovery of a correct means of predicting the future numerically, tables of compound interest.[26] As experience was gained with numbers, it was realized that calculation was not an intellectual feat reserved for thinkers of genius but (using Arabic numerals with a zero) could be reduced to simple rules and taught to children. These rules—modern-day primary school arithmetic—were taught (in Italian) to merchants' sons in the commercial arithmetic schools of the Italian city-states.[27] The better teachers in these schools went further, developing much of modern high school mathematics, in particular algebra, in the sense of problem solving using the manipulation of unknown numbers represented by letters.[28] As with modern textbook authors on algebra, the Italian writers had the custom of illustrating their techniques in a number of unrealistic "real life" examples to which the techniques "apply." Simple number problems arising from dice occasionally make an appearance.[29]

There occur in a commentary on Dante, possibly from the late fourteenth century, some remarks on the throwing of three dice, which have been described as "the earliest mention of probabilities in world literature" (meaning, of course, factual probabilities).[30] It is stated that the lowest and highest throws, 3 and 18, can occur in only one way (three 1s and three 6s, respectively) and that the same is true of the next to highest and next to lowest, 4 and 17: 4 can only be made up of two 1s and a 2 and 17 of two 6s and a 5. This is not correct if the problem is one about probability, since 4 and 17 occur three times as often as 1 and 18. Either the commentator is making a true remark about combinatorics, in which case he knows nothing about relative frequencies in repeated throws, or he is trying to obtain a result about relative frequencies and does not know what combinatorial fact is relevant. Earlier medieval works had enumerated correctly the fifty-six different outcomes (in the natural sense of partitions) for throws of three dice, suggesting the first purely combinatorial interpretation.[31] But the Dante commentary hints at the second, relative frequency, interpretation by adding that the outcomes 1, 2, 17, and 18 are called "hazards" because they appear rarely and that this happens because they are made up in fewer ways than the numbers from 3 to 16: "It is to be observed that the dice are square and every face turns up, so that a number which can appear in more ways must occur more frequently."[32] One of the earlier works hinted at the same interpretation by correctly distinguishing between partitions (*punctaturae*) and "fallings" (*cadentiae*) and correctly counting both for three dice; thus the outcome 3 has one partition and one falling but 4 has one partition and three fallings (1, 1, 2; 1, 2, 1; and 2, 1, 1).[33]

Out of this context, with mathematical thought about dice in existence, but only barely, there suddenly appears the statement, and the exact solution, of a difficult problem on the correct division of stakes in an interrupted game of chance. The problem is one of those solved in 1654 by Fermat and Pascal that constitute their claim to have founded mathematical probability. The trivial cases of an interrupted game are when each of two players needs the same number of points to win. In that case, the stake should obviously be divided equally. The simplest nontrivial case is when one player needs two points to win and the other one. The answer as to how the stakes should be fairly divided if the game is interrupted at that point is not easy, as Pascal and Fermat discovered. The next simplest case is when one player needs three points and the other one. This is the problem solved in an anonymous Italian mathematical manuscript of about 1400. It is not entirely easy to follow, but on examination the reasoning is correct:

> Two men play chess, and [each] bets one ducat on [being the first to win] three games. It happens that the first man wins 2 games from the second. I ask: if the game proceeds no further, how much of the second man's ducat will the first have won? Suppose that the first man has won x from the second in the first game; you must see that he has won the same amount in the second game as in the first, by the same reasoning. Therefore he will have won another x, and so will have won $2x$ in two games. So the second man, the loser, will still have out of his ducat 1 ducat minus $2x$. It is to be known that if the one who has lost two games were to win two further games from his companion, neither would have won anything from the other. Now let us suppose that the second man begins by winning one game from the first; I say that he wins in this game the 1 ducat minus $2x$ that the first had won. The reason is that if the one who had won the first two games had won the third game too, he would have won from the first [read: second] man the whole remaining part of his ducat, and so conversely the second wins from the first 1 ducat minus $2x$, so he wins back 1 ducat minus $2x$ out of the stake that the first player had won from the second, that is, $2x$. There will then remain to the first man $4x$ minus 1 ducat. So the second man before he wins his second game will have 2 ducats minus $4x$. Now observe that for the first man (who has won two games), that if the second man has won the [next?] two games he must have won the third game; there would be

TABLE 11.1. DIVISION OF STAKE IN CHESS GAME

	Two wins to A	Two wins to A, then one to B
Amount of B's ducat won by player A	$2x$	$2x - (1 - 2x) = 4x - 1$
Amount of B's ducat won by player B	$1 - 2x$	$(1 - 2x) + (1 - 2x) = 2 - 4x$

nothing he [the second] had not won [back], in all reason, of what the first had in his [the second's] ducat. If the first had won that third game he would have won 2 ducats minus $4x$, so likewise that is what the second [actually] wins from the first. Now we are supposing the second man wins his second [that is, the fourth] game; therefore he has won from the first 2 ducats minus $4x$, and also he must have won back everything the first had won, because they have won two games each. Now look at how much the second wins from the first in his second [the fourth] game: he wins 2 ducats minus $4x$. Now [in the equation $4x - 1$ $= 2 - 4x$] we should add one ducat to each side, and will have on one side $4x$ and on the other $3 - 4x$; then add $4x$ to each side, and there will be $8x$ equal to 3 ducats. Then divide the number by [the number of] xs, which gives $\frac{3}{8}$ for the value of x. This is what the first man wins in the first game, and in the second game he wins a further $\frac{3}{8}$ ducats which makes $\frac{6}{8}$, that is, $\frac{3}{4}$. This is how much the first man has won when they play no further than two games; and thus you proceed in similar situations.[34]

Table 11.1 may make the course of the reasoning clearer. The calculations follow the arrows, and the quantities in the last column are equated. Some further commentary is required:

— The answer is correct. After A has won two games, B has to win three games in a row to win, the probability of which is $\frac{1}{8}$. So B should take $\frac{1}{8}$ of the total stake of two ducats, that is, $\frac{1}{4}$ of a ducat. Thus A wins $\frac{3}{4}$ of a ducat from B, as the text says.[35] The answer for the division in the simpler case when A has won two games and B one is also correct.

— In this reasoning, a good deal depends on the principle that if a player wins two games in a row, his winnings in each game are equal.

This is especially important in moving from the third game to the fourth. The principle is not obviously true, and the author gives no argument for it. But it can be justified, if "winnings" in a game is taken to mean the difference in the expected payoff of the winner after and before the game.[36]

— The x in the text translates "c." in the manuscript, which stands for *cosa* (thing) and is the standard notation for an unknown in algebra.

— The author does not mention anything like chance or probability. All he uses is the implicit assumption that it is as easy for one player to win a point as it is for the other. This means that he effectively treats the problem as if it were a game of chance, whereas chess is a game of skill (though there was such a thing as chess played with moves determined by dice).[37] The difference is that in a game of chance each player is in an equal position in every game, whereas in a game of skill previous wins to one player indicate he is more skillful and, hence, more likely to win the next game.

— There follows in the text an attempt to apply the same reasoning to a more complicated case. The reasoning goes wrong very early, in a way that suggests a lack of grasp of the principle involved in solving the easier case but perhaps is simply a sign of defeat in the face of complexity.

There is no evidence that anyone but the author ever read the manuscript. But the question at least is mentioned in two known fifteenth-century Italian mathematical manuscripts.[38]

Cardano

Discussion of the problem of dividing the stakes in an interrupted game continued in the arithmetic books intended for merchants. Fra Luca Pacioli's *Summa de Arithmetica* of 1494 is not a superior work of its kind, but it had the all-important advantage of being printed. It treats the problem simply as an exercise in proportion: the stakes should be divided proportionally to the gains already made.[39] Pacioli thus has no notion of the probabilistic aspects of the problem. Indeed, he gives the same rule for games of skill and horse races. Even he mentions the most obvious problem for this formula: that it gives the whole stake to one player when the other has not won any points. This objection was raised again by Tartaglia, who concluded, "The resolution of such a question is rather judicial than rational, so that however one tries to resolve it, one will find there matter for litigation."[40] It was left to Cardano not only to partially solve the problem but also to realize that there was a genuine mathematical problem admitting of an answer.

The son of a lawyer, Gerolamo Cardano eventually acquired a doctorate in medicine after much opposition from his many enemies. In middle life he was a physician of international fame, traveling to Scotland to cure the Archbishop of St. Andrews of asthma. On his return journey, he was impressed by the young King Edward VI of England, for whom he cast a horoscope predicting a long life, shortly before the king's death at sixteen.[41] His readiness to make such a falsifiable prediction contributed to his later reputation as a charlatan. More justly deserving of such a reputation is his invention of a divinatory art based on the position of facial warts.[42] He is more rational in his style of argument in the interpretation of dreams. Interpretation, he says, is conjectural, but it is none the worse for that, since the same is true in medicine and astrology. He argues that Ptolemy's astrology has the same scientific status as the books of Hippocrates and Galen on prognostic. His method supplies a list of dream symbols that may have multiple meanings. An old man dancing signifies folly, indecency, or death. "Riding a mule alludes to a boring enterprise, with little risk or profit."[43] He also fits patterns from life to the events in the dream; for a man of the word, dreaming of cutting a man's throat may indicate he has refuted someone's book.[44] The style is recognizably Jungian. When almost seventy years old, he fell foul of the Inquisition; he was jailed for some months and forbidden to publish. The charges are not known with certainty, but they may have involved a horoscope of Christ and a book in praise of Nero.[45]

Cardano's addictions included, besides pseudosciences, gambling and mathematics. In the 1540s, he was involved in the solution of the cubic equation, the first major discovery in Western mathematics since Oresme.[46] Though he was not the actual solver, his work on this difficult question places him among the best mathematicians of the century. The combination of this with a much more common sixteenth-century accomplishment, vast experience in gambling, gave Cardano the ideal skills for work on probability. The financial results were disappointing, however.[47]

His first successful book was his *Practica Arithmetica* of 1539; it is one of the most influential books on the subject. One chapter is "On the errors of Fra Luca in arithmetic." These include his "very manifest error, that a child could see" in dividing stakes in interrupted games. Cardano considers a game in which the first to gain nineteen points wins, and the first player has eighteen points and the second nine. Pacioli's rule would give the second player a third of the stake, which is obviously too much. Again, if the first to gain nineteen wins, and the first player has two points and the second none, Pacioli's rule would give the whole stake to the first, whereas it is clear that his real advantage is very small, considering how far it is from the end of the game.[48] It is clear that Cardano has grasped the central point: that what matters is the events that have to happen for the various players to win, not what has happened thus far.

The rule he proposes in place of Pacioli's satisfies this requirement and also gives a greater advantage to a player who needs very few points to win. He says one should divide the stakes in the ratio of the "progressions" of the number of points each player needs to win. The progression of a number is the sum of the numbers up to it; thus the progression of 4 is 1 + 2 + 3 + 4 = 10. For example, if the first to reach 10 wins, and one player has 7 points and the other has 9, the stake should be divided in the ratio 1 + 2 + 3 to 1, that is, 6 to 1. He argues, "For the demonstrative reason for this is that if, after the division, the game were begun again, the players would have to stake the same as they received at the time of stopping." The notion that there should be demonstrative reasoning at all is significant, as is the argument from symmetry. The language recalls that later used by Huygens. The actual reasoning offered to justify the rule of ratios of progressions in this example is more like the recursive arguments of Pascal. "Someone says, 'I wish to play on the condition that you need three in a row to win, but I win if I win one.' He who must win three games stakes 2 ducats; how much has the other to stake? I say he must stake 12 ducats. The reason is, that if he had to play one game it would suffice to stake 2, and if two games, he would have to stake three times this; for in winning two games simply, he would win 4 ducats, but he carries on, with the peril of losing the second after having won the first, therefore he should be paid triple. And if to three games, then six times, since the difficulty is doubled, therefore he has to stake 12."[49] Cardano has not arrived at the right answer; however, he understands the question well and has found one of the essentially correct methods of reasoning about it.

The above occupies two pages in a successful book. Besides this Cardano wrote a substantial manuscript on games of chance, *De Ludo Aleae*. It seems to have been written in the 1520s and revised in the 1560s, but it was not published until the collected edition of Cardano's works of 1663, that is, after it had been superseded by the work of Pascal, Fermat, and Huygens.[50] It is a confusing work; it is often not revised well enough to make the author's intention clear, and there remain in it sections explicitly contradicted by later ones. On the throws of one die he writes that "one half of the total number of faces is always equality; thus a given point will appear in three throws, for the total circuit is completed in six, or again one of three given points will turn up in one throw. For example, I can as well throw one, three or five as two, four or six. The wagers are therefore laid in accordance with this equality if the die is honest, and if not they are made so much the larger or smaller in proportion to the departure from true equality."[51] Equality should be thought of as something like equity. The phrase "the total circuit is completed in six" is not clear but obviously has something to do with indicating the whole space of possibilities.

There follows a chapter on why gambling was condemned by Aristotle, then the matter of throws of two dice is taken up. He correctly counts the combinations of two dice. "In the case of two dice, there are six throws with like faces, and fifteen combinations with unlike faces, which when doubled gives thirty, so that there are thirty-six throws in all, and half of these possible results is eighteen. As for the throws with unlike faces, they occur in pairs in the circuit of eighteen to equality, so equality for such a throw consists of nine casts; and therefore the throw $(1, 1)$ can in reason appear and not appear in eighteen throws; and similarly for $(2, 2)$ and $(3, 3)$." The phrase "can in reason appear and not appear" is the closest Cardano comes to equal probability. It seems he believes, in modern language, that since the chance of getting $(1, 1)$ in a single throw is $1/36$, the chance of getting $(1, 1)$ in eighteen throws is $1/2$. This is not correct, but it is a natural mistake. Cardano continues: "But $(1, 2)$ can come up in two ways, therefore there is equality in nine throws; if more frequently or rarely it is from fortune."

But then he realizes that the reasoning is not satisfactory and that the mistake arises somehow from the possibility of having more than one of the successful throws. "The whole circuit is not wrong, except that in one there can be repetition, even two or three times. Accordingly, this knowledge is based on conjecture and is approximate, and the reckoning is not exact in these details; yet it happens in the case of many circuits that the matter falls out very close to conjecture."[52]

The case of three dice is then treated. The reasoning has the same correct results and the same mistakes as for two. He says, incorrectly, that since the chance of a 1 in one throw is $1/6$, the chance of at least one 1 in three throws is $1/2$.[53] But two chapters later he obtains the correct answer to the same question, namely $91/216$, and puts forward the correct method, that of counting the cases for and against. His statement is, for once, perfectly clear: "I have wished this matter not to lie hidden because many people, not understanding Aristotle, have been deceived, and with loss. So there is one general rule, namely, that we should consider the whole circuit, and the number of those casts which represents in how many ways the favorable result can occur, and compare that number to the remainder of the circuit, and according to that proportion should the mutual wagers be laid so that one may contend on equal terms."[54]

He also considers how to deal with repetitions of events. After a false start, he finds the correct answer. Though expressed somewhat clumsily mathematically, it is equivalent to the rule that to find the probability of repetitions of an event one must take the powers of its probability. The examples Cardano gives make it clear he understood this in full generality.[55]

Later, in a discussion on methods of cheating, occurs the only mention of "probable" or "likely": "When the die is thrown straight with such an impetus and such a number of points exposed above that it is likely (*verisimile*) that the point which we wish will come uppermost."[56]

Gamblers and Casuists

The connection between cheating and knowledge of probability in Cardano suggests the hypothesis that, among the many cheaters at dice through the centuries, there must have been some with a reasonable empirical understanding of probabilities. Obviously, any such experts will have been too busy making money to reduce their understanding to writing, so the hypothesis is hard to confirm. One piece of evidence does, however, appear in an English Elizabethan booklet on how to cheat at dice (or "warning against dice play," as such works are always called). It contains the following exchange on whether a certain large loss was a result of cheating, indicating a basic ability to know what would be a rare event if there were no cheating. "But his luck was bad; the like falleth scarcely once in a hundred years. That is but one doctor's opinion. I see it betide every day, though not in this so large a proportion."[57]

There follows a description of how exactly to cheat with biased dice. For the game *novem quinque*, where it is desirable to avoid a throw of 5 or 9 with two dice, one uses one pair of dice slightly elongated along the 4-3 axis, making a throw of 4 or 3, and hence a total of 5 or 9, almost impossible. "So long as a pair of barred cater-treys be walking on the board, so long can ye cast neither 5 nor 9, unless it be, by a great mischance, that the roughness of the board, or some other stay, force them to stay and run against their kind; for without cater-trey ye wot that 5 or 9 can never fall."[58] But it would be suspicious if 5 and 9 never came up, so one occasionally substitutes a pair with the opposite bias to allay suspicion, being careful to bet little in this case. A later Elizabethan book on cheating at cards gives numerical odds (not necessarily intended as an exact estimate) for the dupe's conviction that he is about to win: "The cony [dupe], upon this, knowing his card is the third or fourth card, and that he has forty to one against the barnacle, pawns his rings, if he have any, his sword, his cloak."[59]

The first known genuine probability calculation in a published work intended for gamblers is in Cotton's *Compleat Gamester* of 1674, in which it is shown that a throw of 7 with two dice is more likely than either 6 or 8. This is of course later than the mathematical work of Fermat, Pascal, and Huygens, though whether influenced by it is not easy to say.[60] On the whole, however, one learns from such accounts of cheating that matters of the biasing of dice and the marking of cards are much more important than any knowledge of the outcomes of fair games.

The late Scholastic legal and moral writers also undertook some serious though brief discussion of games of chance. As seen in the last chapter, games of chance were included in the category of aleatory contracts, along with insurance and annuities. Santerna's book discusses merchants' bets with insurance, and Soto's *De Iustitia et Iure* compares the risks in insurance contracts with a fair game. Lessius holds that sometimes players tacitly consent to allow small frauds, so that one need not keep as strictly to equality as in commercial contracts. For example, although one may not in general play a game in which, because of one's skill, one is morally certain of winning (because there must be equality in the condition of the players), it is permitted to raise the stake without revealing that one has a winning hand. Escobar, Pascal's opponent from the *Provincial Letters*, argues that games of chance are generally permitted. He holds that equality of the stakes is normally required but that a player may stake less money licitly if his potential winnings are greater. This seems to mean, though Escobar does not explicitly say it, that what should be equal for the two players is the stake multiplied by the payoff (when their chances of winning are equal).[61]

The *Summa* of the casuist Diana has an interesting remark on calculating the amount of restitution owed by someone who has won a game of chance by cheating. "If someone was declared the winner because it was his turn, when the players had equal points, then he is obliged to make restitution not only of the whole he has won, but over and above that of the amount he would have lost; because that was the whole loss of which his cheating was the cause."[62] This is logically a simple instance of the problem of just division of the stakes in an interrupted game.

Galileo's Fragment

Galileo wrote a fragment on dice, not published until centuries later. In it he correctly explains why, when three dice are thrown, the totals 10 and 11 appear more often than the totals 9 and 12, even though all these numbers can be written as the sum of three of the numbers 1 to 6 in the same number of ways (partitions).[63] There are many remarkable things about this fragment. Though it is the first known proper treatment of probabilities in dice throwing, except for the confusing and unpublished work of Cardano, Galileo writes as though these matters are very obvious and well known. He implies that it is beneath the consideration of a mathematician to write about them and that he is only condescending to do so because he is paid to by his patron.

Then there is the matter of his reference to experience. He writes, "It is known that long observation has made dice players consider 10 and 11 to be more advantageous than 9 and 12." This is again the first clear reference to gamblers collecting any experience of long-run frequencies in games of chance. Yet it has been plausibly argued that the two probabilities are much too nearly equal ($^{27}/_{216}$ against $^{25}/_{216}$) for the difference to be noticeable against the expected background of noise.[64] There is some suspicion that Galileo was here inventing his experimental evidence. It would not be the first time.

Galileo correctly solves the problem by explaining that some "ways" of making up numbers are made up of more "throws" than others. For example, a 4 and two 3s is easier than three 3s, because to get three 3s each die must come up 3, while to get a 4 and two 3s there are three possibilities: 4, 3, 3 or 3, 4, 3 or 3, 3, 4. His probabilistic concepts are expressed by the phrases "more easily and more frequently" ("Some [numbers] are more easily and frequently made than others"; *più facilmente e più frequentemente*) and "indifferently" ("A die has six faces, and when thrown it can indifferently fall on any of these"). No words like "probable" or "likely" appear in this work, and there is no suggestion that Galileo saw any connection between these matters of factual probability and his work on the probability of scientific hypotheses.

De Méré and Roberval

How long was history to wait until there was someone who was interested in probability in dice and at the same time had the mathematical skills necessary to quantify it? Finally, in the 1650s, there appeared at least six such people, Pascal, Fermat, Huygens, Caramuel, Roberval, and de Méré.

"A problem about games of chance proposed to an austere Jansenist by a man of the world was the origin of the calculus of probabilities."[65] Thus legend. But the Jansenist Pascal did not originate mathematical probability, since the "man of the world," the Chevalier de Méré, did not merely propose the problem but calculated the solution. We know this because Pascal says so.

In their series of letters of 1654, Fermat and Pascal discuss two problems, both suggested by, or at least known to, the Chevalier de Méré. The first of these is known as the "problem of points." It is the old problem of the just division of the stake in an interrupted game of chance. The second is the "problem of dice" (*parti des dés*). If one agrees to throw a certain number in a given number of throws, does one have the advantage? For example, if one agrees to throw a 6 in four throws (of a single die) does the thrower or his opponent have the advantage (in modern language, a probability greater than $\frac{1}{2}$ of winning)?

The first letters in the exchange are unfortunately lost, and the following remarks in the first letter from Pascal are all the information that survives about the preexisting state of the problems: "I have seen solutions of the problem of the dice by several persons, as M. le Chevalier de Méré, who proposed the question to me, and by M. de Roberval also. M. de Méré has never been able to find the just value of the problem of the points nor has he been able to find a method of deriving it, so that I found myself the only one who knew this proportion."[66] Later in the same letter, Pascal says of de Méré, "He tells me then that he has found an error in the numbers for this reason: If one undertakes to throw a 6 with a die, the advantage of undertaking to do it in four is as 671 is to 625. If one undertakes to throw double 6's with two dice the disadvantage of the undertaking is 24. But nonetheless, 24 is to 36 (which is the number of faces of two dice) as 4 is to 6 (which is the number of faces of one die). This is what was his great scandal which made him say haughtily that the theorems were not consistent and that arithmetic was demented."[67]

It is clear, then, that before Pascal arrived on the scene, de Méré knew the following facts (expressed in modern language):

— The probability of throwing at least one 6 in four throws is $\frac{671}{1296}$. The modern way to calculate this is as $1 - (\frac{5}{6})^4$, but Fermat, Huygens and Caramuel all give ways of deriving this figure which are clumsier if perhaps more natural for the novice. No method is, however, completely easy.

— The probability of throwing at least one double 6 in twenty-four throws of two dice is less than ½, while the probability of doing so in twenty-five throws is greater than ½. To know this one must presumably calculate the two probabilities, which are respectively $1 - (^{35}/_{36})^{24}$, or 0.4914, and $1 - (^{35}/_{36})^{25}$, or 0.5055. To do that one must understand that to calculate probabilities for two dice one counts numbers of outcomes out of 36. (It has been occasionally suggested that de Méré used long-run gambling experience to guess this fact. But it would take a great deal of carefully controlled experience to distinguish probabilities so close, and there is no suggestion that either de Méré or anyone else made such observations or even noted explicitly that long-run relative frequency was relevant to what was being calculated.)[68]

— The problem of points is harder if a solution of any generality is sought.

To know even the first of these facts is to be considerably adept in mathematical probability. It follows that de Méré's claim to being the real founder of the subject is as good as anyone's, irrespective of any mistakes he may have made. What he claimed is contained in a letter to Pascal, probably of 1658 or 1659, though possibly retouched later. He said, "You know that I discovered things in mathematics so rare that the most learned of the ancients said nothing of them, and the best mathematicians of Europe were surprised at them. You have written on my discoveries, as have M. Huygens, M. de Fermat and many others who admired them."[69]

   What little further information we have about de Méré's contribution comes from inquiries Leibniz made in the 1690s. Leibniz says that he heard that it was "a certain gentleman gambler, but of an extraordinary penetration, who first considered it, and even conceived some idea of it before the learned came into it." He asks for more details. His informant names de Méré but says only that he "gave occasion to" the research.[70] On the basis of this information, Leibniz says that he "almost laughed" at the airs de Méré gave himself in the letter to Pascal but at the same time says, "It is true however that the Chevalier had some extraordinary ability, even for mathematics. . . . Being a great gamer, he made the first openings on the estimation of wagers, which led to the beautiful thoughts *de Alea* of Messieurs Fermat, Pascal and Huygens."[71]

Did de Méré possess the ability Leibniz describes? He gives no other evidence of ability in mathematics, strictly speaking. But he did later write a pamphlet on the game of hombre, arguing for an ability to give a connected exposition of games.[72] More important, if the problem of dice is fundamentally not one of pure mathematics but of mathematical modeling, or translating a physical situation into mathematical language, then de Méré has some claim to possessing the requisite ability. The question that is the main subject of his letter to Pascal is another problem of choosing a correct mathematical model: it is whether a line consists of a finite or an infinite number of points, that is, whether space is infinitely divisible. Pascal believes that it is infinitely divisible and that this is a demonstrable geometrical fact. He writes to Fermat that de Méré "is not a geometer (which is, as you know, a great defect) and he does not even comprehend that a mathematical line is infinitely divisible and he is firmly convinced that it is composed of a finite number of points."[73] Pascal is wrong; the question of the divisibility of physical space, which is what de Méré is considering, is a contingent one and a priori geometrical arguments to the contrary are without exception invalid.[74] De Méré's reasoning for his conclusion, based on the atomism of Epicurus and Lucretius, is by no means without merit.

However, the mention of Roberval as having also solved the problem of dice suggests that such methods were widely known among the mathematically literate. Roberval was a professional mathematician of considerable range and originality. Furthermore, he knew at least one fact of mathematical probability at least seven years earlier. It appears in an unpublished discourse on philosophy, written no later than 1647. There Roberval opposes to the Cartesian emphasis on the certainty of knowledge a mixture of Aristotelian ideas and what could be called a commonsense positivism. He argues that, as the senses are not infallible, it is necessary to act on the basis of signs, such as the shape and color only of an apple. "I say of a proposition that it is believable and likely (*vray semblable*) when, though not infallible, it has more appearances and signs than its contrary. . . . When there are more signs for one thing than for another, one should conclude for the majority of signs if they are equally considerable. One must believe that one thing will happen rather than another when it has more natural possibilities or when the like has happened more often, as in throwing three dice one should believe, and it is likely, that one will get 10 rather than 4, because 10 can be made up in more ways than 4."[75] Notable is the explicit connection between "likely" as used of propositions and the number of "natural possibilities" in throwing dice. The fact about dice that he asserts is of the kind known to Galileo, and even Cardano, but unlike them Roberval sees it as an explicitly probabilistic fact about what is likely to happen and what should be believed.

Roberval, like de Méré, suffered from a bad press. Leibniz was informed that he did not think such problems susceptible of demonstrative reasoning and himself says that Roberval "could not or did not wish to understand" what Pascal and Fermat had done.[76] If this refers to the problem of dice, it would contradict what Pascal says much closer to the time, but it is reasonable if it refers to the problem of points or just division, which could be regarded as a fundamentally legal problem, not solvable purely by mathematics. It therefore does not detract from any claim of Roberval's to have discovered the fundamental ideas of the subject. It is true that Roberval seems never to have claimed anything for himself in the matter, although he was rather known for making late priority claims for his mathematical discoveries. Still, he would not necessarily have wanted to draw attention to a topic in which his rivals had made him look stupid.

One other person has been named as making a contribution. This is Damien Mitton, a friend of de Méré and like him a man of the world and an Epicurean. The opponent of the religious worldview in Pascal's *Pensées* is sometimes given his name. Leibniz's informant believes Mitton "was also part of this discovery with his friend the chevalier." There are no further details.[77]

While Fermat and Pascal, and later writers, solved much more difficult problems than the problem of dice, that problem is the one that resembles the basics of the subject as it is now taught and practiced. It involves the two basic methods: finding probabilities for simple events by counting the different outcomes; and finding probabilities for combinations of events by multiplying the probabilities of the individual events. The answer, then, to the question, Who founded the mathematical theory of probability? must remain A person or persons unknown.

The Fermat-Pascal Correspondence

The first surviving letter of the famous correspondence—actually, only an extract—contains a reply from Fermat to Pascal. The problem is a version of the problem of dice, but it is not clear whether the two correspondents have in mind exactly the same question. The letter includes a quotation from a previous letter, now lost, which is the earliest statement of the problem: "But you proposed in the last example of your letter (I quote your very terms) that if I undertake to find the 6 in eight throws and if I have thrown three times without getting it, and if my opponent proposes that I should not play the fourth time, and if he wishes me to be justly treated [literally: impartial, *désintéresser*], it is proper that I have $^{125}/_{1296}$ of the entire sum of our wagers."[78] It is to be noted that the word "impartial" takes the entire weight of the concept of probability in the problem (and it does so again in Fermat's reply). Unnaturally from a modern point of view, even the problem of dice has been posed as one of a fair division of stakes.

Fermat thinks Pascal's answer is wrong, given a fair interpretation of the problem, although there is another interpretation on which it could be called correct. He begins by saying, "If I undertake to make a point with a single die in eight throws, and if I agree after the money is put at stake, that I shall not cast the first throw, it is necessary by my theory that I take $^1/_6$ of the total sum to be impartial because of the aforesaid throw. And if we agree after that that I shall not play the third throw, I should, for my share, take the sixth of the remainder that is $^5/_{36}$ of the total." And so on. What Fermat is saying has no connection with the announced problem of throwing a 6 in eight throws; the taking of a share in the stake for forgoing a throw seems to be the only way he and Pascal have of conceptualizing a probability. The next letters are not a reply to this first surviving one, and there remains an impression that at this stage Fermat and Pascal are still not entirely sure what the question is.[79]

Most of the rest of the correspondence deals with the problem of points. A solution is sought to the general problem: What is the just division of the stake if the game is interrupted when one player needs a points to win and the other b points? Fermat and Pascal propose different methods for calculating the answer. The difference between their methods is a purely mathematical one and has no bearing on the conceptions the two have of probability, a matter on which they say almost nothing. Fermat's method is the "method of combinations." It is to find the maximum number of throws necessary to complete the game (it is always $a + b - 1$); then to list all the 2^{a+b-1} possible outcomes of these games and to count how many of these result in a win for each player. The stake is divided accordingly. Pascal explains:

> If two players, playing in several throws, find themselves in such a state that the first lacks *two* points and the second *three* of gaining the stake, you say it is necessary to see in how many points the game will be absolutely decided.
>
> It is convenient to suppose that this will be in *four* points, from which you conclude that it is necessary to see how many ways the four points

may be distributed between the two players and to see how many combinations there are to make the first win and how many to make the second win, and to divide the stake according to that proportion. I could scarcely understand this reasoning if I had not known it myself before; but you also have written it in your discussion. Then to see how many ways four points may be distributed between two players, it is necessary to imagine that they play with dice with two faces (since there are but two players), as heads and tails, and that they throw four of these dice (because they play in four throws). Now it is necessary to see how many ways these dice may fall. That is easy to calculate. There can be *sixteen*, which is the second power of *four*; that is to say, the square. Now imagine that one of the faces is marked *a*, favourable to the first player. And suppose the other is marked *b*, favourable to the second. Then these four dice can fall according to one of these sixteen arrangements:

a	a	a	a	a	a	a	a	b	b	b	b	b	b	b	b
a	a	a	a	b	b	b	b	a	a	a	a	b	b	b	b
a	a	b	b	a	a	b	b	a	a	b	b	a	a	b	b
a	b	a	b	a	b	a	b	a	b	a	b	a	b	a	b
1	1	1	1	1	1	1	2	1	1	1	2	1	2	2	2

And because the first player lacks two points, all the arrangements that have two *a*s make him win. There are 5 of these arrangements. Therefore it is necessary that they divide the wager as 11 is to 5.[80]

This is indeed a method of finding a solution for any given *a* and *b*, although it does not give any formula for general *a* and *b*. It is also not completely clear that it is right to count, as if on a par, fictitious cases that cannot occur and real ones. In the example just given, if *aa* occurs, the first player has then won, and the game stops in two throws; why then are we counting all of the outcomes *aaaa*, *aaab*, *aaba* and *aabb*, none of which will occur?

Pascal says that Roberval made this very objection. The objection is not correct, but it is prima facie reasonable and can come only from someone who has real insight into the difficulty of such problems. Pascal says he explained the matter to Roberval with a symmetry argument, to the effect that if the players agreed to continue to four throws, it would make no difference. "Is it not clear that the same gamblers, not being constrained to play the four throws, but wishing to quit the game before one of them has attained his score, can without loss or gain be obliged to play the whole four plays, and that this agreement in no way changes their condition? . . . It certainly is convenient to consider that it is absolutely equal and indifferent to each whether they play in the natural way of the game, which is to finish as soon as one has his score, or whether they play the entire four throws. Therefore, since these two conditions are equal and indifferent, the division should be alike for each."[81]

The phrases "without loss or gain" and "equal and indifferent" here are the closest Pascal ever gets to the notion of equiprobable. Fermat says, however, on the same point, that "the consequence, as you have well called it, 'this fiction,' of extending the game to a certain number of plays serves only to make the rule easy and (according to my opinion) to make all the chances equal (*rendre tous les hasards égaux*); or better, more intelligibly, to reduce all the fractions to the same denomination."[82] Much later argument of doubtful validity about "principles of indifference" and "equal ignorance" tries to explain why various chances should be regarded as equal. Fermat surpasses them all in breathtaking naivete in thinking that it is simply a mathematical convenience for producing fractions with the same denominators. It is the remark of the purest of pure mathematicians.

Pascal then claims that Fermat's method of combinations will give the wrong answer in the case of three players. If one tries to solve it using fictitious extra games, the order in which the players win the games is important, which is not the case for two players. Pascal thinks that *combinations* means, by definition, that one is taking no account of order and hence concludes that Fermat is mistaken. But Fermat replies that is not what he means.[83]

We now come to Pascal's general method of solving the problem of points, his alternative to Fermat's "method of combinations." Here he presents it for the simplest nontrivial case, in which one player needs one point to win, and the other two:

> This is the way I go about it to know the value of each of the shares when two gamblers play, for example, in three throws, and when each has put 32 pistoles at stake:
>
> Let us suppose that the first of them has *two* [points] and the other *one*. They now play one throw of which the chances are such that if the first wins, he will win the entire wager that is at stake, that is to say 64 pistoles. If the other wins, they will be *two* and *two* and in consequence, if they wish to separate, it follows that each will take back his wager, that is to say 32 pistoles.
>
> Consider then, Monsieur, that if the first wins, 64 will belong to him.

If he loses, 32 will belong to him. Then if they do not wish to play this point, and separate without doing it, the first should say "I am sure of 32 pistoles, for even a loss gives them to me. As for the 32 others, perhaps I will have them and perhaps you will have them, the risk is equal (*le hasard est égal*). Therefore let us divide the 32 pistoles in half, and give me the 32 of which I am certain besides." He will then have 48 pistoles and the other will have 16.[84]

Pascal then considers the next hardest case, when one player has two points and the other none (so the first needs one to win, the second needs three). He argues in the same way: if in the next throw, the first wins, he will win the whole 64, while if he does not, they will be back to the previous case.

The second case is exactly the one solved in the Italian manuscript of 1400. The method of solution is also exactly the same: reduce to the simpler case and divide the stake equally when there is an equal chance of each winning it. Pascal expresses himself more clearly, mainly because he has solved the simpler case separately first, but he adds nothing essential to the method except the justifying phrase, "The risk is equal."

Next Pascal attempts a more general formula. It would be perhaps most natural to consider the general case in which one player needs a points to win and the other b points. In fact he attempts the simpler, but almost as difficult, problem in which the first player to reach n points wins, and one player has one point and the other none. Pascal sends the correct formula to Fermat (for the case $n = 8$, but the generality is clear).[85] To find it requires some difficult mathematics with numbers of combinations. At this time, Pascal was in the process of composing a *Treatise on the Arithmetical Triangle*, which explains his results on numbers of combinations. It has an appendix, "The use of the arithmetic triangle to determine the division of the stake between two players." Here the problem is finally solved in complete generality, with a proof by mathematical induction.[86] The answer is that if one player needs a points and the other b points, the stake should be divided in the ratio:

$$\sum_{r=a}^{a+b-1} {}^{a+b-1}C_r \quad : \quad \sum_{r=0}^{a-1} {}^{a+b-1}C_r.$$

(Here the notation ${}^nC_r$ means the number of combinations of n things taken r at a time; that is, the number of r-element subsets of a set of n elements. Pascal does not use such a notation, but his expression of it in words is clear.) It is an excellent piece of pure mathematics. There is

no suggestion that Fermat did anything similar, and Huygens also only
solves the problem for particular small values of *a* and *b*. Its importance
for probability theory lies mainly in giving the confidence to manipulate combinatorial numbers that is needed in all problems concerning
large discrete probability spaces.

It is also of interest that Pascal, able to give a more considered exposition than in the letters of what the calculations are about, does not
mention anything like the modern interpretations of probability in
terms of either long-run frequencies or uncertainty, but expresses himself wholly in legal terms. What is being calculated precisely is a legal
or moral right:

> The money that the players have put at stake no longer belongs to them,
> for they have renounced ownership of it; but they have received in re-
> turn the right to receive whatever chance (*le hasard*) can give them. . . .
> And in that case [of an interrupted game], the distribution of what ought
> to be given to each must be so proportioned to what they have the right
> to hope from fortune that each of them would find it entirely equitable
> (*égal*) to either take what is assigned to him or to continue the adventure
> of the game; and this just distribution is called the division.
>
> The first principle for discovering how one should make the division
> is this:
>
> If one of the players is in a situation such that, whatever happens, a
> certain sum will belong to him in case of either a loss or a win, without
> chance being able to take it away from him, then he should make no di-
> vision of it, but take it all. . . .
>
> The second is this. If two players are in a situation such that, if one
> wins, a certain sum belongs to him, and if he loses, it belongs to the
> other; and if the game is of pure chance and there are as many chances
> for one as for the other, and consequently no more reason why one
> should win rather than the other, then if they wish to separate without
> playing, and take what legitimately belongs to each, the division is that
> they divide the sum that is at stake in halves, and take one each.[87]

Fermat, by contrast, seems not fundamentally interested in problems
of chances as such. He is more reliable in obtaining answers than Pascal but appears to regard the subject matter itself as just another source
of questions in number theory—and not very interesting ones at that.
Only Pascal understands that a completely new topic has been discovered. There is a short account of his view of what he has found, which
survives in an address listing his researches that he gave to an academy
of mathematics in Paris in 1654:

> A treatment quite new and on a matter completely unattempted, namely, the combinations of chance in games subject to it, what is called in French, *"faire les parties des jeux,"* where the uncertainty of fortune is so restrained by the equity of reason, that each of two players can always be assigned exactly what is rightly due. This must be the more inquired into by reason, in that it can be little investigated by trial. For the doubtful outcomes of the lot are rightly attributed rather to fortuitous contingency than to natural necessity. So the matter has hitherto wandered in uncertainty; but now what has been rebellious to experiment has not been able to escape the dominion of reason. For we have reduced it to art so securely, through Geometry, that, participating in [Geometry's] certainty, it now goes forth daringly, and, by thus uniting the demonstrations of mathematics to the uncertainty of chance, and reconciling what seem contraries, it can take its name from both sides, and rightly claim the astonishing title: the Geometry of chance (*aleae Geometria*).[88]

(Calling the subject "geometry" was actually an idea of de Méré, who suggested the name "mobile Geometry.")[89] Pascal seems to believe that the subject he has discovered has nothing to do with anything observational, such as long-run relative frequencies. His calculations are not about the "doubtful outcomes of the lot," because there is no necessity in them. They are about legal claims: what is calculated is "what is rightly due" (*quod iure competit*).

Fermat, at the end of the correspondence, tries without much success to divert Pascal's interest into questions of pure number theory. These are not meant to have any relation to probability, but in retrospect, one of them does. Fermat sends Pascal the "proposition" that all numbers of the form $2^{2^n} + 1$ are prime. This is true, he says, for $n = 1$, 2, 3, and 4, after which the numbers are too large to calculate easily. And "this is a property whose truth I will answer to you. The proof of it is very difficult and I assure you that I have not yet been able to find it fully. I shall not set it for you to find unless I come to the end of it."[90] Despite the evidence, and Fermat's unparalleled insight into what thinking is likely to lead to a proof in number theory, the proposition is false.

The last letter to Fermat is dated October 27, 1654. It thus belongs to the end of what Pascal's pious biographers called his "worldly" period. On a sheet of paper known as the Memorial, dated November 23, 1654, and found after Pascal's death stitched into the lining of his coat, is the record of his final conversion. It includes the words, "FIRE. God of Abraham, God of Isaac, God of Jacob, not of the philosophers and scholars. Certainty. Certainty. Feeling. Joy. Peace."[91] The word that appears twice here indicates the future direction of Pascal's thought, culminating in the *Pensées*.

Pascal and Fermat did briefly resume contact. In 1656 Pascal sent Fermat a "gambler's ruin" problem. In this problem, two players have different probabilities of winning a single point—in Pascal's example the ratios of these probabilities is 9 to 5. Each time a player wins, one point is transferred from his total to his opponent's total. The winner is the first to be 12 points ahead of the other. The advantage of each player is sought, that is, their probabilities of winning. The problem differs from the earlier ones that Pascal and Fermat had considered in that the game can continue indefinitely, and it is presumably for this reason that Pascal thought Fermat would have difficulty if he tried to solve it as usual by enumerating the outcomes. Nevertheless, Fermat sent back the correct numerical answer immediately. Neither Pascal nor Fermat gives his reasoning.[92]

Huygens' *Reckoning in Games of Chance*

Huygens, then aged twenty-six and little known, visited Paris in 1655 on his way to the University of Angers to purchase a doctorate in law. He learned of Fermat and Pascal's achievements, through discussions with the Parisian mathematicians Roberval, Mylon, and Carcavy, who formed an informal academy of sciences. He did not visit Pascal, believing him to have withdrawn into purely religious concerns.[93] On returning to Holland, Huygens worked on a general method of solving dice problems. He decided the answer lay in algebra.[94] He sent the following challenge problem to Roberval and Mylon, apparently hoping that his methods would produce a solution that the French would not be able to emulate: "When I play with another with two dice, with the rule that I win as soon as I throw 7 points, and he wins as soon as he throws 6, who has the advantage, and how much?"[95]

Roberval did not reply. Mylon sent back a wrong answer that relied on an inept argument concerning "composition of reasons" and "subtraction of reasons."[96] The honor of France was saved when Fermat sent the correct answer, along with some challenges about probability as applied to card games and a few comments on how difficult these problems were, as they required different difficult techniques for each one. The old dog does not show the pup any of his tricks.[97] Carcavy is overwhelmed and explains that Fermat is the wonder of the age and the equal of the ancients. Huygens, being the equal of Fermat, is not as impressed and tells Carcavy that he has already discovered a general method for such problems and has written it in a book soon to be published. There follow solutions to the problems set by Fermat.[98] Huygens uses his algebraic method, adding complaints that some of the problems that Fermat set need too much calculation to get the final answer, even though the method is clear. Carcavy reports that Pascal (then very busy with the *Provincial Letters*) has also found the correct answer to Huygens' challenge problem and agrees that Huygens' general method is correct. Pascal does however send a problem that he believes cannot by solved by Huygens' method (the rest of the letter describes Pascal and Fermat's treatment of the "gambler's ruin" problem). Huygens' reply explains how his method is in fact able to solve such problems.[99]

The book he refers to is a small treatise, in Dutch, *On Reckoning in Games of Chance*. A Latin translation was published in 1657 and the Dutch original in 1660.[100] It is the first published work on mathematical probability. In it Huygens complains that the French have hidden their methods and proposes his own foundation for the subject. The reasoning is interesting, as it goes much further than Pascal and Fermat in providing a foundation for the symmetry arguments involved.

> Although in games determined solely by chance the outcomes are uncertain, there is always a fixed value for how much one has for winning over losing. Thus if someone bets on throwing 6 points on the first throw of a die, it is uncertain if he will win or lose, but how much more likely (*quanto verisimilius*) he is of winning his bet than of losing it is something determinate that can be subject to calculation. Similarly, if I play against someone with the winner the first to win three, and I have already won one, it is uncertain which of the two will win. But how much my expectation (*expectatio*) is, and how much his, can be estimated exactly by calculation, and one knows consequently how much greater, if we interrupt the game, the part of the stake rightly mine exceeds his. One can calculate also at what price it would be fair (*aequum*) to sell my place to someone who wished to continue the game in my place. . . .
>
> I take it as a foundation that in a game the chance (*sortem seu expectatio* [in the Dutch, simply *kansse*]) that one has towards something is to be estimated as such that, if one had it, one could procure the same chance in an equitable game (*aequo conditione certans* [the Dutch adds: that is, a game where no loss is offered to anyone]). For example, if someone hides from me 3 shillings in one hand and 7 in the other, and gives me the choice of taking either hand, I say this is worth the same to me as if I were given 5 shillings. For if I have 5 shillings, I can again arrive at having an equal chance (*aequam expectationem*) of getting 3 or 7 shillings, and that by an equitable (*aequo*) game.

> Proposition 1. If I can get *a* or *b*, of which either is equally easy for
> me, my expectation should be said to be worth $(a + b)/2$.[101]

The device used is thus the same as in Lessius, the estimation of the
monetary value of a chance by a fictitious sale.

Let us attend to the vocabulary Huygens uses. (He remarks that he
wrote in Dutch because Latin lacks the relevant words. It is hard to
know what to make of this, and he seems to have changed his mind
later.)[102] The language *aequo conditione certans* recalls Cardan's *aequale
conditione certent*; since Huygens did not know Cardan's work, this concept may be referred to the general conceptions about the subject. What
Huygens is evaluating is what would now be called the expectation, that
is, the probability multiplied by the payoff, but that is not *his* meaning
of the word *expectatio*. Huygens calls the quantity he is evaluating "what
the expectation is worth," so his meaning of "expectation" is closer to
simply "the chance." The word was used standardly in Latin in connection with hopes; Aquinas uses *expectatio* to mean a hope that "pertains to the cognitive powers," that is, is well founded.[103] That Huygens
is thinking primarily in terms of probabilities rather than (what we call)
expectations appears also from his explanation of these results in a letter to Carcavy. He is happy to use the phrase "the number of chances"
(*nombre des hazards*): "If the number of chances one has for getting *b* is
p, and the number of chances one has for getting *c* is *q*, this is worth as
much as if one had

$$(bp + cq)/(p + q)."[104]$$

This formula (proposition 3 in his book) is what Huygens uses to solve
all problems whatsoever concerning chance. He never counts cases or
uses the combinatorial numbers, even when doing so would be easier.
The formula does make explicit the worth of chances (modern "expectation") which in Pascal and the manuscript of 1400 had been present,
but hidden by the reliance on symmetry.

The first problem Huygens solves is the simplest problem of points,
where one player has won two points and the other one, and the winner
is the first to three. His reasoning is the same as Pascal's, with the addition that he invokes his first proposition to justify dividing the expectation in half when the players have equal chances of winning: "An equal
chance of having *a* [the stake] or $\frac{1}{2} a$, is, according to the first proposition, worth the average, that is, $\frac{3}{4} a$."[105] The next propositions solve the
next hardest cases and simple cases for three players. It is explained how
to reduce more complicated cases to simpler cases, but there is no at-

tempt to find a general formula. Then de Méré's problem of dice is
solved. So far, there is no advance on Fermat and Pascal. Next Huygens
considers the problem that had received incorrect answers from his correspondents. The game proceeds as follows: Two dice are thrown; I win
if 7 comes up, my opponent wins if 10 comes up, otherwise we divide
the stake equally. Different problems arising from this game are treated;
the reason it created difficulties is that the chances of 7 and 10 are not
equal. This allows Huygens' formula

$$(bp + cq)/(p + q)$$

to exhibit its advantages over methods that use only intuitive symmetry.

The treatise concludes with five harder exercises, some due to Fermat and Pascal, which attracted attention from those writing on probability until the end of the century. They include two on drawing from
an urn and one on drawing from a pack of cards, revealing an initial attempt to extend the applicability of the subject.

Caramuel

What Huygens only alludes to in explaining the nature of probability
is expanded into considerable detail in the next printed work on mathematics and dice. This is Caramuel's *Mathesis Biceps*, printed on his epis-
copal press in 1670. It is a huge work that takes a wide view of mathematics. It includes, for example, the theory of manned flight, both aerial
and interplanetary. It is certainly the most complete account of the
mathematical sciences published up to that time.[106] Of interest here is
the long section on dice. It is preceded by another on combinatorics,
which contains derivations of combinatorial formulas of considerable
generality. Combinatorics is useful, according to Caramuel, not only in
such expected fields as music but also in areas not normally thought of,
such as jurisprudence. The section on dice depends on Huygens. Indeed
it includes the full text of Huygens' short book (wrongly attributed to
the Danish astronomer Longomontanus).[107] Caramuel claims that his
basic ideas were written before seeing Huygens' treatise. The introduc-
tory material that Caramuel supplies is considerably more elementary
than anything in Huygens or, for that matter, in Fermat and Pascal and
explains more clearly both the basic mathematics and the conceptual
motivation. It is thus valuable historically, giving an insight into the
thinking of the pioneers of mathematical probability, especially in view
of the sketchy nature of what the others say. The connection of probability considerations with moral theology, only briefly mentioned by the
others, is much clearer in Caramuel:

"In games and lots which depend solely on chance, equality should be completely preserved."

This is a most certain and secure proposition. But how can the pure Theologian know whether equality is preserved in a contest? Therefore, we must necessarily have recourse to mathematics. Consider the following syllogism:

For equality to be preserved, it is necessary that payment correspond to risk. So, those who expose themselves to equal risk, put down equal money; to unequal risk, unequal money, such that he who is faced with greater risk should put down less money; and with less risk, more money.

But here and now among certain players, equality is preserved (either the risk is equal or it is compensated for).

Therefore the game is licit, and neither is obliged to make restitution.

Concerning the first premise, which pertains to Moral Theology, there are only too many generalities written in books. But the pure theologian can never examine the second premise. Indeed, neither can the arithmetician, unless he understands some recondite theory of numbers, which almost all those who teach in the schools and serve the republic do not know. Note also, that in a contest one should determine not only whether there is risk, but also how great it is, because the amount of money to be exposed depends on the amount of risk. There is a good example of this proposed by some: If someone hides in one hand 3 shillings and in the other 7, and gives me the choice of hand, this is worth the same as if he gave me 5. For in this case we have no way of measuring any inequality of risk, and it would be going in blind to make a bet on the number of shillings. And so, since there is no more reason for one than the other, I strike the medium between the two extremes, and say the difference should be divided into equal parts: the choice is worth the same as if he gave me 5. And also with dice: the equality and inequality of hope and risk can be known, and determined exactly. . . . In storms the ancients invoked the two divinities, Castor and Pollux; I here invoke the help of two divinities, Theology and Combinatorics, the first to cast light on the first premise of the syllogism, the latter to examine the second premise, and measure the grades of risk.[108]

Caramuel deals first with the throws of two dice and lists the ways of getting each of the possible outcomes from 2 to 12. The language is much the same as we now use; it is probably the first explanation of the subject moderately comprehensible to an ordinary reader: "2 can only be thrown in one way, namely, 1 and 1. 3 can be given by 1 and 2 or 2 and 1. Therefore it can come out in two ways. . . . 7 by 1 and 6, 2 and 5, 3 and 4, 4 and 3, 5 and 2, 6 and 1. Therefore 6 ways. . . . Hence it is clear that the best number (that is, the one that can come out in most ways) is 7."[109]

He has yet another explanation of the foundation of the subject in terms of fair contracts, where modern writers would use either equal probabilities or equal long-run outcomes. "A moral conclusion: To know what you should expose, and what you can with a safe conscience require, begin the computation this way: See how much you would win, if you rendered yourself secure, and how much that security would stand to you. If in that case you lose as much as you win, the game is equal and just; it is unequal and unjust if you win more than you lose, or vice versa."

What "rendered yourself secure" means is that you make bets that cover all outcomes.

In a game with four (implicitly equiprobable) outcomes, one plays against four opponents and bets with the first that the first outcome will appear, with the second that the second will appear, and so on. "'Therefore, because justice requires equity (that is, equality) in the game, he should win as much as he loses. But he wins from one, and loses to three. Therefore, he ought to pay to the three as much as he receives from the one.' Whatever happens, he pays out three, so he should receive three for the win."[110] This is Caramuel's basic way of conceptualizing the measurement of risks in the simplest case in which they are not symmetrical (in this case, when, as we would say, the probability of a win is ¼). He does not, strictly, give a number to a probability or risk. But neither does Pascal, Fermat, or Huygens. Like them, he uses symmetry arguments on contracts, dividing a nonsymmetrical contract into a number of simple contracts in which the fair payouts are obvious by symmetry. Plainly, his method resembles the later method of calculating probabilities by counting the equiprobable cases contained in an outcome, but it adds a justification for doing so. On the same page, however, he does begin to give numbers directly to the probabilities (called "hopes" and "risks"). He prints a table for a game of two dice (see table 11.2). The hope and peril, he says, are just the number of ways in which you win or lose.[111]

Caramuel next deals with the problem of points. Strangely, his example is closer to that in the medieval manuscript on the interrupted game than it is to Huygens or Pascal: it treats the case of one aureus staked by each player, in which one has two points to win and the other three, and includes reference to a fourth game. Caramuel says nothing

TABLE 11.2. PROBABILITIES IN A GAME OF TWO DICE

Number on the Dice	Hope	Risk	Therefore You Pay	Therefore You Receive
2	12	1	35	1	35
3	11	2	34	1	17
4	10	3	33	1	11
5	9	4	32	1	8
6	8	5	31	1	6⅕
7	7	6	30	1	5

about the history of the problem except to note that gamers do not know the answer. The solution is essentially as in Huygens, but the latter's justification by expectations is omitted, and talk of rights is inserted.

As in Huygens, the next most difficult cases are solved by reducing to simpler ones. Caramuel adds, "It is clear one can proceed to infinity": if one player has 1 to win and the other n, then the latter must put down 2^n against 1. A game with three players is then attempted through a complicated division of it into three contracts, each between two players. Then there is consideration of what to do when players cannot begin a game at all but have decided who gets the advantage of playing first. The last problem in Huygens, in which the players have unequal chances of winning each point, is incorrectly solved by Caramuel, although Huygens had given the right answer.[112]

Caramuel then justifies in terms of rights a result equivalent to the multiplication rule for independent probabilities, with reasoning close to Huygens'. Caramuel derives from this the correct value for the chance of throwing at least one 5 in two throws of dice as $\frac{1}{6} + \frac{5}{6} \times \frac{1}{6} = \frac{11}{36}$ and gives a formula for greater numbers of throws. This contrasts, he says, with the "vulgar opinion" that to throw a 5 in two throws is worth $\frac{2}{6}$, in three throws $\frac{3}{6}$, and so on.[113]

Something quite unlike anything in Huygens occurs in Caramuel's analysis of the Genoese lottery. This was a game, dating probably from the early seventeenth century, similar to modern lotto games: one guessed which 5 numbers out of the first 95 (or 100) numbers would come up in a draw and won prizes for having one, two, three, four, or five numbers agreeing with the ones drawn. Caramuel gives the payoffs on a bet of 10 gold pieces as respectively 10, 100, 3,000, 15,000, and 100,000 gold pieces. He calculates extensively and concludes the game is an iniquitous contract, especially for the higher payoffs. His figures were later criticized by Nicholas Bernoulli, whose own results, however, differ little from Caramuel's. What is perhaps more remarkable is that the original setters of the payoff scale, presumably without advice from mathematicians, obviously had a reasonable feel for how safe it would be to offer huge prizes for the rare draws.[114]

What Caramuel is doing is linked to the earlier casuists not only by his constant talk of rights and equity but also by his quoting the remark noted above from Diana's *Summa* on the amount of restitution owed by someone who has won a game of chance by cheating. Caramuel adds a conclusion that is more general but more lax: "One who makes a profit by cheating does not make satisfaction by restoring what he won, but is obliged to pay only as much as in equity (that is, from the equality of condition and the game) the other would have won."[115]

CHAPTER TWELVE

Conclusion

This is a Whig history of mentalities, a story of the Advance of Knowledge as the forces of Reason roll back the frontiers of ignorance. As such, it does not exactly need a conclusion, as it records the gradual discovery of preexisting intellectual terrain in a more or less rational order. Generally, a new idea in probability is seen to replace an older one because it is a better idea. In mental technologies, one idea can fly better than another, just as one plan for heavier-than-air aviation can work and one not.

On the other hand, this is not an Enlightenment history, in which a heap of perfectly formed propositions, previously hidden in darkness, are one by one revealed to the light. Such a conception may be appropriate to, say, number theory, in which the entities studied are perfectly clear and the difficulty lies simply in finding out which of the propositions about them are the true ones. Probability is not like that but is more like law, or psychoanalysis, in which there are confused conceptions that work in practice, up to a point, and progress lies in clarifying those conceptions while keeping them grounded in reality. In such cases, there is a need for the historian to set out what the situation was before and after a transition in ideas and to explain how it occurred.

It is possible to regard the history of ideas as a kind of Brownian motion, ideas being discovered and conceptions changing in a purely random order. Possible, but uninviting, since historians of ideas, like any other theorists, dream of finding order in the flux of experience, of imposing unifying theoretical perspectives on their subject. At one time or another, a variety of engines have been postulated to drive the development of ideas, reference to which would explain why ideas developed as they did and not otherwise. Suggestions for the role include social context, economic forces, the Zeitgeist, national character, class struggle, the necessary unfolding of the Hegelian World Spirit. Diverse as they are, all these entities share a common problem, over and above sheer implausibility, as attempted explanations of ideas: ideas are mental entities, and it is unclear how the proposed causes act on minds. How is social context, for example, represented mentally, so that it can generate or affect ideas? Without an answer to that question, the theories are no better than ones postulating astral or paranormal influence. It is as if one were to take literally the metaphor involved in saying that Newton and Leibniz could invent the calculus simultaneously because the relevant ideas were "in the air." It is the same as the problem with invoking economic rivalry as the cause of the First World War. It is a purely magical pseudoexplanation, unless there are answers to questions like: Were those motivated by greed the ones with the power to declare war? Are the motives revealed in the private letters of the politically powerful in fact economic or substantially influenced by people whose motives were economic? And so on. Once such questions are asked, one can hope to look at the balance of motives in real people's minds and see economic explanation as only one of various possible motivations. Actions, like ideas, are the products of human minds, and proposed explanations of them must be given a way of acting psychologically.

One might fear being left with the theory—with which surely most historians of ideas must have a secret sympathy—that ideas are discovered when great thinkers are dictated to directly by their Muses. Still, even if that were true, and thinkers indeed depended on inspiration for their answers to questions, it might be possible to explain reasonably their choice of questions. If there is a most plausible case for direct inspiration from on high, it is Mozart, but it may still be explicable why he is Viennese classical. Fortunately, there is another perspective on the causes of the development of ideas that is more concordant with what we know about how the world works, though it may not appear so at first sight. It is the theory of the genesis of ideas in Plato's *Meno*, as reinterpreted by Galileo. In the *Meno*, an untutored slave boy, under questioning by Socrates and with the help of a diagram, is induced to discover that to draw a square with area twice the area of a given square, one must use as the side of the new square the diagonal of the given one.[1] The knowledge that the boy discovers is not explicit in either the questions, the diagram, the boy's preexisting knowledge, or even in the sum of all three. Plato draws the conclusion that the boy must have remembered it. Plato mentions a further theory: that this knowledge was explicit in a previous life but hidden from memory before birth; but he emphasizes that the important conclusion is that there is an answer to Meno's challenge (How can one recognize a correct definition of, for example, virtue, unless one already knows what virtue is?). In general terms, there must be latent knowledge, which is somehow accessible with appropriate cues. Galileo was not misled by the fairytale of past lives. In his *Dialogue Concerning the Two Chief World Systems*, Salviati is encouraging Simplicio to realize that a stone released from a sling moves off tangentially:

> *Salviati:* The unravelling depends upon some data well known and believed by you just as much as by me, but because they do not strike you, I shall cause you to resolve the objection by merely recalling them.
> *Simplicio:* I have frequently studied your manner of arguing, which gives me the impression that you lean toward Plato's opinion that *nostrum scire sit quoddam reminisci* [our knowing is a kind of reminiscence]. . . .
> *Salviati:* Well, then, what is its motion?
> *Simplicio:* Let me think a moment here, for I have not formed a picture of it in my mind.
> *Salviati:* Listen to that, Sagredo; here is the *quoddam reminisci* in action, sure enough. Well, Simplicio, you are thinking a long time.
> *Simplicio:* So far as I can see, the motion received on leaving the notch can only be along a straight line. . . .[2]

What Galileo means is that Simplicio has an implicit knowledge of dynamics and that this knowledge is memory—a kind of superposition and automatic generalization of the experiences of (this) life. The task of thinking—in this case, of the thought experiments Galileo suggests—is to make this implicit knowledge explicit, in propositions.

Investigations of the development of ideas in small children show that there is indeed a parallel process of the becoming explicit of ideas. It is further found that it is difficult and that it is not a one-step process. Insight into what has been learned implicitly is not suddenly revealed in full but appears as the mind iteratively redescribes the information it already has. One can distinguish, for example, a first stage, in which one can recognize a zebra; a second, in which one can recognize the analogy between a zebra and a zebra crossing without being able to explain the analogy in words; and a third, in which the respects in which the two are analogous can be stated in words.[3]

So it is with the historical development of ideas. The dredging of the subconscious for its knowledge is long-drawn-out and painful. Seeing the history of probability as an example of the gradual explicitation of concepts that were already implicitly present permits an explanation of various matters that could hardly be explained otherwise. First, we review what is known from modern psychology about the connection between talk about probabilities and actual reasoning about and behavior under risk. In the light of that, the early history of probability as described in this book is considered, and the nature of its achievements is made precise. Then we are in a position to attempt an answer to the old chestnut, Why was the mathematical theory of probability not discovered earlier? The diagnosis is confirmed by comparing the history of probability with that of two closely parallel topics, motion and perspective. These parallels suggest, too, the general utility of a cognitive development approach to the history of ideas. Finally, some explanation is attempted of what might seem to have been biases in the story: the prominence in it of the Scholastics and of law. The strength of Scholasticism lies in conceptual analysis, that is, in bringing confused ideas to clarity. Law, on the other hand, acts to prevent oversimplification in conceptual analysis by constantly providing "hard cases" that challenge the adequacy of existing categories and answers.

Subsymbolic Probability and the Transition to Symbols

That the vast majority of probabilistic inferences are unconscious is obvious from considering animals, for it is not just the human environment that is uncertain but the animal one in general. To find a mechanism capable of performing probabilistic inference (as distinct from talking about it), one need look no further than the brain of the rat, which generates behavior acutely sensitive to small changes in the probability of the results of that behavior.[4] Naturally so, since the life of animals is a constant balance between coping adequately with risk or dying. Foraging, fighting, and fleeing are activities in which animal risk evaluations are especially evident; in general, the combining of uncertain information from many sources is of the essence of brainpower in the higher animals.[5] Some further light on what the brain does is cast by simple artificial neural nets, whose behavior after training on noisy data can be interpreted as implicit estimates of probabilities.[6] These animal and machine studies confirm in the most direct possible way that, to behave probabilistically, it is not necessary to have anything like explicit estimates of probabilities or ways of talking about them.

The human species inherited the mammal brain, with these abilities already loaded and in automatic use. In human life, the only certainties, proverbially, are death and taxes, and of these the time and amount, respectively, are quantities rarely known. The "Iceman" discovered in the Alps in 1991 was certainly one who took a calculated risk. It is clear at least in principle how evolutionary pressures select for rational techniques of risk management.[7] The roads continue to select against those whose evaluation of risk is below par. There are psychological studies that show how much of cognition generally is intuitive statistics. Even such a basic operation as the discrimination of stimuli (for example, in deciding whether two sounds are the same pitch or not) is a probabilistic process of extracting a signal from a noisy background. And perceiving and remembering both involve unconscious testing of hypotheses on the basis of imperfect correlations.[8] Very nearly all uncertain inference is unconscious, performed at the subsymbolic level by the neural net architecture of the brain. But the species did not inherit from its animal forebears a language for discussing its abilities. Consequently, there is no history of it. The brain being biodegradable, there is no archaeology of it either.

Since this book describes the historical coming to consciousness of notions of probability, there is a special interest in the many studies that show how the boundary between unconscious risk assessments and the conscious linguistic assessments of them is neither transparent nor totally impermeable. Human subjective assessments of risk expressed in words are reasonably accurate in many circumstances.[9] Indeed, in business forecasting of such variable quantities as stock prices, human judgmental forecasting is still generally comparable to the best statistical methods (and it is possible to say which statistical methods it resembles).[10] Child development studies show the gradual development of reasonable risk estimates in words.[11]

However, there are some strange features of the brain's implementation of probabilistic reasoning that result in systematic deviations from rationality. Estimates are age-specific, for example.[12] They rely on mental models, or prototypes, leading to such problems as overconfidence in estimates and oversensitiveness to the order of presentation of evidence.[13] The relationship between words used to express risk estimates and what the brain is really doing is a problematic one: driver behavior is related more closely to objective risk than to stated risk, and learning of relative frequencies is often best in the absence of clumsy attempts to make conscious statements about them.[14] On the other hand, having a theory can help bring order into data and correct mistakes in it.[15] Of interest in connection with the relation between numerical and purely linguistic expressions are experiments that show a reasonable consistency between phrases like "very likely" and the fuzzy numerical estimates of probabilities that they mean.[16] Since people often use words in preference to numbers in discussing risks, to avoid committing themselves to accuracy they do not possess, it is fortunate that probabilistic words are reasonably well calibrated.

Kinds of Probability and the Stages in Discovering Them

The book began by distinguishing between logical and factual probability and was organized around that distinction. However, there are four factors that confuse the issue further, making it more difficult to decide what to look for in early authors.

First, actual people's partial belief in, or uncertainty or doubt about, propositions and the support they give one another (sometimes called subjective probability) has some relation to logical probability but is not the same as it; indeed, psychological experiments show consistent discrepancies between the two, though the deviations are not too large in normal circumstances.[17] It is for this reason that purely rhetorical studies are excluded from the present book. The subject matter is what ought to persuade, not what does.

Second, actual relative frequencies, or proportions in populations, have some relation to factual probability but are not identical to it. (In a long sequence of throws of a die, the relative frequency of a 6 is *probably* close to the factual probability of the die's falling 6 but may not be actually close.) This is why the book contains nothing about a subject that might be expected, combinatorics. Many of the mathematical exercises on elementary probability take a form like, "What is the probability of choosing at least three hearts on a wet Sunday?" The solution involves simply *counting* how many ways some outcome can happen; probability enters only in the assumption that the elementary ways are equiprobable. This connection between probability and counting dates only from the work of Pascal, Fermat, and Huygens in the 1650s. A considerable amount of the mathematics of counting existed before that time, but it had no connection with probability.[18] Mersenne, for example, saw no relation between probability and his extensive combinatorial investigations in music.[19] Therefore it is omitted.

Third, there are connections between logical and factual probability, in that if the outcome of an experiment has a factual probability other than 0 or 1, the experimenter is rationally uncertain of the outcome. Hence it will sometimes be unclear which kind of probability an author means or whether he is aware of any distinction. In a subject like insurance, the connection between the two concepts is in its nature close.

Fourth, the *mathematical* theory of probability can be, and usually has been, developed without reference to what kind of probability is being spoken of. This is because the mathematical theory of, say, dice throwing depends only on the *symmetry* or *equiprobability* of certain outcomes, allowing equal numbers to be assigned to them. One can say that there is symmetry between all of the six possible outcomes of a single throw of a die, and hence they each have probability ⅙, without needing to say in what respect they are symmetrical or what kind of probability is being considered—whether one is discussing properties of the die or uncertainty about the outcome.

Another matter of controversy concerning probability is whether all probabilities should be given numbers. Factual probability is essentially numerical, certainly. And the standard mathematical theory of probability treats only of probabilities which are numerical. But Keynes, in his classic *Treatise on Probability*, argued at length that not all logical probabilities should have numbers. Even if they all do have numbers in principle, no convincing way has been discovered of actually assigning a number to, for example, the probability of the steady-state theory of the universe on present evidence—or even such a simple case as the probability that the next ball chosen from an urn will be black, given that all of the twenty balls already chosen have been black. English law, while admitting that proof beyond reasonable doubt is a concept of probability, has refused to allow a number to be assigned to that probability.[20] This should caution us against supposing that, because the concept of probability before Pascal was mostly nonnumerical, it was therefore primitive or in some way inadequate.

The emphasis, then, is on looking not so much for anticipations of what was new in the seventeenth century (that is, numerical quantification) but rather for what was being said about the probabilistic phenomena that became quantified—and those that did not. Some of the fragments of nondeductive logic that one can expect to find before probability was given numbers are as follows:

— *The recognition that there are such things as rationally persuasive arguments that are not necessary.* The subject matter of nondeductive logic was discovered once there was a classification of arguments into the necessary, the probable, and the persuasive but sophistical. These distinctions were fully explicit in Aristotle. The subject of rationally persuasive methods must also be distinguished from irrational methods of reaching decisions in doubtful cases—guessing, throwing a coin, divination, casting lots, trial by ordeal, or taking the opinion of the person next ahead in the pecking order. These distinctions were made by the pre-Socratics and again in twelfth-century Europe.

— *Arguments from presumption* (or *default* or *nonmonotonic* reasoning, as it is called in artificial intelligence).[21] In reasoning about real situations, a computer or brain does not waste time computing all the logically possible things that may have happened but *presumes* various things to be so, unless there is reason to believe the contrary. One presumes objects are in the position one left them, unless one receives contrary information; the law presumes ownership of goods is unchanged, unless there is evidence otherwise. It is unclear whether numbers ought to be attached to the force of such arguments; the majority opinion among artificial intelligence researchers is against doing so. This notion was explicit in ancient Jewish and Roman law and was developed in the medieval theory of presumptions.

— *Gradations in the probability of arguments.* In temperature, for example, classificatory (hot, warm, very cold) and comparative (hotter than) concepts are possible without any numerical scale; so it is with probability. Noteworthy is the recognition that there are arguments that lead almost to certainty but not quite (proof beyond reasonable doubt; morally certain conclusions), while some arguments amount only to half-proof or produce conclusions merely more probable than their opposites. "More probable" was common in the Attic orators; an "almost certain" standard of proof in criminal cases is found in the Talmud and the *Digest*; "half-proof" and "more probable than its opposite" are in medieval law and philosophy; "morally certain" is late medieval.

Then there are qualitative versions of particular kinds of probabilistic arguments:

— *The proportional syllogism* or *statistical syllogism*.[22] The numerical version of the proportional syllogism is as follows: The argument from "x percent of As are Bs. This is an A" to "This is a B" has probability x percent. (If one knows that x percent of As are Bs, then it is rational to be x percent certain that a given A—about which one knows nothing else relevant—is a B.) Nonnumerical versions, such as, "Most As are Bs, so this A is probably a B," and "The vast majority of As are Bs, so this A is very likely a B," can be expected. (Arguments of the "vast majority" form still have a crucial place in science, in statistical mechanics.)[23] The "for the most part" arguments in Aristotle, the Talmud, and ancient medicine are of this type. There was no attempt to refine the numerical proportions involved, at least explicitly. The late Scholastics did however understand in principle the connection between numerical insurance and annuity rates and the proportion of ships that returned or annuitants who died.

— *Combination of evidence.* Several more or less independent pieces of evidence tending toward the same conclusion can support that conclusion strongly, even if the individual pieces are weak. This principle was mentioned in ancient physiognomics and understood by ancient commentators on law such as Quintilian and in medieval law. The technique of averaging noisy measurements of a quantity to obtain a more accurate value was known to Hipparchus and later astronomers.

— *Statistical significance arguments.* If on a certain hypothesis a certain result would be unlikely, then the occurrence of that result tells against the hypothesis. Of this kind are arguments ruling out the hypothesis of chance: a sequence of 1,000 heads when tossing a coin is possible but very unlikely if the tossing were random, so if such a result occurs, one rules out the hypothesis of chance and looks for some explanation of the regularity, such as a bias in the coin. Such arguments occurred in Aristotle, Cicero, the Talmud, and occasionally thereafter. There was never any attempt to quantify *how* unlikely the result was on the hypothesis.

— *Verification of a consequence arguments.* The verification of a (nontrivial) consequence renders a hypothesis more likely. Polya calls this "the fundamental inductive pattern."[24] For example, if a stabbing has occurred, the discovery of bloodstains on the clothing of the accused makes his guilt more probable. If a scientific theory implies that an experiment should have a certain result, and that result is found, then the theory is rendered more probable. It is notoriously difficult to attach numbers to the probabilities in such arguments, but it is easy to make some comparative judgments, such as that ruling out as unlikely an alternative explanation makes the original hypothesis more likely. This was understood implicitly by arguers in courts from ancient Greek times, but the principle seems never to have been clearly stated.

An especially important variety of verification of a consequence argument is the following:

— *Inductive arguments*, or arguments from the observed to the unobserved, that is, from sample to population. If all observed As have been Bs, this lends support to the hypothesis that all As are Bs, and to the hypothesis that the next A to be observed will be a B. The larger the sample, the better supported the inference, though again no convincing way has appeared of attaching numbers to inductive arguments in general. Arguments of this kind are found in ancient medicine, and in the Talmud. The practical and theoretical problems with them are discussed by ancient, medieval, and early modern philosophers.

— *Arguments from analogy*, that is, inductive arguments in which the sample and population consist of properties, not individuals. One argues that because two things share a number of properties, they are likely to share a further one. The idea is mentioned in Aristotle and ancient logic, but little developed.

Why Not Earlier?

It has become usual when writing about the history of probability to speculate on the reasons for the late development of a theory of probability. The evidence presented in this book goes some way toward answering the question but does not seem wholly sufficient. First, there are various questions that need to be separated for clarity. These and the tentative answers to them are listed here.

Q: Why did no theory of probability develop before the seventeenth century?

A: A theory of probability did develop before the seventeenth century. The confirmation of universal generalizations by their instances, the use of nonconclusive arguments in law and medicine, and so on, were standard. See the present work, passim.

Q: Even if there was a logical or epistemic concept of probability, why did the factual or aleatory concept not develop before the seventeenth century?

A: The factual concept of probability did develop before the seventeenth century. Even in ancient times it was realized that a chance event that can happen easily enough in one or two cases is unlikely to keep happening many times, and by 1600 people had been discussing insurance for centuries. "Aleatory" means the same in (ancient and medieval) Roman law as it does now.

Q: Why did no quantitative or numerical concept of factual probability develop before the seventeenth century?

A: A numerical concept of factual probability did develop before the seventeenth century. Insurance and life annuities involve exactly the assignment of a numerical price to a risk, which is a probability at least partly factual; a number of writers said so explicitly.

Q: Why was there no mathematical theory of games of chance before the seventeenth century?

A: There were rudimentary mathematical statements about games of chance before then; except for a difficult-to-read example in a long-lost manuscript, they were however almost entirely wrong.

Q: Why was there no widely known, correct mathematical theory of dice throwing, as we know it today, before the seventeenth century?

A: The answer remains less than entirely clear, but there is one principal explanation, and two lesser ones, that seem to provide a reasonably adequate account. First, as the answer to the last question suggests, there is something very difficult mathematically in the early stages of the mathematical theory of probability. To obtain correct answers, one must have clear either the distinction between permutations and combinations (or partitions and ways of falling) or the concept of independence of events, at least sufficiently to know when probabilities should be multiplied. These notions are very difficult.[25] Consider Cardano: he was among the best mathematicians of his century and had an interest in and experience with gambling. Yet his attempts to solve quite simple problems still mix good ideas with bad, and his exposition is almost impossible to follow.

The improvement in mathematics across the board in the century after Cardano must be accounted a crucial ingredient in the development of the mathematical theory of probability. The scientific revolution of the seventeenth century might almost equally well be called the mathematical revolution. Experimentation was important, but even there one of the advances was the systematic extension of *measurement* to such quantities as time, speed, pressure, and probability. The initial reduction of data to numerical form enabled the discovery of patterns in them like the laws of motion and the creation of mathematical theories of continuity (the calculus) and of probability. The medievals knew about the idea of quantitative laws but never succeeded with reasonably accurate measurement (except, like the ancients, in astronomy).[26]

In fact, the obvious difference in the seventeenth century is the growth of a basic mathematical culture in general. Culture here has two aspects, matter and form, and both were obviously in a new phase at the time in question. The form means the group of people who knew each other, mostly through correspondence via Mersenne. They asked each other questions, set challenges, attacked supposed answers, and praised correct solutions—in other words, provided the motivation for advances that present-day researchers take for granted. Science *can* be done by lone geniuses, but it is not the normal way. And the shared matter of the mathematical culture increased enormously. Elementary mathematical tools like decimal notation, negative numbers, algebraic symbolism, graphs on the Cartesian plane, and logarithms were only tentatively known here and there before 1600 but were common property by 1650. Without them, it is a struggle to do mathematics using essentially only natural language and geometrical diagrams; with them, everything becomes possible. Applied mathematics especially benefits from such graphical and calculatory tools, and many things can naturally be seen in mathematical terms that were before only hinted at in words. Why this happened, and happened so suddenly, remains to be explained. But the explanation has nothing to do with probability, since it happened in all fields equally.

Huygens provides the most direct evidence on how exactly the possession of a suite of mathematical abilities, especially in modeling, helped in creating mathematical probability. He refers to Archimedes in explaining that applied mathematics needs some "translation axioms" to relate the real world to the mathematics.[27] (This is of course exactly what he provided for probability.) The subject he discusses is floating bodies, in which the axioms in question are made reasonable largely by a symmetry principle (the weight of a floating body equals the weight of water it displaces), as is the case with his work on probability (a game with an equal chance of winning three and seven shillings is worth five shillings). Interestingly, the question of the equilibrium of weights on balances—even more obviously a matter of symmetry—had been the subject of a controversy between Pascal's father and Roberval and Fermat in 1636.[28]

It has been argued that there is a close parallel between Huygens' work on probability and what he did a little earlier on the conservation of momentum in collisions of elastic bodies;[29] again the explanatory weight is carried by a symmetry principle (equal elastic bodies colliding head-on with equal speeds recoil with equal speeds; one then divides unequal bodies into parts, as one does with unequal chances). This much is inference, but Huygens does write that algebra is of great assistance in dealing with probability (and he uses algebraic symbols like a and b to represent the worth of games and masses).

Huygens in many places shows himself easily able to slip between an algebraic and a verbal/geometric description of a problem.[30] This is an ability necessary for much applied mathematics, rare enough even today, and acquired only with effort and with the support of a mathematical tradition. In the letter that forms the preface to his book on probability, he writes, "You propose *inter alia* to show, by the range of topics treated, the breadth of the field to which our excellent art of algebra extends; I do not doubt but that the present piece on the subject of calculation in games of chance will be able to achieve that end. In fact, the more difficult it looked to determine through reason what is uncertain and subject to chance, the more remarkable should that science appear which could reach such a result." In another letter he asserts that dice problems are not hard for those "who know the principles of this calculation and a little algebra."[31]

Pascal's tastes did not run to algebra, but they did include a deep understanding of symmetry principles. Probability was not the first subject matter he reduced to mathematical order by the use of symmetry, as is clear from his *Treatise on the Equilibrium of Liquids*. It will be recalled that the only probabilistic concepts in the Fermat-Pascal correspondence were "to be impartial" (*désintéresser*) and "equal" (hazards or chances). In explaining Roberval's mistake about chances, Pascal considers two situations "absolutely equal and indifferent to each [player]." The parallel is close with his explanation of why a square inch anywhere at a given depth on the inside surface of a vessel enclosing a liquid under pressure will feel the same pressure: "Each part of the vessel bears more or less pressure in proportion to its area, whether that part be opposite the aperture, or to one side of it, or far, or near; for the continuity and fluidity of the water make all those circumstances equal and indifferent (*égales et indifférentes*)." The actual phrase "equal and indifferent" to express the symmetry between different individual cases is a legal one: "Bracton requireth to the making up of a true Iudgement . . . an equall and indifferent acceptation of the persons."[32]

It is true that transferring symmetry principles from the continuous situations of motion and pressure to the discrete case of probability is nontrivial, but Pascal was no stranger to original thinking about the relation of the continuous and the discrete either: his calculating machine was exactly a device to "reduce to regulated movement all the operations of arithmetic," that is, to represent digital arithmetic by continuous motions.[33]

Perhaps such abilities represent overkill, given the problem. If Huygens or Pascal or Fermat could create a unified theory out of probability, could not some of the lesser lights in the same milieu have grappled with a few of the easier problems? In fact, we saw that this is exactly what did happen in the cases of Roberval and de Méré, who solved some easier problems but whose confusions on harder ones were treated with condescension by the great.

The short answer to the question, Why not sooner? is, then, that mathematics is difficult; applied mathematics doubly so. There are two lesser explanations.

First, dice throwing is not very interesting. Of the first three thinkers who were able to calculate the probabilities of the outcomes of multiple throws correctly—Galileo, Fermat, and Pascal—the first two seem to have had very little real interest in the subject. None of the three devoted more than a tiny fraction of his time to the subject. Huygens hints that the reader of his book may at first glance find the subject trivial but defends himself by saying that it has interest as pure mathematics, as indicated by the fact that famous mathematicians in France are studying it. Leibniz too is apologetic, saying that the principal use of dice problems is to practice for more important things.[34] Nor is the theory directly useful; no one has ever heard of a mathematician winning money at the gaming tables by knowing about probability.

This is confirmed by the strange fact that the problem that motivated thought on the subject was far from what we would think of as useful: the question of the division of the stake in an interrupted game is a very unobvious place to start. Is there not even something paradoxical in thinking first of a *moral* question in connection with gaming? Further, the various applications of the mathematical theory to law, politics, and so on in the seventeenth and eighteenth centuries were all fakes ("misapplications of the calculus of probabilities which have made it the real opprobrium of mathematics," says Mill).[35] There is hardly a convincing case of the mathematical theory aiding in the evaluation of a hypothesis until the work of Fisher in the 1920s, or at least Pearson in the 1890s. In modern school mathematics it is understood that questions about probability expressed in terms of dice and coins are simply excuses for questions in combinatorics. Those who teach these exercises are not supposed to encourage their students to go out and play dice for money.

There is a deeper reason why dice throwing is uninteresting, even if one is concerned with uncertainty. Dice are an extremely bad model for the great majority of cases of reasoning under uncertainty. In general, when in doubt, the problem is to take into account and weigh matters of history, like the skill of players or horses, the testimony of witnesses, the reliability of market rumors, or past relative frequencies. But in dice, the outcomes are equally probable exactly because history and circumstances are irrelevant. The spin of the die wipes all memory of the past and all influences of the surroundings. The problems with attempting to model partial knowledge by partitions of ignorance has dogged the history of applications of probability all along and is at least partly responsible for the "misapplications" of probability in law and the like. It is still present in, for example, the unreality of game-theoretic prisoner's dilemma models in economics, in which one tries to model a situation in which players are rich in information with one in which they rely on pure guesswork.

A last plausibly suggested obstacle to the development of mathematical probability is the belief that there can be no science of chance, since chance is precisely what escapes science.[36] It is a natural belief and was supported by the authority of Aristotle. Where we tend to think of things that happen "mostly" as due to some chance mechanism or combination of chance mechanisms, Aristotle's conception of chance corresponds more to the modern word "fortuitous": "Chance is also the cause of good things that happen contrary to reasonable expectation, as when, for instance, all your brothers are ugly, but you are handsome; or when you find a treasure that everyone else has missed."[37] By concentrating on the *single* chance event, Aristotle sees chance as something of which there can be no explanation or theory. Whereas mathematical probability links chance with what happens mostly, Aristotle contrasts the two; it is this contrast that produced his argument that the stars must be fixed on a sphere, since "nothing that happens always or mostly can be due to chance."[38] So also "there can be no demonstrative knowledge of chance. What happens by chance is neither a necessary event nor one that happens mostly, but happens differently from either; whereas demonstration is concerned with one or other of these. Every syllogism is from premisses true necessarily or mostly."[39] Aristotle also treats the matter of fallible signs in a way that discourages further inquiry about them: if a sign like a bodily symptom or a weather sign is not followed by what it would ordinarily signify, it is because a more powerful cause has intervened.[40]

The contrast Aristotle sees between chance and what usually happens is heightened rather than lessened by the extraordinary treatment he gives of the one case in which he admits a connection between the two—the occurrence of male and female births. Some offspring take after their parents rather little, sometimes even to the extent of being monstrosities, but at other times only to the extent that some infirmity causes a female to be born. Aristotle does not mention the fact that male and female births keep close to half and half—a fact not easily reconciled with his theory. Indeed he thinks some external influences can cause more females to be born.[41]

"There can be no demonstrative knowledge of chance." What Aristotle means by this is quite correct, but it is obviously a dictum that will discourage any science of random phenomena. It is very difficult to know how this thought affected thinkers, even those who repeated Aristotle's words. There is, however, an interesting remark of Chrysippus that takes dice (actually knucklebones) as a paradigm not of chance but of necessity: "Chrysippus in many places cited as evidence the knucklebone and the balance and many of the things that cannot fall or incline now one way and now another without the occurrence of some cause, that is of some variation either entirely in the things themselves or in their environment, it being his contention that the uncaused is altogether non-existent and so is the spontaneous."[42]

If even dice suggest not chance but causality, there is little prospect of a science of chance arising. One small piece of evidence that the idea was a substantial barrier in later times is Pascal's excitement on discovering a demonstrative science of chance, implying that he finds it natural to think there can be no such science. Pascal's thought elsewhere, in fact, makes some play with the notion of the fortuitous. "Cleopatra's nose: if it had been shorter the whole face of the earth would have been different."[43] The idea that things would have been very different with a small change in initial conditions is part of his strategy of destabilizing the smug view of the worldly that all is much as it seems. Pascal reminds noblemen that their status is an accident of birth. In language that recalls his earlier work on dice, he says: "You are not merely the son of a duke but you find yourself in this world only as the result of an infinite number of fortuitous events (*une infinité des hasards*). . . . Your soul and your body are equally indifferent to the station of a boatman and to that of a duke." Nor is there any reason to be given for many of the conditions of life that we take for granted, such as why the span of human life is a hundred rather than a thousand years.[44] Of course, it is only human calculation that Pascal intends to abase; he believes in a divine plan: "God has not abandoned his elect to caprice and hazard."[45]

Another related speculation has even less evidence to support it but has a certain a priori likelihood. Nearly all writing about chance before modern times was in terms of fortune, fate, the goddess Fortuna, and the Wheel of Fortune.[46] The wheel of fortune is a powerful image, transmuting an understandable *Schadenfreude* at the fall of princes into the warm inner glow of moralizing, while undoubtedly giving a good fit to many empirical cases. It has no tendency to lead to probabilistic thinking about chance, since the "inevitability" of the wheel's turn suppresses such questions as, In how many cases does good fortune turn to bad?, or What is the expected time to a reversal of fortune? So it could be argued that the theory of the wheel of fortune hindered the theory of probability by providing a rival conceptualization of the field. It is hard to find definite evidence for this view. Probabilistic discussions never occur with those of fortune, but whether this is evidence for or against the theory is difficult to say.

It is possible that both the "no science of chance" idea and the Wheel of Fortune are symptoms of a deeper connection between chance and unreason. After all, the *point* of games of chance, or of reference to fortune, is that the outcomes of chance or fortune are unpredictable and uncontrollable. Those cultures that most favor games of chance are the ones that place little value on personal achievement and rely on decision making by divination.[47] Even now, an interest in games of chance is associated with superstition.[48] It seems at the very least that chance is psychologically infertile ground for the play of reason.

One confirmation of the above suggestions as to why mathematical probability developed so slowly is that all alternative explanations that have been thought of are considerably less credible. Suggestions include, with brief reasons against them, the following:

— *The absence of a combinatorial algebra or combinatorial ideas.* Combinatorial ideas were developed but not used in probability except briefly by Galileo. In any case, one needs simply the ability to count to produce the basics of probability as applied to dice. The methods first used successfully by the anonymous writer of 1400 and by Pascal are not exactly combinatorial at all.

— *The superstition of gamblers.* It is certainly true that there is a connection between gambling and superstition and between rationality and not gambling. To that extent, this suggestion agrees with the "gambling is not interesting" suggestion above—that is, not interesting to the kind of person capable of making conceptual advances. But gamblers can also be hardheaded. Cardano is an example. In any case, probability theory was developed by people who were, in almost all cases, not gamblers. Any psychological explanation ought to refer to the psychology of the kind of people who did eventually develop the theory, rather than of those who did not.

— *The absence of a notion of chance events.* Roman law had a perfectly good notion of chance events, clearly expressed, and people from many fields knew about it.

— *Moral or religious barriers to the development of the idea of randomness or chance.*[49] When there were moral or religious barriers to an idea familiar to the lawyers—for example, usury—that was precisely when there was most discussion of the idea, in order to demarcate the legitimate from the illegitimate. The example of usury is also a reminder that whatever moral or religious barriers there may have been, the followers of Mammon, godless, greedy, and calculating to a man, were quite capable of ignoring or circumventing them. It is not any lack of an idea of chance that has to be explained but the failure of that idea to become mathematical.

— *The necessity of capitalism to ask questions about random events.*[50] The random events that capitalists wanted to ask about were connected with insurance and annuities, and they received answers sufficient for their needs without requiring mathematical theory about dice. Dice playing, on the other hand, is not capitalism, and the people who finally did ask numerical questions about it, like Galileo's patron the grand duke of Tuscany and de Méré, were interested in the useless, not the useful. In any case, the period of the rise of capitalism, say 1350–1650, is exactly the time of the gap to be explained.

— *The necessity of the Renaissance development of the idea of signs or uncertain evidence in the lower sciences to a theory of probability.*[51] Most of the present book is concerned with tracing the existence, indeed the ubiquity, of discussion of uncertain evidence from the Greeks to the time of Pascal. No era lacked this concept, except those that lacked the general conceptual apparatus of civilization. In any case, the explanation says nothing about dice, the theory of which was not connected with uncertainty until both theories were substantially developed.

— *The gap before the Renaissance between theory and practice and the resulting "habit of mind which made impossible the construction of theoretical hypotheses from empirical data."*[52] To some degree, this suggestion is in agreement with the ideas on applied mathematics and modeling expressed above. To the extent that it makes a more general claim, it appears to be wrong. While engineering was indeed a much less theoretical discipline in ancient and medieval times than it is now, there was still a great deal of progress. Some of it was unwritten, but there were plenty of practical written guides on how to farm, erect buildings, draw in perspective, add and multiply, do accounting, and so on. Extracting theory from empirical data was also what law did; very well, from the point of view of concepts relevant to probability, and often, even excessively, in writing. Again, what is lacking is not theory but mathematical theory.

There is one aspect of the emergence of probability, however, that remains unaccounted for. Soon after Pascal and Fermat's work on dice,

there appeared the work of Graunt, Petty, Hudde, and de Witt on tables of mortality. This is conceptually much simpler than the work on
dice; all it requires is noticing a relative frequency in a population and
asserting that it will continue. Mathematically, the only requirements
are a population (that is, a set), and a relative frequency in it (that is, a
ratio of numbers). It is surely not much to ask. The ancient medical writers had no problem with talking about what happened "mostly" or "as
often as not." Even in the ancient world, and certainly in the fourteenth
century, the commercial world was full of records and used numbers
everywhere. To set a premium for an insurance one must, at least un-
consciously, have observed a relative frequency. Yet there was no serious
numerical study of such things until 1660, despite the apparently obvious commercial advantages. The reasons for this remain obscure. Perhaps a set and a ratio are so simple, and at the same time abstract, that
they are easy to overlook. (Why did set theory not develop until the late
nineteenth century?[53] Or market surveying by business until this century? And the difficulty biologists had in accepting Mendel's work and
thinking in terms of populations, even in 1900, is well known.)

The absence of averages—except for a few examples concerned with
astronomical observations[54]—is remarkable; taking the average seems
to us an unavoidable idea when looking at any table of numbers. It is the
same with proportions; while there are ratios between lengths and between numbers in Aristotle and Euclid, it is hard to find anything like a
precise numerical percentage, like "49 percent of Romans would vote
for Caesar." Amazingly, even after Fermat, Pascal, and Huygens invented the mathematical theory of probability, they were apparently
quite unaware that the probabilities they had calculated had anything to
do with relative frequencies or averages. Pascal explicitly says that what
he is calculating is "what is rightly due" and that this can be "little investigated by trial; for the doubtful outcomes of the lot are rightly attributed rather to fortuitous contingency than to natural necessity."[55]
Aristotle could hardly have spoken better. Huygens, too, though his use
of the word "expectation" may lead the modern reader to think he is
dealing in averages, speaks solely of what a hope is worth in the single
case. (It is true that Galileo and, up to a point, Cardano have some notion of long-term averages in dice throwing.) Again, the work on mortality rates of around 1660 was hardly used in the next hundred years.[56]
There is something to be explained, but some mystery remains as to
what the true explanation can be.

There is still a mystery too as to what gamblers thought they were
doing before mathematicians explained it to them and whether there

was any lore equivalent to estimates of probability. It is known that some ancient dice were symmetrical and some were biased; surely some people knew the difference? It seems that ancient games involved rather less repeated throwing than modern ones.[57] But still it is surely impossible to play a game like backgammon without acquiring a good idea of the probabilities of winning from various positions. Odds at racecourses are good predictors of performance on the track.[58] That is, people can easily convert experience to numerical odds without training in any theory of probability.

The occasional phrase like "twenty to one" in Shakespeare and earlier suggests a kind of vernacular quantification of the racecourse-odds kind. Modern children between the ages of about five and eight acquire without instruction an ability to reason well about chances depending on relative frequencies and random choices though they do not explain their reasoning well in words.[59] Presumably medieval adults had similar abilities. The question is whether any of this knowledge was expressed in words, written or unwritten. Undoubtedly there was much unwritten lore on practical subjects in ancient and medieval times: what is written on cathedral building is far less than is sufficient to build one. On the other hand, there is much unexplored writing, especially from the fourteenth to the seventeenth centuries. When historians are equipped with suitable optical character recognition and text retrieval tools and can do their work as it should be done, it will be possible to discover with some certainty what was actually known. In the meantime, speculating without adequate evidence would be an offense against that very rationality whose discovery the present book has chronicled.

Two Parallel Histories

Another reason to believe that the problem with discovering mathematical probability lies in the mathematics is the parallel between the history of probability and that of two other abstract and loosely mathematical concepts, continuity and perspective.

Aristotle treats the subject of continuous variation, including motion, but quantification proved difficult.[60] Ancient languages contain no unit of speed, such as miles per hour, nor are any other rates quantified. The study of continuity reached a peak in the fourteenth century in the work of Oresme and others. Even such unlikely figures as Olivi and Scotus contributed.[61] Their work was largely qualitative (which here means quantitative but not exactly numerical, as when one says that the speed of a certain body increases linearly with time, or when one distinguishes between speed and acceleration but does not measure them with any units). Thereafter, little happened until Galileo, whose experiments found that the speed of falling bodies did increase linearly with time and whose discovery of the pendulum allowed adequate measurements of small time intervals. Descartes, Fermat, Pascal, and Huygens studied curves and their tangents, and the areas of curved figures, developing the difficult mathematics of limits. They applied the results to mechanics, and their achievements led to a general science of the continuous, the calculus of Newton and Leibniz.[62]

The parallel with the development of probability could hardly be closer. Even the individuals involved are the same. Aristotle and other ancients made various correct qualitative remarks on probability, but little was achieved in the way of quantification. The mathematical development reached a peak with Oresme, who knew, for example, that a large number is unlikely to be a cube, because cubes thin out among numbers. Thereafter, little happened until Galileo, who achieved correct quantitative results on dice. Fermat, Pascal, and Huygens then created the full mathematical theory, while Descartes, Newton, and Leibniz made at least some correct statements on the subject.

In the years of greatest achievement in both areas, the middle of the seventeenth century, we can trace the same people developing a community in which standards of mathematics are high enough to support breakthroughs. Roberval understands new things about tangents before 1640 and discusses them with Pascal (senior) and Fermat.[63] Around 1642, Hobbes submits an idea on comparing the arc length of a parabola and a spiral to both Mersenne and Roberval; Mersenne writes to Fermat on Roberval's extension of the idea; the problem is further discussed by Pascal (junior).[64] The young Huygens impresses Mersenne by refuting Caramuel's opinion that the distance traveled by a falling body is proportional to the time elapsed.[65] All these events have close parallels, as we have seen, in the emergence of mathematical probability.

The parallel continues to the present day. As in the case of probability, it has proved most difficult in artificial intelligence to formalize what "everyone knows" about motion.[66] And in psychology, consistent patterns of expectation about motion are found on eliciting subjects' "intuitive physics" or "naive physics" by asking them to imagine what motion would occur in various circumstances. Some subjects give the correct answers, but a high proportion, even those educated in physics, persist in such errors as expecting curved motion to continue in a curve when released.[67] The authors of these studies claim that all the medieval mistakes about the motion of projectiles and circular impetus can be found among present-day college students. This suggests, among other things, that the medievals were doing their physics by the same kind of imaginative reasoning from experience as the psychologists are now studying. Again, the parallel is close with the intuitive understanding of probability.

We now apply an argument from analogy. Since the history of probability and the history of continuity share so many features, they probably share the same explanation for the time at which they successfully became mathematical. The secret must be looked for in the mathematical culture that provided the skills to mathematize recalcitrant concepts.

The history of optics and perspective does not afford quite so close a parallel, but the differences are themselves of interest. The most important difference is that the science underlying optics, geometry, became explicit much earlier than those underlying probability and motion. A good deal of theory relevant to perspective is in Euclid's *Optics*, but it is not applied to drawing pictures. There are isolated Roman wall paintings that are obviously drawn by some kind of perspective construction.[68] But, as in the *Digest*, the theory lying behind the applications does not survive, and it is impossible to tell how explicit it was. Experimenters of the thirteenth and fourteenth centuries sought the law of refraction, that is, the relation between the angle at which a light ray strikes a surface between, say, air and water, and the angle at which it emerges on the other side of the surface. The medieval perspectivists knew what they were looking for, as is clear from their wrong answers.[69]

The correct answer, Snell's law, was not discovered until about 1620 and became widely known only through the work of Huygens. Again, the difficulty seems to be mainly mathematical. The seventeenth century had a familiarity with measurement and with the crucial notion of the sine of an angle that the medievals did not. Meanwhile, in painting, Giotto invented, or reinvented, certain "empirical perspective" methods for giving an illusion of depth to a painted surface, such as drawing lines that are in reality parallel as converging toward a point and drawing circles as ellipses.[70] How consciously aware of his rules he was is impossible to tell, but a few of them appear later in the fourteenth century in textbooks and are evident from the construction lines on walls.[71] Finally, shortly after 1400, general explicit rules of foreshortening were discovered.[72] (The time and place is the same as those of the first manuscript on mathematical probability and the first discussions of insurance.) It is clear from Ghiberti's account of the discovery of perspective that the Scholastic treatises on optics were crucial in the final step of making explicit the painters' rules.[73]

In the early seventeenth century, the cast of characters already noticed working on probability and continuity were also the ones who created modern optical theory. Galileo's knowledge of optics allowed him to build a telescope; Kepler discovered the focusing of the lens of the eye on the retina; Roberval wrote on optics; Fermat's principle and Huygens' principle are still crucial to optics.

There are other subjects that are well seen in terms of the gradual coming to explicitness of implicit knowledge. The point of view of Auerbach's *Mimesis*, for example, is that much of the history of literature consists in a gradual colonizing of human experience of life, both inner and outer, as a possible subject for literary treatment. Ancient literature, for example, was able to represent the life of ordinary people only in the comic mode, but later literature gradually learned to express the heroic and tragic in ordinary life. Further examples are found in the next section.

The final sections of this chapter attempt to explain two unusual features of the story told in this book: the prominence in it of the Scholastics and of law. In each case, the explanation lies not in anything unique to probability but in much wider relations between the Scholastics and law and the development of ideas. Just as one's understanding of why nails rust is advanced by knowing that all iron things rust, so one's understanding of the role of the Scholastics and law in probability is improved by noticing the ubiquity of the two in all medieval and early modern advances in ideas.

The Genius of the Scholastics and the Orbit of Aristotle

Each generation has to learn anew the importance of Aristotle and the Scholastics in the history of ideas. Each generation is as surprised as the one before, for everyone approaches the Aristotelians through certain myths. Even now, the only factoid "known" to many people about the medieval theologians is that they debated how many angels could dance on the head of a pin. This ridiculous libel has been repeated at least since Chillingworth without it ever having been found in any medieval text.[74] So a priori unlikely a myth can sustain itself only through the support of a context of surrounding more general myths, such as that Aristotle did science without observation, that the Scholastics refused to look through the first telescopes, and the Renaissance myth generally. There are reasons why these myths are so impervious to refutation. The reasons are somewhat different for Aristotle, for Thomas Aquinas, and for the later Scholastics.

The influence of Aristotle on the development of thought, though widely recognized, is underrated. We are all in his orbit. ("Aristotle's works are full of platitudes in much the same way as Shakespeare's *Hamlet* is full of quotations.")[75] Everything is in Aristotle somewhere—at least in potency but often in actuality. And the reason we underrate his contribution is, of course, precisely that it *has* become platitudinous: we forget about it for the same reason that we forget about the air we breathe. Anything that has become background, or context, or tradition is no longer salient, sometimes no longer represented symbolically at all. The *Meno* theory really is true of what we have learned from Aristotle: we have forgotten that we learned it, but it is still there, waiting to surprise us when we are induced to remember it. But there is another reason that we do not notice our Aristotelianism. Aristotle is a philosopher with more respect than most for "what seems so to all, or to most, or to the wise." His philosophy has none of the paradoxes repugnant to common sense that render the thought of other "great philosophers" so memorable.

Thomas Aquinas occupies a happier position than any other Scholastic, since he is Philosopher By Appointment to the Catholic Church and thus has a corps of followers dedicated to enhancing his memory. Hence his major works are readily available in modern languages, books and articles on his thought appear regularly, there are indexes to what he thought on any topic, and so on. Perhaps there is some tendency on the part of his followers to exaggerate his importance (of works on the history of probability, those of Byrne and Breny could be accused of such a bias).[76] And perhaps there is a tendency of those *extra ecclesiam* to overcompensate. But in general Aquinas can be said to be known as well as he deserves.

The contributions of the Scholastics after Aquinas, and especially those later than Ockham and Scotus, have been the most grossly neglected. Writing only in Latin, attacked by every major thinker from Descartes to the Encyclopedists, and reviled even by Thomists as "decadent Scholastics," they have had few friends. Yet we have seen in almost every chapter of this book the contributions they made to every aspect of probability—evidence in law and moral theology (and "probabilism" and "moral certainty") and to understanding aleatory contracts. Their contributions were much wider than this, and in many cases they are "known" to experts in the history of particular fields. But they are not generally known outside each field and, hence, are invisible in the mass.

Mathematics

Mathematics is usually thought to be a model of absolute clarity of thought achieved with the use of symbols and, hence, no place for

Scholastics to be delving. But a prerequisite of most mathematics is analysis of concepts confusedly present in experience, and in this the Scholastics were masters. Boyer's book on the development of calculus gives an adequate treatment of the work of Oresme and the Merton school on graphs and the analysis of continuity and motion, and the material is well displayed in Grant's *Source Book in Medieval Science*.[77] Analysis of infinities is also suited to the conceptual methods of the Scholastics; Albert of Saxony pairs off infinite sets "in the imagination" in the same style as Cantor.[78] Yet mathematics is still commonly perceived to have gone through a dark age from the third century to the sixteenth.

Legal Theory

Modern law—especially Anglo-American law, which does not depend directly on the Roman texts—shows many evidences of its centuries of development under the Scholastic intellectual *oikumene*. Its basic apparatus includes general conceptions of rights, natural justice, and equality, as well as Scholastic methods of comparison of texts and logical argument for and against propositions.[79]

The process by which this came about has been most closely studied for the law of contract. The largely unexplained Roman rules were given explanations in terms of Aristotelian concepts of the nature of contract, and the Spanish Scholastics of the sixteenth century were responsible for a synthesis, Aristotelian in conception and Roman in detail, involving a casuistry of will, consent, fraud, mistake, duress, and so on. Later legal theorists up to the present retained the Scholastic language and distinctions, though ignoring as far as possible the Aristotelian philosophy.[80] International law, not being under the control of any king, legislature, or court, has a character more suited to the abstract discussion of the Scholastics than the law of individual states. It was studied in the fourteenth century when it was recognized that Italian city-states were de facto independent, whatever the pretensions of the empire. The history of international law has had many decades to absorb the inclusion of Vitoria's and Suarez's works, in English, in Scott's *Classics of International Law*, but credit for those Scholastics' ideas is still often given to later thinkers like Grotius and Pufendorf.[81]

Political Theory

Notions of constitutional government, the restraints of custom and tradition, and the subtle relations between legitimacy of government and popular consent are medieval achievements in both theory and practice.[82] The reasonableness of Scholastic discussions of these matters is in some ways only now being attained again, after many unhappy detours through extremist theories of divine right of kings, utopianism, theocracy, revolution, and class war. The one-sidedness of such theories accounts for their memorability—and hence the corresponding difficulty of remembering more nuanced theories—as well as for their disastrous consequences in practice.

Simplified counterfactual situations are a staple of Scholastic analysis in all fields. In political theory, they include the long-running scenario of a "state of nature" in which all are "born free and equal."[83]

Economics

In economic history, the idea that the serious development of theory began with the seventeenth-century Mercantilists (perhaps with Aristotle deserving some passing mention) was exploded by Schumpeter and others with their examination of the Scholastics of, especially, the sixteenth century.[84] But the idea can hardly be said to be common property even now. Concepts like demand, capital, labor, utility, scarcity, and the cost of lost opportunity were developed at length by the Scholastics, based on Aristotle's remarks.

It is difficult to evaluate the contribution of the business community to these notions, for lack of evidence, but at present it seems that the explicit names and distinctions were due more to the Scholastics' need for precision in ethical and legal discussions than to any strictly business requirements. In any case, as we saw in chapter 10, the distinction between ethics and business was once not so rigidly drawn. Oresme's monetary theory is only one of the connections between the Scholastics and more applied and mathematical investigations into business.[85]

Psychology

The general medieval concentration on the inner life encouraged by Augustine, enforced confession, and Aristotle's De anima is reflected in the large number of texts on psychological subjects.[86] One of the longest-running themes in the history of psychology is faculty psychology as advanced by Avicenna and his Western Scholastic followers. Faculty psychology holds that human information processing is decomposable into largely distinct tasks like sensation, comparison of different sensory modalities, memory, and so on, which are undertaken by different "faculties" located in different regions of the brain. It was the natural way of organizing psychological data for many centuries.[87] Its (Latin) language is the way we still express our "folk" psychological explanations of behavior in terms of aspirations, sensations, imagination, actions, motives, emotions, and so on.[88] While the approach was disparaged in the heyday of behaviorism, its revival under various names has been one vigorous strand in more recent cognitive psychology.[89]

Philosophy

Since Gilson's studies of Descartes' Scholastic vocabulary, there has been much work on the dependence of the founders of modern philosophy on the conceptual apparatus of Scholasticism.[90] The Scholastics controlled the well-funded university posts, the academic presses, and the curricula; their logic and philosophy were a central part of what every educated person had to learn.[91] The Rationalists and Empiricists who attacked the Scholastics so forcefully provide perfect examples of the old topos of students and teachers disagreeing heatedly on answers but exhibiting a unity of questions, unspoken assumptions, and concepts. The fiercest opponents at, so to speak, the species level are found to be united at the genus level. Descartes and Locke underwent particularly heavy exposure to Scholasticism in their formative years, and the questions they ask, and the vocabulary of their answers, stray remarkably little from their teachers' practice. Gassendi very pointedly asks why the Cartesian ego, having doubted everything and put aside all prejudice and tradition, is still spouting Scholastic terminology.[92]

Linguistics

Grammar was a foundation stone of medieval education, and the inherited texts were from the twelfth century subjected to the same process as legal texts—of glossing and commentary in search of the philosophical principles underlying them. Many of the concerns of modern semiotics on the different ways in which words can have meaning are visible in medieval debates about signs.[93] Vocabulary is one of the great hidden successes of Scholasticism. The present-day international language, in which this book is written, is a blend of Old English, Norman French, and Scholastic Latin; and it is the vast inheritance of Scholastic vocabulary, mostly still carrying its medieval meaning, that bears the weight of academic discussion on all subjects.[94] Consider the current section in general, or the present sentence in particular, or examine adjacent portions of the text for a variety of natural and artificial examples of Scholastic abstract vocabulary.

Physics

The situation is partly but not wholly different for physics. The thesis of Duhem, that much of early modern physics was anticipated by the

Merton school and associated Scholastics in the fourteenth century, is
well known and has been the object of much discussion. Despite some
qualifications, it has largely stood the test of time.[95] The careful studies
of Wallace on Galileo's early writings have established how much
Galileo owed to Scholastic physics.[96] But physics is not the subject in
which the Scholastics show to best advantage. While the medieval aver-
sion to observation and experience in general is much exaggerated, and
while the idea of mathematized science was understood, it is undeniable
that *controlled* experiment and accurate measurement were largely absent from medieval science.[97] The medievals supported their physical
theories with "experience," but they usually meant common experience,
"what everyone knows." Without careful measurement and experiment,
the conceptual analysis that is the strong point of the Scholastic method
could make only limited progress in physics. Galileo's addition of those
ingredients to the recipe of science makes his reputation for originality
entirely deserved. Still, purely conceptual work, combined with everyday experience, can result in *some* good physics as well as good social science or law. Statics, the subject of Duhem's original work, can progress
a good way purely on symmetry considerations.[98] And the distinction
between velocity and acceleration, necessary for any serious work on
motion, is purely a matter of concepts.

NONE OF THIS IS TO DENY that there are areas of knowledge in which
the Scholastic method is inappropriate. Chemistry, pharmacology, and
metallurgy are examples: knowledge about the effects of chemicals and
drugs and the strength of materials is so heavily empirical that looking
into oneself and analyzing concepts will not make much progress, however good a memory one has.

It is not an accident that these disciplines were undeveloped before
very modern times. What knowledge there was in them was more a mass
of engineering expertise than a scientifically grounded body of facts. In
particular, they lacked a methodology for improving themselves.

The Place of Law in the History of Ideas

It remains to survey and explain the wide influence law has had in the
development of ideas up to early modern times. Again, the situation that
became evident in the history of probability is part of a greater whole
and can only be understood in that wider context.

"Ignorance of the law is no excuse," according to a universal legal
maxim.[99] It is asking a great deal, both of legal insiders and of those
without. It demands that ordinary persons have a general understand-

ing of legal principles, at least as far as their own affairs are concerned. The *Digest* and the Talmud are huge storehouses of concepts, and to be required to have even a sketchy idea of them is a powerful stimulus to learning abstractions.

What is less obvious, but equally important, is that the maxim imposes a heavy burden on the law itself. Legal concepts must be, in some sense, comprehensible at large. No formulas, no flow charts, no diagrams. Despite a common impression to the contrary, the law cannot possibly be a tangle of esoteric rules that invariably need resort to a lawyer to understand or to have understood on one's behalf. Since the point of the law is to order ordinary affairs, the language in which the rules are expressed must be substantially that of ordinary life. *Responsibility* and *reasonable care*, for example, must mean what they do to ordinary people, if those people are to take them as a guide to their behavior. Of course, in particular cases, the affairs being ordered may themselves be technical and understood only by experts, in which case the applicable law can also be highly technical, as with taxation law. But criminal law cannot be like that.

To regulate conduct in a domain, then, law is engaged in a cooperative exercise in developing the concepts of the domain. The domain experts, such as merchants and their accountants in the case of commercial law, will have practices and customs concerning appropriate actions in different circumstances, but they may not be explicit or may be expressed in a partial or misleading way. They may have been simply actions taken in response to some particular circumstance and copied in circumstances felt for unstated reasons to be similar. Only when disputes arise and the lawyers are called in does the need arise to debate in words what concepts are applicable. (If I am bequeathed an estate and its "fruits," do these include children born to slaves on the estate? Do they include the right to sue for thefts that occurred before I took possession?[100] The merchants are charging different rates when the enterprise involves danger: is there something, a "risk," for which they are charging? If so, what kind of thing is this? Does it make sense to buy and sell it separately? Is the charge for it "interest"? Or again: couples are living together as if married: Are they really married if they have promised to be so but have not gone through a ceremony? Is consent sufficient? Is it necessary? Informed consent? Forced consent? If force is a problem, are threats "force"?)

There is a close parallel in the modern theory of building computerized expert systems, which are intended to clone experts' knowledge of a domain into a system of rules in a software package. There are well-known difficulties in the knowledge elicitation process, whereby the builder of the system quizzes the domain experts on their knowledge, and the same difficulties obstruct the legal process's attempts to codify new areas. Typically, the experts' practice is more subtle than their stated account of it; they believe they are following simple statable rules, but the rules they offer do not give the correct behavior, and the experts have to be asked again for their qualifications and unstated conditions.[101] (Legal scholars will be reminded of Lord Mansfield's advice to a new colonial judge: "Give your judgement, but give no reasons. As you are a man of integrity, sound sense, and information, it is more than an even chance that your judgements will be right; but, as you are ignorant of the law, it is ten to one that your reasons will be wrong.")[102] There are also major problems with the meanings of the terms used in the rules: they are typically fuzzy, and the way borderline cases are handled is difficult for the experts to explain in words. The upshot is that the system builder—and for the same reason the jurist—must be involved in actually building the conceptual model of the domain. Then the refinement of the concepts takes place as individual cases arise that test the adequacy of the model.

A number of these themes come together in some remarks of Guicciardini, which explain why legal thinking is a natural matrix for a wide range of conceptual advances in applicable reasoning. "Common men find the variety of opinions that exists among lawyers quite reprehensible, without realizing that it proceeds not from any defects in the men but from the nature of the subject. General rules cannot possibly comprehend all particular cases. Often, specific cases cannot be decided on the basis of law, but must rather be dealt with by the opinions of men, which are not always in harmony. We see the same thing happen with doctors, philosophers, commercial arbitrators, and in the discourses of those who govern the state, among whom there is no less variety of judgement than among lawyers."[103] Law, that is, is a model for reasoning in all those areas in which there is necessarily a balancing of opinions.

This helps explain why the originators of mathematical probability were all either professional lawyers (Fermat, Huygens, de Witt, Leibniz) or at least the sons of lawyers (Cardano, Pascal) and so will have had some contact with at least the broad concepts of legal thought. Bacon and Copernicus, among the peripheral cast, were also lawyers, Montaigne, a judge; Valla, a notary; Machiavelli and Arnauld, the sons of lawyers; and Petrarch, Rabelais, Luther, Calvin, Donne, and Descartes, former law students.[104] That people with law degrees are famous for achievements outside law is a sign of certain circumstances concerning the interpenetration of the legal and the extralegal that obtained in medieval and early modern times but that do not hold today. Law occupied then the total space now occupied by all the social sciences, both in general perceptions of the system of knowledge and in curricula. So a law degree was the only university training available for someone intending a career in any kind of administration. The England of 1500 had an administration entirely effected through the legal system, with both central and local government being set up as courts and staffed by lawyers; personal secretaries and the like were also lawyers, so that "lawyer" was virtually synonymous with "man of affairs."[105] In one of the few computerized studies of a large medieval data set, it is shown that an Oxford degree in civil law was a passport to success in a wide range of administrative careers.[106] Copernicus's training in canon law was to prepare him for a career in the management of estates.[107] This meant that law graduates tended to control writing on all subjects except in the specialized domains of medicine and theology (although one must take account also of the generalist training available in arts).

So bureaucracy was informed with the legal ethos. A fundamental weakness of Machiavelli's analysis of politics is that he cannot understand how bureaucracies, both church and state, retain power in the face of rulers' *Realpolitik.* Here as elsewhere (accessible) knowledge is power, in the long run. Medieval bureaucracies, staffed by legal and financial experts and relying on records, were well able to survive changes of ruler.[108] Their background in law, especially canon law, meant too that their agenda was informed by moral-legal concepts of rights, which could differ systematically from the natural views of rulers. It is hard to imagine the individualistic knightly ethos issuing in such orderly precursors of modern bureaucratic projects as the medieval poor laws, but it is natural enough from a canon law perspective.[109]

From the perspective of the ordinary person, law also took up a greater proportion of the horizon than is at first obvious. Much of the access of illiterate people to writing was via the semilegal profession of scribe or notary.[110] Compulsory confession, as explained in chapter 4, meant that everyone submitted his life's doings to a miniature court and was encouraged to think of himself in terms of the internal forum—or court—of conscience. Secular law too was everywhere. In Elizabeth I's time, there were forty thousand cases in the court of chancery alone, most requiring ordinary witnesses to answer sophisticated questions like, "Do you not know or have you not credibly heard or are you not fully persuaded in your conscience that it was the true intent, will and meaning of the said Nicholas Bristowe, deceased, that . . . ?"[111] A good indication of how deeply law pervaded medieval thought is the assumption made in nonlegal literature that readers will understand law. Shakespeare's knowledge is substantial if not deep.[112] There is even more in Chaucer; the "Parson's Tale," for example, is based on the theory of homicide in Raymond of Peñafort's *Summa of Cases of Conscience*. The *Canterbury Tales* incorporate substantial legal material on clandestine marriage and on contracts. Nearly as much can be said of almost any other Western medieval author.[113]

It is clear, then, that law contributed more to the growth of abstract ideas than is generally recognized. The Talmud, the *Digest*, and the medieval lawyers provide a huge source of concepts that have been incorporated at a barely conscious level in our ways of thinking.

So much for theory. Let us review some areas in which law has been a prime motive force in the development of concepts, other than the areas of evidence and aleatory contracts discussed in earlier chapters.

Arithmetic

The main reason for learning arithmetic has always been to deal with the measurement of legal obligations, such as taxes or debts owing, the division of profits, and the like. When money was invented, all such problems were expressed in terms of it, and the point of money is to measure contracts of sale. Problems of simple proportion, the staple of applicable calculations beyond simple additions, arise in conversion between units, the division of profits, and the determination of commission and in turn require facility with the representation and manipulation of large numbers. These developments are all evident in ancient Greek financial documents.[114]

Later, times measured to the hour first appear, outside monasteries, in contracts.[115] Accuracy in dating can have legal implications, so there was attention in legal contexts to exact dating (though the law tended to retain archaic methods of expressing the result, like regnal years and roman numerals).[116] The problem of the accuracy of measurement in contracts that resolves the courtroom drama in *The Merchant of Venice* nicely illustrates the point made above about the fuzzy nature of the concepts in legal rules. Shakespeare may have his law wrong in the case in point, but the underlying problem, about the legal implications of inaccuracy, had been considered by lawyers in connection with the latitude allowed in the determination of the just price.

Negative numbers are most naturally interpreted as debts; they so appear in Indian, Islamic, and medieval Latin arithmetic. Fractions, little considered by the Greeks, appear in the same texts in connection with problems of dividing inheritances and the profits of shared enter-prises.[117] The difficult concepts of rates and ratios are most naturally ap-plicable in mixed legal and accounting contexts. There are ancient con-tracts referring to, for example, expenses per acre.[118] The English "per," now the usual term for forming the name of any rate, appears in the six-teenth century in such phrases as *per diem*, *per annum*, and *per cent* (as early as vernacular versions like "in the hundred"), always concerning matters of wages, customs duties, and the like.[119]

In modern times, it may be true that law is chosen as a profession by intelligent people who do not like figures, but in earlier times, there were not the alternative employment opportunities involving numbers in science and technology. So mathematically able lawyers were not so rare as now; Fermat stands as an example. In any case, modern practic-ing lawyers know that the division between an upper class of lawyers and an inferior caste of accountants is more theoretical than practical, in that a large proportion of bread-and-butter legal work involves sorting through accounts.

Geometry

The supposed origin of geometry in surveying for redetermining prop-erty rights after the Nile flood may have more than a touch of myth about it.[120] But if nothing else it reflects Greek notions of what would count as a natural use of geometry. Later legal thought again saw exac-titude of land measurement as outside its domain, though of course ad-vice on the matter would be crucial in real cases. An exception is Bar-tolus's *On Rivers*, which applies Euclidean constructions to the problem of land ownership when a river changes course; it remains one of the few legal treatises to contain geometrical diagrams.[121] Even diagrammatic reasoning has a partly legal origin, one of the commonest medieval di-agrams being the tree of consanguinity, a kind of generic family tree used to explain the legal effects of blood relationships.[122]

Political Theory

Political theory is one area in which legal sources are well recognized. Even in the days when history was said to be mainly concerned with kings and battles, the traditional pageant of English monarchs and their doings was always interrupted for the high point of Magna Carta. The general background of constitutionalism that the modern Western po-litical order takes for granted—the complex of ideas including the rule of law, balance of powers, due process, freedom from the arbitrary will of despots, laws made in consultation with those affected, representation, the expectation that laws promulgated will be known to and generally obeyed by those concerned—is as difficult to create and sustain as it is essential. Its theory and practice are both legacies of medieval law.[123]

Talk of rights that one possesses naturally and ought to be able to have enforced in law originates largely in medieval law, informed by Scholastic thought.[124] The political ideal of equality owes much to the concept of equality before the law, which was upheld, with sundry grave qualifications, in medieval law.[125]

Of particular significance for social organization was the invention of the concept of corporations, or legal persons—entities like companies, guilds, monasteries, and universities—that lay between individuals and the whole state and had their own legal existence. They could own property, sue, make contracts, negotiate, make their own internal ordinances, and generally pursue their own interests. As always, there are some origins in Roman law, in the notion of partnerships.[126] There was an interesting debate about how entities such as legions or peoples can stay the same although their members change; it was one of the rare instances in which the *Digest* admitted taking up ideas from Greek philosophy. "Likewise, if a ship had been repaired so often that no plank remained the same as the old had been, it was nevertheless considered the same ship. For if anyone thought that a thing became a different one when its parts were changed, it would follow from this reasoning that we ourselves would not be the same as we were a year ago, because, as the philosophers said, the extremely tiny particles of which we were made up daily left our bodies and others came from outside to take their place."[127]

But corporations themselves are largely a development of medieval law. (By contrast, in Islamic law "the whole concept of institution is missing.")[128] Such bodies provide a kind of space for the individuals within them to develop their own interests, with a relative freedom from interference from outside. From the point of view of the history of ideas, such freedom is especially necessary when the ideas under development are those regarded with suspicion by the outside world, such as Aristotelian philosophy and science in medieval universities and loan instruments in medieval business.

Psychology and Psychiatry

The remark of the *Digest* that "the expression 'according to the law' is to be understood as referring to the intention of the laws as well as to their express statements" illustrates the need to perform inference about internal mental states to determine legal meaning, responsibility, duty, and so on.[129] On the necessity of intent for criminal responsibility, Plato had already written that "voluntary and involuntary wrongs are recognized as distinct by every legislator who has ever existed in any society."[130]

For medical or religious thought about insanity, crude concepts may suffice; but for legal thought about who may be held legally responsible for actions, some subtlety is called for. While medical thought on mental illness was mostly fanciful, the legal need to deny the validity of contracts, wills, and so on made by the insane produced rational considerations about their responsibility.[131] Legal ideas on insanity have perhaps been less subject to changes of fashion than those in medical and religious circles.

The concentration on the inner life (as described in chapter 4), especially the sins therein, gave an impetus to conceiving psychology legalistically, in terms of cases of conscience. If conscience is to be identified with the superego, and the passions with the id, then the rational ego that mediates between them and resolves their conflicting demands must essentially perform casuistry.

Ethics

Even before the medieval canon lawyers criticized the deposit of law in ethical terms, law itself had developed a sophisticated appreciation of what are now regarded as ethical concepts. Among many examples, a particularly clear one is the subtle analysis of responsibility in Roman law. Compensation is payable for the wrongful killing of someone's slaves or cattle. But for killing to be wrongful, there must be true responsibility. So "he who kills a robber is not liable, that is if he could not otherwise avoid danger." Accidental killing is also a defense, but one must distinguish between accident and negligence. "If anyone while engaged in sport or practising with javelins should kill your slave while he is passing by, a distinction arises; for if this was done by a soldier while in camp, or where such practice ordinarily takes place, he cannot be charged with negligence; but if anyone else should commit such an act, he is guilty of negligence." If someone trimming a tree cuts a branch and it falls on the road and kills your slave, he is liable; but if he gives warning, and the slave ignores it, he is not liable. Again, intention requires knowledge: "Things which have to be done intentionally cannot be done except in true and certain knowledge."[132] It is clear that the decision making is being driven by a developed extralegal conception of moral responsibility, not a formalist or precedent-based legalism.

Much the same lessons could be drawn from the primacy that Roman law gives to consent in marriage: later canon law made much of this, but the basic idea that "not copulation but consent creates marriage" is Roman.[133] Roman contract theory does not deal in a general notion of consent but does recognize mistake, fraud, and duress as able to invalidate a contract.[134]

Both Greek and Roman law had a doctrine of equity, a kind of moral standard by which the strict application of law could be bent if its straightforward application in a particular case seemed too harsh.[135] The idea that the existing law implements, to some degree, a higher moral law is of course common in Christian legal theorists, but it is also found in Cicero and in the *Digest*.[136] The *Digest* in fact begins with a declaration on the subject: "The law is the art of goodness and fairness. Of that art we [jurists] are deservedly called the priests. For we cultivate the virtue of justice and claim awareness of what is good and fair."[137] Many of the rules in the last title of the *Digest* go some way toward implementing those ambitious claims: "Someone is free of blame who knows about something but cannot prevent it"; "Always in [marital] unions one must consider not only what is lawful but what is honourable"; "By the law of nature it is fair that no-one become richer by the loss and injury of another."[138]

Anthropology

The beginnings of the transition from the tales of marvels in strange lands, in the style of Mandeville's *Travels*, to the serious analysis of the humanity and customs of the native tribes of newly discovered regions occur in the juristic-moral debate on the Spanish conquest of America. A crucial issue was whether the Indians had a legal status like idiots, who might therefore need civilized masters. Vitoria and Las Casas argue that the Indians had a developed civilization and hence a right to rule themselves.[139] But not all is sweetness and light. Eurocentric perceptions of tribal life as inferior, and of slavery as an acceptable institution, also have roots in medieval law.[140]

Hermeneutics

In law there is a pressing need for reasonable solutions to problems arising from the conflict of possible interpretations, from the change in historical circumstances since texts were written, from the need to fill in gaps in the text, and from the possible disharmony between authorial intention and the public rules of language.[141] The *Digest* takes a position on the interpretation of language that seems sensitive to postmodernist concerns, without capitulating to them. "When a statement is ambiguous, the person uttering it is not saying both the things that it might be interpreted as saying but only that which he intends to say. Thus, if someone says something different from what he intended, he neither says what the words mean, because that is not what he intends to say, nor says what he intends, because his formulation is wrong."[142]

Knowledge Organization and Information Technology

Classifying, summarizing, and indexing information is a difficult intellectual operation in its own right. Aristotle's *Categories* may be an excellent metaphysical division of being, but the Roman jurists' division into persons, things, or actions is perhaps more relevant to normal life.[143] The division of assertions into matters of fact and matters of law is again a crucial distinction at the highest level of abstraction.[144] As an exercise in excerpting and rationally arranging a huge body of ideas, the *Digest* and its summary versions, the *Code* and the *Institutes*, are perhaps unsurpassed until modern encyclopedias. If there is a rival, it is the other great legal encyclopedia, the Talmud with its commentaries. In Gratian's *Decretum* and its predecessors in codifying canon law, the effort of classifying centuries of scattered Patristic comments and papal decretals is a necessary preparation for the task of reconciling their contradictions and extracting general principles from them.

Document authentication has always been a concern of law, as is clear from the huge number of ancient seals. Later (as we saw in chapter 7) it led to the development of the first critical historical methods. The Inquisition was in the lead in finding methods of recording and accessing large quantities of documents, in its efforts to determine which heretics were relapsed.[145]

It is the concern of law with record taking that makes legal documents such a rich resource for the modern historiography of ideas, particularly on ideas related to ordinary life. Almost all of the surviving information about medieval marriage and sexual practices, for example, is in decisions on legal cases—a biased sample, naturally, but the only one we have.[146] Social historians have realized that the status of medieval and Renaissance women and children can be best investigated through legal texts.[147] It is a less biased source, because driven by real cases, than alternatives like troubadour poetry. Much the same can be said for peasants and other groups who did not contribute much directly to the written record. Jews, of course, left written records, but again questions about their relation to Christian society must largely be answered from legal records.[148]

The documentary and tradition-oriented nature of law is also a reason why it has been a remarkably good preservative of ideas. While discoveries in science and engineering, for example, have been easily lost at certain periods and rediscovered only with difficulty, the legal tradition has been much less broken, and its texts have been kept alive through constant use. Again, there are disadvantages, as mistakes like slavery and torture proved very difficult to suppress. In the legal area especially, in which theory and practice are so closely connected, practice exerts a beneficial effect on theory, but mistakes in theory are costly.

Argument and Logic

Plutarch's remarkable essay "On the E at Delphi" contemplates a range of possible meanings of the sacred letter displayed in the shrine. The first suggestion is that *E* means "if," which is said to be a uniquely human expression of the awareness of the interconnectedness of things and the possibility of inference from one to another.[149] It is a notion that hides enormous complexity behind its apparent simplicity.[150] It is well known in the psychological literature that humans have great difficulty with "if" in an abstract setting. A simple conditional like "If a card has a vowel on one side, then it has an even number on the other" tends to make the human mind seize up, when questions are asked as to what evidence is relevant to the truth of the conditional. It is found, however, that performance miraculously improves if the conditional is cast in a semilegal context, or permission schema, like "If a letter is sealed, it must carry a 20-cent stamp."[151] It has been maintained that this is best explained in terms of the evolution of specialized reasoning schemas for social contracts.[152] Whether the connection between logic and the legal realm is quite as old as that suggests, it is certainly true that legal contexts have always provided a natural milieu for conditionals.

Legal rules provide the typical cases of the need for some complexity in combining logical terms like *if, and, or, except, unless,* and so on. Ulpian's rules for the pricing of life annuities (considered in chapter 10) have a complicated "if . . . else if" structure, for example, and many other rules of the *Digest* have similarly complicated conditions.[153] As for *all,* there is, besides the implied generality of the rules of law in the last title of the *Digest,* an extended discussion of the fulfillment of rules whose antecedents involve an *all* statement.[154] Modal notions are well handled: what to do if someone attaches an impossible condition to a will or one that refers to the future? And "there is no obligation to impossibilities."[155] Counterfactual reasoning is discussed by the *Digest* in inquiring about the liability for profit that would have been made if things had proceeded correctly.[156]

As to logical relations, Justinian's original call for the composition of the *Digest* demanded that "there is to be no place for any antinomy (as it is called, from old times, by a Greek word), but there is to be total concord, total consistency."[157] It was easier said than done but was certainly a stimulus for logical thinking by later generations who wished to resolve apparent contradictions in the laws. According to Plutarch, the use of logic for this purpose is divinely intended: "When the god gives out ambiguous oracles, he is promoting logical reasoning as indispensable for those who are to apprehend his meaning aright."[158] Legislators have certainly more than filled the gap left by the cessation of the oracles. The Roman jurists also use the logical language of genus and species and mention fallacies such as the *sorites*.[159] It appears that what there was of conscious generalizing from cases, using definitions, distinctions, and rules, came from the contact the earlier Roman jurists had with Greek dialectic and grammatical theory.[160]

There is, then, some justice in Leibniz's "amusing but true paradox, that there are no authors whose manner of writing resembles the style of geometers more than the ancient Roman jurists . . . they are admirable for their consistency and applications of logic and they reason with such simple clarity and with such exact subtlety that they put our philosophers to shame in even the most philosophical matters which they are often obliged to treat."[161]

THE CONTRIBUTIONS OF LAW to the discovery of ideas, it is clear, have been substantial. But they have been little appreciated for the most part. And they have been obscured by none so much as by legal theorists. The commonest views of the nature of law, contradictory though they are among themselves, all conspire to draw attention away from the law's sensitivity to facts about the world, whether particular or general. This is most obvious with crude positivist theories that hold that law is the more or less arbitrary utterance of rules by a king or legislature. For what is to restrain arbitrary rules? Certainly not mere facts about the world: the legal maxim that the impossible cannot be commanded is perhaps the only vestige of a restriction on the will of the prince.[162]

On the other hand, theories that law arises from custom or tradition, though in principle open to debate about the formative influence of the state of the world on custom, in practice emphasize the arbitrariness of tradition, regarded as a set of practices arising out of the need to resolve disputes. And custom is thought of as essentially social, even tribal, and hence not the kind of process leading to cutting-edge conceptual discoveries—as might have been expected if the custom of merchants instead of tribal custom had been taken as the paradigm. Again, the view of law that is a central myth of the English common law tradition, that new law is "discovered" by judges as new borderline cases explore the interstices between existing rules, is little better. It concentrates on the logical relations between existing pieces of law as the matters to be discovered, not on the characteristics of the new cases. There is a legal pretence that there is really no new law but that existing law is sufficient in principle to cover all cases, actual and potential, and that the role of the judge is to discern what the law always was. That must implicitly mean that all concepts are already available in the law and hence, again, that there is little role for law in the advance of ideas. In any case, there is a general suspicion that the "discovery" is a fiction and that there is no matter of fact as to what the correct decision is.[163]

Legal training also concentrates on the internals of law, neglecting students' futures in evaluating facts. The concentration of legal education on disputed questions of law in upper courts is not obviously a good preparation for the straightforward applications of law to disputed matters of fact that make up the vast bulk of work in a normal legal practice.[164] In the Anglo-American legal tradition, attention is further diverted away from facts by the use of the jury as a black-box fact evaluator, into whose workings it is unseemly to inquire.

Conclusion and Moral

In some ways, the story of probability told here resembles the history of number theory. The plot is simply Whig; there is progress, an accumulation, sometimes slow and sometimes fast, of knowledge about probability. Progress is immensely difficult, as it is in any abstract field. But once achieved, successes are rarely lost. And there are few outright errors. Probability differs from number theory in important ways, nevertheless. In the case of probability, advance is doubly difficult, as it must coordinate formulations in words and symbols with real phenomena and also with intuitive and vague preexisting understandings of the phenomena. What is discovered about probability must be in tune with what competent intuitive evidence evaluators are doing with legal, historical, and scientific evidence.

There is a darker subplot, with a moral. If certainty is demanded when it is not available, there is a strong temptation to resort to extreme measures to achieve it—or to some false appearance of it, using torture to convert half-proofs to full proofs, resorting to alleged divine interventions like ordeals or lots, relying on "faith" instead of reasoning in religious matters. The costs of each of these deviations into irrationality have been severe. The cost of repeating those errors, in those reaches of the humanities that have traded in their birthright of rationality for a mess of postmodernist pottage, will be no less severe. If it comes to be believed that the deployment of evidence in support of conclusions is just a rhetorical cover for "power relations," there is no defense possible against those who wish to impose their agendas on the academy by naked force.

The Survival of Unquantified Probability

O ne gains the impression from most histories of probability that the subject since the time of Pascal has been entirely a part of mathematics. Or at least, the story is taken to be one of the gradual colonization by mathematical methods of areas of thought that require reasoning with uncertainty. On this view, the period before 1654 is prehistory, in which there was a tangled undergrowth of poorly formulated ideas, which were of no interest once true quantified probability was discovered.

This epilogue serves as an antidote, surveying briefly the persistence of the mostly unquantified notions that form the subject matter of this book. It will be seen that, in general, in the fields in which these ideas were developed, probability is still far from fully quantified. Nor is this a bad thing, since there is every reason to suppose that new attempts at mathematization would be as unsuccessful as previous ones.

The *Port-Royal Logic*

The first attempt to put together the old and the new ideas on probability is a peculiar hybrid, which stands as a warning of incompatibilities in the proposed union. Arnauld and Nicole's *Logic* of 1662, commonly known as the *Port-Royal Logic*, transmitted to the future a number of ideas on how the new mathematical ideas should apply to real cases of reasoning. The relevant chapters are believed to have been written by Arnauld with Pascal's advice.[1] One example is of believing notarized contracts have not been deceitfully postdated simply "because it is certain that in 1,000 contracts 999 have not been postdated. It is incomparably more probable that the contract I see is one of the 999 rather than the single one of the 1,000 which is postdated."[2] The question of evidence for notaries' signatures had been considered by Baldus, for example. Arnauld lives in a time when it is beginning to be considered desirable to add numbers, however spurious. The fact that the numbers are a purely decorative addition to an older style of argument shows that applying mathematical probability to legal reasoning will not be as straightforward as it appears.

There follows some discussion of historical evidence, belief in miracles, and of what is likely to happen in the future. It proceeds without numbers, along the lines of the various writers of the sixteenth and early seventeenth centuries on these subjects. Something new, inspired by the work on dice, occurs in the last chapter, in which probability in games is used to illustrate the importance of expectations when making decisions. A game in which one stakes one coin and wins nine may be fair if losing is "nine times more probable" than winning. Some more spurious numbers then appear, in a version of the modern "How likely is it that monkeys would type *Hamlet?*" argument: "It would be sheer folly to bet even ten coppers against 10,000 gold pieces that a child arranging at random a printer's supply of letters would compose the first twenty lines of Virgil's *Aeneid.*"[3] The chapter, and the book, conclude with a version of Pascal's wager.

Leibniz's Logic of Probability

The most explicit and well-known discussion of probabilities in the legal and logical sense is in Leibniz's *New Essays* of 1704. He holds that

> the *investigation of degrees of probability* would be very important, that we are still lacking in it, and that this lack is a great defect in our Logic. For when we cannot decide a question absolutely, we might still determine the degree of likelihood from the data, and can consequently judge reasonably which side is the most likely. And when our moralists (I mean the wisest ones, such as the present-day General of the Jesuits) join the "safest" with the "most probable" and even prefer the safe to the probable, they are not in fact far removed from the most probable; for the question of *safety* is the same as that of the small probability of an evil to be feared. The fault of the moralists, lax on this point, has been in good part due to a too limited and inadequate notion of *the probable*, which they have confused with Aristotle's *endoxon* or *opinable*. . . . And when Copernicus was nearly alone in his opinion, it was still incomparably more *probable* than that of all the rest of mankind. Now I do not know but that *the art of estimating verisimilitudes* would not be more useful than a good part of our demonstrative sciences, and I have often thought about it.[4]

He approves the older legal theorists as well as the moral theologians, while attempting to connect the old and the new:

> Jurisconsults in treating the proofs, presumptions, conjectures and indices have said a number of good things on this subject [of degrees of

assent], and have entered into some considerable detail. They begin with *notoriety.* . . . There are *proofs more than half complete* . . . Beyond this there are many degrees of *conjecture* and *indices* . . . the entire *form of juridical procedures* is in fact nothing but a species of Logic applied to questions of law. Physicians also have a number of degrees and differences in their *signs* and *indications* which may be seen among them.

The Mathematicians of our day have begun to estimate chances in connection with gambling games. . . . I have more than once said that we should have a *new kind of Logic* which would treat of degrees of probability, since Aristotle in his *Topics* has done nothing less than that. . . . But he did not take the trouble to give us a necessary balance to weigh probabilities and to form solid judgements accordingly.[5]

It is obvious enough that Leibniz has taken these ideas from the kind of authors treated in this book. It is possible to tell in some cases which writers exactly Leibniz read. In a note of 1688 he remarks that it would be good to excerpt Menochio and Mascardi on proofs.[6] These two gave, as we saw in chapter 3, the most complete account of the legal theory of half-proofs and presumptions. In his *Art of Combinations* of 1666 Leibniz refers to another work of Menochio in support of the idea that judicial matters should be left as little as possible to the discretion of the judge. The idea immediately following this reference is that jurisprudence is similar to geometry in compounding cases out of simples, from which it follows that combinatorics is very useful in law.[7] This is a very Caramuelian idea—indeed, the kind of bizarre idea of Caramuel that only someone like Leibniz could take seriously. Leibniz does in fact mention Caramuel's *Mathesis Audax* in the plan for the *Art of Combinations*.[8] A related idea, or perhaps the same one taken to an extreme, is Leibniz's thought that his Universal Characteristic, designed to solve all problems of inference by mechanical calculation, should be able to handle "estimating grades of probability and the status of proofs, presumptions, conjectures and indices" as well.[9]

Leibniz also casts light on the discovery of mathematical probability, by trying to explain more clearly than Huygens what makes a game just. "A game is just if there is the same reason for hope and fear on each side. In a just game a hope is sold for what it is worth, because it is just to sell something for what it is worth, and the price of the hope is as much as the fear. Axiom: If players deal similarly so that no difference can be assigned between them, except what consists in the outcome, there is the same reason for hope and fear."[10] The point of this axiom is to allow symmetry arguments and thus introduce the mathematics.

"If several outcomes are equally easy, and in some outcomes I will have the thing, and in the others lack it, the estimation of the hope will be the proportion the thing has to the whole, which is as the number of outcomes which are favourable to the number of all outcomes." This is the first appearance of the modern way of introducing probability as a ratio of "favourable" outcomes to all outcomes. The phrase "estimation of the hope" takes us back to the origin of the whole project in the *Digest*.

Leibniz represents, however, not only the coming together of legal and mathematical probability but also their divergence. There was an old legal saying, known to Leibniz, "reasons are not to be counted, but weighed."[11] The fundamental problem in trying to apply mathematical probability to evidence in law (or for that matter, in science) is that there seems to be no set of equiprobable basic alternatives. The end of the phase of optimism about applying mathematics to legal reasoning may be taken to be the correspondence of 1703 between Leibniz and James Bernoulli. Bernoulli asks Leibniz for legal examples to which one could apply the determination of probabilities a posteriori (that is, determining probabilities by experience, as in estimating the likelihood of death from mortality tables). Leibniz is unable to supply anything; he replies that such calculations are not usually necessary, as one can simply enumerate cases.[12] But since he had never produced any real legal cases solvable by such a priori quantification either, even to his own satisfaction, his claim is empty.

To the Present

This last section reviews what has happened in the areas in which we saw, in earlier chapters, that probability developed.

The Law of Evidence

By 1700 law had served its purpose for the mathematical theory of probability. The service was never returned. Legal probability has continued to exist, and it is accepted in legal theory that such notions as proof beyond reasonable doubt involve probability. But all attempts to quantify the concept have been resisted. While writers on mathematical probability of the eighteenth and early nineteenth centuries offered their services, and even justified their subject for its applications in law,[13] the legal profession, both Continental and English, continued to speak as if dice had never been studied. Thus Sir William Blackstone's monumental *Commentaries on the Laws of England* of 1765–69 uses the unreconstructed language of violent and probable presumptions and half-proof, with references to Roman law, to explain the grading of evidence.[14] Standard works such as Gilbert's *Law of Evidence* (1754), Starkie's *Practical Treatise of the Law of Evidence* (1824), Best's *Principles of the Law of Evidence* (1849), Sir James Fitzjames Stephen's *Digest of the Law of Evidence* (1876), and Wigmore's *Treatise on the Anglo-American System of Evidence in Trials at Common Law* (1904–5) discuss evidence in terms Baldus would have understood without difficulty. English law kept the doctrine of presumptions that Coke and Blackstone had given it. Best's *Treatise on Presumptions* of 1844, on which his popular *Principles of the Law of Evidence* was largely based, refers briefly to Bartolus and Baldus and extensively to Menochio and Mascardi.[15] It is fair to say that English law remained somewhat unhappy with these foreign grafts. A notable 1937 article says, "Every writer of sufficient intelligence to appreciate the difficulties of the subject-matter has approached the subject of presumptions with a sense of hopelessness and has left it with a feeling of despair."[16]

From around 1770, English law adopted the phrase "proof beyond reasonable doubt" (originally defined as equivalent to "moral certainty") for the standard of proof required in a criminal case.[17] But the status of the rule caused confusion. Even when the rule was believed to be understood, no number became attached to it. Instead, attempts to explain it have been purely linguistic, as in the 1947 case of *Miller v. Minister for Pensions*, in which Lord Denning declared that in a criminal charge, "that degree is well settled. It need not reach a certainty, but it must carry a high degree of probability. Proof beyond a reasonable doubt does not mean proof beyond the shadow of a doubt. The law would fail to protect the community if it admitted fanciful possibilities to deflect the course of justice. If the evidence is so strong against a man as to leave only a remote possibility in his favour, which can be dismissed with the sentence, 'of course it is possible but not in the least probable,' the case is proved beyond reasonable doubt, but nothing short of that will suffice."[18]

In 1981 there was a judicial reminder that "the concept of 'probability' in the legal sense is certainly different from the mathematical concept; indeed, it is rare to find a situation in which the two usages coexist."[19] Even in the United States, where quantification and experts are most welcome, the inroads of mathematical probability into law have been minimal.[20]

Not only reasonable doubt but also that antique conception the "reasonable man" became a staple of English law. Given a statement like, "The reasonable man, then, to whose ideal behaviour we are to look as the standard of duty, will neither neglect what he can forecast as prob-

able, nor waste his anxiety on events that are barely possible" (from Pollock's *Law of Torts* of 1895),[21] A. P. Herbert's caricature begins to look
like only a small exaggeration:

> No matter what may be the particular department of human life which
> falls to be considered in these Courts, sooner or later we have to face
> the question: Was this or was it not the conduct of a reasonable man?
> Did the defendant take such care to avoid shooting the plaintiff in the
> stomach as might reasonably be expected of a reasonable man? (*Moocat
> v. Radley* (1883) 2 Q.B.) Did the plaintiff take such precautions to inform
> himself of the circumstances as any reasonable man would expect of an
> ordinary person having the ordinary knowledge of an ordinary person
> of the habits of wild bulls when goaded with garden-forks and the per-
> sistent agitation of red flags? (*Williams v. Dogbody* (1841) 2 A.C.)
>
> I need not multiply examples. It is impossible to travel anywhere or
> to travel for long in that confusing forest of learned judgements which
> constitutes the Common Law of England without encountering the
> Reasonable Man. He is at every turn, an ever-present help in time of
> trouble, and his apparitions mark the road to equity and right. There
> has never been a problem, however difficult, which His Majesty's judges
> have not in the end been able to resolve by asking themselves the simple
> question, "Was this or was it not the conduct of a reasonable man?" and
> leaving that question to be answered by the jury.[22]

Moral Theology

It is perhaps not surprising that moral theology and canon law found
little use for quantification. Disputes on probabilism continued much as
they always had. In 1662 there appeared at Louvain *Saul Ex-rex* by the
Irish Jansenist John Sinnigh, with an extreme attack on probabilism:
"Whoever delivers himself to the act of opinion without urgent necessity exposes himself spontaneously and uselessly to the danger of sin-
ning." One may not act on a probable opinion, "even the most probable among probables."[23]

In 1665 the Sorbonne condemned certain lax propositions. The
strongly worded censure condemned one writer who "has tried to infect
the hearts of the faithful with all the ordures of writers of his kind, by
using some sort of probability called extrinsic probability."[24] One of the
condemned propositions, however, asserted the infallibility of the pope,
leading to a major diplomatic incident between France and the Holy
See. In 1665–66 Pope Alexander VII, though favorably disposed toward
probabilism, condemned some of the more extreme laxist propositions,

including number 26, "When litigants have equiprobable opinions for them, the judge may accept money for giving sentence in favour of one rather than the other," and number 27, "If a book is by a recent modern author, one can consider its opinion probable, as long as it is not agreed that it has been rejected as improbable by the Apostolic See."[25]

In 1679 the more rigorist Innocent XI condemned sixty-nine laxist propositions, including one that asserted that probability, no matter how tenuous, is always sufficient to excuse conduct, and another that asserted the probability of the proposition that unborn children have no rational souls.[26] The antiprobabilist *Foundation of Moral Theology, that is, a Theological Treatise on the Right Use of Probable Opinions* by the general of the Jesuits, Tirso Gonzalez, appeared in 1694 (this is the work to which Leibniz refers).[27] Theoretical laxism was thereafter a spent force. Practical laxism, of course, is not so easily disposed of.

The next century, when the fortunes of the Catholic Church reached a low ebb with the Enlightenment and the suppression of the Jesuits, saw a moderate compromise agreed on and a waning of interest in the whole topic. Concina's antiprobabilist *History of Probabilism and Rigorism* of 1743 became the definitive view of the history of the affair, but the standard expression of the accepted answer was Saint Alphonsus Liguori's *On the Moderate Use of Probable Opinion*, which defended a system of equiprobabilism, almost indistinguishable from probabilism.[28]

Thereafter one hears little about either moral theology or canon law in general intellectual discourse. Gladstone, however, gave some attention to these questions. Though attracted to Catholicism, he took grave exception to its doctrine of probabilism, which, he says, "will be found to threaten the very first principles of duty . . . if universally received and applied, it would go far to destroy whatever there is of substance in moral obligation."[29]

Anglican moral theology was a buoyant enterprise in the late seventeenth century, and Boyle, for example, consulted clerics to calm his scruples over doubts about vows and the meaning of "following the safer path."[30] Not much more was heard of such matters after 1700, though as late as 1947 an Anglican academic moral theologian wrote the remarkable sentence, "An opinion is either so slightly more probable than another as to be doubtfully more probable and therefore really only equally probable, and in that case it is certain that the law which it supports is doubtful; or it is, however slightly, yet certainly more probable than another, and in that case it is notably and clearly more probable and therefore amounts to a practical certainty and makes the law practically certain."[31]

The new Catholic Code of Canon Law in 1917 retained some of the old language, declaring that "A presumption is a probable conjecture of an uncertain thing," and using the concepts of half-proof and moral certitude. And in 1942 an allocution of Pope Pius XII asserted that moral certitude lies between absolute certainty and simple probability and admits of degrees.[32]

Evaluation of Scientific Theories

Has the quantification of probability helped in the evaluation of uncertain evidence in science? In the restricted cases in which statistical tests apply, it has, but for more general theory evaluation, it seems not. Even Huygens, one of the discoverers of dice theory, had no use for it in this context, saying instead that one may obtain "a degree of probability which very often is scarcely less than complete proof" when the deductions of theory agree with many existing phenomena and correctly predict new ones.[33] Newton, who solved a few problems concerning dice, though he never spent much time on the topic, followed his famous pronouncement that "hypotheses are not to be regarded in experimental Philosophy" with these words, which appear to mean that hypotheses *are* to be regarded, if they are highly probable: "Although the arguing from Experiments and Observations by Induction be no Demonstration of general Conclusions; yet it is the best way of arguing which the Nature of Things admits of, and may be looked upon as so much the stronger, by how much the Induction is more general. And if no Exception occur from Phaenomena, the Conclusion may be pronounced generally."[34]

When Darwin, for example, speaks about probability it can hardly be supposed he has some quantitative theory in mind.[35] Even in mathematics, the evaluation of evidence for conjectures was regarded as non-quantitative in Polya's *Mathematics and Plausible Reasoning* of 1954.[36]

And in other cases in which the evidence to be evaluated is strictly numerical, as in evaluating business prospects given financial information, it appears that human judgment is competitive with the most sophisticated mathematical methods.[37] The same is largely true of testing computer programs for reliability.[38] For decisions on the most important projects, like the safety of space travel or the prospects for global warming, it is still the custom to rely on the opinion of experts to evaluate risks. They are known to make mistakes, but no method has been found to replace them.[39]

When the time came to actually implement probabilistic reasoning mechanically in artificial intelligence, the existing purely quantitative theory was found wanting. There is very much a "back to square one" feeling about the whole "defeasible reasoning" project in artificial intelligence. To perform medical diagnosis with an expert system, one needs to understand some basic things about probabilistic inference, to which the standard theory of probability does not, in general, give answers. Especially, one needs to know how to combine pieces of evidence that tend toward the same conclusion. One needs to know what to decide when evidence bearing on a conclusion conflicts (illustrated in artificial intelligence by the "Nixon diamond": Nixon is a Republican and Republicans are generally not pacifists; Nixon is a Quaker and Quakers are generally pacifists; what should we presume about Nixon?). One needs to know how to chain defeasible links: if p supports q and q supports r, under what circumstances does p support r? It has so far proved impossible to imitate a convincingly large portion of real human reasoning under uncertainty, and what formalism will be required to do so is still a matter for debate; there is no lack of approaches.[40]

The Social Sciences

Gibbon speaks for the mentality still characteristic of the social sciences: "As soon as I understood the principles, I relinquished for ever the pursuit of Mathematics; nor can I lament that I desisted before my mind was hardened by the habit of rigid demonstration so destructive of the finer feelings of moral evidence which must however determine the actions and opinions of our lives."[41] Modern defenses of the rationality of historical inferences continue to rely on an objective concept of probability, but do not demand it be quantified.[42]

The notion of probability found in the evaluation of historical evidence pervades Jane Austen's novels on persuasion, sense, and prejudice. The crucial chapter 36 of *Pride and Prejudice*, in which Elizabeth Bennet is forced to make humiliating changes to her beliefs in response to the new evidence contained in Mr. Darcy's letter, is a tour de force of the careful reappraisal of a belief system that has been based on a large body of evidence. As Elizabeth "weighed every circumstance with what she meant to be impartiality—deliberated on the probability of each statement . . . reconsidering events, determining probabilities," she came to see that her previous opinions of Darcy's and Wickham's characters were based on vanity and were now insupportable. Austen goes through each piece of evidence carefully, explaining its relation to the whole ("How could she deny that credit to his assertions, in one instance, which she had been obliged to give in the other?—He declared himself to have been totally unsuspicious of her sister's attachment;—and she could not help remembering what Charlotte's opinion had always been.—Neither could she deny the justice of his description of Jane").[43]

Economics, the modern social science that most makes a fetish of quantification, still tends to regard some probabilities as too vague to quantify (or insure against) though too important to ignore.[44]

Philosophy

Philosophical discussions of the nature of probability have sometimes begun with dice theory—but certainly not always. Locke's *Essay Concerning Human Understanding* has a concept more or less identical to the old legal one.[45] Much the same is true of the probability used in Bishop Butler's *Analogy of Religion*, famous for its saying, "probability is the very guide of life."[46] And in Hume's *Enquiry Concerning Human Understanding*, section 10, "On Miracles," argues that a miracle is a breach of a law of nature, and a law of nature has the highest possible degree of probability and so cannot be defeated by the probability of any testimony, given the known fabrications of marvels. Similar uses of probabilistic argument in modern philosophy, such as Swinburne's in *The Existence of God* (1979), require no equations.

The "ordinary language" concept of probability owes much more to these philosophers than to mathematicians. Johnson's *Dictionary* of 1755 defines *probability* as "Likelihood; appearance of truth; evidence arising from the preponderance of argument: it is less than moral certainty" (illustrated with quotations from Locke, Hooker, Wilkins, Tillotson, and Law).[47] The standard modern dictionaries like the *Oxford English* and *Merriam Webster's* also give the evidential meaning of probability as the principal one, followed by the mathematical ones.

Modern philosophical discussions of the nature of probability arise principally from Keynes' *Treatise on Probability* of 1921, which argues that all probability is logical and that it is not always possible to quantify it.[48] Carnap and the later Bayesian philosophers of science did believe that logical probability could be given numbers in principle, but they rested almost all the explanatory weight on the comparative principles that could be derived from the axioms.[49]

In his classic 1947 defense of induction using probability theory, *The Ground of Induction*, D. C. Williams wrote that "Western civilization— any hopeful, humanitarian, knowledgeable and right culture—depends on induction not merely in its parts, to justify its particular scientific inquiries and political inventions. It depends on induction altogether and in principle. . . . In the political sphere, the haphazard echoes of inductive skepticism which reach the liberal's ear deprive him of any rational right to champion liberalism, and account already as much as anything for the flabbiness of liberal resistance to dogmatic encroachments from the left or the right. The skeptic encourages atavistic rebellion with, 'Who are we to say?' and puts in theory the forms and methods of democracy on a level with the grossest tyranny."[50]

Contrary to Williams' fears, civilization and democracy survived. In the universities, however, and especially in the humanities faculties, the values of rationality remain under threat. In the philosophy of science, the progressive skepticism about the objective bearing of experimental evidence on theory, from Popper through Kuhn and Lakatos to Feyerabend, undermined the defenses science might have mounted against its irrationalist enemies.[51] With science deprived—in the eyes of theorists in the humanities—of reliance on evidence, the way was open for the even more extreme attacks from sociologists of knowledge and postmodernists, all eager to give accounts of scientific theory change independent of rational responses to new evidence.[52] Leading proponents of the "social construction of science" movement write, "Something can only become 'evidence' within the framework of an agreed theoretical understanding of nature. For the sociologist, the question is how that agreement was reached and how it is sustained. These (social) processes must be presupposed before talk of 'evidence' makes sense."[53] To begin with what social negotiations might take to be evidence (in quotation marks), instead of what actually is evidence, is to miss the point of science entirely. A fortiori, disciplines unfortunate to be caught institutionally in the humanities, like history, have suffered even more sustained attack from the forces of unreason.[54]

Alleged historical studies of theory development have been prominent in supporting these various attempts to "privilege" (as the jargon has it) irrational causes of belief change over rational reasons for it. Perhaps a historical survey of methods of evidence evaluation can play a small part in reason's fight back.

Review of Work on Probability
before 1660

The literature on probability before Pascal is dominated by Ian Hacking's *The Emergence of Probability* (Cambridge, 1975). Entertainingly written and with a sound philosophical viewpoint, it well deserved its popular success. As history, it left something to be desired with regard to the ratio of evidence to conclusions. In particular, it claimed that there was very little concept of evidential support in the literature before 1660. Hacking writes that "until the end of the Renaissance, one of our concepts of evidence was lacking: that by which one thing can indicate, contingently, the state of something else." He also claims that "a probable opinion [in the Renaissance] was not one supported by evidence, but one which was approved by some authority, or by the testimony of respected judges." To reach these conclusions, it was necessary to ignore everything written about evidence in law and almost all medieval and early modern writing in Latin. These lacunae in Hacking's evidence are substantial, given that law is the central discipline that deals with nonconclusive evidence and that Latin was the international language of learning.

A number of reactions to Hacking's book, pointing out the overwhelming evidence against his theses, formed the basis of the present book: D. Garber and S. Zabell, "On the emergence of probability," *Archive for History of Exact Sciences* 21 (1979): 33–53; R. Brown, "History versus Hacking on probability," *History of European Ideas* 8 (1987): 655–73; J. van Brakel, "Some remarks on the prehistory of the concept of statistical probability," *Archive for History of Exact Sciences* 16 (1976–77): 119–36; J. Kassler, "The emergence of probability reconsidered," *Archives internationales d'histoire des sciences* 36 (1986): 17–44; C. Howson, "The prehistory of chance," *British Journal for the Philosophy of Science* 29 (1978): 274–80; M. Ferriani, "Storia e 'preistoria' del concetto di probabilità nell'età moderna," *Rivista di filosofia* 69 (1978): 129–53; H. Breny, "Les origines de la théorie des probabilités," *Revue des questions scientifiques* 159 (1988): 453–77.

Contemporary with Hacking is O. Sheynin, "On the prehistory of the theory of probability," *Archive for History of Exact Sciences* 12 (1974):

97–141. Earlier works of similar broad scope are P. Hagstroem, *Les préludes antiques de la théorie des probabilités* (Stockholm, 1932); S. Sambursky, "On the possible and probable in ancient Greece," *Osiris* 12 (1956): 35–48, reprinted in *Studies in the History of Statistics and Probability*, ed. M. Kendall and R. Plackett (London, 1977), 2:1–14.

A separate source of evidence are works that reveal many kinds of probabilistic concepts in Jewish legal literature. They are N. Rabinovitch, *Probability and Statistical Inference in Ancient and Medieval Jewish Literature* (Toronto, 1973); A. Hasofer, "Some aspects of Talmudic probabilistic thought," *Proceedings of the Association of Orthodox Jewish Scientists* 2 (1969): 63–80; A. Hasofer, "Random mechanisms in Talmudic literature," *Biometrika* 54 (1967): 316–21, repr. in *Studies in the History of Statistics and Probability*, ed. E. Pearson and M. Kendall (London, 1970), 1:39–43; A. Hasofer, "The Talmud on expectation," *Milla wa-Milla* (*The Australian Bulletin of Comparative Religion*), no. 13 (Dec. 1973): 33–39; R. Ohrenstein and B. Gordon, "Risk, uncertainty and expectation in Talmudic literature," *International Journal of Social Economics* 18 (1991): 4–15; O. Sheynin, "Stochastic thinking in the Bible and the Talmud," *Annals of Science* 55 (1998): 185–98. Islamic parallels can be found in W. Hallaq, "On inductive corroboration, probability and certainty in Sunni legal thought," in *Islamic Law and Jurisprudence*, ed. N. Heer (Seattle, 1990), 3–31.

Recent reviews of the topic after integration of this material are A. Crombie, *Styles of Scientific Thinking in the European Tradition* (London, 1994), pt. 6; A. Crombie, "Contingent expectation and uncertain choice: Historical contexts of arguments from probabilities," in Crombie, *Science, Art and Nature in Medieval and Modern Thought* (London, 1994); L. Daston, "Probability and evidence," in *Cambridge History of Seventeenth-Century Philosophy*, ed. D. Garber and M. Ayers (Cambridge, 1998), ch. 31; A. Garrett, "The history of probability theory," in *Maximum Entropy and Bayesian Methods: Boise, Idaho, USA, 1997*, ed. G. Erickson, J. Rychert, and C. Smith (Dordrecht, 1998), 223–38.

The medieval theory of legal evidence, argued earlier to be the central strand in the development of (logical) probability, has been little studied, especially in recent decades. J.-P. Lévy's little-noticed *La hierarchie des preuves dans le droit savant du moyen-âge* (Paris, 1939) covers many of the twelfth- and thirteenth-century developments, but the later culmination of the theory in Baldus has not been discussed. The work in the area has been mainly aimed at legal historians and has made little outside impact. See also J.-P. Lévy, "Le problème de la preuve dans les droits savants du moyen âge," *Recueils de la société Jean Bodin pour l'his-*

toire comparative des institutions 17 (1965): 137–67; R. van Caenegem, "The law of evidence in the twelfth century," in *Proceedings of the Second International Congress of Medieval Canon Law, 1963,* ed. S. Kuttner and J. Ryan (Vatican City, 1965), 297–310; R. van Caenegem, "Methods of proof in Western medieval law," in van Caenegem, *Legal History: A European Perspective* (London, 1991), 71–113; A. Giuliani, "The influence of rhetoric on the law of evidence and pleading," *Juridical Review* 7 (1962): 216–52; F. Sinatti D'Amico, *Le prove giudiziarie nel diritto longobardo* (Milan, 1968); A. Gouron, "Aux racines de la théorie des présomptions," *Rivista internazionale di diritto commune* 1 (1990): 99–109; R. Fraher, "'Ut nullus describatur reus prius quam convincatur': Presumption of innocence in medieval canon law?" in *Proceedings of the Sixth International Congress of Medieval Canon Law,* ed. S. Kuttner and K. Pennington (Vatican City, 1985), 493–506; G. Donatuti, *Le presumptiones iuris in diritto romano* (Perugia, 1930). There is something on the fourteenth century in W. Ullmann, "Medieval principles of evidence," *Law Quarterly Review* 62 (1946): 77–87; and W. Ullmann, "Some medieval principles of criminal procedure," *Juridical Review* 59 (1947): 1–28. A rather distant view of the material is available in English in B. Shapiro, *"Beyond Reasonable Doubt" and "Probable Cause"* (Berkeley, 1991). Other connections with later developments are drawn in the predecessor to the present work: J. Franklin, "The ancient legal sources of seventeenth century probability," in *The Uses of Antiquity,* ed. S. Gaukroger (Dordrecht, 1991), 123–44. The volumes by K. Gross on proof in canon law—*Das Beweistheorie im canonischen Prozess* (Lausanne, 1880)—show that as usual the massive German scholarship of a century ago was there first. Also see J. Brundage, "Proof in canonical criminal law," *Continuity and Change* 11 (1996): 329–39; and R. Motzenbäcker, *Die Rechtsvermutung im kanonischen Recht* (Munich, 1958).

The ideas on proof in law in various earlier cultures were treated by a number of articles in the *Recueils de la société Jean Bodin pour l'histoire comparative des institutions* of 1964–65. See H. Pirenne, "La preuve dans la civilisation de l'Egypte antique," 16 (1964): 9–42; R. Boyer, "La preuve dans les anciens droits du proche-orient," 16 (1964): 61–87; B. Cohen, "Evidence in Jewish law," 16 (1964): 103–15; G. Sautel, "Les preuves dans le droit grec archaïque," 16 (1964): 117–60; C. Préaux, "La preuve à l'époque hellenistique principalement dans l'Egypte grecque, 16 (1964): 161–222; G. Pugliese, "La preuve dans le procès romain de l'époque classique," 16 (1964): 277–348; J.-P. Lévy, "L'évolution de la preuve, des origines à nos jours," 17 (1965): 9–70; L. Rocher, "The theory of proof in ancient Hindu law," 18 (1965): 325–71. Also see G. Ries,

"Altbabylonische Beweisurteile," *Zeitschrift der Savigny-Stiftung für Rechtsgeschichte, Romanistische Abteilung* 106 (1989): 56–80.

Probabilism and moral certainty in moral theology were studied in excellent prewar French Catholic works: T. Deman, "Probabilisme," in *Dictionnaire de Théologie Catholique*, vol. 13, pt. 1 (Paris, 1936), cols. 417–619; J. de Blic, "Barthélémy de Medina et les origines du probabilisme," *Ephemerides Theologicae Lovanienses* 7 (1930): 46–83, 264–91; M. Gorce, "A propos de Barthélémy de Medina et du probabilisme," *Ephemerides Theologicae Lovanienses* 7 (1930): 480–82; E. Amann, "Laxisme," pt. 2, *Dictionnaire de Théologie Catholique*, vol. 9, pt. 1 (Paris, 1926), cols. 42–86; O. Lottin, "Le tutiorisme du treizième siècle," *Recherches de théologie ancienne et médiévale* 5 (1933): 292–301. More recent are I. Kantola, "Probability and Moral Uncertainty in Late Medieval and Early Modern Times" (dissertation, Schriften d. Luther-Agricola-Gesellschaft, vol. 32, Helsinki, 1994); and I. Kantola, "Some modern aspects of medieval probability," *Publications: Laboratoire de la pensée ancienne et médiévale, Université d'Ottawa* 1 (1995): 520–28. Earlier works are A. Schmitt, *Zur Geschichte der Probabilismus* (Innsbruck, 1904); J. Ternus, *Zur Vorgeschichte der Moralsysteme von Vitoria bis Medina* (Paderborn, 1930); U. Lopez, *Thesis probabilismi ex Sancto Thoma demonstrata* (Rome, 1937); and D. Concina, *Della storia del probabilismo e del rigorismo* (Venice, 1743). The reaction to these works by the mid-Atlantic anglophone hegemony in the history of ideas may be imagined.

The relations of Pascal with casuistry have been noted, but many scholars have been misled, as Hacking was, by the propaganda of his *Provincial Letters*, without looking at the works of the casuists attacked. More informative on Puritan casuistry is D. Bellhouse, "Probability in the sixteenth and seventeenth centuries: An analysis of Puritan casuistry," *International Statistical Review* 56 (1988): 63–74. See also D. Bell, "Pascal: Casuistry, probability, uncertainty," *Journal of Medieval and Early Modern Studies* 28 (1998): 37–50; and N. Rescher, "The philosophers of gambling," in *An Intimate Relation: Studies in the History and Philosophy of Science*, ed. J. Brown and J. Mittelstrass (Dordrecht, 1989), 203–20.

Much has been written on Pascal's wager from various points of view, but the most complete work on its history is L. Blanchet, "L'attitude religieuse des jésuites et les sources du pari de Pascal," *Revue de métaphysique et de morale* 26 (1919): 477–516, 617–47.

A certain amount has been written about Aristotle's *Rhetoric* and the notion of probability in Plato and the Greek orators—but not normally with an eye for arguments that are genuinely probable as opposed to rhetorically effective. See for example G. Kennedy, *The Art of Persuasion in Greece* (London, 1963); C. Kuebler, *The Argument from Probability in Early Attic Oratory* (Chicago, 1944); J. Smith, "Plato's myths as 'likely accounts,'" *Apeiron* 19 (1988): 24–42; S. Nadler, "Probability and truth in the *Apology*," *Philosophy and Literature* 9 (1985): 198–202; T. Lewis, "Parody and the argument from probability in the *Apology*," *Philosophy and Literature* 14 (1990): 359–66; W. Grimaldi, "*Semeion, tekmerion, eikos* in Aristotle's *Rhetoric*," *American Journal of Philology* 101 (1980): 383–98; S. Raphael, "Rhetoric, dialectic and syllogistic argument: Aristotle's position in *Rhetoric* I–II," *Phronesis* 19 (1974): 153–67; J. le Blond, *Eulogos chez Aristote et l'argument de convenance* (Paris, 1938); E. Madden, "Aristotle's treatment of probability and signs," *Philosophy of Science* 24 (1957): 167–72; M. Mignucci, "*Hōs epi to poly* et nécessaire dans la conception aristotelicienne de la science," in *Aristotle on Science: The Posterior Analytics*, ed. E. Berti (Padua, 1981), 173–203; M. Winter, "Aristotle, *hōs epi to polu* relations, and a demonstrative science of ethics," *Phronesis* 42 (1997): 163–89; K. Schoen, *Die Scheinargumente bei Lysias* (Paderborn, 1918); W. Fairchild, "Argument from probability in Lysias," *Classical Bulletin* 55 (1979): 49–54; and G. Preti, "Sulla dottrina della *semeion* nella logica stoica," *Rivista critica di storia della filosofia* 1 (1956): 5–16.

There are some works on the treatment of experimental errors in astronomy but few that analyze how astronomers or other scientists evaluated theories. On experimental errors in astronomy, see O. Sheynin, *The History of the Theory of Errors* (Egelsbach, 1996); O. Sheynin, "The treatment of observations in early astronomy," *Archive for History of Exact Sciences* 46 (1993): 153–92; O. Sheynin, "J. Kepler as a statistician," *Bulletin of the International Statistical Institute* 46 (1975), bk. 2: 341–54; I. Schneider, "Wahrscheinlichkeit und Zufall bei Kepler," *Philosophia Naturalis* 16 (1976): 40–63; R. Plackett, "The principle of the arithmetic mean," in *Studies in the History of Statistics and Probability*, ed. E. Pearson and M. Kendall (London, 1970), 1:121–26; O. Gingerich, "Kepler's treatment of redundant observations," in *Internationales Kepler-Symposium, Weil-der-Stadt, 1971*, ed. F. Krafft et al. (Hildesheim, 1973), 307–14; A. Hald, "Galileo's statistical analysis of astronomical observations," *International Statistical Review* 54 (1986): 211–20; G. Hon, "On Kepler's awareness of the problem of experimental error," *Annals of Science* 44 (1987): 545–91; G. Hon, "Is there a concept of experimental error in Greek astronomy?" *British Journal for the History of Science* 22 (1989): 129–50; and G. Lloyd, "Observational error in later Greek science," in G. Lloyd, *Methods and Problems in Greek Science* (Cambridge, 1991).

Theory evaluation in other sciences is discussed in P. De Lacy and E. De Lacy, "Ancient rhetoric and empirical method," *Sophia* 6 (1938): 523–30; M. McVaugh, "The nature and limits of medical certitude at fourteenth-century Montpellier," *Osiris* 2d ser. 6 (1990): 62–84; and E. Sylla, "Galileo and probable arguments," in *Nature and Scientific Method,* ed. D. Dahlstrom (Washington, D.C., 1991), 211–34.

The only study of randomization in technology is S. Stigler, "Eight centuries of sampling inspection: The trial of the Pyx," *Journal of the American Statistical Association* 72 (1977): 493–500. Carneades' probabilistic theory has attracted considerable attention: see J. Allen, "Academic probabilism and Stoic epistemology," *Classical Quarterly* 44 (1994): 85–113; R. Bett, "Carneades' *pithanon:* A reappraisal of its role and status," *Oxford Studies in Ancient Philosophy* 7 (1989): 59–94; M. Burnyeat, "Carneades was no probabilist" (unpublished); I. Schneider, "The contributions of the sceptic philosophers Arcesilas and Carneades to the development of an inductive logic," *Indian Journal of History of Science* 12 (1977): 173–80; M. Burnyeat, "The origins of non-deductive inference," in *Science and Speculation*, ed. J. Barnes et al. (Cambridge, 1982), 193–238; and D. Sedley, "On signs," in *Science and Speculation*, 239–72.

And there are works that analyze probabilistic concepts in particular philosophers, notably Aristotle and Aquinas. But except for those on Descartes, they are not dense in material genuinely relevant to probability. See J. Lennox, "Aristotle on chance," *Archiv für Geschichte der Philosophie* 66 (1984): 52–60; J. Junkersfeld, *The Aristotelian-Thomistic Conception of Chance* (Notre Dame, 1945); E. Byrne, *Probability and Opinion: A Study in the Medieval Presuppositions of Post–Medieval Theories of Probability* (The Hague, 1968); A. Gardeil, "La certitude probable," *Revue des sciences philosophiques et théologiques* 5 (1911): 327–68, 441–85; M. de Gandillac, "De l'usage et de la valeur des arguments probables dans les questions du Cardinal Pierre d'Ailly sur le 'Livre des Sentences,'" *Archives d'histoire doctrinale et littéraire du moyen âge* 8 (1933): 43–91; L. Cohen, "Some historical remarks on the Baconian conception of probability," *Journal of the History of Ideas* 41 (1980): 219–31; E. Curley, "Certainty: Psychological, moral, and metaphysical," in *Essays on the Philosophy and Science of René Descartes*, ed. S. Voss (New York, 1993), 11–30; S. Voss, "Scientific and practical certainty in Descartes," *American Catholic Philosophical Quarterly* 67 (1993): 569–85; and J. Morris, "Descartes and probable knowledge," *Journal of the History of Philosophy* 8 (1970): 303–12.

Some scattered works on the language of chance and probability (on which more work may be productive) are D. Bellhouse and J. Franklin,

"The language of chance," *International Statistical Review* 65 (1997): 73–85; G. Shipp, "'Chance' in the Latin vocabulary," *Classical Review* 15 (1937): 209–12; and T. Deman, "Notes de lexicographie philosophique médiévale: *probabilis*," *Revue des sciences philosophiques et théologiques* 22 (1933): 260–90.

Some substantial studies of the general spread of probable arguments and "reasonableness" in the seventeenth century, especially in England, include B. Shapiro, *Probability and Certainty in Seventeenth-Century England* (Princeton, 1983); H. van Leeuwen, *The Problem of Certainty in English Thought, 1630–1690* (The Hague, 1963); M. Ferreira, *Scepticism and Reasonable Doubt: The British Naturalist Tradition in Wilkins, Hume, Reid and Newman* (Oxford, 1986); R. Serjeantson, "Testimony and proof in early-modern England," *Studies in History and Philosophy of Science* 30 (1999): 195–236; and B. Nelson, "Probabilists, anti-probabilists and the quest for certitude on the 16th and 17th centuries," in B. Nelson, *On the Roads to Modernity*, ed. T. Huff (Totowa, 1981), 115–21. Related issues can be found in D. Patey, *Probability and Literary Form* (Cambridge, 1984).

A rather disconnected stream of works on the problem of induction has made little impression on the urban myth that the problem was discovered by Hume: see J. Milton, "Induction before Hume," *British Journal for the Philosophy of Science* 38 (1987): 49–74; J. Stocks, "Epicurean induction," *Mind* 34 (1925): 185–203; J. Weinberg, *Abstraction, Relation and Induction* (Madison, 1965); E. Bos, "Pseudo-Johannes Duns Scotus über Induktion," in *Historia philosophiae medii aevi*, ed. B. Mojsisch and O. Planta (Amsterdam, 1991), 1:71–103. (Hacking writes that "the problem of induction could not have arisen much before 1660, for there was no concept of inductive evidence in terms of which to raise it.") The seventeenth-century Jesuit achievements on induction are known from S. Knebel, "Necessitas moralis ad optimum (III): Naturgesetz und Induktionsproblem in der Jesuitenscholastik während des zweiten Drittels des 17. Jahrhunderts," *Studia Leibnitiana* 24 (1992): 182–215; S. Knebel, "Die früheste Axiomatisierung des Induktionsprinzips: Pietro Sforza Pallavicino SJ (1607–1667)," *Salzburger Jahrbuch für Philosophie* 41 (1996): 97–128; and S. Knebel, "Agustin de Herrera: A treatise on aleatory probability," *Modern Schoolman* 73 (1996): 199–264.

Aleatory contracts such as annuities and insurance have been little studied recently, and not from the point of view of probabilistic argument, although Coumet pointed out their importance for the mathematization of probability by Pascal, Fermat, and Huygens. See E. Coumet, "La théorie du hasard est-elle née par hasard?" *Annales* 25

(1970): 574–98. Brief notices of the ancient legal background are in J. Thomas, "*Venditio hereditatis* and *emptio spei*," *Tulane Law Review* 33 (1958–59): 541–50; H. Guitton, "Le droit romain en face de l'aléa," *Revue française de recherche opérationelle* 7 (1963): 194–95; and E. Betti, "Periculum: problemi del rischio contrattuale in diritto romano classico e giustinianeo," *Jus* 5 (1954): 333–84. Other relevant information is in the early chapters of P. Bernstein, *Against the Gods: The Remarkable Story of Risk* (New York, 1996).

Works on the origins of insurance include G. de Ste. Croix, "Ancient Greek and Roman maritime loans," in *Debits, Credits, Finance and Profits*, ed. H. Edey and B. Yamey (London, 1974), 41–59; P. Millett, "Maritime loans and the structure of credit in fourth-century Athens," in *Trade in the Ancient Economy*, ed. P. Garnsey, K. Hopkins, and C. Whittaker (Berkeley, 1983), 36–52; G. Calhoun, "Risk in sea-loans in ancient Athens," *Journal of Economic and Business History* 2 (1930): 561–84; C. Hoover, "The sea loan in Genoa in the twelfth century," *Quarterly Journal of Economics* 40 (1925): 499–529; G. Stefani, *Insurance in Venice from the Origins to the End of the Serenissima* (Trieste, 1958), vol. 1; F. Edler de Roover, "Early examples of marine insurance," *Journal of Economic History* 5 (1945): 172–200; H. Nelli, "The earliest insurance contract: A new discovery," *Journal of Risk and Insurance* 39 (1972): 215–20; L. Boiteux, *La fortune de mer: le besoin de sécurité et les débuts de l'assurance maritime* (Paris, 1968); F. Melis, *Origini e sviluppi delle assicurazione in Italia* (Rome, 1975); J. Heers, "Le prix de l'assurance maritime à la fin du moyen âge," *Revue d'histoire économique et sociale* 37 (1959): 7–19; M. del Treppo, "Assicurazioni e commercio internazionale a Barcellona nel, 1428–1429," *Rivista storica italiana* 69 (1957): 508–41, 70 (1958): 44–81; A. Garcia i Sanz and M.-T. Ferrer i Mallol, *Assegurances i canvis marítims medievals a Barcelona* (Barcelona, 1983); K. Nehlsen-von Stryk, *L'assicurazione marittima a Venezia nel XV secolo* (Rome, 1988); A. Jack, *An Introduction to the History of Life Assurance* (London, 1912); S. Passamaneck, *Insurance in Rabbinic Law* (Edinburgh, 1974); R. Smith, "Life insurance in fifteenth-century Barcelona," *Journal of Economic History* 1 (1941): 57–59; M.-T. Ferrer i Mallol, "Intorno all'assicurazione sulla persona di Filipozzo Soldani, nel 1399, e alle attività dei Soldani, mercanti fiorentini, a Barcellona," in *Studi in Memoria di Federigo Melis* (Naples, 1978), 2:441–78; D. Houtzager, *Hollands lijf- en losrenteleningen voor 1672* (Schiedam, 1950); V. Barbour, "Marine risks and insurance in the seventeenth century," *Journal of Economic and Business History* 1 (1928–29): 561–96; P. Santerna, *Tractatus de assecurationibus et sponsionibus* (Venice, 1552); repr. with notes by M. Amzalak and Portuguese, English, and

French translations (Lisbon, 1971); and D. Maffei, *Il giureconsulto portoghese Pedro de Santarém, autore del primo trattato sulle assicurazioni (1488)* (Separata do número especial do Boletim da Faculdade de Direito de Coimbra, Coimbra, 1983), 703–28. None of the works just listed was noticed by historians of probability.

The most significant article on annuities remains the theological work of F. Veraja, *Le origini della controversia teologica sull contratto di censo nel XIII secolo* (Rome, 1960). Also see H.-P. Baum, "Annuities in late medieval Hanse towns," *Business History Review* 59 (1985): 24–48; G. Alter and J. Riley, "How to bet on lives: A guide to life-contingent contracts in early modern Europe," *Research in Economic History* 10 (1986): 1–53; and M. Greenwood, "A statistical mare's nest?" *Journal of the Royal Statistical Society* 103 (1940): 246–48.

Work on lotteries is descriptive rather than concerned with analyzing concepts, with the exception of D. Bellhouse, "The Genoese lottery," *Statistical Science* 6 (1991): 141–48. Also see R. Kinsey, "The role of lotteries in public finance" (Ph.D. diss., Columbia Univ., 1959); F. Endemann, *Beiträge zur Geschichte der Lotterie und zum Heutigen Lotterierechte* (Bonn, 1882); R. Woodhall, "The English state lotteries," *History Today* 14 (July 1964): 497–504; and R. Johnson, "The lotteries of the Virginia Company, 1612–1621," *Virginia Magazine of History and Biography* 74 (July 1966): 259–90. Little has been written about early modern gambling, and it is likely that there is still much to discover in obscure sources. But see J. Ashton, *The History of Gambling in England* (1898; repr., Montclair, N.J., 1969); D. Bellhouse, "The role of roguery in the history of probability," *Statistical Science* 8 (1993): 410–20. The Scholastic analyses of the probabilistic concepts in aleatory contracts is best covered in J. Noonan, *The Scholastic Analysis of Usury* (Cambridge, Mass., 1957); Noonan does not, however, attend to the two authors who went furthest in that direction, Olivi and Alexander. See also A. Lapidus, "Information and risk in the medieval doctrine of usury during the thirteenth century," in *Perspectives on the History of Economic Thought*, ed. W. Barber (1991), 5:23–38.

On dice throwing and its mathematics, the best-known book is F. David, *Games, Gods and Gambling* (London, 1962); a similar but inferior work is L. Maistrov, *Probability Theory: A Historical Sketch* (New York, 1974). David's book is very informative but is written under the philosophical influence of Jerzy Neyman, who aggressively promoted the view that there is no such thing as logical probability, inductive reasoning, or nondeductive logic and hence that all probability is factual. David's book is thus strictly confined to stochastic phenomena like dice.

For descriptive works on dice games, see E. Tylor, "On American lot-games, as evidence of Asiatic intercourse before the time of Columbus," supp., *International Archives for Ethnography* 9 (1896): 56–66; repr. in *The Study of Games*, ed. E. Avedon and B. Sutton-Smith (New York, 1971), 77–93; H. Lamer, "Lusoria tabula," in *Paulys Realencyclopädie der Classischen Altertumswissenschaft*, band 13, 2 (Stuttgart, 1927), cols. 1900–2029; L. Zdekauer, *Il gioco d'azzardo nel medioevo italiano* (Firenze, 1993); and W. Tauber, *Das Würfelspiel im Mittelalter und in der frühen Neuzeit* (Frankfurt, 1987). Discussions of the significance of dice games for probability are in R. Ineichen, *Würfel und Wahrscheinlichkeit: stochastisches Denken in der Antike* (Heidelberg, 1996); and R. Ineichen, "Der schlechte Würfel—Ein selten behandeltes Problem in der Geschichte der Stochastik," *Historia Mathematica* 18 (1991): 253–61.

Of the surveys on the main line of development through Cardano, the most prominent is O. Ore, *Cardano, the Gambling Scholar* (Princeton, 1953). Also see E. Coumet, "Le problème des partis avant Pascal," *Archives internationales d'histoire des sciences* 17 (1965): 244–72; M. Kendall, "The beginnings of a probability calculus," *Biometrika* 43 (1956): 1–14, repr. in *Studies in the History of Statistics and Probability*, ed. E. Pearson and M. Kendall (London, 1970), 1:19–34; R. Ineichen, "Dante-Kommentare und die Vorgeschichte der Stochastik," *Historia Mathematica* 15 (1988): 264–69; I. Schneider, "Why do we find the origin of a calculus of probabilities in the seventeenth century?" in *Probabilistic Thinking, Thermodynamics and the Interaction of the History and Philosophy of Science: Proceedings of the 1978 Pisa Conference on the History and Philosophy of Science*, ed. J. Hintikka, D. Gruender, and E. Agazzi (Dordrecht, 1981), 3–24; I. Schneider, ed., *Die Entwicklung der Wahrscheinlichkeitstheorie von den Anfangen bis 1933: Einfuhrungen und Texte* (Darmstadt, 1988); M. de Mora Charles, *Los inicios de la teoria de la probabilidad: siglos XVI y XVII* (Erandio, 1989); and I. Todhunter, *A History of the Mathematical Theory of Probability* (Cambridge, 1865).

The medieval manuscript on the interrupted game is noticed only by two writers: L. Toti Rigatelli, "Il 'problema delle parti' in manoscritti del XIV e XV secolo," in *Mathemata: Festschrift für Helmuth Gericke*, ed. M. Folkerts and U. Lindgren (Stuttgart, 1985); and I. Schneider, "The market place and games of chance in the fifteenth and sixteenth centuries," in *Mathematics from Manuscript to Print, 1300–1600*, ed. C. Hay (Oxford, 1988), 220–35.

Pascal's and Fermat's discovery of mathematical rules for probability has been described many times, as well as by David. Especially illuminating on the mathematical aspects are A. Edwards, "Pascal and the problem of points," *International Statistical Review* 50 (1982): 259–66; and A. Edwards, "Pascal's problem: The 'gambler's ruin,'" *International Statistical Review* 51 (1983): 73–79; A. Edwards, *Pascal's Arithmetical Triangle* (London, 1987). See also O. Ore, "Pascal and the invention of probability theory," *American Mathematical Monthly* 67 (1960): 409–19; P. Raymond, *De la combinatoire aux probabilités* (Paris, 1975); P. Dupont, "Concetti probabilistici in Roberval, Pascal e Fermat," *Rendiconti del Seminario Matematico dell'Università e del Politecnico di Torino* 34 (1975–76): 235–45; H. Freudenthal, "Huygens' foundations of probability," *Historia Mathematica* 7 (1980): 113–17; L. Thirouin, *Le hasard et les règles: Le modèle du jeu dans la pensée de Pascal* (Paris, 1991); I. Schneider, "Christiaan Huygens's contribution to the development of a calculus of probabilities," *Janus* 67 (1980): 269–79; E. Coumet, "Sur 'le calcul ès jeux de hasard' de Huygens: Dialogues avec les mathématicians français (1655–1657)," in *Huygens et la France* (Paris, 1982), foreword by R. Taton; P. Dupont and C. Silvia Roero, "Il trattato 'De ratiociniis in ludo aleae' di Christiaan Huygens con le 'Annotationes' di Jakob Bernoulli ('Ars Conjectandi,' parte 1) presentati in traduzione italiana, con commento storico-critico e risoluzioni moderne," *Memorie della Accademia delle Scienze di Torino, Classe di Scienze Fisiche, Matematiche e Naturali* ser. 5, 8, fasc. 1–2 (1984); and P. Holgate, "The influence of Huygens' work in dynamics on his contribution to probability," *International Statistical Review* 52 (1984): 137–40.

Notes

Preface

1. J. Butler, intro. to *The Analogy of Religion*, § 4, in *The Works of Joseph Butler*, ed. W. E. Gladstone (Oxford, 1896), 1:5.

2. Introductions to this distinction are found in I. Hacking, *The Emergence of Probability* (Cambridge, 1975), ch. 2; and D. Stove, *Probability and Hume's Inductive Scepticism* (Oxford, 1973), ch. 1.

3. J. Keynes, *Treatise on Probability* (London, 1921) (all probability is logical); J. Neyman, "Outline of a theory of statistical estimation based on the classical theory of probability," *Philosophical Transactions of the Royal Society of London* A 236 (1937): 333–80 (all probability is factual); R. Carnap, *Logical Foundations of Probability* (London, 1950) (neither factual nor logical probability is reducible to the other).

4. D. Stove, *Popper and After: Four Modern Irrationalists* (Oxford, 1982); repr. as *Anything Goes* (Sydney, 1998).

5. For the situation in historiography see K. Windschuttle, *The Killing of History: How a Discipline Is Being Murdered by Literary Critics and Social Theorists* (New York, 1997), esp. ch. 7.

6. W. Twining, *Rethinking Evidence: Exploratory Essays* (Oxford, 1990), esp. ch. 4; also see symposium in *Hastings Law Journal* 49 (2–4) (1998).

Chapter 1: The Ancient Law of Proof

1. J. Jaynes, *The Origin of Consciousness in the Breakdown of the Bicameral Mind* (Boston, 1976), 149.

2. H. Pirenne, "La preuve dans la civilisation de l'Egypte antique," *Recueils de la société Jean Bodin pour l'histoire comparative des institutions* 16 (1964): 9–42, at 22; cf. E. Cuq, *Etudes sur le droit babylonien* (Paris, 1929), 72, 352.

3. C. Desroches-Noblecourt, *Tutankhamen* (Harmondsworth, 1965), 34; A. McDowell, *Jurisdiction in the Workmen's Community of Deir el-Medîna* (Leiden, 1990), 166, 198–200; further in E. Bedell, "Criminal Law in the Egyptian Ramesside Period" (Ph.D. diss., Brandeis Univ., 1973); T. Peet, *The Great Tomb Robberies of the Twentieth Egyptian Dynasty* (Oxford, 1930).

4. Pirenne, "La preuve dans la civilisation," 38; cf. Exodus 22:9.

5. McDowell, *Jurisdiction*, 111, 137; A. Breasted, *Ancient Records of Egypt* (Chicago, 1906), 4: § 673; S. Allam, "De la divinité dans le droit pharaonique," *Bulletin de la société française d'égyptologie* 68 (Oct. 1973): 17–30.

6. *Code of Hammurabi* §§ 131–32, in *Babylonian Laws*, ed. and trans. G. Driver and J. Miles (Oxford, 1952–55), 2:53; commentary at 1:283–84; cf. § 2, §§ 106, 107, 9; Exodus 22:11; R. Boyer, "La preuve dans les anciens droits du proche-orient," *Recueils de la société Jean Bodin pour l'histoire comparative des institutions* 16 (1964): 61–87, sec. 2; and Cuq, *Etudes sur le droit babylonien*, 453–55.

7. G. Sautel, "Les preuves dans le droit grec archaïque," *Recueils de la société Jean Bodin pour l'histoire comparative des institutions* 16 (1964): 117–60, at 147.

8. G. Ries, "Altbabylonische Beweisurteile," *Zeitschrift der Savigny-Stiftung für Rechtsgeschichte, Romanistische Abteilung* 106 (1989): 56–80.

9. Sautel, "Les preuves dans le droit grec archaïque"; cf. G. Glotz, *L'ordalie dans la Grèce primitive* (Paris, 1904); but see M. Gagarin, *Early Greek Law* (Berkeley, 1986), 29–30, 43 n. 66.

10. B. Cohen, "Evidence in Jewish law," *Recueils de la société Jean Bodin pour l'histoire comparative des institutions* 16 (1964): 103–15, at 111.

11. *The Law Code of Gortyn*, ed. and trans. R. Willetts (Berlin, 1967), col. 10 at 49; cf. col. 9 at 47; Plato, *Gorgias* 471e, 472a; Aristotle, *Politics* 1268b41–1269a3; Plutarch, *Life of Cato the Younger* 19; cf. G. Pugliese, "La preuve dans le procès romain de l'époque classique," *Recueils de la société Jean Bodin pour l'histoire compararive des institutions* 16 (1964): 277–348, at 319–20.

12. Deuteronomy 17:6, 19:15; Numbers 35:30; see B. Jackson, *Essays in Jewish and Comparative Legal History* (Leiden, 1975), chs. 6–7; Matthew 18:16; John 8:17; 2 Corinthians 13:1; 1 Timothy 5:19.

13. Philo, *Special Laws*, bk. 4:54; cf. 4:59–61; *The Confusion of Tongues* 140; cf. E. Goodenough, *The Jurisprudence of the Jewish Courts in Egypt* (Amsterdam, 1968), 188, 193–94, 252.

14. Talmud, Sanhedrin 37a–b; all quotations from the Talmud are from *The Babylonian Talmud*, trans. under editorship of I. Epstein (London, 1935); cf. N. Rabinovitch, *Probability and Statistical Inference in Ancient and Medieval Jewish Literature* (Toronto, 1973), 110; but see Talmud, Kiddushin 80a.

15. Sanhedrin 9b; but in civil cases, see Baba Metzia 1.10.

16. Jerusalem Talmud, Yevamot 16:7; *Encyclopaedia Judaica* (Jerusalem, 1971–72), s.v. "agunah."

17. Talmud, Gittin 2b.

18. Talmud, Gittin 28a, Hullin 10b; cf. Rabinovitch, *Probability and Statistical Inference*, 112; Boyer, "La preuve dans les anciens droits du proche-orient," 80–81.

19. Talmud, Bava Batra 5a–b; Rabinovitch, *Probability and Statistical Inference*, 114, 43.

20. Rabinovitch, *Probability and Statistical Inference*, 115.

21. Talmud, Pesahim 9a; Rabinovitch, *Probability and Statistical Inference*, 42.

22. Bava Batra 158a; cf. 93a; Rabinovitch, *Probability and Statistical Inference*, 76.

23. C. Morrison, "Some features of the Roman and the English law of evidence," *Tulane Law Review* 33 (1959): 577–94.

24. *Corpus iuris civilis*, Digest 22, tit. 3, law 2; all translations from the *Digest*

are from *The Digest of Justinian*, ed. T. Mommsen, trans. A. Watson (Philadelphia, 1985).

25. *Digest* 48.19.5; similar in Aristotle(?), *Problemata* 951a37–951b8.

26. *Code* 4.19.25; cf. 3.28.34; 5.9.10.6; 8.38.14.2.

27. *Digest* 22.5.15.1.

28. Ibid., 22.5.3.

29. Ibid., 48.18.1.16; 22.5.21.

30. Ibid., 22.5.3.2.

31. J.-P. Lévy, "L'évolution de la preuve, des origines à nos jours," *Recueils de la société Jean Bodin pour l'histoire comparative des institutions* 17 (1965): 9–70, at 36 n. 1.

32. *Digest* 22.5.21.3; cf. 22.5.1.2; 22.3.13.

33. H. Jolowicz and B. Nicholas, *Historical Introduction to the Study of Roman Law*, 3d ed. (Cambridge, 1992), 443–44; M. Kaser, *Das römische Zivilprozessrecht* (Munich, 1966), 484–86.

34. *Code* 4.20.8; also see 4.20.3; *Digest* 22.5.12; 48.18.20; cf. *Theodosian Code* 11.39.3; W. Ashburner, *The Rhodian Sea-Law* (Oxford, 1909), lxxxix–xc.

35. Cicero, *Topics* xx.74; cf. Cicero, *De Partitione Oratoria* xxiv.117–18; *Digest* 22.5.21.2; A. Greenidge, *The Legal Procedure of Cicero's Time* (New York, 1971), 479–81; E. Peters, *Torture* (New York, 1985), 18–36.

36. *Digest* 48.18.1.1; cf. *Novels* 90.1; C. Préaux, "La preuve à l'époque hellenistique principalement dans l'Egypte grecque," *Recueils de la société Jean Bodin pour l'historie comparative des institutions* 16 (1964): 161–222, at 211.

37. *Code* 9.41.8.1.

38. *Digest* 48.18.1.23–25.

39. Ibid., 48.18.5; 48.18.10.1.

40. Ibid., 30.96; 35.1.36.1; cf. 32.64; 50.5.56; 50.10.1.1; 29.1.32; *Code* 6.21.4; 6.25.2.

41. *Digest* 12.4.6.

42. Ibid., 23.4.26.

43. Ibid., 37.10, 43.4.3.

44. Ibid., 4.2.23; also see L. Donatuti, *Le presumptiones iuris in diritto romano* (Perugia, 1930), esp. 4–5; G. Donatuti, "Le 'presumptiones iuris' come mezzi di svolgimento del diritto sostanziale romano," *Rivista di diritto private* 3 (1933): 161–206.

45. *Digest* 34.3.28.3.

46. Ibid., 41.3.44.4.

47. Ibid., 50.17.114; compare rules at 1.3.3, 1.3.4.

48. *The Minor Law Books*, trans. J. Jolly (Oxford, 1889), pt. 1, xvii (*The Sacred Books of the East*, ed. F. Müller, vol. 33).

49. Barth, quoted in R. Lingat, *The Classical Law of India*, trans. J. Derrett (Berkeley, 1973), 102.

50. *The Laws of Manu*, trans. G. Bühler (Oxford, 1886), 264 (*Sacred Books of the East*, vol. 25); cf. S. Radhakrishnan and C. Moore, eds., *A Source Book in In-*

dian Philosophy (Princeton, 1957), 368–69; P. Mahalanobis, "The foundations of statistics," *Sankhya* 18 (1957): 183–94.

51. *Laws of Manu,* 267.

52. Ibid., 269.

53. *Apastamba,* 2.5.11.2, trans. G. Bühler (Oxford, 1879), 124 (*Sacred Books of the East,* vol. 2).

54. Intro. to *Narada* 1.71–72, in J. Jolly, *Minor Law Books,* pt. 1, 23; see L. Rocher, "The theory of proof in ancient Hindu law," *Recueils de la société Jean Bodin pour l'histoire comparative des institutions* 18 (1965): 325–71.

55. Intro. to *Narada* 2.28–30, in Jolly, *Minor Law Books,* 30–31, but see 248; 2.31, at 31; 3.8, 3.11 at 341; cf. *Brihaspati,* in Jolly, *Minor Law Books,* pt. 1, 294.

56. *Narada* 1.143, at 78; some similar discussions in S. Vidyabhusana, *A History of Indian Logic* (Calcutta, 1921), 56, 94–97, 370.

57. *Narada* 1.153 at 81; 1.172–76 at 85–86; 1:189 at 89; 1:192 (90); cf. *Brihaspati* 7.18 at 301.

58. *Narada* 1.229 at 95–96; cf. *Brihaspati* 5.12–13 at 296; 7.35 at 303–5.

59. *Narada* 1.177–88 at 86–89; 12.98–100 at 185.

Chapter 2: The Medieval Law of Evidence

1. R. Eggleston, *Evidence, Proof, and Probability,* 2d ed. (London, 1983), chs. 9–10.

2. See refs. in J. Winkler, "Roman law in Anglo-Saxon England," *Journal of Legal History* 13 (1992): 101–27; cf. G. Nicolaj, *Cultura e prassi di notai preirneriani: alle origini del rinascimento giuridico* (Milan, 1991).

3. R. Coleman, "Reason and unreason in early medieval law," *Journal of Interdisciplinary History* 4 (1974): 571–91.

4. *Capitularia Regum Francorum* in *Monumenta Germaniae Historica, Leges in quarto,* sec. 2, vol. 1, no. 134 (816), cap. 1, 268, quoted in J. Nelson, "Dispute settlement in Carolingian West Francia," in *The Settlement of Disputes in Early Medieval Europe,* ed. W. Davies and P. Fouracre (Cambridge, 1986), 45–64, at 47; cf. 58; and cf. Shakespeare, *Richard II* 1.1.

5. I. Wood, "Disputes in late fifth- and sixth-century Gaul: Some problems," in Davies and Fouracre, *Settlement of Disputes,* 1–22; P. Wormald, "Charters, law, and the settlement of disputes in Anglo-Saxon England," in Davies and Fouracre, *Settlement of Disputes,* 149–68; P. King, *Law and Society in the Visigothic Kingdom* (Cambridge, 1972), 102–3.

6. T. Schieffer, *Die Urkunden Lothars I. und Lothars II.* (Berlin, 1966), 165; see *Monumenta Germaniae Historica, Diplomatum Karolinorum,* vol. 3, band 3; J. Niermeyer, *Mediae Latinitatis Lexicon Minus* (Leiden, 1976), s.v. "verus."

7. S. Chin, *Proof by Witnesses in Canon Law* (Taipei, 1971), ch. 4; F. Herrmann and B. Speer, "Facing the accuser: Ancient and medieval precursors of the confrontation clause," *Virginia Journal of International Law* 34 (1994): 481–552, at 500–502.

8. *Monumenta Germaniae Historica, Concilia,* vol. 3 (Hanover, 1984), 117, 283, 471.

9. *Decretales Pseudo-Isidorianae et Capitula Angilramni*, ed. P. Hinschius (Aalen, 1963), 449; cf. *The Collection in Seventy-Four Titles: A Canon Law Manual of the Gregorian Reform*, trans. J. Gilchrist (Toronto, 1980), 108.

10. Notably H. Berman, *Law and Revolution* (Cambridge, Mass., 1983); C. Morris, *The Discovery of the Individual, 1050–1200* (New York, 1972); B. Stock, *The Implications of Literacy* (Princeton, 1983); P. Wolff, *The Awakening of Europe*, trans. A. Carter (New York, 1969); N. Cantor, *Medieval History: The Life and Death of a Civilization* (New York, 1963); C. M. Radding, *A World Made by Men* (Chapel Hill, 1985); R. Southern, *Scholastic Humanism and the Unification of Europe* (Oxford, 1995), vol. 1; summary in M. Colish, *Medieval Foundations of the Western Intellectual Tradition* (New Haven, 1997), ch. 12. These all build on C. Haskins, *The Renaissance of the Twelfth Century* (Cambridge, Mass., 1927); M. Bloch, *Feudal Society*, trans. L. Manyon (London, 1961); and E. Rosenstock-Huessy, *Die Europäischen Revolutionen* (Jena, 1931).

11. J. Franklin, "The Renaissance myth," *Quadrant* 26 (11) (Nov. 1982): 51–60.

12. K. Clark, *Civilisation* (London, 1969), 17.

13. Ibid., 33.

14. Bloch, *Feudal Society*, 60; see R. Lopez, *The Commercial Revolution of the Middle Ages* (Englewood Cliffs, 1971).

15. See J. Smith, *Medieval Law Teachers and Writers* (Ottawa, 1975), 5; F. de Zulueta and P. Stein, *The Teaching of Roman Law in England Around 1200* (London, 1990), ch. 1; S. Kuttner, "The revival of jurisprudence," in *Renaissance and Renewal in the Twelfth Century*, ed. R. Benson and G. Constable (Oxford, 1982), 299–323; P. Stein, *Roman Law and European History* (Cambridge, 1999), ch. 3.

16. *Expositio* to the *Liber Papiensis*, Wido 6, quoted in Radding, *World Made by Men*, 180; and in C. Radding, *The Origins of Medieval Jurisprudence* (New Haven, 1988), 87.

17. *Expositio*, Rothar 1, quoted in Radding, *World Made by Men*, 183; on inferring intention see A. Kiralfy, "Taking the word for the deed: The mediaeval criminal attempt," *Journal of Legal History* 13 (1992): 95–100; see also F. Sinatti D'Amico, *Le prove giudiziarie nel diritto longobardo* (Milan, 1968).

18. B. Shapiro, *"Beyond Reasonable Doubt" and "Probable Cause"* (Berkeley, 1991), 200–241.

19. Berman, *Law and Revolution*, 582 n. 3; survey of their output in P. Weimar, "Die legistische Literatur der Glossatorenzeit," in *Handbuch der Quellen und Literatur der neueren europäischen Privatrechtsgeschichte*, ed. H. Coing (Munich, 1973), pt. 1, sec. 2; H. Lange, *Romisches Recht im Mittelalter*, vol. 1 (Munich, 1997); more generally, E. Cortese, *Il rinascimento giuridico medievale* (Rome, 1992).

20. J. Franklin, "Mental furniture from the philosophers," *Et Cetera* 40 (1983): 177–91; on the first century of disputations in law, see A. Belloni, *Le questioni civilistiche del secolo XII: Da Bulgaro a Pillio da Medicina e Azzone* (Frankfurt, 1989); B. Lawn, *The Rise and Decline of the Scholastic "Quaestio disputata"* (Leiden, 1993); Southern, *Scholastic Humanism*, 1: chs. 7–9.

21. Rabelais, *Pantagruel* 2, ch. 5.

22. Placentinus, *Summa codicis* 4, tit. 19, 151; see J. Baldwin, "The intellectual preparation for the canon of 1215 against ordeals," *Speculum* 36 (1961): 613–36, at 616 n. 22; cf. W. Ullmann, "Medieval principles of evidence," *Law Quarterly Review* 62 (1946): 77–87, at 78; W. Ullmann, "Reflections on medieval torture," *Juridical Review* 56 (1944): 123–37, at 124.

23. *Digest* 29.3.1.2; *Code* 4.19.5.

24. [Irnerius], *Summa, Code* 4, tit. 20; see J.-P. Lévy, *La hierarchie des preuves dans le droit savant du moyen-âge* (Paris, 1939), 107, 109, 110 n. 14.

25. A. Gouron, "Aux racines de la théorie des présomptions," *Rivista internazionale di diritto commune* 1 (1990): 99–109, at 103.

26. Azo, *Lectura super codicem* 1 tit. 4 (Turin, 1966), 32; ibid., 4 tit. 19, 283.

27. Azo, *Brocarda* (Turin, 1967), 122; cf. Augustine, *De Doctrina Christiana* 2 ch. 30; S. Caprioli, "Tre capitoli intorno alla nozione di *Regula Iuris* nel pensiero dei glossatori," *Annali di storia del diritto* 5–6 (1961–62): 221–374, at 354.

28. Azo, *Lectura super codicem* 2 tit. 22, 123.

29. Gouron, "Aux racines de la théorie des présomptions," 105; H. Kiefner, "Qui possidet dominus esse praesumitur," *Zeitschrift der Savigny-Stiftung für Rechtsgeschichte. Romanistische Abteilung* 79 (1962): 239–306.

30. Azo, *Lectura super codicem* 4 tit. 19; H. Jolowicz, *Roman Foundations of Modern Law* (Oxford, 1957), 102 n. 6; Placentinus, *Summa codicis* 4 tit. 19 (Turin, 1962, 150).

31. Lévy, *La hierarchie des preuves*, 108.

32. Azo, *Lectura super codicem* 4 tit. 1, 254; cf. I. Fasolus, *De summariis cognitionibus*, ed. L. Wahrmund (Aalen, 1962), 21; I. Andreae, *Novella commentaria in secundum decretalium*, tit. 19, rubric (Turin, 1963), fol. 109A.

33. *Code* 4.21.19; Lévy, *La hierarchie des preuves*, 112.

34. *Digest* 22.5.3.2; Lévy, *La hierarchie des preuves*, 115; contrary tendencies in F. Herrmann, "The establishment of a rule against hearsay in romano-canonical procedure," *Virginia Journal of International Law* 36 (1995): 1–51.

35. Lévy, *La hierarchie des preuves*, 122, 124, 126; Azo, *Lectura* 4 tit. 28, 308.

36. *Corpus iuris civilis, Code* 6.33.3.

37. William of Drogheda, *Summa aurea*, ed. L. Wahrmund (Aalen, 1962), 142.

38. Voltaire to Damilaville, May 3, 1763, quoted in J.-P. Lévy, "L'évolution de la preuve, des origines à nos jours," *Recueils de la société Jean Bodin pour l'histoire comparative des institutions* 17 (1965): 9–70, at 39 n.1; cf. A. Allard, *Histoire de la justice criminelle au seizième siècle* (Aalen, 1970), 269–70.

39. J. Langbein, *Torture and the Law of Proof* (Chicago, 1976), 5; R.-M. Sargent, "Scientific experiment and legal expertise: The way of experience in seventeenth-century England," *Studies in History and Philosophy of Science* 20 (1989): 19–46, at 27.

40. Berman, *Law and Revolution*, chs. 9–14; D. Kelley, *The Human Measure* (Cambridge, Mass., 1990), ch. 7.

41. Gratian, *Decretum* pt. 2 causa 2 q. 4 can. 2, in *Corpus iuris canonici*, ed. A. Friedberg (Leipzig, 1879), 1: col. 466; refs. on Gratian's sources on the two-witness rule in R. Fraher, "Preventing crime in the High Middle Ages," in *Popes, Teachers, and Canon Law in the Middle Ages*, ed. J. Sweeney and S. Chodorow (Ithaca, 1989), 212–33, at 216 n. 17.

42. Aquinas *Summa theologiae* II-II q. 70 art. 2 ad 3.

43. Rufinus, *Summa decretorum* to causa 32 q. 1 cap. 2, ed. H. Singer (Aalen, 1963), 476; further refs, in J. Brundage, *Law, Sex and Christian Society in Medieval Europe* (Chicago, 1987), 321 n. 304 and 410 n. 449; on violent presumptions see *Decretals of Gregory IX* 1 tit. 9 cap. 5, in Friedberg, *Corpus iuris canonici*, 2: col. 104; 2: tit. 19 cap. 3 col. 307; 2: tit. 20 cap. 27 col. 324; 2: tit. 23 caps. 10 and 12 cols. 355–56. Full discussions in R. Motzenbäcker, *Die Rechtsvermutung im kanonischen Recht* (Munich, 1958); see also W. Bryson, "Witnesses: A canonist's view," *American Journal of Legal History* 13 (1969): 57–67; further on presumptions in the early canonists and their connections with Ciceronian rhetoric in S. Kuttner, *Gratian and the Schools of Law* (London, 1983), ch. 9, 771–76, reporting A. Lang, "Rhetorische Einflüsse auf die Behandlung des Prozesses in der Kanonistik des 12. Jahrhunderts," in *Festschrift Eduard Eichmann* (Paderborn, 1940), 69–97; A. Giuliani, "The influence of rhetoric on the law of evidence and pleading," *Juridical Review* 7 (1962): 216–52, at 231–32.

44. *Corpus iuris civilis*, Code 9.9.30, and appended extract from *Novels* 117.15.

45. P. Boskoff, "Quintilian in the late Middle Ages," *Speculum* 27 (1952): 71–78.

46. *Digest* 4.4.3.1; 41.3.44.4; 19.1.13.3.

47. Ibid., 22.3.24; alleged by Tancredus, in Pillius, Tancredus, Gratia, *Libri de iudiciorum ordine* (Aalen, 1965), 259; but note "most open proofs of violence," *Digest* 4.2.23.

48. *Digest* 48.18.1.

49. J. de Lignano, *Super Clementina "Saepe,"* ed. L. Wahrmund (Aalen, 1962), 9.

50. 1 Kings 3:16–28; John of Salisbury, *Policraticus* 5, ch. 14; *Decretals of Gregory IX* 2 tit. 23 cap. 2, in Friedberg, *Corpus iuris canonici* 2: col. 353; cf. Rabelais, *Pantagruel* 2, ch. 14.

51. In the later glossators, Azo, *Lectura super codicem* 4 tit. 19, 283; cf. 9 tit. 22, 699; and Damasus, *Summa de ordine iudiciario*, ed. L. Wahrmund (Aalen, 1962), 42; Accursius, *Glossa in "Digestum vetus"* (Turin, 1969), 644; many refs. in canon law up to 1860 in K. Gross, *Das Beweistheorie im canonischen Prozess* (Lausanne, 1880), 2:98 n. 19.

52. Thomas Aquinas, *In IV Sententiarum* dist. 9 q. 1 art. 5 sol. 2, in Thomas Aquinas, *Opera omnia* (Rome, 1570), vol. 7.

53. Morris, *Discovery of the Individual*, 127–33; cf. R. Hanning, *The Individual in Twelfth Century Romance* (New Haven, 1977), Afterword; G. Silano, "Of sleep and sleeplessness: The papacy and law, 1150–1300," in *The Religious Roles of the Papacy*, ed. C. Ryan (Toronto, 1989), 343–61; I. Robinson, *The Papacy,*

1073–1198 (Cambridge, 1990), ch. 5, also 144; see generally R. Moore, *The Formation of a Persecuting Society* (Oxford, 1987).

54. *Decretals of Gregory IX* 2 tit. 20 cap. 28, in Friedberg, *Corpus iuris canonici* 2: cols. 324–25.

55. E. Kemp, *Canonization and Authority in the Western Church* (Oxford, 1948), 105–6; *Decretals of Gregory IX* 2 tit. 22 cap. 10, in Friedberg, *Corpus iuris canonici* 2: cols. 350–52; cf. J.-P. Lévy, "Le problème de la preuve dans les droits savants du moyen âge," *Recueils de la société Jean Bodin pour l'histoire comparative des institutions* 17 (1965): 137–67, at 154.

56. *Decretals of Gregory IX* 3 tit. 2 cap. 8 col. 456; cf. Placentinus, *Tractatus de accusationibus publicorum iudiciorum* tit. 2, in *Tractatus universi iuris* (Venice, 1584) 11: pt. 1, fols. 2r–5r, at fol. 2v; *Assize of Clarendon*, c. 14, in *Select Charters and Other Illustrations of English Constitutional History*, ed. W. Stubbs (Oxford, 1881), 144–45; E. Peters, "Wounded names: The medieval doctrine of infamy," in *Law in Medieval Life and Thought*, ed. E. King and S. Ridyard (Sewanee, 1990), 43–89.

57. Fraher, "Preventing crime," 222–25; J. Brundage, *Medieval Canon Law* (London, 1995), 144–46.

58. *Decretals of Gregory IX* 5, tit. 39, cap. 44, in Friedberg, *Corpus iuris canonici* 2: col. 908; repeated in Raymond of Peñafort, *Summa de matrimonio* tit. 21 n. 2, col. 984 (Rome, 1978).

59. Gratian, *Decretum* pt. 2 causa 30 q. 5, in Friedberg, *Corpus iuris canonici* 1: cols. 1104–8; see A. Esmein, *Le mariage en droit canonique* (New York, 1968), 1:189–202; further refs. in C. Donahue, "'Clandestine' marriages in the later Middle Ages," *Law and History Review* 10 (1992): 315–22.

60. *Decretals of Gregory IX* 2 tit. 19 cap. 4, in Friedberg, *Corpus iuris canonici* 2: col. 307; more direct methods of evidence on this question in R. Helmholz, *Marriage Litigation in Medieval England* (Cambridge, 1974), 88.

61. *Decretals of Gregory IX* 2 tit. 20 cap. 2 col. 324; cf. 3 tit. 43 cap. 3 col. 649; 4 tit. 3 cap. 3 col. 680; cf. R. Helmholz, *Canon Law and the Law of England* (London, 1987), 148–49.

62. *Decretals of Gregory IX* 2 tit. 23 cap. 14 col. 357; cf. W. Wakefield, *Heresy, Crusade and Inquisition in Southern France, 1100–1250* (Berkeley, 1974), 257; discussion of "vehement" in Johannes Teutonicus, *Apparatus glossarum in Compilationem tertiam*, ed. K. Pennington (Vatican City, 1981), 275 (*Monumenta iuris canonici, series A: Corpus Glossatorum*, vol 3).

63. M. Russell, "Hired champions," *American Journal of Legal History* 3 (1959): 242–59; D. Ireland, "First catch your toad: Medieval attitudes to ordeal and battle," *Cambrian Law Review* 11 (1980): 50–61.

64. A. Barthélemy, "Diversité des ordalies médiévales," *Revue historique* 280 (1988): 3–25.

65. Baldwin, "The intellectual preparation for the canon of 1215 against ordeals," 628–29; J. Baldwin, *Masters, Princes, and Merchants: The Social Views of Peter the Chanter and His Circle* (Princeton, 1970), 326–29; cf. J. Gaudemet, "Les ordalies au moyen âge," *Recueils de la société Jean Bodin pour l'histoire compara-*

tive des institutions 17 (1965): 99–135; Radding, *World Made by Men*, 9; R. Koss, "'Swindling' 'justice': The *judicium dei* in medieval French literature," in King and Ridyard, *Law in Medieval Life*, 231–43.

66. Peter the Chanter, *Verbum abbreviatum*, 78, in *Patrologia Latina*, ed. J. Migne, 205: col. 231, partly quoted in R. Bartlett, *Trial by Fire and Water* (Oxford, 1986), 86.

67. R. Helmholz, "Crime, compurgation, and the courts of the medieval church," *Law and History Review* 1 (1983): 1–26, at 13–20; R. Helmholz, *The Spirit of Classical Canon Law* (Athens, Ga., 1996), 156–59; *The Law of Hywel Dda*, ed. D. Jenkins (Llandysul, 1986), xxxiii, 94–95.

68. Langbein, *Torture and the Law of Proof*, 6; P. Hyams, "Trial by ordeal: The key to proof in the early common law," in *On the Laws and Customs of England*, ed. M. Arnold et al. (Chapel Hill, 1981), 90–126; R. van Caenegem, "The law of evidence in the twelfth century," *Proceedings of the Second International Congress of Medieval Canon Law, 1963*, ed. S. Kuttner and J. Ryan (Vatican City, 1965), 297–310; R. van Caenegem, "Methods of proof in Western medieval law," in R. van Caenegem, *Legal History: A European Perspective* (London, 1991), 71–113; R. van Caenegem, "Reflexions on rational and irrational modes of proof in medieval Europe," *Legal History Review* 58 (1990): 263–79; Bartlett, *Trial by Fire and Water*, 135–43, with bibliography, 167–69; cf. review of Bartlett, *Trial by Fire and Water*, by K. Pennington, *Journal of Ecclesiastical History* 39 (1988): 263–66.

69. G. Hall, ed., *The Treatise on the Laws and Customs of England, Commonly Called Glanvill* (London, 1965), 28; see R. Helmholz, "The early history of the grand jury and the canon law," *University of Chicago Law Review* 50 (1983): 613–27.

70. Hall, *Glanvill*, 171, 173–74; repeated in Scots law in *Regiam Majestatem* 4, ch. 1, ed. T. Cooper (Edinburgh, 1947), 249.

71. R. Groot, "The jury of presentment before 1215," *American Journal of Legal History* 24 (1982): 1–24; *Curia Regis Rolls*, 5:64; M. Kerr, R. Forsyth, and M. Plyley, "Cold water and hot iron: Trial by ordeal in England," *Journal of Interdisciplinary History* 22 (1992): 573–95.

72. D. Stenton, ed., *Rolls of the Justices in Eyre for Yorkshire, 1218–1219* (1937), xl; R. Groot, "The early-thirteenth-century criminal jury," in *Twelve Good Men and True: The Criminal Trial Jury in England, 1200–1800*, ed. J. Cockburn and T. Green (Princeton, 1988), 3–35.

73. See generally J. Heath, *Torture and English Law* (Westport, 1982), ch. 1; Peters, *Torture*, 190; similarity of early torture and later ordeals in M. Ruthven, *Torture: The Grand Conspiracy* (London, 1978), 47, 50–51.

74. Placentinus, *Tractatus de accusationibus publicorum iudiciorum* tit. 6, in *Tractatus universi iuris* 11: pt. 1, fols. 2r–5r, at 4r; cf. Vacarius, *Liber Pauperum*, summarized in Heath, *Torture and English Law*, 21; Frederick II, *Constitutiones regni Siciliae*, in P. Fiorelli, *La tortura giudiziaria nel diritto commune* (Rome, 1953–54), 1:86; Brundage, *Medieval Canon Law*, 95.

75. The example from [Cicero] *Rhetorica ad Herennium* 2.vii.10.

76. Langbein, *Torture and the Law of Proof*, 9; but note low English conviction rates: Helmholz, "Crime, compurgation, and the courts of the medieval church," 19–21; B. Hanawalt, *Crime and Conflict in English Communities, 1300–1348* (Cambridge, Mass., 1979), 54–62.

77. Various refs in Fiorelli, *La tortura giudiziaria nel diritto commune*, 2:23–33.

78. *Decretals of Gregory IX* 2 tit. 23, in Friedberg, *Corpus iuris canonici* 2 cols. 354–55; Innocent IV, *Commentaria super libros quinque decretalium* (Frankfurt, 1968), fol. 281; conflict of presumptions in *Decretals of Gregory IX* 4 tit. 1 cap. 30 col. 672; cf. Hostiensis, *In Quartum decretalium librum commentaria* tit. 5 cap. 6 (Turin, 1965), fol. 17.

79. G. Durand, *Speculum judiciale* (Aalen, 1975), 628, 151; cf. R. Millar, "Memory verses of the Romano-canonical procedurists," *Juridical Review* 67 (1955): 284–97, at 290.

80. Lévy, *La hierarchie des preuves dans le droit savant du moyen-âge*, 115, 125; Durand, *Speculum judiciale*, 736–40, 625; cf. [anon] *Ordo judiciarius "Scientiam,"* ed. L. Wahrmund (Aalen, 1962), 62–64; Hostiensis, *In Secundum decretalium librum commentaria* tit. 23 cap. 10 fol. 122A; Ioannes Andreae, *Novella commentaria in secundum decretalium* tit. 23 cap. 8; Baldus, *In decretalium* 2 tit. 23 cap. 10 (Turin, 1971), fol. 178A.

81. Hostiensis, *In Secundum decretalium librum commentaria* tit. 20 cap. 32 fols. 93A–94; cf. Ioannes Andreae, *Novella commentaria in secundum decretalium* tit. 20 cap. 32 fols. 141A–142; Baldus, *In decretalium* 2 tit. 20 cap. 32 fols. 224A–225.

82. R. Fraher, "Conviction according to conscience," *Law and History Review* 7 (1989): 23–88, at 37–39.

83. *Bracton on the Laws and Customs of England*, trans. S. Thorne (Cambridge, Mass., 1979), 3:371, 378, 4:330, 2:386, 404; cf. C. Güterbock, *Bracton and His Relation to Roman Law*, trans. B. Coxe (Philadelphia, 1866), 156–57; *Selected Passages from the Works of Bracton and Azo*, ed. F. Maitland, Publications of the Selden Society 8 (London, 1895); H. Robinson, "Tancred, Raymond and Bracton," *English Historical Review* 59 (1944): 376–84; F. Schulz, "Bracton and Raymond de Peñafort," *Law Quarterly Review* 61 (1945): 286–92; T. Plucknett, *Early English Legal Literature* (Cambridge, 1958), 53; H. Jakobs, *De similibus ad similia bei Bracton und Azo* (Frankfurt, 1996).

84. Shapiro, *"Beyond Reasonable Doubt" and "Probable Cause,"* 207–8.

85. J. Weber, "The birth of probable cause," *Anglo-American Law Review* 11 (1982): 155–67; J. Bellamy, *Crime and Public Order in England in the Later Middle Ages* (Toronto, 1973), 102–3; 3 Edward I, cap. 15, in *Statutes at Large* (London, 1811), 1:48.

86. Shapiro, *"Beyond Reasonable Doubt" and "Probable Cause,"* ch. 2; on the Continental system in English church courts, C. Donahue, "Proof by witnesses in the church courts of Medieval England," in Arnold et al., *On the Laws and Customs of England*, 127–58.

87. R. Fraher, "The theoretical justification for the new criminal law of the

High Middle Ages: 'Rei publicae interest, ne crimina remaneant impunita,'"
University of Illinois Law Review (1984): 577–95.

88. W. Ullmann, *The Medieval Idea of Law* (New York, 1946), 122–23; further on the line between direct and inferred knowledge in Bartolus, *De testimoniis*, in *Opera omnia* (Venice, 1615),10: fols. 158r–164r, at 159v–163v.

89. Ullmann, *Medieval Idea of Law*, 133, 123–25; Ullmann, "Medieval principles of evidence," 83; Fraher, "Conviction according to conscience," 52–53.

90. W. Ullmann, "Some medieval principles of criminal procedure," *Juridical Review* 59 (1947): 1–28, at 16–17; Ullmann, *Medieval Idea of Law*, 123–24.

91. Ullmann, "Some medieval principles of criminal procedure," 86; cf. L. Daston, *Classical Probability in the Enlightenment* (Princeton, 1988), 43.

92. Ullmann, "Reflections on medieval torture," 128–29.

93. Bartolus, *De quaestionibus*, in *Opera omnia* 10: fols. 178v–180v, at 179r.

94. *Opera omnia* 10: fols. 179v–180r.

95. A. Sheedy, *Bartolus on Social Conditions in the Fourteenth Century* (New York, 1942), 100–102.

96. Levy, *La hierarchie des preuves*, 111, 112, 114, 125, 127–29.

97. W. Ullmann, "Baldus' conception of law," *Law Quarterly Review* 58 (1942): 386–99; N. Horn, "Philosophie in der Jurisprudenz der Kommentatoren: Baldus philosophus," *Ius Commune* 1 (1967): 104; see generally H. Canning, intro. to *The Political Thought of Baldus de Ubaldis* (Cambridge, 1987); N. Horn, *Aequitas in den Lehren des Baldus* (Cologne, 1968); K. Pennington, "The consilia of Baldus de Ubaldis," *Legal History Review* 56 (1988): 85–92.

98. Baldus, *Code* 9.2.7, in *Opera omnia* (Venice, 1577) 8: fols. 212v–213r; cf. 4.19.5, 6: fol. 37v; Baldus, *In decretalium* 2 tit. 23, rubric (Turin, 1971), fol. 244.

99. Baldus, *Code* 6.15.4, in *Opera omnia* 7: fol. 42v; further, with comparison of judge to doctor, 4.19 rubric, in *Opera omnia* 6: fol. 35r; reference to Cynus is Cynus of Pistoia, *In codicem* 2.19(20).9 (Turin, 1964) 1: fol. 91v.

100. Quoted in Durand, *Speculum Judiciale*, 740; cf. Baldus, *Consilia* 2 cons. 394 (Venice, 1575), 5 at fol. 106; 5 cons. 447, 5 at fols. 118r–120r.

101. Baldus, *In decretalium*; cf. 246; Baldus, *Consilia* 3 cons. 364, 3 at fol. 102v; cf. H. Lange, "Die Consilien des Baldus de Ubaldis," *Akademie der Wissenschaften und der Literatur: Abhandlungen der Geistes- und Socialwissenschaftlichen Klasse*, no. 12 (1973): 44, 46; contrast Quintilian's "circumstances," *Institutio oratoria* 5.10.103–4.

102. Baldus, *Code* 4.19.25 at fol. 45v.

103. Baldus, *Digest* 22.3.7, 2 at fol. 177r; cf. *Code* 6.30.19, 7 at fol. 103v.

104. Baldus, *Consilia* 5 cons. 515, 5 at fol. 138v.

105. Baldus, *In decretalium* 2 tit. 18, quoted in Lévy, *La hierarchie des preuves*, 124, 127.

106. Baldus, *Consilia* 1 cons. 37, 1: fol. 13r–v.

107. T. Walsingham, *Historia Anglicana*, ed. H. Riley, Rolls Series (London, 1863–64), 1:464, selection in R. Dobson, *The Peasants' Revolt of 1381* (London, 1970), 177; see A. Harding, "The revolt against the justices," in *The English Ris-*

ing of 1381, ed. R. Hilton and T. Aston (Cambridge, 1984), 165–93; A. Myers, ed., *English Historical Documents, 1327–1485* (New York, 1969), 4:127–40; cf. Shakespeare, *Henry VI, Part II*, 4.2.73.

108. Ioannes Andreae, *Novella commentaria in secundum decretalium* tit. 23 cap. 9 fol. 178A; cf. H. Kiefner, "Semel malus semper praesumitur malus," *Zeitschrift der Savigny-Stiftung für Rechtsgeschichte. Romanistische Abteilung* 78 (1961): 308–54.

109. *Decretals of Gregory IX* 5 tit. 7 cap. 15, in Friedberg, *Corpus iuris canonici* 2 col. 789.

110. B. Gui, *Practica Inquisitionis*, see *Manuel de l'inquisiteur*, ed. G. Mollat (Paris, 1926–27), selection in *Introduction to Contemporary Civilization in the West*, 2d ed., ed. J. Buchler et al. (New York, 1954), 1:281–90, at 282; cf. E. Privat, ed., *Bernard Gui et son monde* (Toulouse, 1981), 253–64.

111. *Corpus iuris civilis*, Code 1.5.2.1; repeated in Innocent IV, *In V libros decretalium commentaria* 5 tit. 3 cap. 6 (Frankfurt, 1968), fol. 498v; later in G. de Villadiego, *Tractatus contra haereticum pravitatem*, in *Tractatus universi iuris* (Venice, 1584) 11: pt. 2, fols. 32r–42r, at 37r.

112. Herrmann and Speer, "Facing the accuser."

113. N. Eymerich and F. Peña, *Le manuel des inquisiteurs*, trans. L. Sala-Molins (Paris, 1973), 219; see also E. van der Vekene, *Bibliographie des Directorium Inquisitorum des Nicolaus Eymerich* (Luxembourg, 1961).

114. Eymerich and Peña, *Le manuel des inquisiteurs*, 216; *Liber Sextus* 5 tit. 2 caps. 5 and 8, in Friedberg, *Corpus iuris canonici* 2: cols. 1071–72; cf. Fraher, "Preventing crime," 226.

115. J. Given, "The inquisitors of Languedoc and the medieval technology of power," *American Historical Review* 94 (1989): 336–59, at 354–55.

116. *Liber Sextus* 5 tit. 2 cap. 8, ibid.

117. A. Patschovsky, *Die Anfänge einer ständigen Inquisition in Böhmen: ein Präger Inquisitoren-Handbuch aus der ersten Hälfte des 14. Jahrhunderts* (Berlin, 1975), 153, 172, 176, 212, 218–19.

118. Gui, *Practica Inquisitionis*, 287; but see E. Peters, *The Magician, the Witch, and the Law* (Philadelphia, 1978), 133; cf. R. Mentzer, "Heresy proceedings in Languedoc, 1500–1560," *Transactions of the American Philosophical Society* 74 (1984): pt.5, 69–70.

119. Eymerich and Peña, *Le manuel des inquisiteurs*, 207–8, 218; cf. E. Vacandard, s.v. "Inquisition," *Dictionnaire de théologie catholique*, vol. 7, pt. 2 (Paris, 1927), cols. 2016–68, at col. 2038.

120. Talmud, Sanhedrin 30b, 41a, 69a; Baba Kama 70b; Jackson, *Essays in Jewish and Comparative Legal History*, 201; cf. Lévy, "Le problème de la preuve," 158.

121. G. Lizerand, *Le Dossier de l'affaire des Templiers* (Paris, 1923), 62–71, summarized in Heath, *Torture and English Law*, 33; M. Barber, *The Trial of the Templars* (London, 1978), 106.

122. Barber, *Trial of the Templars*, 148–49; cf. M. Michelet, *Le procès des templiers* (Paris, 1841), 1:262.

123. Michelet, *Le procès des templiers*, 1:185; cf. R. Sève and A. Chagny-Sève, *Le procès des templiers d'Auvergne* (Paris, 1986), 84.

124. Given, "Inquisitors of Languedoc," 352.

125. M. Barber, "Lepers, Jews and Moslems: The plot to overthrow Christendom in 1321," *History* 66 (1981): 1–17; C. Ginzburg, *Ecstasies*, trans. R. Rosenthal (London, 1991), pt. 1, ch. 1.

126. J. Langbein, *Prosecuting Crime in the Renaissance* (Cambridge, Mass., 1974), 149.

127. Maimonides, *Sefer HaMitzvot*, negative commandments 290; also trans. C. Chavel (London, 1967), 2: 269–70, quoted in N. Rabinovitch, *Probability and Statistical Inference in Ancient and Medieval Jewish Literature* (Toronto, 1973), 111; cf. ibid., 135–36; and cf. Maimonides, *Guide of the Perplexed* pt. 1 ch. 32, trans. S. Pines (Chicago, 1963), 68.

128. Maimonides, *Mishneh Torah* 4 tr. 2, ch. 12.10.

129. A. Hulsewé, *Remnants of Han Law* (Leiden, 1955), 1:74–79.

130. M. Shapiro, *Courts: A Comparative and Political Analysis* (Chicago, 1981), ch. 4.

131. *T'ang-Yin-Pi-Shih: Parallel Cases from Under the Pear Tree*, trans. R. van Gulick (Leiden, 1956), 56–58; but see D. Buxbaum, "Some aspects of civil procedure and practice at the trial level in Tanshui and Hsinchu from 1789 to 1895," *Journal of Asian Studies* 30 (1971): 255–79, at 269–70.

132. "Chih-Yüan Hsin-Ko (1291)," in P. Ch'en, *Chinese Legal Tradition under the Mongols* (Princeton, 1979), 154.

133. A. Conner, "The law of evidence during the Ch'ing Dynasty" (Ph.D. diss., Cornell Univ., 1979), 121, 142–43, 193; C. Boxer, ed., *South China in the Sixteenth Century* (London, 1953), 19, 201–2; T. Huff, *The Rise of Early Modern Science: Islam, China and the West* (Cambridge, 1993), 262–75.

134. D. Kaiser, *The Growth of Law in Medieval Russia* (Princeton, 1980), ch. 5.

135. M. Lippman, S. McConville, and M. Yerushalmi, *Islamic Criminal Law and Procedure: An Introduction* (New York, 1988), 69–72; J. Coulson, *A History of Islamic Law* (Edinburgh, 1964), 124–28; J. Schacht, *The Origins of Mohammedan Jurisprudence* (Oxford, 1950), 187–88; W. Hallaq, "On inductive corroboration, probability, and certainty in Sunni legal thought," in *Islamic Law and Jurisprudence*, ed. N. Heer (Seattle, 1990), 3–31, at 7–9, and on grades of probability, 19 n. 44; M. el-Awa, *Punishment in Islamic Law* (Indianapolis, 1982), 124–33.

Chapter 3: Renaissance Law

1. H. Kelly, *The Matrimonial Trials of Henry VIII* (Stanford, 1976), 107, 151, 61, 122–23, 153, 148–49.

2. Ibid., 176–79; G. Bedouelle and P. le Gal, eds., *Le "divorce" du Roi Henry VIII* (Geneva, 1987), pt. 2; E. Surtz and V. Murphy, eds., *The Divorce Tracts of Henry VIII* (Angers, 1988).

3. Kelly, *Matrimonial Trials*,176, 235–37, 298.

4. W. Roper, *Life of Sir Thomas More*, in *Two Early Tudor Lives*, ed. R. Sylvester and D. Harding (New Haven, 1962), 246; J. Derrett, "The trial of Sir Thomas More," in *Essential Articles for the Study of Thomas More*, ed. R. Sylvester and G. Marc'hadour (Hamden, 1977), 55–78, at 67; cf. R. Marcus, "The Tudor treason trials: Some observations on the emergence of forensic themes," *University of Illinois Law Review* (1984): 675–704, at 703 n. 181.

5. L. Smith, *Treason in Tudor England: Politics and Paranoia* (London, 1986), esp. ch. 2.

6. T. Kyd, *Spanish Tragedie* 3.4.9.

7. J. Heath, *Torture and English Law* (Westport, 1982), 76–77, 144, 285 n. 236; also Statute 33 Henry VIII (1541), cap. 23, in *Statutes at Large*, 2:190–91.

8. T. Howell, ed., *A Complete Collection of State Trials* (London, 1816), 1, cols. 1053–54, 1061; B. Shapiro, *"Beyond Reasonable Doubt" and "Probable Cause"* (Berkeley, 1991), 212–13.

9. Howell, *Complete Collection of State Trials*, col. 1070.

10. Lord Sankey in *Woolmington v. Director of Public Prosecutions* [1935], Appeal Cases 462, at 481; see R. Cross, *The Golden Thread of the English Criminal Law* (Cambridge, 1976); J. Mortimer, *Rumpole and the Golden Thread* (Harmondsworth, 1983).

11. [William Cecil, Lord Burleigh?], *A Declaration of the Favourable Dealing of Her Majesties Commissioners, Appointed for the Examination of Certaine Traytours, and of Tortures Unjustly Reported to be Done Upon Them for Matter of Religion*, in Heath, *Torture and English Law*, 235–39, at 238.

12. A. Steuart, ed., *The Trial of Mary Queen of Scots* (Edinburgh, 1923), 44; cf. ibid., 72; Howell, *Complete Collection of State Trials*, 1: col. 1183; cf. A. Smith, *The Babington Plot* (London, 1936), 226–27.

13. *The briefe of the Lord Chancellors first speech*, in *Holinshed's Chronicles of England, Scotland and Ireland*, ed. H. Ellis (London, 1807–8), 4:930.

14. *A summarie report of the second speach uttered by the Speaker of the Lower House*, in *Holinshed's Chronicles*, 936–38, at 937; cf. L. Gallagher, *Medusa's Gaze: Casuistry and Conscience in the Renaissance* (Stanford, 1991), ch. 1; the "proofs" set out in *A True Copie of the Proclamation*, in *Holinshed's Chronicles*, 941–43.

15. W. Lambarde, *Eirenarcha* (London, 1581, repr. London, 1972) 2, ch. 15, at 453; M. Dalton, *The Country Justice* (London, 1635), 336–38; see Shapiro, *"Beyond Reasonable Doubt" and "Probable Cause,"* 131–32, 145, 152.

16. For example, P. Farinacci, *Tractatus de testibus* (Frankfurt, 1598, 1606, Venice, 1603, 1609); G. Struve, *Disputatio juridica De indiciis cui annectitur quaestio de proba per aquam frigidam sagarum* (Jena, 1666, 1683, 1687, 1714); A. Matthaeus, *De probationibus liber* (Leiden, 1678, Groningen, 1739).

17. Leibniz, *Textes inédits*, ed. G. Grua (Paris, 1948), 2:548; cf. Leibniz, *De arte combinatoria*, in *Philosophical Papers and Letters*, trans. L. Loemker (Chicago, 1956), 1:133; Viscount Stair, *Institutions of the Law of Scotland* (Edinburgh, 1693) 2 tit. 12.7; 4 tit. 45.9, ed. D. Walker (Edinburgh, 1981), 548, 1011.

18. W. Best, preface to *A Treatise on Presumptions of Law and Fact, with the Theory and Rules of Presumptive or Circumstantial Proof* (Philadelphia, 1845); further refs. in Shapiro, *"Beyond Reasonable Doubt" and "Probable Cause,"* 300 n. 32, 325 n. 127, 216, 233, 235, 327 n. 157; cf. *The Canon Law of the Church of England: Being the Report of the Archbishops' Commission on Canon Law* (London, 1947), 56–57.

19. G. Menochio, *De praesumptionibus, coniecturis, signis et indiciis*, vol. 1 (Venice, 1587–90), q. 4 fol. 4v; q. 5 fol. 5r; q. 32 fol. 24v; q. 6 fol. 5v.

20. Ibid., q. 7 fols. 6r–7v; cf. A. Allard, *Histoire de la justice criminelle au seizième siècle* (Aalen, 1970), 262–64.

21. Menochio, *De praesumptionibus*, q 29 fols. 23r–v; q. 41 fols. 32r–33v.

22. G. Mascardi, *Conclusiones probationum omnium* or *De probationibus*, vol. 1 (Venice, 1584–88, 1593, Frankfurt, 1593, etc; references are to Venice, 1584 edition), q. 14 fols. 29v–30v; q. 15 fols. 30v–31r; qq. 11–13 fols. 28r–29v.

23. Allard, *Histoire de la justice criminelle*, 443.

24. Mascardi, *Conclusiones probationum omnium*, 1: fols. 84v–85r concl. 60; fol. 85r–v concl. 61; 2: fols. 72v–73r concl. 677; fol. 330r–v concl. 995; 3: fol. 183v concl. 1224; similar lists in shorter treatises: A. Alciati, *Tractatus de presumptionibus* (Lyons, 1561), in A. Alciati, *Opera* (Basel, 1582), 4: cols. 677–894; Lancelottus, *Institutiones iuris canonici* 3 tit. 14 § 12, quoted in W. Best, *The Principles of the Law of Evidence*, 5th ed. (London, 1970), § 21, 19.

25. *Constitutio Criminalis Carolina*, secs. 18, 22, 24–30, in J. Langbein, *Prosecuting Crime in the Renaissance* (Cambridge, Mass., 1974), 261–308, at 272–75.

26. F. Bruni, *De indiciis et tortura*, in *Tractatus universi iuris*, 11: pt. 1 fols. 246r–260v, at 246v; cf. Allard, *Histoire de la justice criminelle*, 256–57.

27. H. Beinart, ed., *Records of the Trials of the Spanish Inquisition in Ciudad Real* (Jerusalem, 1974), 1:86; see H. Beinart, *Conversos on Trial: The Inquisition in Ciudad Real* (Jerusalem, 1981), 127–85; a different opinion in H. Kamen, *The Spanish Inquisition* (New Haven, 1997), 62–63.

28. Grillandus, *De quaestionibus et tortura*, in *Tractatus universi iuris* (Venice, 1584) 11: pt. 1 fols. 294r–298r, at 294v–295v.

29. J. de Damhouder, *Practica rerum criminalium* (Louvain, 1555, Würzburg, 1641), cap. 36 fols. 105 and 107; cap. 39 fol. 122; on the book's popularity see Allard, *Histoire de la justice criminelle*, 464–69; more restrained rules for the inquisition in R. Helmholz, *Roman Canon Law in Reformation England* (Cambridge, 1990), 17–18.

30. B. Pullan, *The Jews of Europe and the Inquisition of Venice, 1550–1670* (Oxford, 1983), 133.

31. L. Stern, "Inquisition procedure and crime in fifteenth-century Florence," *Law and History Review* 8 (1990): 297–308.

32. L. Stern, *The Criminal Law System of Medieval and Renaissance Florence* (Baltimore, 1994), 39, 31–32; cf. T. Dean, "Criminal justice in mid-fifteenth-century Bologna," in *Crime, Society and the Law in Renaissance Italy*, ed. T. Dean (Cambridge, 1994), 16–39, at 20–21.

33. G. Marañón, *Antonio Pérez*, trans. C. Ley (London, 1954), 251, 265.

34. Gratian, *Decretum* pt. 2 C. 26 q. 5 c. 12, in *Corpus iuris canonici*, ed. A. Friedberg (Leipzig, 1879), 1: cols. 1030–31; also in *Witchcraft in Europe, 1100–1700*, ed. A. Kors and E. Peters (London, 1972), 29–31; Peters, *The Magician, the Witch, and the Law*.

35. Bartolus, *Consilia*, in *Opera omnia* (Venice, 1615) 10:183–211, cons. 3, 184.

36. P. Ariès, *The Hour of Our Death*, trans. H. Weaver (New York, 1981), ch. 3; J. Huizinga, *The Waning of the Middle Ages* (Harmondsworth, 1955), ch. 11.

37. H. Kramer and J. Sprenger, *The Malleus Maleficarum*, trans. M. Summers (New York, 1971), pt. 1 qq. 3, 5, 9 pp. 25, 35, 58; pt. 3 q. 34 p. 269; W. Stephens, "Witches who steal penises: Impotence and illusion in *Malleus Maleficarum*," *Journal of Medieval and Early Modern Studies* 28 (1998): 495–530.

38. Kramer and Sprenger, *Malleus Maleficarum*, pt. 2 q. 1 p. 111.

39. Ibid., pt. 1 q. 6 pp. 41–47.

40. Ibid., 112–13.

41. Ibid., pt. 3 q. 2 p. 208; q. 4 p. 209; q. 5 p. 210.

42. Ibid., pt. 3 q. 6 p. 211.

43. Ibid., 212.

44. Ibid., 213; cf. G. Matthews, *News and Rumor in Renaissance Europe* (New York, 1959), 142.

45. G. Quaife, *Godly Zeal and Furious Rage* (London, 1987), 201–4; J. Janssen, *History of the German People*, trans. A. Christie (London, 1910), 16:286–88; further refs. in R. Kieckhefer, *European Witch Trials* (Berkeley, 1976), 162.

46. E. Walter, "Nature on trial," *Boston Studies in the Philosophy of Science* 84 (1984): 295–321; J. Finkelstein, "The ox that gored," *Transactions of the American Philosophical Society* 71 (1981), pt. 1; E. Cohen, *The Crossroads of Justice* (Leiden, 1993), 110–33.

47. C. Otten, ed., *A Lycanthropy Reader: Werewolves in Western Culture* (Syracuse, 1986), readings 8, 9, 11, 12; M. Summers, *The Werewolf* (London, 1933), ch. 3; P. Barber, *Vampires, Death, and Burial* (New Haven, 1988), ch. 2.

48. Kramer and Sprenger, *Malleus Maleficarum*, pt. 3 q. 7 pp. 213–14.

49. Ibid., q. 8 p. 215.

50. Ibid., q. 11 pp. 219–20.

51. Ibid., q. 15 pp. 227–28.

52. Ibid., q. 22 p. 243.

53. Ibid., q. 14 p. 226.

54. Ibid., q. 16 pp. 231–32.

55. Ibid., q. 19 p. 237.

56. Ibid., 239–40.

57. Ibid., pt. 3 qq. 23–25 pp. 245–50.

58. Ibid., q. 31 p. 260.

59. H. Lea, *Materials toward a History of Witchcraft* (New York, 1957), 1:356–57.

60. H. Lea, *A History of the Inquisition of the Middle Ages* (New York, 1955),

3:545; cf. P. Jobe, "Inquisitorial manuscripts in the Bibliotheca Apostolica Vaticana: Preliminary handlist," in *The Inquisition in Early Modern Europe*, ed. G. Henningsen and J. Tedeschi (De Kalb, 1986), 32–53, at 48; F. Bavoux, *Hantises et diableries dans la terre abbatiale de Luxeuil d'un procès d'Inquisition, 1592, à l'épidémie démoniaque de 1628–1630* (Monaco, 1956); M. Kunze, *Highroad to the Stake*, trans. W. Yuill (Chicago, 1987), 360–64; E. Monter, *Witchcraft in France and Switzerland* (Ithaca, 1976), 92–99; the evidence in the case of the devils of Loudun described in R. Rapley, *A Case of Witchcraft: The Trial of Urbain Grandier* (Montreal, 1998), 169–74.

61. F. Ponzinibio, *Tractatus de lamiis*, in *Tractatus universi iuris* 11: pt. 2 fols. 350r–56r; Allard, *Histoire de la justice criminelle*, 285–89, 469–79.

62. Luther, quoted in Janssen, *History of the German People*, 16:273; cf. J. Todd, *Luther: A Life* (New York, 1982), 5; H. Grisar, *Martin Luther: His Life and Work*, adapted by F. Eberle (New York, 1971), 492; J. Haustein, *Martin Luthers Stellung zum Zauber- und Hexenwesen* (Stuttgart, 1990).

63. J. Bodin, *De la Démonomanie des Sorciers* (Paris, 1580) 4, ch. 5, selection in Kors and Peters, *Witchcraft in Europe*, 213–15; cf. C. Baxter, "Jean Bodin's *De la Démonomanie des Sorciers:* The logic of persecution," in *The Damned Art*, ed. S. Anglo (London, 1977), 76–105, at 98–100.

64. H. Midelfort, *Witch Hunting in Southwestern Germany, 1562–1684* (Stanford, 1972), 190–91.

65. R. Scribner, "Witchcraft and judgement in reformation Germany," *History Today* 40 (4) (Apr. 1990): 12–19.

66. R. Scot, *The Discoverie of Witchcraft* 2, chs. 3–5 (Arundel, 1964), 41–43; cf. R. West, *Reginald Scot and Renaissance Writings on Witchcraft* (Boston, 1984), 94–96; G. Gifford, *A Dialogue Concerning Witches and Witchcraftes* (London, 1931), selection in B. Rosen, *Witchcraft in England, 1558–1618* (Amherst, 1969), 158–60.

67. James VI and I, *Demonologie* (Edinburgh, 1597, rcpr. Edinburgh, 1966), 80; see S. Clark, "King James's *Demonologie:* Witchcraft and kingship," in Anglo, *Damned Art*, 156–81; C. Larner, *Witchcraft and Religion* (Oxford, 1984), ch. 1.

68. H. Paul, *The Royal Play of Macbeth* (New York, 1950), 75–130, 255–74; G. Wills, *Witches and Jesuits: Shakespeare's Macbeth* (New York, 1995), ch. 3.

69. W. Raleigh, preface to *History of the World*, in *The Works of Sir Walter Raleigh* (New York, 1964), 2:ii; cf. 2 ch. 23 sec. 4 at 4:676; 2 ch. 14 sec. 1 at 4:446; James VI, *Demonologie*, 62.

70. J. Teall, "Witchcraft and Calvinism in Elizabethan England," *Journal of the History of Ideas* 23 (1962): 21–36.

71. W. Perkins, *A Discourse of the Damned Art of Witchcraft*, ch. 7, in *The Works of William Perkins*, ed. I. Breward (Abingdon, 1970), 604; cf. M. MacDonald, *Witchcraft and Hysteria in Elizabethan London* (London, 1991), 29.

72. Gifford, *Dialogue Concerning Witches and Witchcraftes*, fols. H3–4, L1–2.

73. Shapiro, *"Beyond Reasonable Doubt" and "Probable Cause,"* 52–54, 164–67, ref. esp. to J. Cotta, *The Trial of Witchcraft* (London, 1616); T. Cooper, *The Mys-*

terie of Witchcraft (London, 1617); Dalton, *Country Justice*; R. Bernard, *Guide to Grand Jury Men in Cases of Witchcraft* (London, 1627).

74. D. Hall, *Witch-Hunting in Seventeenth Century New England* (Boston, 1991), 229.

75. G. Henningsen, *The Witches' Advocate* (Reno, 1980), 347.

76. E. Monter, *Frontiers of Heresy* (Cambridge, 1990), ch. 12; Lea, *History of the Inquisition in Spain*, 4: ch. 9; Kamen, *Spanish Inquisition*, 270–75.

77. Henningsen, *Witches' Advocate*, 225–37.

78. Salazar de Frias, selection in Kors and Peters, *Witchcraft in Europe*, 340–41; see Henningsen, *Witches' Advocate*, chs. 11–12.

79. J. Tedeschi, "Inquisitorial law and the witch," in *Early European Witchcraft*, ed. B. Ankarloo and G. Henningsen (Oxford, 1990), 83–118.

80. E. Monter and J. Tedeschi, "Towards a statistical profile of the Italian inquisitions, sixteenth to eighteenth centuries," in Henningsen and Tedeschi, *Inquisition in Early Modern Europe*, 131–57, at 155 n. 61; cf. J. Tedeschi, "The Roman inquisition and witchcraft: An early seventeenth-century 'instruction' on correct trial procedure," *Revue de l'Histoire des Religions* 200 (1983): 163–88; R. Martin, *Witchcraft and Inquisition in Venice, 1550–1650* (Oxford, 1989), 178.

81. Montaigne, *Essais* 3 ch. 11, in *Great Books of the Western World*, trans. C. Cotton (Chicago, 1952), 25:500–501; the Augustine reference is *City of God* 18 ch. 18.

82. Montaigne, *Essais* 2 ch. 5 p. 176.

83. P. Reilly, "Friedrich von Spee's belief in witchcraft," *Modern Language Review* 54 (1959): 51–55; cf. H. Zwetsloot, *Friedrich von Spee und die Hexenprozesse* (Trier, 1954); T. Cornelis van Stockum, *Friedrich von Spee in de heksen processen* (Amsterdam, 1949); J.-F. Ritter, *Friedrich von Spee* (Trier, 1977), 49–66.

84. F. von Spee, *Cautio Criminalis* (1631), selection in Kors and Peters, *Witchcraft in Europe*, 351–57.

85. J. Fortescue, *De Laudibus Legum Angliae*, ch. 20, ed. and trans. S. Chrimes (Oxford, 1942), 43, 45.

86. Ibid., 65; see generally J. Bellamy, *Criminal Law and Society in Late Medieval and Tudor England* (Gloucester, 1984), ch. 3.

87. T. Twiss, *The Black Book of the Admiralty*, Rolls Series, 4 vols. (London, 1871–76); see K. Nörr, "Bologna and the Court of Admiralty: A Latin text in the Black Book," *Proceedings of the Seventh International Congress of Medieval Canon Law, Cambridge, 1984*, ed. P. Linehan (Vatican City, 1988), 475–83.

88. Heath, *Torture and English Law*, ch. 3; C. Hall, "Some perspectives on the use of torture in Bacon's time and the question of his 'virtue,'" *Anglo-American Law Review* 18 (1989): 289–321, at 300–303; H. Kelly, "English kings and the fear of sorcery," *Mediaeval Studies* 39 (1977): 206–38, at 213, 231.

89. J. Selden, *Table-Talk*, ed. E. Arber (London, 1905), 113; see W. Holdsworth, *History of English Law*, 3d ed. (London, 1945), 5:185–87.

90. A. Cobban, *The Medieval English Universities* (Berkeley, 1988), 239–42;

T. Aston, "Oxford's medieval alumni," *Past and Present* 74 (1977): 3–40, at 28–30; B. Levack, *The Civil Lawyers in England, 1603–1641* (Oxford, 1973).

91. Statute 27 Henry VIII, cap. 4, in *Statutes at Large* 2:111–12; also Statute 28 Henry VIII, cap. 15, in *Statutes at Large* 2:140–41.

92. J. Wigmore, "Required numbers of witnesses: A brief history of the mathematical system in England," *Harvard Law Review* 15 (1901): 83–108, at 100–101; J. Bellamy, *The Tudor Law of Treason* (London, 1979), 152–54; J. Hill, "The two-witness rule in English treason trials," *American Journal of Legal History* 12 (1968): 95–111, at 101.

93. Wigmore, "Required numbers of witnesses," 102; Bellamy, *Tudor Law of Treason,* 77–78, 155; Hill, "The two-witness rule," 104; Marcus, "Tudor treason trials," 697.

94. Marcus, "Tudor treason trials," 702 n. 174; Howell, *Complete Collection of State Trials,* 2: col. 15.

95. N. Williams, *Sir Walter Raleigh* (London, 1962), 187; cf. Howell, *Complete Collection of State Trials,* 2: col. 18; E. Coke, *Third Part of the Institutes* 137; Hill, "The two-witness rule," 107–10.

96. Shapiro, *"Beyond Reasonable Doubt" and "Probable Cause,"* 193.

97. Coke, *First Part of the Institutes* 6b; cf. *English Reports* 77: 534; 145: 73; B. Shapiro, *Probability and Certainty in Seventeenth-Century England* (Princeton, 1983), 177–78; later quotations in Dalton, *Country Justice,* 303; Shapiro, *"Beyond Reasonable Doubt" and "Probable Cause,"* 157–58.

98. Coke, *Third Part of the Institutes* 25; cf. ibid., 137; *English Reports* 145: 40.

99. Coke, *First Part of the Institutes* 79b, 264a; cf. ibid., 125b, 232b, 283a; Coke, *Second Part of the Institutes* 137; Coke, *Third Part of the Institutes* 210; Coke, *Fourth Part of the Institutes* 279; *English Reports* 79: 877; 81: 36; *English Reports* 77: 370, 471; Howell, *Complete Collection of State Trials,* 2: col. 18.

100. Coke, *First Part of the Institutes* 373a–b; cf. *English Reports* 77:208; for the Latin originals of some of these, see Helmholz, *Canon Law and the Law of England,* 178, 197; Mascardi, *Conclusiones probationum omnium* 2: fol. 138r concl. 788; *Digest* 2.4.5; for the presumption based upon flight, see Shakespeare, *Macbeth* 3.4.9.

101. *English Reports* 77:727, 872; cf. 145:48.

102. Coke, *First Part of the Institutes* 254a; see J. Martin, *Francis Bacon, the State and the Reform of Natural Philosophy* (Cambridge, 1992), 92–95.

103. *English Reports* 80:1022; cf. Coke, *Third Part of the Institutes* 29.

104. Hall, "Some perspectives on the use of torture in Bacon's time," 304.

105. *Slade v. Morley* (1602), in *English Reports* 76:1074–79; K. Teeven, "Seventeenth century evidentiary concerns and the Statute of Frauds," *Adelaide Law Review* 9 (1983–85): 252–56; K. Teeven, "Problems of proof and early English contract law," *Cambrian Law Review* 15 (1984): 52–72; A. Simpson, "The place of Slade's case in the history of contract," *Law Quarterly Review* 74 (1958): 381–96; J. Baker, "New light on Slade's case," *Cambridge Law Journal* 29 (1971): 51–67, 213–36.

106. Coke, *Second Part of the Institutes* proeme, vi, quoted in Levack, *Civil Lawyers in England*, 152.

107. Shapiro, *Probability and Certainty in Seventeenth-Century England*, 190.

108. Ibid., 180.

109. Shapiro, *"Beyond Reasonable Doubt" and "Probable Cause,"* ch. 1; T. Waldman, "Origins of the legal doctrine of reasonable doubt," *Journal of the History of Ideas* 20 (1959): 299–316.

110. T. Deman, s.v. "Probabilisme," *Dictionnaire de théologie catholique*, 13: pt. 1 (Paris, 1936), cols. 417–619, at col. 473; cf. col. 475; C. Michalski, *La philosophie au XIVe siècle* (Frankfurt, 1969), 253.

111. *Decretals of Gregory IX* 4 tit. 1 cap. 15; cf. cap. 28, in Friedberg, *Corpus iuris canonici* 2: cols. 666–67, 671; J. Noonan, *Power to Dissolve* (Cambridge, Mass., 1972), ch. 1; Brundage, *Medieval Canon Law*, 166; cf. *Digest* 4.2.6; *Digest* 50.17.184.

112. Hooker, *Laws of Ecclesiastical Polity*, 4 ch. 7 at 183; preface at 28; 2 ch. 7 at 118; cf. W. Southgate, *John Jewel and the Problem of Doctrinal Authority* (Cambridge, Mass., 1962), 140; Cotta, *Triall of Witchcraft*; Shapiro, *Probability and Certainty in Seventeenth-Century England*, 202.

113. Shapiro, *"Beyond Reasonable Doubt" and "Probable Cause,"* ch. 1, esp. 22.

Chapter 4: The Doubting Conscience and Moral Certainty

1. Aristotle, *Nicomachean Ethics*, trans. W. Ross, revised by J. O. Urmson, in *The Complete Works of Aristotle*, ed. J. Barnes (Princeton, 1984), 1094b12–27; cf. 1140a31–b2.

2. C. Cochrane, *Christianity and Classical Culture* (New York, 1957), ch. 11; C. Taylor, *Sources of the Self: The Making of Modern Identity* (Cambridge, Mass., 1989), ch. 7.

3. C. Morris, *The Discovery of the Individual, 1050–1200* (New York, 1972), 86–95; M. Radding, *A World Made by Men* (Chapel Hill, 1985), 224–36, 242; S. Spence, *Texts and the Self in the Twelfth Century* (Cambridge, 1996).

4. K. Clark, *Civilisation* (London, 1969), 56; Morris, *Discovery of the Individual*, 79–86, 107–20; C. Ferguson, "Autobiography as therapy: Guibert de Nogent, Peter Abelard, and the making of medieval autobiography," *Journal of Medieval and Renaissance Studies* 13 (1983): 187–212; J. Benton, "Consciousness of self and perceptions of individuality," in *Renaissance and Renewal in the Twelfth Century*, ed. R. Benson and G. Constable (Oxford, 1982), 262–95; S. Bagge, "The *Autobiography* of Abelard and medieval individualism," *Journal of Medieval History* 19 (1993): 327–50; G. Bynum, *Jesus as Mother* (Berkeley, 1982), ch. 3; K. Harris, "Portraying the prophet: The role of autobiography in the writings of Hildegard of Bingen," *Tjurunga* 56 (1999): 5–25; C. Lewis, *The Allegory of Love* (Oxford, 1936), ch. 1; P. Dronke, *Medieval Latin and the Rise of the European Love Lyric* (Oxford, 1968), 1: ch. 5; R. Hanning, *The Individual in Twelfth Century Romance* (New Haven, 1977); Abelard, *Ethics*, ed. and trans. D. Luscombe (Oxford, 1971), 15, 33; P. King, "Abelard's intentionalist ethics," *Modern School-*

man 72 (1995): 213–31; cf. H. Berman, *Law and Revolution* (Cambridge, Mass., 1983), 598 n. 47.

5. G. Frank, *The Medieval French Drama* (Oxford, 1954), ch. 8.

6. *De vera et falsa poenitentiae*, in *Patrologia Latina*, ed. J. Migne, 40: cols. 1114–30, at cols. 1129–30.

7. Robert of Flamborough, *Liber Poenitentialis*, ed. J. Firth (Toronto, 1971), 65; cf. Raymond of Peñafort, *Summa de poenitentia* 3 tit. 34 n. 30, ed. X. Ochoa and A. Diez (Rome, 1976), col. 827.

8. M. Braswell, *The Medieval Sinner: Characterization and Confession in the Literature of the English Middle Ages* (New Brunswick, 1983); A. Murray, "Confession as a historical source in the thirteenth century," in *The Writing of History in the Middle Ages*, ed. R. Davies and J. Wallace-Hadrill (Oxford, 1981), 275–322; P. Biller and A. Minnis, *Handling Sin: Confession in the Middle Ages* (York, 1998); J. le Goff, *Your Money or Your Life* (New York, 1988), 11–12; Protestant continuations in C. Slights, "Notaries, sponges and looking-glasses: Conscience in early modern England," *English Literary Renaissance* 28 (1998): 231–46.

9. H. Coing, "English equity and the *denunciatio evangelica* of the canon law," *Law Quarterly Review* 71 (1955): 223–41.

10. R. Fraher, "Conviction according to conscience," *Law and History Review* 7 (1989): 23–88; B. Shapiro, *"Beyond Reasonable Doubt" and "Probable Cause"* (Berkeley, 1991), 11–13, 260 n. 32.

11. *Digest* 28.4.3; 34.5.21.1; 50.17.56; 50.17.192.1; A. Berger, "In dubiis benigniora," *Seminar* 9 (1951): 36–49.

12. Augustine, *De Sermone Domini in Monte* 2 ch. 18, in *The Lord's Sermon on the Mount*, trans. J. Jepson (Westminster, Md., 1948), 147; adopted by canon law in *Decretals of Gregory IX* 5 tit. 41 cap. 2, in *Corpus iuris canonici*, ed. A. Friedberg (Leipzig, 1879) 2: col. 927; cf. Aquinas, *Summa theologiae* II-II q. 60 art. 4.

13. *Decretals of Gregory IX* 5 tit. 27 cap. 5 cols. 828–29.

14. O. Lottin, *Psychologie et morale aux XIIe et XIIIe siècles* (Gembloux, 1949), 3:675.

15. O. Lottin, "Le tutiorisme du treizième siècle," *Recherches de Théologie Ancienne et Médiévale* 5 (1933): 292–301, at 293 n. 5; cf. P. Glorieux, *La littérature quodlibétique, 1260–1320* (Le Saulchoir, 1925), 183, 265.

16. Lottin, "Le tutiorisme du treizième siècle," 295 nn. 10, 11.

17. Albert the Great, *Summa "De creaturis"* pt. 2 q. 72 a.2, in Albert the Great, *Opera omnia*, ed. A. Borgnet (Paris, 1890–99), 35:600–601.

18. Ibid., pt. 2 q. 53 a.1, at 35:447; Albert the Great, *Summa de bono*, quoted in Lottin, "Le tutiorisme du treizième siècle," 298 n. 22; repeated by Aquinas, *Summa theologiae* II-II q. 2 art. 1; I q. 79 art. 9 ad 4; Aquinas, *Disputed Questions on Truth* q. 14 art. 1; a similar phrase in Ghazali, in W. Hallaq, "On inductive corroboration, probability, and certainty in Sunni legal thought," in *Islamic Law and Jurisprudence*, ed. N. Heer (Seattle, 1990), 19 n. 44.

19. Aquinas, *In IV Sententiarum* dist. 38 q. 1 ad 3; q. 1 ad 6; Aquinas, *Summa theologiae* II-II q. 32 art. 5 ad 3; cf. III q. 83 art. 6 ad 5; Aquinas, *Commentary on*

Aristotle's "De Caelo" 1 lectio 22; I. Kantola, *Probability and Moral Uncertainty in Late Medieval and Early Modern Times* (Helsinki, 1994), 49–50; U. Lopez, *Thesis probabilismi ex Sancto Thoma demonstrata* (Rome, 1937); T. Richard, *Le probabilisme moral et la philosophie* (Paris, 1922).

20. F. Russell, *The Just War in the Middle Ages* (Cambridge, 1975), 226, 47, 87 n. 3, 229.

21. J. Gerson, *De praeparatione ad missam*, in *Oeuvres Complètes*, ed. P. Glorieux (Paris, 1973), 9:35–50, at 37; cf. D. Brown, *Pastor and Laity in the Theology of Jean Gerson* (Cambridge, 1987), 69; the beginning of the laxist rot, according to P. Fagnani, *Commentaria in quinque libros decretalium* (Venice, 1729), 1:32.

22. T. Deman, s.v. "Probabilisme," *Dictionnaire de théologie catholique*, 13: pt. 1 (Paris, 1936), cols. 417–619, at cols. 442–43.

23. D. Stove, *Popper and After: Four Modern Irrationalists* (Oxford, 1982), 29–31; repr. as *Anything Goes* (Sydney, 1998), 65–67.

24. J. Nider, *Consolatorium timoratae conscientiae*, in Kantola, *Probability and Moral Uncertainty*, 112 n. 7; and quoted in Deman, "Probabilisme," col. 446.

25. Nider, *Consolatorium timoratae conscientiae*, in Kantola, *Probability and Moral Uncertainty* 113 n. 8; Gerson's opinion in Deman, "Probabilisme," col. 443.

26. St. Antoninus, *Summa Moralis* (Verona, 1740); cf. R. de Roover, *San Bernardino of Siena and Sant' Antonino of Florence: The Two Greatest Economic Thinkers of the Middle Ages* (Boston, 1967); J. Kirshner, "Raymond de Roover on Scholastic economic thought," in *Business, Banking and Economic Thought in Late Medieval and Early Modern Europe*, ed. R. de Roover (Chicago, 1974), 15–36.

27. Deman, "Probabilisme," cols. 447–48.

28. T. Tentler, *Sin and Confession on the Eve of the Reformation* (Princeton, 1977), 35–36.

29. J. de Blic, "Barthélémy de Medina et les origines du probabilisme," *Ephemerides Theologicae Lovanienses* 7 (1930): 46–83, 264–91, at 47–50.

30. M. Gorce, "A propos de Barthélémy de Medina et du probabilisme," *Ephemerides Theologicae Lovanienses* 7 (1930): 480–82.

31. Deman, "Probabilisme," cols. 449–50.

32. R. Tawney, *Religion and the Rise of Capitalism* (Harmondsworth, 1938), 110.

33. Adrian, *Quaestiones Quodlibeticae* II q. 2 (Paris, 1527); de Blic, "Barthélémy de Medina," 54; P. Haggenmacher, *Grotius et la doctrine de la guerre juste* (Geneva, 1983), 196.

34. J. Scott, *The Spanish Origin of International Law: Francisco de Vitoria and His Law of Nations* (Oxford, 1934); J. Ternus, *Zur Vorgeschichte der Moralsysteme von Vitoria bis Medina* (Paderborn, 1930).

35. B. Hamilton, *Political Thought in Sixteenth-Century Spain* (Oxford, 1963), 114–34; A. Pagden, *Spanish Imperialism and the Political Imagination* (New Haven, 1990), ch. 1.

36. F. de Vitoria, *De Indis* q. 2 art. 4, in F. de Vitoria, *Political Writings*, ed. A. Pagden and J. Lawrance (Cambridge, 1991), 269–71.

37. F. de Vitoria, *De Iure Belli* q. 2 art. 3, in Vitoria, *Political Writings*, 311–12;

de Blic, "Barthélémy de Medina," 54; Scott, *Spanish Origin of International Law*, 224–28.

38. F. de Vitoria, *Commentary* on [Aquinas' *Summa theologiae*] I-II q. 19 art. 5 (lecture notes of a course of 1539), in de Blic, "Barthélémy de Medina," 55.

39. M. Cano, *Commentary* on [Aquinas' *Summa theologiae*] I-II q. 19 art. 5, in de Blic, "Barthélémy de Medina," 58.

40. De Blic, "Barthélémy de Medina," 59–60.

41. B. de Medina, *Expositio in I-II S. Thomae* (Salamanca, 1577), q. 19 art. 5 and 6; de Blic, "Barthélémy de Medina," 67–72; Kantola, *Probability and Moral Uncertainty*, 124–30; on Soto, see D. Soto, *De dubio et opinione*, in Ternus, *Vorgeschichte der Moralsysteme von Vitoria bis Medina*, 47–67, discussed in Kantola, *Probability and Moral Uncertainty*, 117–23.

42. These two meanings distinguished by an anonymous follower of Medina in de Blic, "Barthélémy de Medina," 78, 80.

43. Deman, "Probabilisme," cols. 478, 473, 477.

44. F. Suarez, *De Legibus* 1, ch. 9.11, in F. Suarez, *Selections from Three Works* (Oxford, 1944), 2:113; cf. Hamilton, *Political Thought in Sixteenth-Century Spain*, 49.

45. F. Suarez, *De triplici virtute theologica: De caritate* (Coimbra, 1621), disp.13 sec. 6.12, in Suarez, *Selections from Three Works*, 2:836; Kantola, *Probability and Moral Uncertainty*, 134–40.

46. J. Keynes, *Treatise on Probability* (London, 1921), ch. 6; see L. Cohen, "Twelve questions about Keynes' concept of weight," *British Journal for the Philosophy of Science* 37 (1986): 263–78; R. O'Donnell, "Keynes' weight of argument and Popper's paradox of ideal evidence," *Philosophy of Science* 59 (1992): 44–52.

47. Suarez, *De triplici virtute theologica: De caritate*, disp.13 sec. 6.2–3, in *Selections from Three Works*, 2:828–29.

48. Ibid., disp.13 sec. 4.10, in *Selections from Three Works*, 2:822–23.

49. H. Grotius, *Rights of War and Peace* 2, ch. 26 (London, 1682), 430; J. Johnson, *Ideology, Reason, and the Limitation of War* (Princeton, 1975), 220; B. Vermeulen, "Grotius on conscience and military orders," *Grotiana* n.s. 6 (1985): 3–19; for Grotius's dependence on the Scholastic Lessius, see L. Lessius, *Commentarius in Secundam Secundae Divi Thomae*, in J. Bittremieux, ed., *Lessius et le droit de la guerre* (Brussels, 1920); Haggenmacher, *Grotius et la doctrine de la guerre juste*, 217–18, 294, 495; B. Meulenbroek, *Briefwisseling van Hugo Grotius* (The Hague, 1966), 5:194.

50. H. Grotius, *De iure praedae commentarius*, trans. G. Williams (Oxford, 1950), 1:xiii–xv; see Haggenmacher, *Grotius et la doctrine de la guerre juste*, 200–203.

51. Grotius, *De iure praedae commentarius*, 1:78–79.

52. J. de Silhon, "Lettre à l'Evesque de Nantes," in N. Faret, *Receuils de lettres nouvelles* (Paris, 1627), quoted in W. Church, *Richelieu and Reason of State* (Princeton, 1973), 168.

53. J. de Silhon, *Le Ministre d'Estat, avec le veritable usage de la Politique moderne* (Paris, 1631), quoted in Church, *Richelieu and Reason of State*, 263–64.

54. *Ministre*, in Church, *Richelieu and Reason of State*, 266.

55. Silhon, "Lettre," quoted in Church, *Richelieu and Reason of State*, 169–70.

56. T. Hobbes, *Leviathan*, pt. 1, ch. 13, W. Molesworth, ed., *The English Works of Thomas Hobbes* (London, 1839), 3:113; on Hobbes on the probable reasoning of Aristotle's *Rhetoric* see J. Harwood, ed., *The Rhetorics of Thomas Hobbes and Bernard Lamy* (Carbondale, 1986), 9, 14, 39.

57. Hobbes, *Leviathan* pt. 1 ch. 14 at 3:124–25.

58. Hobbes, *De Cive* ch. 2 par 11 at 2:21; cf. ch. 13 sec. 7; and T. Hobbes, *Elements of Law* pt. 1 ch. 15 sec. 10.

59. J. Azorius, *Institutionum moralium pars prima* 1 ch. 10 (Cologne, 1602), 17; P. Sarpi et al., *Tractatus de interdicto sanctitatis papae Pauli V*, in *Controversiae memorabilis inter Paulum V. pontificem Max. et Venetos* (San Vincentiana, 1607), first pagination, pp. 169–242, at 180–81; see J. Sommerville, *Thomas Hobbes: Political Ideas in Historical Context* (New York, 1992), 54.

60. Hobbes, *Leviathan* pt. 2 ch. 26 (3:276).

61. H. Warrender, *The Political Philosophy of Hobbes* (Oxford, 1957), 115–18.

62. M. Nedham, *The Case of the Commonwealth of England Stated* (London, 1650) (*Goldsmiths'-Kress Library of Economic Literature*, item 1179), 6, 37; cf. A. Ascham, *Of the Confusions and Revolutions of Governments*, 2d ed. (London, 1649) (*Goldsmiths'-Kress Library of Economic Literature*, item 1035), 99, 121.

63. Books listed in H. Hurter, *Nomenclator Literarius Theologicae Catholicae* (Innsbruck, 1907), cols. 590–603, 880–94, 1185–1202.

64. Deman, "Probabilisme," cols. 484–86; also in G. Vasquez, *Commentaria in II* 67: iv, quoted in A. Jonsen and S. Toulmin, *The Abuse of Casuistry: A History of Moral Reasoning* (Berkeley, 1988), 167; "extrinsic" for authorities is from Cicero, *Topics* iv.24, xix.72–73; cf. I. Hacking, *The Emergence of Probability* (Cambridge, 1975), 79.

65. Azorius, *Institutiones morales pars prima* 2.17.7, quoted in A. Malloch, "John Donne and the casuists," *Studies in English Literature* 2 (1962): 57–76, at 67.

66. Deman, "Probabilisme," col. 487; T. Tamburini, *Theologia moralis* 1 cap. 3 (Venice, 1726), 13.

67. Deman, "Probabilisme," col 492.

68. Jonsen and Toulmin, *Abuse of Casuistry*, 372 nn. 10–11.

69. Deman, "Probabilisme," col. 499.

70. Ibid., cols. 502–3.

71. L. Gallagher, *Medusa's Gaze: Casuistry and Conscience in the Renaissance* (Stanford, 1991); C. Slights, *The Casuistical Tradition in Shakespeare, Donne, Herbert, and Milton* (Princeton, 1981); Spanish parallels in M. Alvarez, "El probabilismo y el teatro español del siglo XVII" (Ph.D. diss., New York Univ., 1982).

72. J. Donne, *Biathanatos* (New York, 1930), 30; see Slights, *Casuistical Tradition*, 141, 147.

73. E. Gosse, ed., *The Life and Letters of John Donne* (London, 1899), 1:174; see Malloch, "John Donne and the casuists," 59; Slights, *Casuistical Tradition*,

140; C. Cathcart, "Doubting conscience: John Donne and the tradition of casuistry" (Ph.D. diss., Vanderbilt Univ., 1968); on Donne's contacts with the Scholastic tradition more generally, M. Ramsay, *Les doctrines médiévales chez Donne* (London, 1917); T. Eliot, *The Varieties of Metaphysical Poetry*, ed. R. Schuchard (New York, 1993), 67–92.

74. H. Garnett, *A Treatise of Equivocation*, quoted in P. Zagorin, *Ways of Lying* (Cambridge, 1990), 195; cf. 200–201; cf. E. Rose, *Cases of Conscience* (Cambridge, 1975), 84, 88.

75. T. Howell, ed., *A Complete Collection of State Trials* (London, 1816), 2: col. 219; *A True and Perfect Relation of the Proceedings at the Several Arraignments of the Late Most Barbarous Traitors* (London, 1606), sig. Fff3, quoted in Zagorin, *Ways of Lying*, 197.

76. C. McIlwain, ed., *The Political Works of James I* (Cambridge, Mass., 1918), 113.

77. J. Donne, *Pseudo-Martyr* (London, 1610), 230–31; Malloch, "John Donne and the casuists," 62; the suggestion taken up by a Catholic in T. Preston, alias R. Widdrington, *A New-Yeares Gift for English Catholikes* (1620, repr. Menston, 1973), 185–87.

78. F. Bacon, "An advertisement touching the controversies of the Church of England," in *The Letters and the Life of Francis Bacon*, ed. J. Spedding (London, 1861–72), 1:92; cf. Rose, *Cases of Conscience*, ch. 11.

79. H. Hammond, *Of Resisting the Lawful Magistrate upon Colour of Religion* (London, 1643), 23; M. Sampson, "Laxity and liberty in seventeenth-century English political thought," in *Conscience and Casuistry in Early Modern Europe*, ed. E. Leites (Cambridge, 1988), 72–118, at 106.

80. *A Most Learned, Conscientious, and Devout Exercise . . . By Lieutenant-Generall Crumwell: As it was faithfully taken . . . by Aaron Guerdon*, in Sampson, "Laxity and liberty in seventeenth-century English political thought," 111.

81. W. Perkins, *Discourse of Conscience* (London, 1596), and *Whole Treatise of the Cases of Conscience* (London, 1608), both in *William Perkins: His Pioneer Works on Casuistry*, ed. T. Merrill (Nieuwkoop, 1966); see Shapiro *"Beyond Reasonable Doubt" and "Probable Cause,"* 15–16; a more extreme Puritan casuistry in W. Ames, *Medulla SS. Theologiae*, trans. as *The Marrow of Sacred Divinity* (London, 1623); *De conscientia et eius iure, vel casibus*, trans. as *Conscience, with the Power and Cases Thereof* (Amsterdam, 1975).

82. H. McAdoo, *The Structure of Caroline Moral Theology* (London, 1949); T. Wood, *English Casuistical Divinity during the Seventeenth Century with Special Reference to Jeremy Taylor* (London, 1952).

83. *The Whole Works of the Right Rev. Jeremy Taylor*, ed. R. Heber (London, 1822), 12:36–37.

84. Ibid., 75.

85. Ibid., 90.

86. Ibid., 88–89.

87. Biographical details in J. Velarde Lombraña, *Juan Caramuel: vida y obra*

(Oviedo, 1989); H. Hernández Nieto, *Las ideas literarias de Caramuel* (Barcelona, 1992); D. Pastine, *Probabilismo ed enciclopedia* (Florence, 1975); P. Bellazzi, *I. Caramuel Lobkowitz* (Vigevano, 1982); L. Ceyssens, "Autour de Caramuel," *Bulletin de l'Institut historique belge de Roma* 33 (1961): 329–410; R. Cenal, "Juan Caramuel: su epistolario con Atanasio Kircher," *Revista de filosofia* 12 (1953): 101–47; J.-A. Tadesi, *Memorie della vita di Giovanni Caramuele* (Venice, 1760); P. Pissavino, ed., *Le meraviglie del probabile: Juan Caramuel 1606–1682: Atti del convegno internazionale di studi, Vigevano . . . 1982* (Vigevano, 1990); summaries in V. Oblet, s.v. "Caramuel," *Dictionnaire de théologie catholique*, 2: pt. 2 (1932), cols. 1709–12; Hurter, *Nomenclator Literarius*, 4: cols. 604–10.

88. J. Caramuel, *Rationalis et realis philosophia* (Louvain, 1642), 60–62.

89. D. Fernández Diéguez, "Un matemático español del siglo XVII: Juan Caramuel," *Revista Matemática Hispano-americana* 1 (1919): 121–27, 178–89, 203–12, at 178–79.

90. J. Caramuel, *Theologia regularis* (Bruges, 1638; Frankfurt, 1648; Venice, 1651; Lyons, 1665).

91. Caramuel, *Theologia regularis* disp. 6 art. 1 (Venice, 1651), 40–43; cf. M. Mersenne, *Questions inouyes* q. 15, ed. A. Pessel (Paris, 1985), 45–46.

92. Caramuel, *Theologia regularis* art. 3 (44); Caramuel's own theory on how to recognize a joke, in J. Caramuel, *Grammatica audax*, pt. 1.31 (Stuttgart, 1989), 14.

93. Caramuel, *Theologia regularis*, 46–48.

94. J. Caramuel, *Theologia moralis fundamentalis* (Frankfurt, 1652) fundamentum 11 § 265, 132–33.

95. Ibid., § 315, 152–53.

96. Ibid., fundamentum 60, 623–24.

97. Ibid., fundamentum 10, 98; fundamentum 60, 627; fundamentum 11, § 273, 137.

98. Pascal, *Provincial Letter 7*, in Pascal, *Pensées and the Provincial Letters*, trans. T. M'Crie (New York, 1941), 416; Pascal, *Les Provinciales*, ed. L. Cognet (Paris, 1965), 130–31.

99. Pascal, *Pensées and the Provincial Letters*, 393; Pascal, *Les Provinciales*, 102–3.

100. E. Amann, s.v. "Laxisme," *Dictionnaire de théologie catholique*, 9, pt. 1 (1926), cols. 42–86, at col. 54.

101. Caramuel, *Theologia moralis fundamentalis* fundamentum 11, 133.

102. Pastine, *Probabilismo ed enciclopedia* 211–16; also Velarde Lombraña, *Juan Caramuel*, 152–54.

103. *Oeuvres complètes de Huygens* (La Haye, 1888–) 1:562–64; 2:262, 295, 500, 505; *Correspondance du P. Marin Mersenne*, ed. C. de Waard (Paris, 1972), 12:124–25.

104. J. Vernet, "Copernicus in Spain," in *The Reception of Copernicus' Heliocentric Theory*, ed. J. Dobrzycki (Dordrecht, 1973), 271–91, at 275–77; Fernández Diéguez, "Un matemático español del siglo XVII: Juan Caramuel," 179;

Pastine, *Probabilismo ed enciclopedia*, 266, 273; J. Martinez de Prado, *Observationes circa Theologiam fundamentalem DD. Ioannis Caramuel* (Alcala de Henares, 1656), 69; B. Nelson, "Probabilists, anti-probabilists and the quest for certitude on the 16th and 17th centuries," *Proceedings of the Tenth International Congress on the History of Science* (Paris, 1965), 269–73, at 271.

105. Vincenzo Baron, quoted in Velarde Lombraña, *Juan Caramuel*, 152 n. 218.

106. Martinez de Prado, *Observationes circa Theologiam fundamentalem DD. Ioannis Caramuel*, 60, 57, 112, 121, 69; Caramuel on the advantages of defending infallibility as only probable when converting Lutherans in P. Gassendi, *Opera omnia* (Lyons, 1658), 6:465–67.

107. Ceyssens, "Autour de Caramuel," 349 n. 3.

108. Fagnani, *Commentaria in quinque libros decretalium* 1 tit. 2 cap. 5 (Venice, 1729), 1:32–43.

109. P. Fagnani, *De opinione probabili tractatus* (Rome, 1665), esp. 23–25; Caramuel, *Apologema pro antiquissima et universalissima doctrina de probabilitate* (Lyons, 1663); F. Verde, *Theologiae fundamentalis Caramuelis positiones selectae novitatis singularitatis et improbabilitatis frustra appellatae* (Lyons, 1662); Amann, "Laxisme," col. 71.

110. J. Brodrick, *Robert Bellarmine: Saint and Scholar* (Westminster, Md., 1961), ch. 7.

111. J. Kilcullen, *Sincerity and Truth* (Oxford, 1988), 7; R. Knox, *Enthusiasm* (Oxford, 1951), 196.

112. A. Arnauld, *Théologie morale des jesuites*, in *Oeuvres* (Paris, 1779), 29:74–94, at 74.

113. L. Cognet, intro. to Pascal, *Les Provinciales*, xi; Amann, "Laxisme," cols. 44–45; cf. W. Shirer, *The Rise and Fall of the Third Reich* (London, 1962), 1048.

114. Cognet, intro. to Pascal, *Les Provinciales*, xii–xiii; R. Parish, *Pascal's "Provincial Letters": A Study in Polemic* (Cambridge, 1989), ch. 5; L. Ceyssens, "Que penser finalement de l'histoire du jansénisme et de l'antijansénisme?" *Revue d'histoire ecclésiastique* 88 (1993): 108–30.

115. Cognet, intro. to Pascal, *Les Provinciales*, xvi–xviii.

116. Ibid., xxxiii; 71 n. 1.

117. Pascal, *Pensées and the Provincial Letters*, 375; see P. Cariou, *Pascal et la casuistique* (Paris, 1993), ch. 2; D. Bell, "Pascal: Casuistry, probability, uncertainty," *Journal of Medieval and Early Modern Studies* 28 (1998): 37–50.

118. Amann, "Laxisme," cols. 42–43; cf. J. Brodrick, *The Economic Morals of the Jesuits* (New York, 1972), 42.

119. Pascal, *Pensées and the Provincial Letters*, 380–81.

120. Ibid., 381.

121. R. Duchêne, *L'imposture littéraire dans les Provinciales de Pascal* (Aix-en-Provence, 1985), 136–48, 160–84; K. Weiss, *P. Antonio de Escobar y Mendoza als Moraltheologe in Pascals Beleuchtung und im Lichte der Wahrheit* (Freiburg-im-Breisgau, 1911); A. Gazier, *Blaise Pascal et Antoine Escobar* (Paris, 1912); Brodrick,

Economic Morals of the Jesuits, 109–18; Jonsen and Toulmin, *Abuse of Casuistry*, 243–49; D. Descotes, *L'argumentation chez Pascal* (Paris, 1993), 298–99; M. Houle, "The Fictions of Casuistry and Pascal's Jesuit in 'Les provinciales'" (Ph.D. diss., UC San Diego, 1983); review in Parish, *Pascal's "Provincial Letters,"* 58–61.

122. Deman, "Probabilisme," col. 492; C.-A. Saint-Beuve, *Port-Royal* 3 ch. 9, ed. M. Leroy (Paris, 1953–55), 2:129.

123. A. de Escobar y Mendoza, *Liber theologiae moralis* (Lyons, 1644) Proemium, ex. 3 cap. 3, 15–16.

124. Brodrick, *Economic Morals of the Jesuits*, 119.

125. Jonsen and Toulmin, *Abuse of Casuistry*, 246.

126. Pascal, *Pensées and the Provincial Letters*, 442.

127. Ibid., 446; Escobar, *Liber theologiae moralis* tract. 1 ex. 8 cap. 1, 106.

128. Cognet, intro. to Pascal, *Les Provinciales*, xlvi.

129. R. Taveneaux, *La vie quotidienne des jansénistes* (Paris, 1973), 180–83; cf. P. Dear, "Miracles, experiments and the ordinary course of nature," *Isis* 81 (1990): 663–83.

130. Deman, "Probabilisme," col. 515.

131. Amann, "Laxisme," cols. 50–52; Jonsen and Toulmin, *Abuse of Casuistry*, 243.

132. Amann, "Laxisme," col. 82; B. Dolhagaray, s.v. "Fornication," *Dictionnaire de théologie catholique*, 6: pt. 1 (1947), cols. 600–611, at col. 601.

133. Knox, *Enthusiasm*, 181.

Chapter 5: Rhetoric, Logic, Theory

1. J. Jones, *The Law and Legal Theory of the Greeks* (Oxford, 1956), 136–49; on *eikota*, 140 n. 2.

2. Demosthenes, *Oration* 37.48; Plato, *Phaedo* 88d; Aristotle, *Metaphysics* 1000a10.

3. Herodotus, *History* 1, ch. 214; Plato, *Laws* 782d; cf. 839d; Plato, *Critias* 110d; Plato, *Theaetetus* 178e; Aristotle, *Topics* 151a29.

4. S. Sambursky, "On the possible and probable in ancient Greece," *Osiris* 12 (1956): 35–48, at 36–37; also in *Studies in the History of Statistics and Probability*, ed. M. Kendall and R. Plackett (London, 1971), 2:1–14.

5. Sophocles, *Electra* lines 1025–27, trans. R. Jebb, in *The Complete Greek Drama*, ed. W. Oates and E. O'Neill (New York, 1938), 1:530.

6. Thucydides, *Peloponnesian War* 1, ch. 121; cf. Herodotus, *History* 7, ch. 239; Polybius, *Histories* 7.7.4; Antiphon the Sophist, in H. Diels, *Die Fragmente der Vorsokratiker*, 9th ed. (Berlin, 1960), 87B60; Plato, *Euthyphro* 3a; Hippocrates, *On Ancient Medicine* ch. 3.

7. Plato, *Crito* 45d.

8. Plato, *Republic* 564a, trans. P. Shorey (London, 1956), adjusted; cf. *Crito* 53e; *Laches* 190d; *Theaetetus* 152b, 171c–d, 202d; *Protagoras* 311a; *Phaedo* 62e, 67a, 70b, 78c; *Symposium* 200a; *Phaedrus* 232e, 238e, 255e, 258c, 269d; *Philebus* 33b; *Timaeus* 40e; *Laws* 625b, 655d, 691a, 930a; *Republic* 334c, 372d, 407d.

9. See *Phaedo* 81d–82b, 90b; *Phaedrus* 270b, 276c; *Meno* 89b; *Theaetetus* 149c, e, 188d, 203c; *Philebus* 31d, 45b, 63b.

10. *Phaedrus* 273d; but see also *Xenophanes* frag. 35 (Plutarch, *Moralia* 746B), J. Lesher, "Xenophanes' Scepticism," *Phronesis* 23 (1978): 1–21; Sextus Empiricus, *Adversus Mathematicos* 7.110; Philo of Alexandria, *On Rewards and Punishments* 29; etymology of *eikos* discussed in K.-H. Hagstroem, *Les préludes antiques de la théorie des probabilités* (Stockholm, 1932), 22–23.

11. G. Kennedy, *The Art of Persuasion in Greece* (London, 1963), 35–39.

12. Homer, *Iliad* 13 line 68; Homer, *Odyssey* 1 line 215, 3 line 93, 16 lines 470–73; cf. Aeschylus, *Agamemnon* 269, 272–81; cf. Sophocles, *Electra* 774, and Euripides, *Rhesus* 94.

13. Hesiod, frag. 338, in *Fragmenta Hesiodea*, ed. R. Merkelbach and M. West (Oxford, 1967), 168; see Cicero, *Letters to Atticus* 7.18.4; Euripides, *Children of Herakles* lines 179–80; Aristophanes, *Wasps* line 725; Aeschylus, *Eumenides* 431, 435.

14. C. Kuebler, *The Argument from Probability in Early Attic Oratory* (Chicago, 1944), 7–8; cf. G. Kennedy, "The earliest rhetorical handbooks," *American Journal of Philology* 80 (1959): 169–78.

15. Aristotle, *Rhetoric* 1402a17–28; text translations from Aristotle from *The Works of Aristotle*, trans. W. Ross (Oxford, 1910–52), with adjustments; cf. Plato, *Phaedrus* 272e–73c; Kennedy, *Art of Persuasion in Greece*, 58–61, 129–31; R. Bonner, *Lawyers and Litigants in Ancient Athens* (Chicago, 1927), 226–28.

16. Cicero, *Brutus* xii.46; see G. Lloyd, *Magic, Reason and Experience* (Cambridge, 1979), 250–54; G. Lloyd, *Demystifying Mentalities* (Cambridge, 1990), 59–60, ch. 3.

17. Kuebler, *Argument from Probability in Early Attic Oratory*, 2–3, 15; cf. Antiphon, *On the Revolution*, in *The Older Sophists*, ed. R. Sprague (Columbia, S.C., 1972), 204.

18. Plato, *Gorgias* 447c.

19. Plato, *Phaedrus* 267a; Gorgias quoted in Athenaeus, *Deipnosophists* 505d; discussion in R. Wardy, *The Birth of Rhetoric: Gorgias, Plato and Their Successors* (London, 1996), chs. 1–2.

20. Gorgias, *Defence of Palamedes* sec. 9, in Sprague, *Older Sophists*, 56; cf. Cicero, *Topica* xx.76; see A. Long, "Methods of argument in Gorgias' *Palamedes*," in *The Sophistic Movement* (Athens, 1984), 233–41; Kuebler, *Argument from Probability in Early Attic Oratory*, 33, 63; similar in Gorgias, *Encomium of Helen* sec. 5, in Sprague, *Older Sophists*, 51; see Kuebler, *Argument from Probability*, 28–30; E. Schiappa, "Gorgias' *Helen* revisited," *Quarterly Journal of Speech* 81 (1995): 310–24; R. McComiskey, "Gorgias and the art of rhetoric," *Rhetoric Society Quarterly* 27 (4) (1997): 5–24; P. Woodruff, "Rhetoric and relativism: Protagoras and Gorgias," in *The Cambridge Companion to Early Greek Philosophy*, ed. A. Long (Cambridge, 1999), 290–310, at 296–98.

21. Thucydides, *Peloponnesian War* 8, ch. 68; Sprague, *Older Sophists*, 108; Cicero, *On Divination* 2.70.144.

22. Antiphon, *First Tetralogy* B sec. 3–8, see Sprague, *Older Sophists*, 139–141; see Kuebler, *Argument from Probability*, 43–46; M. Gagarin, "The nature of proofs in Antiphon," *Classical Philology* 85 (1990): 22–32; E. Carawan, "The Tetralogies and Athenian homicide trials," *American Journal of Philology* 114 (1993): 235–70.

23. Antiphon, *Speech V* sec. 25–28, 37, 43, 45, see Sprague, *Older Sophists*, 169–74; also see Kuebler, *Argument from Probability*, 53–57.

24. Aristotle, *Rhetoric* 1402a11–12; cf. *Poetics* 1461b12–15.

25. Gorgias, *Defence of Palamedes* sec. 24, see Sprague, *Older Sophists*, 60; cf. Antiphon, *Speech I* sec. 7, see Sprague, *Older Sophists*, 131; M. Untersteiner, *The Sophists*, trans. K. Freeman (Oxford, 1954), 203–4.

26. Plato, *Apology* 17a–18a; T. Lewis, "Parody and the argument from probability in the *Apology*," *Philosophy and Literature* 14 (1990): 359–66, with further refs. esp. S. Nadler, "Probability and truth in the *Apology*," *Philosophy and Literature* 9 (1985): 198–202.

27. Isaeus, *Oration* 4.18, cf. Isocrates, *Oration* 3.53, 18.15, 15.280; Demosthenes, *Oration* 44.38; W. Grimaldi, "*Semeion, tekmerion, eikos*, in Aristotle's *Rhetoric*," *American Journal of Philology* 101 (1980): 383–98, at 397 n. 12.

28. Lysias, *Oration* 7.24, in *Lysias*, trans. W. Lamb (London, 1930), 159.

29. Ibid., 7.38, page 165; cf. 3.23–39; 19.58; W. Fairchild, "Argument from probability in Lysias," *Classical Bulletin* 55 (1979): 49–54; K. Schoen, *Die Scheinargumente bei Lysias* (Paderborn, 1918).

30. Demosthenes, *Oration* 30.37.

31. Aristotle, *Rhetoric* 1356b29–30, also 32–35.

32. Ibid., 1354a15–18; 1355b14–16; 1402a26–28.

33. Ibid., 1355a35–38.

34. Ibid., 1357a22–b1.

35. Ibid., 1356b14–18.

36. Ibid., 1357b26–36; cf. 1393a32–b4.

37. Ibid., 1398a33–b20.

38. Ibid., 1357b18–21; 1401b9–14; cf. Antiphon, *Art of Speaking*, see Sprague, *Older Sophists*, 236; see also E. Madden, "Aristotle's treatment of probability and signs," *Philosophy of Science* 24 (1957): 167–72; S. Raphael, "Rhetoric, dialectic and syllogistic argument: Aristotle's position in Rhetoric I–II," *Phronesis* 19 (1974): 153–67; L. Arnhart, *Aristotle on Political Reasoning* (De Kalb, 1981), 43–46; Grimaldi, "*Semeion, tekmerion, eikos*, in Aristotle's *Rhetoric*."

39. Aristotle, *Rhetoric* 1368a29–31.

40. Ibid., 1394a5–8; cf. 1393a1–8.

41. Ibid., 1397b14–17.

42. Ibid., 1392b24–26 and 31–32.

43. Ibid., 1402b20–1403a1; further at 1403a6–10.

44. See L. Self, "Rhetoric and *phronesis*: The Aristotelian ideal," *Rhetoric and Philosophy* 12 (1979): 130–45.

45. Aristotle, *Rhetoric* 1376a17–23.

46. Ibid., 1376b31–1377a4.

47. Ibid., 1400a5–9.

48. Aristotle, *Posterior Analytics* 91b15–16, 35–36; possibly Aristotle, *Topics* 105a13–16, 157a35.

49. Aristotle, *Prior Analytics* 68b15–29.

50. Ibid., 68b38–69a14.

51. For example, Aristotle, *Posterior Analytics* 90a5–23 and 93a30–b15.

52. Ibid., 77a41, 79a14, 88b12; Aristotle, *Generation of Animals* 721b27–30.

53. Aristotle, *Prior Analytics* 70a4–7.

54. Cf. ibid., 70a33–39; *Topics* 112b1–2; such arguments as a grade of possibility in *On Interpretation* 19a18–23; deductive arguments involving "most" in *Posterior Analytics* 96a8–19; cf. M. Mignucci, "*Hōs epi to poly* et nécessaire dans la conception aristotelicienne de la science," in *Aristotle on Science: The Posterior Analytics*, ed. E. Berti (Padua, 1981), 173–203; M. Winter, "Aristotle, *hōs epi to polu* relations, and a demonstrative science of ethics," *Phronesis* 42 (1997): 163–89.

55. *Topics* 100a30–b25; cf. *On Sophistical Refutations* 165b4; see also J. Evans, *Aristotle's Concept of Dialectic* (Cambridge, 1971), 78–79; J. Le Blond, *Logique et Methode chez Aristote*, 3d ed. (Paris, 1973), 9–16; R. Smith, "Aristotle on the uses of dialectic," *Synthese* 96 (1993): 335–58.

56. Boethius *Topica*, in *Patrologia Latina*, ed. J. Migne, 64: col. 910D.

57. Aristotle, *Topics* 108b12–14, 156b10–18, 157a26–29.

58. Aristotle, *Poetics* 1461b12–15, 17–21; cf. S. Halliwell, *Aristotle's "Poetics"* (London, 1986), 101–6; N. O'Sullivan, "Aristotle on dramatic probability," *Classical Journal* 91 (1995): 47–63.

59. J. Barnes, ed., *The Complete Works of Aristotle* (Princeton, 1984), 2:2431; cf. Plotinus, *Enneads* I.1.12.35, 4.3.14.5.

60. E. Cope, *An Introduction to Aristotle's "Rhetoric"* (London, 1867), 402.

61. [Aristotle], *Rhetoric to Alexander* 1428a25 b32, trans. E. S. Forster, in *The Works of Aristotle*, ed. W. D. Ross, vol. 11, with adjustments.

62. Ibid., 1428b40–1429a10.

63. Ibid., 1429a20–b34.

64. Ibid., 1430b31–1431a3.

65. Ibid., 1431b22.

66. Cicero, *De Inventione* 1.46–48, trans. H. Hubbell (London, 1949), 85–89, with adjustments; cf. 1.29; *De Partitione Oratoria* x.34 and xi.39–40; *Pro Cluentio* 174; *De Republica* 1.xxxviii.59; L. Montefusco, "Cicero's technical treatment of argument in *De Inventione: Omnis autem argumentatio . . . aut probabilis aut necessaria esse debebit* (I.44)," *Rhetorica* 16 (1998): 1–24.

67. S. Brewer, "Exemplary reasoning: Semantics, pragmatics, and the rational force of legal argument by analogy," *Harvard Law Review* 109 (1996): 923–1028.

68. [Cicero], *Rhetorica ad Herennium* 2.v.8–2.vi.9; cf. 2.ii.3; A. Greenidge, *The Legal Procedure of Cicero's Time* (New York, 1971), 482.

69. [Cicero], *Rhetorica ad Herennium* 4.xli.53; cf. 2.vii.11.

70. Quintilian, *Institutio oratoria* 7.2.25; cf. 7.2.29–30; 5.10.19; Aulus Gellius, *Attic Nights* 14.2.20–26.

71. Quintilian, *Institutio oratoria* 5.9.8–12; cf. 2.17.39; 5.12.5; P. Meador, "Minucian, *On Epicheiremes:* An introduction and a translation," *Speech Monographs* 31 (1964): 54–63.

72. Quintilian, *Institutio oratoria* 5.10.15–17; cf. D. Garber and S. Zabell, "On the emergence of probability," *Archive for History of Exact Sciences* 21 (1979): 33–53, at 42–46; on the vocabulary of *probabilis* and *versimilis*, Quintilian, *Institutio oratoria* 4.2.31.

73. Seneca, *Epistulae* 99.12–13.

74. Marius Victorinus, *In De Inventione* 1.29, in *Rhetores Latini Minores*, ed. C. Halm (Frankfurt, 1964), 236; cf. C. Julius Victor, *Ars rhetorica* cap. 8, in Halm, *Rhetores Latini Minores*, 408–9.

75. Boethius, *De topicis differentiis* 1, in Migne, *Patrologia Latina*, 64: cols. 1180C–1181B, trans. E. Stump (New York, 1978), 39–40.

76. Boethius, *De topicis differentiis* 2, in Migne, *Patrologia Latina*, 64: cols. 1183D–1184D.

77. D. Black, *Logic and Aristotle's Rhetoric and Poetics in Medieval Arabic Philosophy* (Leiden, 1990), chs. 1–2.

78. K. Erickson, *Aristotle's Rhetoric* (Metuchen, 1975), 9; see M. Lyons, ed., *Aristotle's "Ars Rhetorica": The Arabic Version* (Cambridge, 1982); J. Watt, "From Themistius to Alfarabi: Platonic political philosophy and Aristotle's *Rhetoric* in the East," *Rhetorica* 13 (1995): 17–41.

79. J. Murphy, "Aristotle's *Rhetoric* in the Middle Ages," *Quarterly Journal of Speech* 52 (1966): 109–15, at 110; cf. *Al-Farabi's Short Commentary on Aristotle's "Prior Analytics,"* trans. N. Rescher (Pittsburgh, 1963), 94; *Al-Farabi's Commentary and Short Treatise on Aristotle's "De Interpretatione,"* trans. F. Zimmerman (London, 1981), 89–90, 92; K. Gyekye, "Al-Farabi on the logic of the arguments of the Muslim philosophical theologians," *Journal of the History of Philosophy* 27 (1989): 135–43, at 138–39.

80. *Avicenna's Treatise on Logic*, trans. F. Zabeeh (The Hague, 1971), 38–39.

81. *Averroes' Three Short Commentaries on Aristotle's "Topics," "Rhetoric," and "Poetics,"* trans. C. Butterworth (Albany, 1977), 75; cf. Averroes, *On the Harmony of Religion and Philosophy*, trans. G. Hourani (London, 1961), 64–65; on induction and dialectic, Averroes, *Three Short Commentaries*, 50; *Averroes' Middle Commentary on Aristotle's "Categories" and "De Interpretatione,"* trans. C. Butterworth (Princeton, 1983), 146–47; also Averroes, *The Decisive Treatise Determining the Nature of the Connection between Religion and Philosophy*, in *Philosophy in the Middle Ages*, ed. A. Hyman and J. Walsh (New York, 1967), 287–306, at 291, 297–301.

82. Black, *Logic and Aristotle's Rhetoric and Poetics*, ch. 5; Hallaq, "On inductive corroboration, probability, and certainty," 3–31; B. Weiss, "Knowledge of the past: The theory of *Tawatur* according to Ghazali," *Studia Islamica* 61 (1985): 81–105.

83. *St. Anselm's "Proslogion,"* trans. M. Charlesworth (Oxford, 1965).

84. A. Luddy, *The Case of Peter Abelard* (Westminster, Md., 1947), 87, 89; R. Naulty, "St. Bernard's lost opportunity," *Compass Theology Review* 16 (2) (1982): 32–41.

85. N. Green-Pedersen, *The Tradition of the Topics in the Middle Ages* (Munich, 1984), 142–43, 161; Garlandus Compotista, *Dialectica*, ed. L. de Rijk (Assen, 1959), 93–94; B. Dod, "Aristoteles latinus," in *Cambridge History of Later Medieval Philosophy*, ed. N. Kretzmann, A. Kenny, and J. Pinborg (Cambridge, 1982), ch. 2.

86. J. Ward, "From antiquity to the Renaissance: Glosses and commentaries on Cicero's *Rhetorica,"* in *Medieval Eloquence*, ed. J. Murphy (Berkeley, 1978), 25–67; J. Murphy, "Cicero's rhetoric in the Middle Ages," *Quarterly Journal of Speech* 53 (1967): 334–41; M. Dickey, "Some commentaries on the *De Inventione* and *Ad Herennium* of the eleventh and early twelfth centuries," *Mediaeval and Renaissance Studies* 6 (1968): 1–41; K. Fredborg, "Petrus Helias on rhetoric," *Université de Copenhague Cahiers de l'Institut du Moyen-Age Grec et Latin* 13 (1974): 31–41; K. Fredborg, "The commentaries on Cicero's *De Inventione* and *Rhetorica ad Herennium* by William of Champeaux," *Université de Copenhague Cahiers de l'Institut du Moyen-Age Grec et Latin* 17 (1976): 1–39; J. Ward, "The date of the commentary on Cicero's *De Inventione* by Thierry of Chartres and the Cornifician attack on the liberal arts," *Viator* 3 (1972): 219–73; K. Fredborg, "The commentary of Thierry of Chartres on Cicero's *De Inventione,"* *Université de Copenhague Cahiers de l'Institut du Moyen-Age Grec et Latin* 7 (1971): 1–36.

87. Pictured in C. Brooke, *The Twelfth Century Renaissance* (London, 1969), 23.

88. Abelard, *Dialectica*, ed. L. de Rijk (Assen, 1956), 277–78; cf. 271–72, 460–61; Green-Pedersen, *Tradition of the Topics in the Middle Ages*, 214, 215 n. 17; O. Bird, "The formalizing of the topics in medieval logic," *Notre Dame Journal of Symbolic Logic* 1 (1960): 138 49; L. Minio-Paluello, *Twelfth Century Logic* (Rome, 1956), 1:14; L. de Rijk, *Logica Modernorum* (Assen, 1962), 69, 542–43, 614.

89. Green-Pedersen, *Tradition of the Topics in the Middle Ages*, 214; on arguments from authority, Abelard, *Dialectica* 438–39.

90. For example, Peter of Spain, *Summulae Logicales*, ed. L. de Rijk (Assen, 1972), 56–57, 90–91, 94; *William of Sherwood's "Introduction to Logic,"* trans. N. Kretzmann (Minneapolis, 1966), 105; R. Bacon, *Summulae Dialectices* III, ed. A. de Libera, in *Archives d'histoire doctrinale et littéraire du moyen-âge* 54 (1987): 171–278, at 220, 234–35, 269; Simon of Faversham, *Quaestiones super libro elenchorum*, ed. S. Ebbesen et al. (Toronto, 1984), 196–97; cf. W. Ong, *Ramus, Method, and the Decay of Dialogue* (Cambridge, Mass., 1958), 61; C. Lafleur, *Quatre introductions à la philosophie au XIIIe siècle* (Montreal, 1988), 241, 281, 283–84, 344, 346; E. Stump, *Dialectic and Its Place in the Development of Medieval Logic* (Ithaca, 1989), 153, 161–62; further refs. in A. Crombie, *Styles of Scientific Thinking in the European Tradition* (London, 1994), 2:1520 n. 72.

91. J. Monfasoni, "Humanism and rhetoric," in *Renaissance Humanism*, ed. A. Rabil (Philadelphia, 1988), 171–235, at 173.

92. J. Murphy, *Rhetoric in the Middle Ages* (Berkeley, 1974), 90–101.

93. Giles of Rome, *Commentaria in rhetoricam Aristotelis* (Frankfurt, 1968), in S. Robert, "Rhetoric and dialectic: According to the first Latin commentary on the *Rhetoric* of Aristotle," *New Scholasticism* 31 (1957): 484–98, at 491, 492 n. 25; also in K. Erickson, ed., *Aristotle: The Classical Heritage of Rhetoric* (Metuchen, 1974), 90–101; J. O'Donnell, "The Commentary of Giles of Rome on the *Rhetoric* of Aristotle," in *Essays in Medieval History Presented to Bertie Wilkinson*, ed. T. Sandquist and M. Powicke (Toronto, 1969), 138–56; further refs. in J. Murphy, *Medieval Rhetoric: A Select Bibliography* (Toronto, 1971), 35–36; cf. Giles of Rome, *De Differentia Rhetoricae, Ethicae et Politicae* pt. 1, trans. J. Miller, in *Readings in Medieval Rhetoric*, ed. J. Miller, M. Prosser, and T. Benson (Bloomington, 1973), 265–68.

94. W. Boggess, "Hermannus Alemannus' rhetorical translations," *Viator* 2 (1971): 227–50; M. Grabmann, "Eine lateinische Übersetzung der pseudo-aristotelischen Rhetorica ad Alexandrum aus dem 13. Jahrhundert," *Sitzungsberichte der Bayerischen Akademie der Wissenschaften* 4 (1931–32): 3–81; G. Lacombe et al., eds., *Aristoteles Latinus: Codices* (Rome, 1939), 1:78–79; cf. J. Hackett, "Moral philosophy and rhetoric in Roger Bacon," *Philosophy and Rhetoric* 20 (1987): 18–40, at 27, 30–31.

95. K. Fredborg, "Buridan's *Quaestiones super Rhetoricam Aristotelis*," in *The Logic of John Buridan*, ed. J. Pinborg (Copenhagen, 1976), 47–59; cf. C. Michalski, *La philosophie au XIVe siècle* (Frankfurt, 1969), 82–83; M. Herrick, "The early history of Aristotle's *Rhetoric* in England," *Philological Quarterly* 5 (1926): 242–57; J. Banker, "The *ars dictaminis* and rhetorical textbooks at the Bolognese university in the 14th century," *Medievalia et Humanistica* 5 (1974): 153–68; S. Karaus Wertis, "The commentary of Bartolinus de Benincasa de Canulo on the *Rhetorica ad Herennium*," *Viator* 10 (1979): 283–310.

96. Albert the Great, *In Lib. I Topicorum* tract. 1 cap. 2, in Borgnet, *Opera omnia* 2:241.

97. Albert the Great, *In Lib. I Topicorum* tract. 1 cap. 5, in Borgnet, *Opera omnia* 2:247–48.

98. Albert the Great, *In Lib. II Posteriorum analyticorum* tract. 3 cap. 7, in Borgnet, *Opera omnia* 2:206–7; cf. *In Lib. I Ethicorum* tract. 4 cap. 3, in Borgnet, *Opera omnia* 7:53; *In Lib. II Priorum analyticorum* tract. 7 cap. 4, in Borgnet, *Opera omnia* 1:795; J. Shaw, "Albertus Magnus and the rise of an empirical approach in medieval science and philosophy," in *By Things Seen*, ed. D. Jeffrey (Ottawa, 1979), 175–85.

99. Aquinas, *Summa theologiae* II-II q. 60 art. 3.

100. Ibid., I-II q. 105 art. 2 ad 8; T. Aquinas, *Super evangelium Johannis* cap. 8 lec. 2, quoted in I. Kantola, *Probability and Moral Uncertainty in Late Medieval and Early Modern Times* (Helsinki, 1994), 36.

101. Aquinas, *Summa theologiae* II-II q. 70 art. 2; cf. E. Byrne, *Probability and*

Opinion: A Study in the Medieval Presuppositions of Post-Medieval Theories of Probability (The Hague, 1968), 203–5; A. Celano, "Peter of Auvergne's Questions on Books I and II of the *Ethica Nicomachea,*" *Mediaeval Studies* 48 (1986): 1–110, at 38.

102. Aquinas, *Summa theologiae* I–II q. 96 art. 1; see *Digest* 1.3.3.

103. Aquinas, *Summa theologiae* II–II q. 70 art. 2; q. 69 art. 2.

104. Ibid., I q. 1 art. 8 ad 2.

105. Ibid., II–II q. 49 art. 1.

106. Ibid., III q. 9 art. 3 ad 2.

107. Aquinas, *Commentary on Aristotle's "Metaphysics"* 4 lectio 4 n. 576; cf. Aquinas, *Commentary on Aristotle's "De Caelo"* 1 lectio 2 n. 14; Ong, *Ramus, Method, and the Decay of Dialogue,* 162.

108. Refs. in Stump, *Dialectic,* 137 n. 10.

109. Aquinas, *Summa theologiae* 1 q. 12 art. 7 ad 3; 1 q. 14 art. 3.

110. Aquinas, foreword to *Commentary on Aristotle's "Posterior Analytics"*; cf. L. Minio-Paluello, *Opuscula: The Latin Aristotle* (Amsterdam, 1972).

111. Aquinas, *Summa theologiae* 111 q. 70 art. 4 ad 2; 111 q. 37 art. 7; T. Deman, "Notes de lexicographie philosophique médiévale: *Probabilis,*" *Revue des sciences philosophiques et théologiques* 22 (1933): 260–90.

112. J. Franklin, "Mental furniture from the philosophers," *Et Cetera* 40 (1983): 177–91; R. Evans et al., "The notion of vernacular theory," in *The Idea of the Vernacular: An Anthology of Middle English Literary Theory, 1280–1520,* ed. J. Wogan-Browne (University Park, Pa., 1998), 314–30.

113. A. Campbell, *The Black Death and Men of Learning* (New York, 1966), 177–78; D. Herlihy, *The Black Death and the Transformation of the West* (Cambridge, Mass., 1997).

114. B. Latini, *Li livres dou tresor* 3 pt. 1 ch. 50, ed. P. Chabaille (Paris, 1863), 539–42; similar in Italian in Boccaccio, *Filostrato* pt. 5 ott. 137.

115. J. Knops, *Etude sur la traduction française de la Morale à Nicomache d'Aristote par Nicole Oresme* (The Hague, 1952); R. Taylor, "Les néologismes chez Nicole Oresme, traducteur du XIVe siècle," *Actes du Xe Congrès International de Linguistique et Philologie Romanes, 1962,* ed. G. Straka (Paris, 1965), 2:589–604; F. Meissner, "Maistre Nicolas Oresme et la lexicographie française," *Cahiers de lexicologie* 40 (1982): 51–66; M. Cantor, *Vorlesungen über Geschichte der Mathematik* (Leipzig, 1913), 2:129; S. Babbitt, "Oresme's *Livre de Politiques* and the France of Charles V," *Transactions of the American Philosophical Society* 75 (1985), pt. 1,10; C. Sherman, *Imaging Aristotle: Verbal and Visual Representation in Fourteenth-Century France* (Berkeley, 1995).

116. N. Oresme, *Le Livre de Ethiques d'Aristote* 2 ch. 2, ed. A. Menut (New York, 1940); N. Oresme, *Le Livre de Politiques d'Aristote,* ed. A. Menut, in *Transactions of the American Philosophical Society* 60 (1970), pt. 6, at 56, 380.

117. R. Higden, *Polychronicon* (with Trevisa's translation), ed. C. Babington and J. Lumby (London, 1865–86), 1:339; similar at 2:71; on Trevisa see D. Fowler, *The Life and Times of John Trevisa, Medieval Scholar* (Seattle, 1995).

118. H. Kurath, ed., *Middle English Dictionary* (Ann Arbor, 1954–), s.v. "likli," "liklihod," "liklines," "probabilite," "probable."

119. R. Pecock, *The Folewer of the Donet* (ca. 1454), ed. E. Hitchcock, Early English Text Series 164 (London, 1924), 70; similar in R. Pecock, *The Repressor of Over Much Blaming of the Clergy*, ed. C. Babington, Rolls Series 19 (London, 1860–61), 42, 77–78, 133; L. Mooney, "A Middle English treatise on the seven liberal arts," *Speculum* 68 (1993): 1037–52, at 1041.

120. J. Hartigan, "Similarity and probability," in *Foundations of Statistical Inference*, ed. V. Godambe and D. Sprott (Toronto, 1971), 305–11.

121. V. Kahn, "Giovanni Pontano's rhetoric of prudence," *Philosophy and Rhetoric* 16 (1983): 16–34; L. Jardine, "Humanism and dialectic in sixteenth century Cambridge," in *Classical Influences on European Culture, AD 1500–1700*, ed. R. Bolgar (Cambridge, 1976), 141–54; L. Jardine, "The place of dialectic teaching in sixteenth century Cambridge," *Studies in the Renaissance* 21 (1974): 31–62; J. McNally, "*Rector et dux populi:* Italian humanists and the relationship between rhetoric and logic," *Modern Philology* 67 (1969): 168–76; D. Patey, *Probability and Literary Form* (Cambridge, 1984), 13, 17–18, 25; C. Noreña, *Juan Luis Vives* (The Hague, 1970), 267–68, 279–80; C. Armstrong, "The dialectical road to truth: The dialogue," in *French Renaissance Studies*, ed. P. Sharratt (Edinburgh, 1976), 36–51, at 42–45; C. Trinkaus, *In Our Image and Likeness* (London, 1970), 1:319; C. Schmitt, ed., *Cambridge History of Renaissance Philosophy* (Cambridge, 1988), 180–84; Melanchthon, *Liber de anima*, in *Philippi Melanchthoni opera quae supersunt omnia* (Halle, 1834–60), 13 col. 166; P. Fabri, *Le grand et vrai art de pleine rhétorique* (Geneva, 1969), 1:95–100; T. Tasso, *Discourses on the Heroic Poem*, trans. M. Cavalchini and I. Samuel (Oxford, 1973), 30.

122. P. Brandes, *A History of Aristotle's "Rhetoric"* (Metuchen, 1989); Herrick, "Early history of Aristotle's *Rhetoric*"; T. Conley, "Some Renaissance Polish commentaries on Aristotle's *Rhetoric* and Hermogenes' *On Ideas*," *Rhetorica* 12 (1994): 265–92; L. Osborn, *The Life, Letters, and Writings of John Hoskyns*, Yale Studies in English 87 (New Haven, 1937), 155; cf. A. Fraunce, *The Lawiers Logike* (London, 1588), fol. 5–6.

123. Patey, *Probability and Literary Form*, 80–82; B. Weinberg, *A History of Literary Criticism in the Italian Renaissance* (Chicago, 1961), 1:6, 8, 33, 391, 508, 531; 2:754–55, 762–64, 889.

124. Savonarola, *On the Division of the Sciences*, quoted in C. Greenfield, *Humanist and Scholastic Poetics, 1250–1500* (Lewisburg, 1981), 256.

125. R. Agricola, *De Inventione Dialectica* 2 ch. 2, in J. McNally, "Rudolph Agricola's *De Inventione Dialectica Libri Tres:* A translation of selected chapters," *Speech Monographs* 34 (1967): 393–422, at 408; see P. Mack, *Renaissance Argument: Valla and Agricola in the Traditions of Rhetoric and Dialectic* (Leiden, 1993), ch. 9.

126. Ong, *Ramus, Method, and the Decay of Dialogue*, 94.

127. R. Wagner, "Thomas Wilson's *Arte of Rhetorique*," *Speech Monographs* 27 (1960): 1–32; probability briefly in the *Logique*, see T. Wilson, *The Rule of Rea-*

son, Conteinying the Arte of Logique, ed. R. Sprague (Northridge, Calif., 1972), 77–82.

128. T. Wilson, *The Arte of Rhetorique* (1553, repr. Amsterdam, 1969), fols. 51–52; cf. R. Lever, *The Art of Reason, rightly termed, Witcraft* (London, 1573), 189–90; T. Blundeville, *The Art of Logike* (London, 1599), 87–88; see Patey, *Probability and Literary Form*, 38, 51.

129. L. Stephen and S. Lee, *Dictionary of National Biography* (London, 1912), s.v. "Thomas Wilson," at 21:605; G. Donaldson, *The First Trial of Mary Queen of Scots* (London, 1969), 220.

130. P. da Fonseca, *Institutionum dialecticarum libri octo* (Cologne, 1605), 372–74; F. de Toledo, *Commentaria, una cum quaestionibus, in universam Aristotelis logicam* (Cologne, 1607), 281–83, 405–8; F. de Oviedo, *Cursus Philosophicus*, 2d ed. (Lyons, 1651), 2:61–62; R. de Arriaga, *Cursus Philosophicus* (Lyons, 1669), 273–74; B. Mastrius and B. Bellutus, *Philosophiae ad mentem Scoti Cursus Integer* (Venice, 1708), 52–56; T. Spencer, *The Art of Logick* (London, 1628), 288–90; cf. P. Dear, *Mersenne and the Learning of the Schools* (Ithaca, 1988), 30; M. Reif, "Natural philosophy in some early seventeenth century Scholastic textbooks" (Ph.D. diss., St. Louis Univ., 1962), 290–94.

131. T. Compton, alias Carleton, *Philosophia Universa* (Antwerp, 1649), 195; *Disputationes in universam Aristotelis logicam* (Salamanca, 1716), 483–84.

132. L. Carbone, *Introductio in logicam* (Venice, 1597), 170–73, discussed in W. Wallace, *Galileo's Logic of Discovery and Proof* (Dordrecht, 1991), 120–29; J. Dietz Moss, *Novelties in the Heavens: Rhetoric and Science in the Copernican Controversy* (Chicago, 1993), 7–9; cf. Wallace, *Galileo's Logical Treatises*, 91.

133. J. Jungius, *Logica hamburgensis* (Hamburg, 1638), 511–12.

134. Eustachio a Sancto Paulo, *Summa Philosophiae Quadripartita* (Paris, 1609), pt. 1:253.

135. M. Smiglecki, *Logica* (Oxford, 1638), 661.

136. *The Material Logic of John of St. Thomas*, trans. Y. Simon et al. (Chicago, 1955), 527–29.

137. F. Burgersdijk (Burgersdicius), *Institutionum logicarum* 1, ch. 31 (Cambridge, 1666) [*Early English Books*, 1641–1700, reel 811, item 3], 97; also see 2, ch. 11, 144; 2, ch. 15, 156.

138. S. Morison, *The Founding of Harvard College* (Cambridge, Mass., 1935), 67; P. Miller, *The New England Mind: The Seventeenth Century* (Cambridge, Mass., 1954), 103, 118, 122; M. Feingold, "The ultimate pedagogue: Franco Petri Burgersdijk and the English speaking academic learning," in *Franco Burgersdijk (1590–1635): Neo-Aristotelianism in Leiden*, ed. E. Bos and H. Krop (Amsterdam, 1993), 151–65.

Chapter 6: Hard Science

1. P. Duhem, *The Aim and Structure of Physical Theory*, trans. P. Wiener (New York, 1962), 183; see D. Gillies, *Philosophy of Science in the Twentieth Century* (Oxford, 1993), ch. 5.

2. S. Stigler, *The History of Statistics* (Cambridge, Mass., 1986), ch. 1.

3. L. Tarán, *Parmenides* (Princeton, 1965), 296–98; A. Coxon, *The Fragments of Parmenides* (Assen, 1986), 229; Aristotle, *On the Heavens* 297b24–31; O. Neugebauer, *A History of Ancient Mathematical Astronomy* (Berlin, 1975), 109–12.

4. Ptolemy, *Almagest* 1 ch. 4, trans. G. Toomer (London, 1984), 40.

5. Diodorus Siculus, *History* 1 ch. 39; cf. Herodotus, *History* 2 ch. 23; Strabo, *Geography* 13 ch. 3.1–2; Plutarch, *Moralia* 951F.

6. Aristotle, *Generation of Animals* 760b30–33; generally in G. Lloyd, "Experiment in early Greek philosophy and medicine," in *Methods and Problems in Greek Science* (Cambridge, 1991), ch. 4.

7. Diodorus Siculus, *History* 3 chs. 36–37; a modern parallel in R. Swinburne, "The paradoxes of confirmation: A survey," *American Philosophical Quarterly* 8 (1971): 318–29, at 326–27.

8. J. Franklin, "Non-deductive logic in mathematics," *British Journal for the Philosophy of Science* 38 (1987): 1–18.

9. Archimedes, *The Method*, supp. in T. Heath, *The Works of Archimedes* (Cambridge, 1912), 13.

10. Aristotle, *On the Heavens* 289b21–28; similar argument with a "likely" conclusion in *Meteorology* 346a23–31.

11. *Parts of Animals* 641b18–24.

12. *Physics* 198b25–199a1; cf. 199b15–25.

13. Ptolemy, *Almagest* 1 ch. 3 at 39.

14. Neugebauer, *History of Ancient Mathematical Astronomy*, 347–555, 577; N. Swerdlow, *The Babylonian Theory of the Planets* (Princeton, 1998); J. Steele, "Eclipse prediction in Mesopotamia," *Archive for History of Exact Sciences* 54 (2000): 421–54; on the Babylonians' fitting of parameters to the data, A. Aaboe, "Observation and theory in Babylonian astronomy," *Centaurus* 24 (1980): 14–35; G. Toomer, "Hipparchus' empirical basis for his empirical mean motions," *Centaurus* 24 (1980): 97–109; Chinese parallels in T. Deane, "Instruments and observation at the Imperial Astronomical Bureau during the Ming Dynasty," *Osiris* 9 (1994): 127–40.

15. A. Jones, "The adaptation of Babylonian methods in Greek numerical astronomy," *Isis* 82 (1991): 441–53; A. Jones, "Hipparchus's computations of solar longitudes," *Journal of the History of Astronomy* 22 (1991): 101–25.

16. J. Kepler, *Astronomia Nova*, trans. W. Donahue (Cambridge, 1992), 115.

17. A. Aaboe, "On the Babylonian origin of some Hipparchian parameters," *Centaurus* 4 (1955): 122–25; G. Toomer, "Hipparchus and Babylonian astronomy," in *A Scientific Humanist: Studies in Memory of Abraham Sachs*, ed. E. Leichty, M. DeJ. Ellis, and P. Gerardi (Philadelphia, 1988), 353–62.

18. Ptolemy, *Almagest* 7 ch. 1 at 321; B. Goldstein and A. Bowen, "The introduction of dated observations and precise measurement in Greek astronomy," *Archive for History of Exact Sciences* 43 (1991): 93–132.

19. Ptolemy, *Almagest* 3 ch. 1 at 132–36.

20. G. Toomer, "The chord table of Hipparchus and the early history of

Greek trigonometry," *Centaurus* 18 (1974): 6–28; N. Swerdlow, "Hipparchus on the distance of the sun," *Centaurus* 14 (1969): 287–305; cf. G. Toomer, "The size of the lunar epicycle according to Hipparchus," *Centaurus* 12 (1967): 145–50; G. Toomer, "Hipparchus on the distances of the sun and the moon," *Archive for History of Exact Sciences* 14 (1974): 126–42; see further O. Sheynin, "The treatment of observations in early astronomy," *Archive for History of Exact Sciences* 46 (1993): 153–92, § 2; W. Hartner, "The role of observations in ancient and medieval astronomy," *Journal for the History of Astronomy* 8 (1977): 1–11; B. Goldstein, "Saving the phenomena: The background to Ptolemy's planetary theory," *Journal for the History of Astronomy* 28 (1997): 1–12; related issues in J. Mancha, "Heuristic reasoning: Approximation procedures in Levi ben Gerson's astronomy," *Archive for History of Exact Sciences* 49 (1998): 1–49.

21. Ptolemy, *Almagest* 3 ch. 1 at 136–37; see R. Plackett, "The principle of the arithmetic mean," *Biometrika* 45 (1958): 130–35; also in *Studies in the History of Statistics and Probability*, ed. E. Pearson and M. Kendall (London, 1970), 1:121–26; cf. G. Hon, "Is there a concept of experimental error in Greek astronomy?" *British Journal for the History of Science* 22 (1989): 129–50, at 141–43; O. Sheynin, *The History of the Theory of Errors* (Egelsbach, 1996).

22. R. Newton, *The Crime of Claudius Ptolemy* (Baltimore, 1977); G. Lloyd, "Observational error in later Greek science," in *Science and Speculation*, ed. J. Barnes et al. (Cambridge, 1982), 128–64, at 138 n. 25, 151; also in Lloyd, *Methods and Problems in Greek Science*, ch. 13; review in N. Hetherington, "Ptolemy: on trial for fraud," *Astronomy and Geophysics* 38 (2) (1997): 24–27; also A. Smith, "Ptolemy's search for a law of refraction," *Archive for History of Exact Sciences* 26 (1982): 221–40.

23. G. Grasshoff, *The History of Ptolemy's Star Catalogue* (New York, 1990), 209–16; O. Gingerich, *The Eye of Heaven* (New York, 1993), 74–80; Sheynin, "The treatment of observations in early astronomy," § 3.4; J. Britton, *Models and Precision: The Quality of Ptolemy's Observations and Parameters* (New York, 1992); M. Riley, "Ptolemy's use of his predecessors' data," *Transactions of the American Philological Association* 125 (1995): 221–50.

24. Ptolemy, *Almagest* 13 ch. 2 at 600–601.

25. Proclus, *In Platonis Timaeum commentaria* 3, ed. E. Diehl (Amsterdam, 1965), 56, quoted in Lloyd, *Methods and Problems in Greek Science*, 161.

26. S. Ebbesen, "Ancient Scholastic logic as the source of medieval Scholastic logic," in Kretzmann, Kenny, and Pinborg, *Cambridge History of Later Medieval Philosophy*, ch. 4.

27. Alexander of Aphrodisias, *On the Principles of the All*, quoted in intro. to Maimonides, *Guide of the Perplexed*, trans. S. Pines (Chicago, 1963), lxix; cf. Simplicius, *In Aristotelis Physica Commentaria*, ed. H. Diels (Berlin, 1882), 18 (on *Physica* 184a16); Maimonides's (over?)-interpretation in *Guide of the Perplexed* pt. 2 chs. 3, 22; on reasoning from "for the most part" premises: Alexander of Aphrodisias, *Commentary on Aristotle's "Prior Analytics"* 1.3 and 1.13, ed. M. Wallies, *Commentaria in Aristotelem Graeca* 2:39–41, 165; see R. Sharples, "Alexander

of Aphrodisias on the compounding of probabilities," *Liverpool Classical Monthly* 7 (1982): 74–75; cf. Aristotle, *Metaphysics* 1064b36; *Prior Analytics* 32b11–14.

28. T. Stiefel, *The Intellectual Revolution in Twelfth-Century Europe* (London, 1985); B. Stock, *Myth and Science in the Twelfth Century* (Princeton, 1972); M. Dreyer, *More mathematicorum: Rezeption und Transformation der antiken Gestalten wissenschaftlichen Wissens im 12. Jahrhundert* (Münster, 1996); Marius, *On the Elements*, ed. and trans. R. Dales (Berkeley, 1977).

29. P. Duhem, *To Save the Phenomena*, trans. E. Doland and C. Maschler (Chicago, 1969); G. Lloyd, "Saving the appearances," in *Methods and Problems in Greek Science*, ch. 11.

30. *Al-Bitruji: On the Principles of Astronomy*, ed. and trans. B. Goldstein (New Haven, 1971); B. Goldstein, "Theory and observation in medieval astronomy," *Isis* 63 (1972): 39–47; G. Saliba, "Theory and observation in Islamic astronomy," *Journal for the History of Astronomy* 18 (1987): 35–43.

31. J. Pecham, *On the Sphere*, quoted in R. Avi-Yonah, "Ptolemy vs al-Bitruji: A study of scientific decision-making in the Middle Ages," *Archives internationales d'histoire des sciences* 35 (1985): 124–47, at 141; "probability" of theses in physics in Buridan, *Physics* 8 q. 12 § 7, quoted in H. Shapiro, *Medieval Philosophy* (New York, 1964), 534; J. Wyclif, *Tractatus de logica*, ed. M. Dziewicki (London, 1899), 3:3; Buridan, *In De Caelo et Mundo* 2 q. 7, in E. Moody, "John Buridan on the habitability of the earth," *Speculum* 16 (1941): 415–25, at 424; in optics, Albert of Saxony, *Quaestiones super quattuor libros de Celo et Mundo* 2 q. 22 (Venice, 1492); Blasius of Parma, *Quaestiones super perspectivam*, in D. Lindberg, *Theories of Vision from al-Kindi to Kepler* (Chicago, 1976), 130–31.

32. E. Grant, ed. and trans., *Nicole Oresme and the Kinematics of Circular Motion: Tractatus de commensurabilitate vel incommensurabilitate motuum celi* (Madison, 1971), 295; cf. L. Thorndike, *History of Magic and Experimental Science* (New York, 1934), 4:169; N. Steneck, *Science and Creation in the Middle Ages: Henry of Langenstein on Genesis* (Notre Dame, 1976), 92, 149.

33. L. Gillard, "Nicole Oresme, économiste," *Revue historique* 279 (1988): 3–39; K. Bales, "Nicole Oresme and medieval social science," *American Journal of Economics and Sociology* 42 (1983): 101–12; *Traité des monnaies (Nicolas Oresme) et autres écrits monétaires du XIVe siècle (Jean Buridan, Bartole de Sassoferrato)*, ed. C. Dupuy (Lyons, 1989); A. Lapidus, "Metal, money and the prince: John Buridan and Nicholas Oresme after Thomas Aquinas," *History of Political Economy* 29 (1997): 21–53.

34. N. Oresme, *De Moneta*, ed. and trans. C. Johnson (London, 1956), 44.

35. Babbitt, "Oresme's *Livre de Politiques* and the France of Charles V," sec. 4; E. Bridrey, *Nicole Oresme* (Geneva, 1978).

36. N. Oresme, *De proportionibus proportionum* and *Ad pauca respicientes*, ed. and trans. E. Grant (Madison, 1966), 246–47.

37. Ibid., 248–51; further on the increase of probability with increasing relative frequency at 252–55; a modern instance of this argument form in E. Jaynes,

"Information theory and statistical mechanics," *Physical Review* 106 (1957): 620–30, at 627.

38. Oresme, *De proportionibus proportionum* and *Ad pauca respicientes*, at 252–53.

39. Ibid., 384–85.

40. Ibid., 303–5, 61–63; N. Meusnier, "À propos de l'utilisation d'une argument probabiliste," in *Nicole Oresme: Tradition et innovation chez un intellectuel du XIVe siècle* (Paris, 1988), 165–77; cf. Thorndike, *History of Magic*, 4:117; more popular anti-astrology arguments from "induction from the past" in N. Oresme, *Livre de Divinacions*, ch. 8, in *Nicole Oresme and the Astrologers*, trans. G. Coopland (Cambridge, Mass., 1952), 70; cf. 74; S. Caroti, "Nicole Oresme's polemic against astrology in his *Quodlibeta*," in *Astrology, Science and Society*, ed. P. Curry (Woodbridge, 1987), 75–93; Thorndike, *History of Magic*, 3: ch. 25.

41. R. Dales, *The Scientific Achievement of the Middle Ages* (Philadelphia, 1973), 153; P. Duhem, *Le système du monde* (Paris, 1958), 8:445–48.

42. H. Poincaré, *Les méthodes nouvelles de la mécanique céleste* (New York, 1957), 3: ch. 26; cf. R. Small, "Incommensurability and recurrence: From Oresme to Simmel," *Journal of the History of Ideas* 52 (1991): 121–37; J. von Plato, "Oresme's proof of the density of rotations of a circle through an irrational angle," *Historia Mathematica* 20 (1993): 428–33.

43. M. Barnsley and S. Demko, eds., *Chaotic Dynamics and Fractals* (New York, 1985).

44. N. Oresme, *Le Livre du ciel et du monde*, ed. and trans. A. Menut and A. Denomy (Madison, 1968), 588.

45. Oresme, *Tractatus* 310–13; J. Kassler, "The emergence of probability reconsidered," *Archives internationales d'histoire des sciences* 36 (1986): 17–44, at 21–29; Kepler's agreement in J. Kepler, *Harmony of the World* 5 ch. 9; see B. Stephenson, *The Music of the Heavens: Kepler's Harmonic Astronomy* (Princeton, 1994), 188.

46. B. Hansen, *Nicole Oresme and the Marvels of Nature* (Toronto, 1985), 361, 97; cf. 393.

47. Ibid., 223, 269–71, 254–57.

48. Oresme, *Livre du ciel et du monde* 2 ch. 25, 532–37; selection in E. Grant, *A Source Book in Medieval Science* (Cambridge, Mass., 1974), 508–9; cf. Buridan, *Quaestiones super libris quattuor De caelo et mundo* 2 q. 22, selections in Grant, *Source Book*, 501.

49. Oresme, *Livre du ciel et du monde*, 536–39; selection in Grant, *Source Book*, 510.

50. Grant, *Nicole Oresme and the Kinematics of Circular Motion*, 132 n. 122.

51. E. Rosen, *Copernicus and the Scientific Revolution* (Malabar, 1984), ch. 4; R. Lemay, "The late medieval astrological school at Cracow and the Copernican system," *Studia Copernicana* 16 (1978): 337–54.

52. N. Copernicus, *De Revolutionibus orbium coelestium*, trans. A. Duncan (Newton Abbot, 1976), 24–25; cf. E. Rosen, trans., *Three Copernican Treatises*, 2d ed. (New York, 1959), 59, 99.

53. Copernicus, *De Revolutionibus orbium coelestium* 51; cf. A. Burdick, "Faithful witness: The empirical method intrudes on medieval medicine," *Sciences* 30 (1) (Jan.–Feb. 1990): 34–35.

54. Copernicus, *De Revolutionibus orbium coelestium* 176–77; see Y. Maeyama, "Determination of the sun's orbit: Hipparchus, Ptolemy, al-Battani, Copernicus, Tycho Brahe," *Archive for History of Exact Sciences* 52 (1998): 13–50.

55. K. Kirk, *Conscience and Its Problems* (London, 1927), 264–65; see also R. Fülöp-Miller, *The Power and Secret of the Jesuits*, trans. F. Flint and D. Tait (New York, 1956), 188.

56. Copernicus, preface to *De Revolutionibus orbium coelestium* 22; M. Kokowski, "Copernicus and the hypothetico-deductive method of correspondence thinking," *Theoria et Historia Scientiarum* 5 (1996): 7–101.

57. See B. Nelson, "The early modern revolution in science: Fictionalism, fideism, and Catholic 'prophetism,'" *Boston Studies in the Philosophy of Science* 3 (1967): 1–39.

58. J. Kepler, *Epitome of Copernican Astronomy* 4 pt. 1 ch. 3, trans. C. Wallis in *Encyclopaedia Britannica Great Books of the Western World*, 16:861; Kepler, *Astronomia Nova*, trans. W. Donahue (Cambridge, 1992), 51.

59. Copernicus, *Epitome of Copernican Astronomy* 4 pt. 2 ch. 5 at 911–13.

60. Ibid., 914–16.

61. See R. Westman, "Kepler's theory of hypothesis and the 'realist dilemma,'" *Studies in History and Philosophy of Science* 3 (1972): 233–64, esp. 249–61; A. Petroni, *I modelli, l'invenzione e la conferma: saggio su Keplero, la rivoluzione copernicana e la "New philosophy of science"* (Milan, 1990).

62. Kepler, *Astronomia Nova*, 286; see G. Hon, "On Kepler's awareness of the problem of experimental error," *Annals of Science* 44 (1987): 545–91, at 564.

63. Hon, "On Kepler's awareness," 551–53.

64. O. Gingerich, "The computer versus Kepler," *American Scientist* 52 (1964): 218–26; O. Gingerich, "Kepler's treatment of redundant observations," in *Internationales Kepler-Symposium, Weil-der-Stadt, 1971*, ed. F. Krafft et al. (Hildesheim, 1973), 307–14; D. Whiteside, "Kepler's planetary eggs," *Journal for the History of Astronomy* 5 (1974): 1–21.

65. O. Sheynin, "J. Kepler as a statistician," *Bulletin of the International Statistical Institute* 46 (1975): 2, 341–54, at 344; I. Schneider, "Wahrscheinlichkeit und Zufall bei Kepler," *Philosophia Naturalis* 16 (1976): 40–63, at 42–44; Sheynin, "Treatment of observations in early astronomy," § 5.3.

66. Kepler, *Astronomia Nova*, 47; cf. Dietz Moss, *Novelties in the Heavens*, 70–73; C. Wilson, "From Kepler's laws, so-called, to universal gravitation," *Archive for History of Exact Sciences* 6 (1969–70): 89–170, at 99–100.

67. Kepler, *Astronomia Nova*, 51; cf. 68.

68. *Keplers Gesammelte Werke* (Munich, 1937–), 8:274; *Mysterium Cosmographicum* cap. xiii, in *Keplers Gesammelte Werke* 1:43; cf. cap. xii at 1:42, trans. A. Duncan (New York, 1981), 149, 137; see N. Jardine, *The Birth of History and Philosophy of Science: Kepler's Defense of Tycho against Ursus* (Cambridge, 1984),

251–52; Schneider, "Wahrscheinlichkeit und Zufall bei Kepler"; J. Field, *Kepler's Geometrical Cosmology* (Chicago, 1988), 64–70.

69. Kepler, preface to *Epitome of Copernican Astronomy* 4, at 847.

70. Ibid., 4 pt. 2 ch. 6 at 918–19; cf. 4 pt. 3 ch. 4 at 950–51.

71. Kepler, *Harmony of the World* 5 ch. 4; see Stephenson, *Music of the Heavens*, 145–54.

72. Kepler, *Harmony of the World* ch. 9, 451.

73. D. Walker, "Kepler's celestial music," *Journal of the Warburg and Courtauld Institutes* 30 (1967): 228–50, at 228, 230; Stephenson, *Music of the Heavens*, 45; J. Kozhamthadam, *The Discovery of Kepler's Laws: The Intersection of Science, Philosophy and Religion* (Notre Dame, 1994), 75–80.

74. Kepler, *Harmony of the World* 5 ch. 9, 466; Stephenson, *Music of the Heavens*, 204.

75. Walker, "Kepler's Celestial Music," 229; cf. J. Field, "Astrology in Kepler's cosmology," in *Astrology, Science and Society*, ed. P. Curry (Woodbridge, 1987), 143–70, at 151.

76. Walker, "Kepler's Celestial Music," 245.

77. Kepler, *Harmony of the World* 5 ch. 9, 490, ch. 10, 492.

78. *Keplers Gesammelte Werke* 1:285; cf. 1:397; selection trans. in O. Gingerich, s.v. "Kepler," *Dictionary of Scientific Biography*, ed. C. C. Gillispie (New York, 1973), 37: 297.

79. M. Caspar, *Kepler*, trans. C. Hellman (New York, 1962), 249–65; C. Rosen, "Kepler and witchcraft trials," *Historian* 28 (1966): 447–50; documents in *Keplers Gesammelte Werke* 12:65–100, esp. 12:72, 93–94, 98.

80. J. Langford, *Galileo, Science, and the Church*, 2d ed. (Ann Arbor, 1971), 152.

81. W. Wallace, *Galileo's Early Notebooks: The Physical Questions* (Notre Dame, 1977), 93–99.

82. Ibid., 268–69, 310–11; cf. W. Wallace, "The certitude of science in late medieval and Renaissance thought," *History of Philosophy Quarterly* 3 (1986): 281–91.

83. Galileo, *Dialogue Concerning the Two Chief World Systems*, trans. S. Drake, 2d ed. (Berkeley, 1967), 53–54; cf. 145, 230; other scientific protests against the weight of authorities in T. Brahe, *Apologetica responsio ad Craigum Scotum de cometis*, in *Opera omnia*, ed. I. Dreyer (Haunia, 1913–29), 4:422; W. Gilbert, *De magnete*, ch. 10, trans. P. Mottelay (New York, 1958), 47.

84. Langford, *Galileo, Science, and the Church*, 60–61; Dietz Moss, *Novelties in the Heavens*, 135–47; R. Blackwell, "Foscarini's defense of Copernicanism," in *Nature and Scientific Method*, ed. D. Dahlstrom (Washington, D.C., 1991), 199–210.

85. J. Brodrick, *Robert Bellarmine: Saint and Scholar* (Westminster, Md., 1961), chs. 7, 10.

86. Galileo, *Dialogue* 408, 443; cf. 185.

87. Ibid., 115; cf. Kepler, *Epitome* 4 pt. 1 ch. 4.

88. Galileo, *Dialogue* 118–19; further probable arguments 120.

89. Ibid., 122–23; cf. 6; Dietz Moss, *Novelties in the Heavens*, 281–83; E. Sylla, "Galileo and probable arguments," in Dahlstrom, *Nature and Scientific Method*, 211–34.

90. Galileo, *Dialogue* 123; cf. 327, 340; Kepler's view on simplicity in *Gesammelte Werke* 14 nr. 166.

91. Galileo, *Dialogue* 464.

92. *Discoveries and Opinions of Galileo*, trans. S. Drake (New York, 1957), 169.

93. E. Grant, "In defence of the earth's centrality and immobility: Scholastic reaction to Copernicanism in the seventeenth century," *Transactions of the American Philosophical Society* 74 (1984), pt. 4, at 58.

94. M. Mersenne, *L'usage de la raison*, trans. in P. Dear, *Mersenne and the Learning of the Schools* (Ithaca, 1988), 32; Mersenne on the probability of an astronomical hypothesis in W. Hine, "Mersenne and Copernicanism," *Isis* 64 (1973): 18–32, at 21–22; also P. Dear, "Marin Mersenne and the probabilistic roots of 'mitigated scepticism,'" *Journal of the History of Philosophy* 22 (1984): 173–206.

95. Hine, "Mersenne and Copernicanism," 25, 30; cf. L. Auger, "Les idées de Roberval sur le système du monde," *Revue d'histoire des sciences* 10 (1957): 226–34, at 227; further on the probability of Copernicanism in J. Wilkins, *The Discovery of a World in the Moone, or a Discourse Tending to prove, That ('tis probable) there may be another Habitable World in that Planet* (London, 1638); J. Wilkins, *A Discourse Concerning a New Planet: Tending to prove, That ('tis probable) our Earth is one of the Planets* (London, 1640); described in Dietz Moss, *Novelties in the Heavens*, 307–28; H. van Leeuwen, *The Problem of Certainty in English Thought, 1630–1690* (The Hague, 1963), 54–55; S. Dick, *Plurality of Worlds* (Cambridge, 1982), 97–105; P. Dear, *Discipline and Experience* (Chicago, 1995), 173–74; much of Wilkins' argument taken from T. Campanella, *Apology for Galileo*, trans. G. McColley (Northampton, 1937), 14, 70–71.

96. I. Newton, *Mathematical Principles of Natural Philosophy*, trans. A. Motte, rev. F. Cajori (Berkeley, 1934), 398.

97. Galileo, *Dialogue Concerning the Two Chief World Systems*, 309; A. Hald, "Galileo's statistical analysis of astronomical observations," *International Statistical Review* 54 (1986): 211–20.

Chapter 7: Soft Science and History

1. G. Gigerenzer et al., *The Empire of Chance: How Probability Changed Science and Ordinary Life* (Cambridge, 1989), esp. ch. 3; S. Stigler, *The History of Statistics* (Cambridge, Mass., 1986), chs. 7–10; T. Porter, *The Rise of Statistical Thinking, 1820–1900* (Princeton, 1986), ch. 9.

2. Aristotle, *Prior Analytics* 70b6–39.

3. [Aristotle], *Physiognomics* 808a13–17, 811b19–20, 812b18–19, 813b7–9.

4. Ibid., 807a1–2.

5. Ibid., 808a2–16; cf. 808b25–26.

6. Ibid., 807a25–26.

7. E. Evans, "Physiognomics in the ancient world," *Transactions of the American Philosophical Society* 59 (1969), pt. 5; T. Barton, *Power and Knowledge: Astrology, Physiognomics, and Medicine under the Roman Empire* (Ann Arbor, 1994); M. Gleason, *Making Men* (Princeton, 1995), chs. 2, 3; V. Tsouna, "Doubts about other minds and the science of physiognomics," *Classical Quarterly* 48 (1998): 175–86; L. Braswell-Means, "A new look at an old patient: Chaucer's Summoner and medieval physiognomia," *Chaucer Review* 25 (1991): 266–75; R. Steele, ed., *Three Prose Versions of the Secreta Secretorum*, Early English Text Society n.s. 74 (London, 1898), 38–39, 118, 216–36; M. Manzalaoui, *Secreta secretorum*, Early English Text Society 276 (London, 1977), pt. 2; Montaigne, *Essais* 3 ch. 12.

8. Theophrastus, *Concerning Weather Signs* paras. 15, 10; cf. para. 24, in *Theophrastus on Plants and Minor Works on Odours and Weather Signs*, trans. A. Hort (London, 1961), vol. 2.

9. Varro, *De Re Rustica* 1.18.8 and 1.4.3–4; Hippocrates, *On Ancient Medicine* ch. 1.

10. Cicero, *De Divinatione* 1.xii.23–1.xiv.25, trans. W. Falconer in *Cicero: De Senestute, De Amicitia, De Divinatione* (London, 1923): 249–51, with adjustments; cf. 1.xlix.111, 1.lv.124–26, and 2.vi.15–16; see D. Garber and S. Zabell, "On the emergence of probability," *Archive for History of Exact Sciences* 21 (1979): 33–53, at 40; A. Crombie, *Styles of Scientific Thinking in the European Tradition* (London, 1994), 2:1303–8; Philo of Alexandria, *The Special Laws* 1.61; Sextus Empiricus, *Adversus Dogmaticos* 5.103–5.

11. Cicero, *De Divinatione* 2.xlvi.97; cf. 1.xlix.109; Manilius, *Astronomica* 1, lines 53–65.

12. Ptolemy, *Tetrabiblos* 1.2, trans. F. Robbins (London, 1940); cf. Aulus Gellius, *Attic Nights* 14.1.5–7, 17–18, 26; A. Long, "Astrology: Arguments pro and contra," in *Science and Speculation*, cd. J. Barnes et al. (Cambridge, 1982), 165–92.

13. O. Sheynin, "J. Kepler as a statistician," *Bulletin of the International Statistical Institute* 46 (1975), 341–54, at 346–47; J. Kepler, preface to *Rudolphine Tables*, trans. O. Gingerich and W. Walderman, *Quarterly Journal of the Royal Astronomical Society* 13 (1972): 360–73, at 360.

14. Cicero, *De Divinatione* 1.x.16; cf. 1.vii.13.

15. Aristotle, *Posterior Analytics* 79a12–16.

16. Hippocrates, *The Art* 7, trans. J. Chadwick and W. Mann, in *Hippocratic Writings*, ed. G. Lloyd (Harmondsworth, 1978), 142; see G. Lloyd, *Revolutions of Wisdom* (Berkeley, 1987), ch. 3.

17. H. van Staden, "Experiment and experience in Hellenistic medicine," *Bulletin of the Institute of Classical Studies* (University of London) 22 (1975): 178–99, at 187; H. Jaeger, "La preuve judiciaire d'après la tradition rabbinique et patristique," *Recueils de la société Jean Bodin pour l'histoire comparative des institutions* 16 (1964): 415–594, at 521–22; J. Longrigg, "Anatomy in Alexandria in the third century B.C.," *British Journal for the History of Science* 21 (1988):

455–88; cf. H. von Staden, *Herophilus: The Art of Medicine in Early Alexandria* (Cambridge, 1989), 433.

18. Hippocrates, *Aphorisms* sec. I.1.

19. Hippocrates, *Prognostic* chs. 2 and 12; cf. *Epidemics* 1 ch. 10.

20. Galen, *Opera omnia*, ed. C. Kühn (Hildesheim, 1965), 17.1:611–13; cf. *Hippocrates*, with English trans. by W. Jones (London, 1923), 1:213–14.

21. Hippocrates, *Aphorisms* sec. 2 nos. 44, 45, sec. 6 no. 32; cf. Aristotle, *Problemata* 862a34–b6, 891a26, 892a1; Theophrastus, *De ventis* c. 57; O. Sheynin, "On the prehistory of the theory of probability," *Archive for History of Exact Sciences* 12 (1974): 97–141; the empirics on concurrence of symptoms in Sextus Empiricus, *Against the Logicians* 1.179.

22. Hippocrates, *Prognostic* ch. 25, trans. in *Hippocratic Writings*, 185.

23. H. Kyburg, "Randomness and the right reference class," *Journal of Philosophy* 74 (1977): 501–21; H. Kyburg, "The reference class," *Philosophy of Science* 50 (1983): 374–97.

24. K. Deichgräber, *Die Griechische Empirikerschule* (Berlin, 1965), 83, 279–81; cf. P. De Lacy and E. De Lacy, "Ancient rhetoric and empirical method," *Sophia* 6 (1938): 523–30; Sextus Empiricus, *Pyrrhonic Hypotyposes* 1.236; R. Hankinson, "Causes and empiricism: A problem in the interpretation of later Greek medical method," *Phronesis* 32 (1987): 329–48; M. Frede, "The empiricist attitude towards reason and theory," *Apeiron* 21 (2) (1988): 79–97; C. Cosans, "Galen's critique of rationalist and empiricist anatomy," *Journal of the History of Biology* 30 (1997): 35–54.

25. A. Favier, *Un médicin grec du IIe siècle après J.-C., précurseur de la méthode expérimentale moderne: Ménodote de Nicomèdie* (Paris, 1906); cf. V. Brochard, *Les sceptiques grecs* (Paris, 1887), 364–68; L. Robin, *Pyrrhon et le scepticisme grec* (Paris, 1944), 190–95.

26. Deichgräber, *Griechische Empirikerschule*, no. 293, 213–14.

27. Galen, *Subfiguratio Emperica*, text in Deichgräber, *Griechische Empirikerschule*, 42–90, trans. in *Outline of Empiricism*, trans. M. Frede, in *Galen: Three Treatises on the Nature of Science* (Indianapolis, 1985), 23–45; see L. Thorndike, "Translations of works of Galen from the Greek by Niccolò da Reggio," *Byzantina Metabyzantina* 1 (1946): 213–35; *Galen on Medical Experience*, ed. and trans. R. Walzer (London, 1944), also in Frede, *Outline of Empiricism*, 49–106.

28. Walzer, *Galen on Medical Experience*, ch. 9, 98–99.

29. Ibid., ch. 12, 107.

30. Ibid., ch. 15, 112–13; cf. Sextus Empiricus, *Adversus Dogmaticos* 2.291 and 2.151–56 and *Pyrrhonic Hypotyposes* 2.102.

31. Walzer, *Galen on Medical Experience*, ch. 7, 96–97, ch. 18, 120.

32. Ibid., chs. 17–18, 115–21; cf. J. Barnes, "Medicine, experience and logic," in Barnes et al., *Science and Speculation*, 24–68.

33. Walzer, *Galen on Medical Experience*, ch. 30, 151–53; Galen, *On the Sects for Beginners* ch. 5, also in Frede, *Outline of Empiricism*, 3–20, at 8–9.

34. Galen, *Outline of Empiricism* ch. 1, in Frede, *Outline of Empiricism*, 23.

35. Galen, *Outline of Empiricism* ch. 2, 24–25.

36. Ibid., ch. 6, 31.

37. Galen, *Outline* ch. 8, 35–36; further on the grading of belief with agreement by authorities, ch. 10, 39.

38. Ibid., ch. 9, 37.

39. Ibid., 37–39; for the Aristotelian "the more and the less" to express continuous variation, see J. Franklin, "Aristotle on species variation," *Philosophy* 61 (1986): 245–52; J. Lennox, "Aristotle on genera, species and 'the more and the less,'" *Journal of the History of Biology* 13 (1980): 321–46.

40. Galen, *De anatomicis administrationibus* 1 ch. 11, in Kühn, *Opera omnia* 2:278, quoted in Lloyd, "Observational error in later Greek science," 152 n. 58; cf. Galen, *On the Therapeutic Method* 1, 2, trans. R. Hankinson (Oxford, 1991), 120–21.

41. Talmud, Ketubot 15a; N. Rabinovitch, *Probability and Statistical Inference in Ancient and Medieval Jewish Literature* (Toronto, 1973), 45; cf. Talmud, Pesahim 9b, Hullin 10b, Berakhot 28a, Yoma 84b; Rabinovitch, *Probability and Statistical Inference*, 39.

42. Hullin 11a, Gittin 2b; Rabinovitch, *Probability and Statistical Inference*, 39; cf. 43.

43. Talmud, Makshirin ch. 2 mishnah 7.

44. Talmud, Bava Batra 93b; Rabinovitch, *Probability and Statistical Inference*, 46; cf. Bava Batra 24a, Ketubot 15a; Rabinovitch, *Probability and Statistical Inference*, 130–31, 51, 65, 80–83, 94–95.

45. Rabinovitch, *Probability and Statistical Inference*, 50, 54.

46. Talmud, Yevamot 37a; Rabinovitch, *Probability and Statistical Inference*, 59–60; cf. Yevamot 119a–b; a "follow the majority" argument defeats a presumption: Kiddushin 80a.

47. Talmud, Eiruvin 97a; Rabinovitch, *Probability and Statistical Inference*, 84; cf. 90; a similar but simpler case at Shabbat 61a,b.

48. Talmud, Yevamot 64b.

49. Talmud, Ta'anith 21a–b; see O. Sheynin, "Stochastic thinking in the Bible and the Talmud," *Annals of Science* 55 (1998): 185–98.

50. Maimonides, Mishneh Torah, Bikkurim 12.21; Rabinovitch, *Probability and Statistical Inference*, 69–70.

51. Mishneh Torah, Bikkurim 11.30; Rabinovitch, *Probability and Statistical Inference*, 67–68; later uses of *near* and *remote* to mean probabilities in Rabinovitch, *Probability and Statistical Inference*, 157.

52. Rabinovitch, *Probability and Statistical Inference*, 72–73.

53. Ibid., 27–28, 93.

54. Thucydides, *Peloponnesian War* 3, ch. 20; W. Wallis and H. Roberts, *Statistics: A New Approach* (New York, 1956), 215; R. Bolzan, "An ancient anticipation of the probability calculus," *Scientia* 107 (1972): 876–78.

55. Thucydides, *Peloponnesian War* 1 chs. 8 and 10; cf. 1 chs. 1, 20, 22.

56. H. Klein, *The World of Measurements* (London, 1975), 66–67; cf. Talmud, Kelim ch. 17 mishnah 6.

57. Jenkins, *Law of Hywel Dda* 3, 180.

58. *Treatise on the New Money*, ca. 1280, in *The De Moneta of Nicole Oresme and English Mint Documents*, ed. C. Johnson (London, 1956), 67.

59. Johnson, *De Moneta of Nicole Oresme and English Mint Documents*, 79–82; S. Stigler, "Eight centuries of sampling inspection: The trial of the Pyx," *Journal of the American Statistical Association* 72 (1977): 493–500; later problems with the process in S. Wortham, "Sovereign counterfeits: the trial of the pyx," *Renaissance Quarterly* 49 (1996): 334–59.

60. *Treatise on the New Money*, in Johnson, *De Moneta of Nicole Oresme and English Mint Documents*, 91.

61. Kepler to Senate of Ulm, July 30, 1627, in M. Caspar and W. von Dyck, *J. Kepler in seinen Briefen* (Munich, 1930), 2:248; see Sheynin, "Treatment of observations in early astronomy," 182.

62. Avicenna, *Canon medicinae* 2.1.2 (Venice, 1608), 1:245–46, quoted in A. Crombie, "Avicenna's influence on the medieval scientific tradition," in *Avicenna: Scientist and Philosopher*, ed. G. Wickens (London, 1952), 84–107, at 103–4 (I am grateful to Br. Michael Naughtin for help in understanding this passage); cf. Galen, *De simplicium medicamentorum temperamentis et facultatibus* 1 ch. 29 and 2 chs. 9 and 21, in Kühn, *Opera omnia*, 11:431–34, 485, 518; Galen, *De compositione medicamentorum* 7 ch. 4, in Kühn, *Opera omnia*, 13:960–62; actual Islamic medical inference from observations in M. Meyerhof, "Thirty-three clinical observations of Rhazes," *Isis* 23 (1933): 321–55, esp. case 1, 333; A. Dietrich, "Islamic sciences and the medieval West," in *Islam and the West*, ed. K. Semaan (Albany, 1980), 50–63.

63. L. Thorndike, *History of Magic and Experimental Science* (New York, 1934), 2:508–13.

64. Bernard of Gordon, *Tractatus de gradibus*, quoted in M. McVaugh, "Quantified medical theory and practice at fourteenth-century Montpellier," *Bulletin of the History of Medicine* 43 (1969): 397–413, at 403.

65. *Arnaldi de Villanova Opera medica omnia, Aphorismi de gradibus*, ed. M. McVaugh (Granada, 1975), 2:176; see M. McVaugh, "The nature and limits of medical certitude at fourteenth-century Montpellier," *Osiris* 2d s. 6 (1990): 62–84.

66. J. Huizinga, *The Waning of the Middle Ages* (Harmondsworth, 1955), ch. 15.

67. M. Foucault, *The Order of Things* (London, 1970), 39.

68. Paracelsus, *Selected Writings*, ed. J. Jacobi, trans. N. Guterman (Princeton, 1951), 122–23; cf. Foucault, *Order of Things*, 26–27.

69. G. Fracastoro, *De contagione* 1 ch. 13, in *Opera omnia* (Venice, 1555), fol. 113v; Hacking, *Emergence of Probability*, 28.

70. F. Bacon, *Sylva sylvarum* no. 402, in *The Works of Francis Bacon*, ed. J. Spedding (London, 1859), 2:476.

71. R. Grant, *Miracle and Natural Law in Graeco-Roman and Early Christian Thought* (Amsterdam, 1952), ch. 4.

72. Sextus Empiricus, *Adversus Mathematicos* 1.266–68.

73. *Mandeville's Travels* chs. 2 and 12.

74. C. de Fleury, *Memoire sur les instruments de la passion de N.-S. J.-C.* (Paris, 1870).

75. J. Russell, *Inventing the Flat Earth* (New York, 1991); A. Boureau, *The Lord's First Night* (Chicago, 1998); J. Franklin, "Heads of pins," *Australian Mathematical Society Gazette* 20 (1993): 127.

76. R. Higden, *Polychronicon* (with Trevisa's translation), ed. C. Babington and J. Lumby (London, 1865–86), 1:16–18.

77. B. Guenée, *Histoire et culture historique dans l'occident médiéval* (Paris, 1980), ch. 4.

78. Guibert of Nogent, *De pignoribus sanctorum* 1 ch. 3, in *Patrologia Latina*, ed. J. Migne, 156: cols. 607–80, at col. 623 (I am grateful to Br. Michael Naughtin for help in understanding this passage).

79. R. Moore, "Guibert of Nogent and his world," in *Studies in Medieval History Presented to R. H. C. Davis*, ed. H. Mayr-Harting and R. Moore (London, 1985), 107–17; B. Stock, *The Implications of Literacy* (Princeton, 1983), 244–51.

80. William of Newburgh, *History of English Affairs* 1, ch. 27, ed. and trans. P. Walsh and M. Kennedy (Warminster, 1988), 115.

81. *The Golden Legend of Jacobus de Voragine*, trans. G. Ryan and H. Ripperberger (New York, 1941), 630.

82. William of Newburgh, prologue to *History of English Affairs* 29–35.

83. L. Delisle, "Notice sur les manuscrits de Bernard Gui," *Notices et extraits des manuscrits de la Bibliothèque Nationale* 27 (1879): 169–455, at 368 n. 4, 370 n. 1 (I am grateful to Joseph McCarthy of Suffolk University for advice on Bernard Gui's works).

84. Quoted in B. Gui, *De fundatione et prioribus conventuum provinciarum Tolosanae et provinciae ordinis praedicatorum*, ed. P. Amargier (Rome, 1961), x–xi.

85. Delisle, "Notice sur les manuscrits de Bernard Gui," 392; cf. B. Guenée, *Between Church and State: The Lives of Four French Prelates in the Late Middle Ages*, trans. A. Goldhammer (Chicago, 1991), 63.

86. Deslisle, "Notice sur les manuscrits de Bernard Gui," 290 n. 1.

87. Ibid., 373 n. 3.

88. D. Maschi, "Accursio precursore del methodo storico-critico nello studio del 'Corpus iuris civilis,'" in *Atti del convegno internazionale di studi accursiani* (Milan, 1968), 2:597–618.

89. *Corpus iuris civilis*, Digest 22.4; Code 4.21, esp. 4.21.19, 4.21.20; Novels 73; Code 4.19.5.

90. Baldus, on Digest 4.21.20 (19), quoted in R. Helmholz, "The origin of holographic wills in English law," *Journal of Legal History* 15 (1994): 97–108, at 107 n. 45.

91. Refs. in K. Nörr, "Procedure in mercantile matters: some comparative aspects," in *The Courts and the Development of Commercial Law*, ed. V. Piergiovanni (Berlin, 1987), 195–201, at 198–99.

92. Helmholz, "The origin of holographic wills in English law," esp. 100.

93. C. Brooke, *Medieval Church and Society* (London, 1971), ch. 5; M. Clanchy, *From Memory to Written Record* (London, 1979), 248–57; E. Brown, "Medieval forgers and their intentions," in *Fälschungen im Mittelalter*, Monumenta Germaniae Historica Schriften 33 (Hanover, 1988–90), 1:109–19; D. Bates, "The forged charters of William the Conqueror and Bishop William of St. Calais," in *Anglo-Norman Durham*, ed. D. Rollason et al. (Woodbridge, 1994), 111–24; Guenée, *Between Church and State*, 143.

94. C. Duggan, *Twelfth-Century Decretal Collections* (London, 1963), 41.

95. C. Cheney, *Innocent III and England* (Stuttgart, 1976), 111.

96. *Decretals of Gregory IX* 2 tit. 22 cap. 6, in *Corpus iuris canonici*, ed. A. Friedberg (Leipzig, 1879) 2: cols. 347–48 (I am grateful to Br. Michael Naughtin for help in understanding this passage); a similar case in England in V. Galbraith, *Studies in the Public Records* (London, 1948), 51–52.

97. Clanchy, *From Memory to Written Record*, 254–55; cf. 323–37.

98. Guenée, *Histoire et culture historique*, 133–34, 145–46.

99. Ibid., 137–38, based on H.-F. Delaborde, "Le procès du chef de saint Denis en 1410," *Mémoires de la Société de l'histoire de Paris et de l'Ile-de-France* 11 (1884): 297–409.

100. Petrarch to Emperor Charles IV, 1355, trans. in P. Burke, *The Renaissance Sense of the Past* (London, 1969), 51; debate on whether Renaissance historiography was more "critical" of sources than medieval in D. Hay, *Annalists and Historians* (London, 1977), ch. 5; E. Cochrane, *Historians and Historiography in the Italian Renaissance* (Chicago, 1981), ch. 1; G. Ianziti, "Leonardo Bruni: first modern historian?" *Parergon* 14 (2) (Jan. 1997): 85–99.

101. L. Valla, *Opera omnia*, ed. E. Garin (Turin, 1962), 1:633; see D. Kelley, *Foundations of Modern Historical Scholarship* (New York, 1970), 41.

102. Valla, *Opera omnia*, 1:693; see J. Seigel, *Rhetoric and Philosophy in Renaissance Humanism* (Princeton, 1968), ch. 5; M.-B. Gerl, *Rhetorik als Philosophie* (Munich, 1974).

103. Valla, *Opera omnia* 1:447; cf. L. Valla, *On Pleasure*, trans. A. Hieatt and M. Lorch (New York, 1977), 9.

104. Valla, *Opera omnia* 1:644; passage trans. in L. Jardine, "Lorenzo Valla and the intellectual origins of humanist dialectic," *Journal of the History of Philosophy* 15 (1977): 143–64, at 155.

105. Valla, *Opera omnia* 1:172; see V. Kahn, "The rhetoric of faith and the use of usage in Lorenzo Valla's *De libero arbitrio*," *Journal of Medieval and Renaissance Studies* 13 (1983): 91–109.

106. Jardine, "Lorenzo Valla and the intellectual origins of humanist dialectic," 153–59; Valla, *Opera omnia* 1:717–19.

107. L. Jardine, "Lorenzo Valla: Academic skepticism and the new human-

ist dialectic," in *The Skeptical Tradition*, ed. M. Burnyeat (Berkeley, 1983), 253–86, at 283 n. 82.

108. *The Treatise of Lorenzo Valla on the Donation of Constantine*, ed. and trans. C. Coleman (Toronto, 1993), 21.

109. D. Maffei, *La donazione di Constantino nei giuristi medievali* (Milan, 1964).

110. William of Ockham, *Breviloquium de potestate papae* 2 ch. 4, ed. L. Baudry (Paris, 1937), 164.

111. R. Pecock, *The Repressor of Over Much Blaming of the Clergy*, ed. C. Babington, Rolls Series 19 (London, 1860–61), 361; J. Levine, "Reginald Pecock and Lorenzo Valla on the Donation of Constantine," *Studies in the Renaissance* 20 (1973): 118–43.

112. Nicholas of Cusa, *De concordantia catholica* 3 ch. 2, in *Opera omnia* (Frankfurt, 1962), 3: fols. 53–54; Nicholas of Cusa, *The Catholic Concordance*, trans. P. Sigmund (Cambridge, 1991), 217–18; discussed in M. Watanabe, *The Political Ideas of Nicholas of Cusa* (Geneva, 1963), ch. 5; connection with Valla in R. Fubini, "Humanism and truth: Valla writes against the Donation of Constantine," *Journal of the History of Ideas* 57 (1996): 79–86.

113. Coleman, *Treatise of Lorenzo Valla*, 35.

114. Ibid., 62–63, 71; cf. 67.

115. Ibid., 85, 91, 95, 131, 139, 131.

116. R. Delph, "Valla grammaticus, Agostino Steuco, and the Donation of Constantine," *Journal of the History of Ideas* 57 (1996): 55–77.

117. Coleman, *Treatise of Lorenzo Valla*, 3.

118. Quoted in L. Janik, "Lorenzo Valla: The primacy of rhetoric and the demoralization of history," *History and Theory* 12 (1973): 389–404, at 396.

119. Ibid., 400, 403.

120. M. Winterbottom, "Fifteenth century manuscripts of Quintilian," *Classical Quarterly* 17 (1967): 339–69, at 356–60; J. d'Amico, *Theory and Practice of Renaissance Textual Criticism* (Berkeley, 1988), 14–17; further refs. in Kelley, *Foundations of Modern Historical Scholarship*, 37 n. 40; cf. C. Trinkaus, *In Our Image and Likeness* (London, 1970), 2:571–76.

121. Poliziano, *Opera* (Basel, 1553), 259; A. Grafton, *Defenders of the Text* (Cambridge, Mass., 1991), 57; E. Kenney, *The Classical Text* (Berkeley, 1974), 5–10.

122. J. Bentley, *Humanists and Holy Writ* (Princeton, 1983), ch. 4.

123. Erasmus, *Annotations to the New Testament*, to I Corinthians 15, in *Opera omnia Des. Erasmi Roterodami* (Hildesheim, 1961–62), 6: col. 742, quoted in J. Bentley, "Erasmus, Jean le Clerc and the principle of the harder reading," *Renaissance Quarterly* 31 (1978): 309–21, at 318.

124. Intro. to *The Discourses of Niccolò Machiavelli*, trans. L. Walker (London, 1950), 74.

125. N. Machiavelli, *The Discourses* 1 ch. 6, ed. B. Crick (Harmondsworth, 1983), 121.

126. C. Schmitt, *Cicero Scepticus* (The Hague, 1972); R. Popkin, *The History of Scepticism from Erasmus to Descartes* (Assen, 1964).

127. F. Patrizzi, *De historia dialogi X* (1560), quoted in J. Franklin, *Jean Bodin and the Sixteenth-Century Revolution in the Methodology of Law and History* (New York, 1963), 97; cf. D. La Russo, "A neo-Platonic dialogue: Is rhetoric an art?" *Speech Monographs* 32 (1965): 393–410, at 396; Schmitt, *Cicero Scepticus*, 754–58.

128. M. Cano, *De locis theologicis* 11 cap. 4, in *Opere* (Padua, 1734), 289; Franklin, *Jean Bodin*, 109; see generally intro. to M. Cano, *L'autorità della storia profana*, trans. A. Biondi (Turin, 1973); there is a slight basis in Augustine, *De fide rerum invisibilium* 1.1–2.

129. *Opere*, 89; Franklin, *Jean Bodin*, 110–12.

130. Franklin, *Jean Bodin*, 142–51, 111 n. 4; cf. C. Vasoli, "Il problema cinquecentesco della 'Methodus' e la sua applicazione all conoscenza storica," *Filosofia* 21 (1970): 137–72; testimony in history discussed in Herbert of Cherbury, *De Veritate* ch. 11, trans. M. Carré (Bristol, 1937), 314–22.

131. Guenée, *Histoire et culture historique*, 154–64.

132. Pierre d'Ailly, ca. 1400, in Grafton, *Defenders of the Text*, 284 n. 71.

133. Grafton, *Defenders of the Text*, ch. 4.

134. Ibid., ch. 7; Copernicus's attempts in the same area in *De Revolutionibus* 3 ch. 11, trans. A. Duncan (Newton Abbot, 1976), 158.

Chapter 8: Philosophy

1. Parmenides, frag. 8, lines 50–61; frag. 1, lines 28–32, in L. Tarán, *Parmenides* (Princeton, 1965), 86, 9; see G. Lloyd, *Magic, Reason, and Experience* (Cambridge, 1979), 78–79; T. Ebert, "Wo beginnt der Weg der Doxa?" *Phronesis* 34 (1989): 121–38; R. Brague, "La vraisemblance du faux: Parménide fr. I, 31–32," in *Etudes sur Parménide*, ed. P. Aubenque (Paris, 1987), 44–68; P. Kelley, "Philodoxy: mere opinion and the question of history," *Journal of the History of Philosophy* 16 (1996): 117–32.

2. Plato, *Theaetetus* 162e; cf. *Phaedo* 92d, *Phaedrus* 229e; M. Untersteiner, *The Sophists*, trans. K. Freeman (Oxford, 1954), 57, 72 n. 23.

3. Plato, *Timaeus* 29c–d, 48d, 59c, 72d; cf. Plotinus, *Enneads* 2.1.6.8, 3.6.12.11; see J. Smith, "Plato's myths as 'likely accounts,'" *Apeiron* 19 (1988): 24–42; A. Ashbaugh, *Plato's Theory of Explanation* (Albany, 1988), ch. 1; G. Lloyd, *Methods and Problems in Greek Science* (Cambridge, 1991), 342–44; K.-H. Hagstroem, *Les préludes antiques de la théorie des probabilités* (Stockholm, 1932), ch. 5.

4. M. Burnyeat, "The origins of non-deductive inference," in *Science and Speculation*, ed. J. Barnes et al. (Cambridge, 1982), 193–238, at 238; cf. G. Preti, "Sulla dottrina della *semeion* nella logica stoica," *Rivista critica di storia della filosofia* 1 (1956): 5–16.

5. Diogenes Laertius, *Lives of Eminent Philosophers* 7.75–76; cf. 107, 116, 130; M. Burnyeat, "Carneades was no probabilist" (unpublished); R. Bett, "Carneades' *pithanon*: A reappraisal of its role and status," *Oxford Studies in Ancient Philosophy* 7 (1989): 59–94, at 62–67; D. Sedley, "On signs," in Barnes et al., *Sci-*

ence and Speculation, 239–72, at 251 n. 31; Aristotelian origins in J. le Blond, *Eulogos chez Aristote et l'argument de convenance* (Paris, 1938); cf. Hippocrates, *The Art* 7; Cicero's translation of *eulogon* as *probabilis* at *De Finibus* 3.58; cf. *Letters to Atticus* 14.22.2.

6. Plutarch, *Moralia* 1071a, 1072c–d.

7. Sextus Empiricus, *Adversus Mathematicos* 7.241–43; further on *pithanon* in the Stoics in Diogenes Laertius, *Lives* 7.190, 199, 200; Plutarch, *Moralia* 1036d–e.

8. Diogenes Laertius, *Lives* 7.162, trans. R. Hicks (London, 1925); cf. 49, 78, 89; Sextus, *Adversus Mathematicos* 7.157; Cicero, *Academica* 2.36.

9. Aristotle, *Metaphysics* 1009b8–12; cf. *On Generation and Corruption* 316a1; Theophrastus, *De Sensu et Sensibilibus* 60.

10. Cicero, *Academica* 2.84; cf. 2.77; see J. Franklin, "Healthy Scepticism," *Philosophy* 66 (1991): 305–24.

11. Diogenes Laertius, *Lives* 7.177; see Athenaeus, *Deipnosophists* 354e.

12. Cicero, *Academica* 2.39, 108.

13. C. Schmitt, "The rediscovery of ancient Skepticism in modern times," in M. Burnyeat, ed., *The Skeptical Tradition* (Berkeley, 1983), 225–51.

14. Sextus Empiricus, *Adversus Mathematicos* 7.187–88, 169–70; *Pyrrhonic Hypotyposes* 1.227–28.

15. *Adversus Mathematicos* 7.176–84, in *Sextus Empiricus*, trans. R. Bury (London, 1935–49), 2: 95–99; H. Tarrant, *Scepticism or Platonism?* (Cambridge, 1985), 9; similar more briefly in Arcesilaus: Sextus Empiricus, *Adversus Mathematicos* 7.158; but see *Pyrrhonic Hypotyposes* 1.232; debate on whether action requires assent in Plutarch, *Moralia* 1122b–d; H. von Arnim, ed., *Stoicorum veterum fragmenta* (Stuttgart, 1964), 2: § 714.

16. Cicero, *Academica* 2.78, 137, 139; in favor of the *reductio* theory, M. Burnyeat, "Carneades was no probabilist"; P. Couissin, "The Stoicism of the New Academy," in Burnyeat, *Skeptical Tradition*, 31–63; against the theory, L. Robin, *Pyrrhon et le scepticisme grec* (Paris, 1944), 90–102; Hagstroem, *Les préludes antiques*, ch. 4; C. Stough, *Greek Scepticism* (Berkeley, 1969), 50–64; I. Schneider, "The contributions of the Sceptic philosophers Arcesilas and Carneades to the development of an inductive logic," *Indian Journal of History of Science* 12 (1977): 173–80; G. Striker, "Sceptical strategies," in *Doubt and Dogmatism*, ed. M. Schofield et al. (Oxford, 1980), 54–83, at 73; J. Glucker, *Antiochus and the Late Academy*, Hypomnemata 56 (Göttingen, 1978), 75–78; A. Long, *Hellenistic Philosophy* (London, 1974), 93, 96–99; A. Ioppolo, *Opinione e scienza* (Naples, 1986), 121–26; M. Frede, *Essays in Ancient Philosophy* (Oxford, 1987), 213–15; for a mixed opinion, see Bett, "Carneades' *pithanon.*"

17. Homer, *Odyssey* 19 line 203; Hesiod, *Theogony* line 27.

18. *Adversus Mathematicos* 7.154.

19. Ibid., 7.160.

20. Ibid., 7.175, trans. in Bury, 2:95.

21. Cicero, *Academica* 2.33–34, 36; trans. Bury, *Sextus Empiricus*, 2:95.

22. Ibid., 2.104–5; P. Meador, "Skeptic theory of perception: A philosophical antecedent of Ciceronian probability," *Quarterly Journal of Speech* 54 (1968): 340–51.

23. Cicero, *Academica* 2.100, trans. H. Rackham, in *Cicero: De Natura Deorum, Academica* (London, 1933), 595; cf. Seneca, *De Beneficiis* 4.33.2–3.

24. Cicero, *De Natura Deorum* 1.12; *Tusculan Disputations* 1.17 and 2.5; *Academica* 2.7–8; *De Officiis* 2.7–8; *De Finibus* 3.72; 5.9; *De Inventione* 1.43.

25. *De Inventione* 1.44; further refs. in H. Merguet, *Lexicon zu den Philosophischen Schriften Ciceros* (Hildesheim, 1961), 3:170–71.

26. For example, D. Armstrong, *What Is a Law of Nature?* (Cambridge, 1983), 52–59; D. Stove, *The Rationality of Induction* (Oxford, 1986), chs. 1, 4.

27. Philodemus, *On Methods of Inference*, ed. and trans. P. De Lacy and E. De Lacy (Naples, 1978), secs. 3 and 4, pp. 92–93; cf. Sextus Empiricus, *Pyrrhonic Hypotyposes* 1.141–4; Lucian, *Hermotimus* 58–61; E. Asmis, *Epicurus' Scientific Method* (Ithaca, 1984), ch. 11.

28. Philodemus, *On Methods of Inference* secs. 6–9, 12, pp. 96–99; cf. Sedley, "On signs," 249.

29. Philodemus, *On Methods of Inference* sec. 11, p. 99.

30. Cf. Philodemus, *Rhetoric*, ed. S. Sudhaus (Leipzig, 1892), 1:246, 264; Philo, *Special Laws* 3.56; Plotinus, *Enneads* 2.1.6.39; more generally, T. Brennan, "Reasonable impressions in Stoicism," *Phronesis* 41 (1996): 318–34.

31. Philodemus, *On Methods of Inference* secs. 20–21, 27, 41, pp. 106–9, 116–17; J. Stocks, "Epicurean induction," *Mind* 34 (1925): 185–203; J. Milton, "Induction before Hume," *British Journal for the Philosophy of Science* 38 (1987): 49–74, at 53–54.

32. Sextus Empiricus, *Pyrrhonic Hypotyposes* 2.195.

33. Ibid., 2.204.

34. J. Weinberg, *Abstraction, Relation, and Induction* (Madison, 1965), 134; Avicenna, *Remarks and Admonitions*, trans. S. Inati (Toronto, 1984), 129; the "example" in other Islamic authors in Hallaq, "On inductive corroboration, probability, and certainty in Sunni legal thought," 3–31, at 5 n. 6.

35. Avicenna, *Kitab al-najat*, quoted in S. Pines, "La conception de la conscience de soi chez Avicenne et chez Abu'l Barakat al-Baghdadi," *Archives d'histoire doctrinale et littéraire du moyen-âge* 21 (1954): 21–98, at 96–97; also in F. Rahman, *Avicenna's Psychology* (London, 1952), 55; cf. *The Propositional Logic of Avicenna*, trans. N. Shehaby (Dordrecht, 1973), 265; B. Kogan, *Averroes and the Metaphysics of Causation* (Albany, 1985), 87–88; further refs. in Weinberg, *Abstraction, Relation, and Induction*, 133 n. 42, 134 n. 45.

36. Aquinas, *Summa theologiae* I–II q. 112 art. 5.

37. References in I. Kantola, *Probability and Moral Uncertainty in Late Medieval and Early Modern Times* (Helsinki, 1994), 24.

38. Aquinas, *Commentary on Boethius' "De Trinitate"* q. 6 a. 1; cf. C. Salutati, *De nobilitate legum et medicinae* cap. 6 (Florence, 1947), 38–48.

39. Aquinas, *Disputed Questions on Truth* q. 8 art. 12, trans. R. Mulligan

(Chicago, 1952), 1:377–78; cf. *Summa Theologiae* I q. 57 art. 3, and II–II q. 60 art. 3 ad 1; a modern revival in K. Popper, *A World of Propensities* (Bristol, 1990).

40. *In Aristotelis librum De memoria et reminiscentia* lectio 1 n. 305, in *In Aristotelis libros De sensu et sensato, De memoria et reminiscentia Commentaria*, ed. R. Spiazzi (Turin, 1949), 88.

41. H. Breny, "Les origines de la théorie des probabilités," *Revue des questions scientifiques* 159 (1988): 453–77, at 465.

42. Aquinas, *Summa theologiae* II-II q. 95 art. 5, body, and ad 2; cf. I q. 115 art. 6; art. 4 ad 3; II-II q. 96 art. 3 ad 1, 2; *De sortibus* cap. 4, in *Opera omnia*, ed. Leonina (Rome, 1882-), 43:229–38, at 233–34; E. Byrne, *Probability and Opinion: A Study in the Medieval Presuppositions of Post-Medieval Theories of Probability* (The Hague, 1968), 210; T. Wedel, *The Medieval Attitude toward Astrology* (New Haven, 1920), 62, 67–71, 124; on Albert, P. Zambelli, *The Speculum Astronomiae and its Enigma* (Dordrecht, 1992), 68, 71.

43. Aquinas, *Summa contra Gentiles* l.3, c. 135, trans. in Kantola, *Probability and Moral Uncertainty*, 40–41; cf. Boethius of Dacia, quoted in Kantola, *Probability and Moral Uncertainty*, 30 n. 16.

44. Aquinas, *Summa theologiae* I q. 63 art. 9.

45. Aquinas, *Disputed Questions on Truth* q. 5 art. 10 ad 7.

46. Nicholas of Cusa, *Catholic Concordance* 3, preface at 206; 2 ch. 4, 58; cf. 208.

47. J. Duns Scotus, *Opus Oxoniense* 1 dist. 3 pt. 1 q. 4 art. 2, in *Opera omnia* (Vatican City, 1954), 3:141–43; trans. in *Duns Scotus: Philosophical Writings*, trans. A. Wolter (Edinburgh, 1962), 109–10; R. McKeon, *Selections from Medieval Philosophers* (London, 1931), 2:327.

48. *Opera omnia* 3:146–47; Wolter, *Duns Scotus*, 114; McKeon, *Selections from Medieval Philosophers*, 331.

49. *Opera omnia* 3:144; Wolter, *Duns Scotus*, 111; McKeon, *Selections from Medieval Philosophers*, 329; J. Duns Scotus, *Metaphysics* 5 q. 3 n. 5, quoted in P. Vier, *Evidence and Its Function According to John Duns Scotus* (St. Bonaventure, 1951), 143–44; cf. E. Bos, "Pseudo-Johannes Duns Scotus über Induktion," in *Historia philosophiae medii aevi*, ed. B. Mojsisch and O. Planta (Amsterdam, 1991), 1:71–103.

50. J. Duns Scotus, *Priorum analyticorum quaestiones* 2 q. 8, in *Opera omnia*, 1:340–41, quoted in A. Crombie, *Robert Grosseteste and the Origins of Experimental Science*, 3d ed. (Oxford, 1971), 168 n. 6; Duns Scotus, *Metaphysics* 1 q. 4 n. 6, quoted in Vier, *Evidence and Its Function*, 149; cf. R. Holkot, *Determinatio* 3, quoted in C. Michalski, *La philosophie au XIVe siècle* (Frankfurt, 1969), 245; Aureolus, *Scriptum super primum sententiarum* q. 1 a. 1, ed. E. Buytaert (St. Bonaventure), 1:135; J. Buridan, *Summa totius Logicae* (Frankfurt, 1965), tract. 6; cf. J. Zupko, "Buridan and Skepticism," *Journal of the History of Philosophy* 31 (1993): 191–221.

51. William of Ockham, *Summa Logicae* pt. 3 tract. 2 cap. 10, ed. P. Boehner (St. Bonaventure, 1974), 523–24; see further L. Baudry, *Lexicon philosophique de Guillaume de Ockham* (Paris, 1958), 119–20.

52. Ockham, *Summa Logicae* pt. 3 tract 1 cap. 1 p. 360; E. Moody, *The Logic of William of Ockham* (London, 1935), 211.

53. Ockham, *Tractatus super libros elenchorum*, quoted in Baudry, *Lexicon philosophique*, 217; cf. Ockham, *Scriptum in librum primum Sententiarum Ordinatio* 1 dist. 2 q. 10, ed. G. Gal (St. Bonaventure, 1969–70), 2:354; Ockham, *Expositio super physicam Aristotelis*, quoted in Baudry, *Lexicon philosophique*, 56.

54. D. Knowles, "A characteristic of the mental climate of the fourteenth century," in *Mélanges offerts à Etienne Gilson* (Toronto, 1959), 315–25.

55. Duns Scotus, *Opus Oxoniense* 1 dist. 42 q. unica, in *Opera omnia*, 6:346.

56. John of Rodington, *On the Sentences* 2, q. 1, quoted in Michalski, *La philosophie au XIVe siècle*, 105.

57. Pierre de Ceffons, *On the Sentences* 1, quoted in Michalski, *La philosophie au XIVe siècle*, 408; cf. Aureolus, *Scriptum super primum sententiarum*, 2:686; Johannes Brammart, quoted in B. Xiberta, *De scriptoribus Scholasticis saeculi XIV ex ordine Carmelitarum* (Louvain, 1931), 435.

58. John de Bassolis, *On the Sentences* dist. 2 q. 3 a. 3, quoted in Michalski, *Philosophie au XIVe siècle*, 183–84; cf. R. Swineshead, *On the Sentences* 1 q. 2, quoted in Michalski, *Philosophie*, 106–7; Andreas de Novo Castro, *On the Sentences* 1, quoted in Michalski, *Philosophie*, 113–16.

59. Hugues de Castro Novo, *On the Sentences* 1, quoted in Michalski, *Philosophie au XIVe siècle*, 112; cf. Stephen Patrington, quoted in L. Kennedy, "Late-fourteenth-century philosophical scepticism at Oxford," *Vivarium* 23 (1985): 124–51, at 137.

60. John of Mirecourt, *On the Sentences* 1 q. 19, quoted in Michalski, *Philosophie au XIVe siècle*, 93; an example with a purely philosophical proposition in Marsilius of Inghen (?), *Quaestiones super VIII libros Physicorum Aristotelis*, quoted in Weinberg, *Abstraction, Relation, and Induction*, 152; *probabiliorem* used in the pronouncements of a church council in H. Denziger and C. Rahner, *Enchiridion Symbolorum* (Barcelona, 1957), 223 (but perhaps meaning as much "approvable" as "probable").

61. Michalski, *Philosophie au XIVe siècle*, 326; cf. R. Brown, "History versus Hacking on probability," *History of European Ideas* 8 (1987): 655–73, at 659, 661.

62. Gregory of Rimini, *On the Sentences* prologue q. 2 a. 4, quoted in Michalski, *Philosophie au XIVe siècle*, 92; cf. G. Leff, *Gregory of Rimini* (Manchester, 1961), 62–66; John Baconis, *On the Sentences* 3 D. 24 q. 2 (Cremona, 1618), 158, quoted in Michalski, *Philosophie*, 92.

63. See for example E. Grant, *A Source Book in Medieval Science* (Cambridge, Mass., 1974), 237–53.

64. Boethius of Dacia, *Quaestiones super librum Topicorum* 2 c. 1 q. 15, ed. N. Green-Pedersen and J. Pinborg (Gad, 1976), 139, quoted in Kantola, *Probability and Moral Uncertainty*, 46 n. 15.

65. John of Dumbleton, *Summa logicae et philosophiae* 1 cap. 29–31; see J. Weisheipl, "The place of John Dumbleton in the Merton School," *Isis* 50

(1959): 439–54, at 450–51; J. Weisheipl, "Ockham and some Mertonians," *Mediaeval Studies* 30 (1968): 163–213, at 200.

66. Nicholas of Autrecourt, *Letters to Bernard of Arezzo*, trans. E. Moody, in *Philosophy in the Middle Ages*, ed. A. Hyman and J. Walsh (New York, 1967), 656–64, at 657–59; also in *Nicholas of Autrecourt: His Correspondence with Master Giles and Bernard of Arezzo*, ed. and trans. L. de Rijk (Leiden, 1994); biography in Z. Kaluza, *Nicolas d'Autrecourt: ami de la vérité* (Paris, 1995).

67. Hyman and Walsh, *Philosophy in the Middle Ages*, 660–63; see J. Weinberg, *Nicolaus of Autrecourt* (New York, 1969), 34; on whether logic is itself fallible, L. Groarke, "On Nicholas of Autrecourt and the law of non-contradiction," *Dialogue* (Canada) 23 (1984): 129–34; T. Stiefel, "The problem of causal inference in the empirical world of Nicholas of Autrecourt," *History of Science* 30 (1992): 295–309.

68. Hyman and Walsh, *Philosophy in the Middle Ages*, 663–64; cf. Brown, "History versus Hacking on probability," 660.

69. Nicholas of Autrecourt, *Exigit Ordo*, ed. J. O'Donnell, *Mediaeval Studies* 1 (1939): 179–280, at 185; also in *The Universal Treatise of Nicholas of Autrecourt*, trans. L. Kennedy, R. Arnold, and A. Millward (Milwaukee, 1973); cf. Weinberg, *Nicolaus of Autrecourt*, 127.

70. Weinberg, *Nicolaus of Autrecourt*, 75; cf. J. d'Eltville, *On the Sentences* 1, D. 2, quoted in Michalski, *Philosophie au XIVe siècle*, 109.

71. N. Oresme, *Le Livre de Politiques d'Aristote*, ed. A. Menut, in *Transactions of the American Philosophical Society* 60 (1970), pt. 6, 243.

72. Nicholas of Autrecourt, *Exigit Ordo* 186.

73. Ibid., 189; see B. Dutton, "Nicholas of Autrecourt and William of Ockham on atomism, nominalism, and the ontology of motion," *Medieval Philosophy and Theology* 5 (1996): 63–85.

74. Nicholas of Autrecourt, *Exigit Ordo* 205.

75. Ibid., 228–29.

76. But see J. Leslie, *Value and Existence* (Totowa, 1979).

77. W. James, *Principles of Psychology* (New York, 1950), ch. 5.

78. Nicholas of Autrecourt, *Exigit Ordo* 203; Weinberg, *Nicolaus of Autrecourt*, 121.

79. Nicholas of Autrecourt, *Exigit Ordo* 204; on the connection of probability and belief, 279; on Ockham's razor and onus of proof arguments, 222–23, 228.

80. Ibid., 187.

81. H. Rashdall, "Nicholas de Ultricuria, a medieval Hume," *Proceedings of the Aristotelian Society* 7 (1906–7): 1–27, at 6–11.

82. K. Tachau, *Vision and Certitude in the Age of Ockham* (Leiden, 1988), 378–79, 281–82 n. 87; W. Courtenay, "Covenant and causality in Pierre d'Ailly," *Speculum* 46 (1971): 94–119, at 101 n. 23, 102 n. 28.

83. *The Chronicle of Jean de Venette*, selections in *The Black Death*, ed. W. Bowsky (New York, 1971), 17; see further Campbell, *Black Death and Men of*

Learning, esp. ch. 6, 158, 160; L. Eldredge, "The concept of God's absolute power at Oxford in the later fourteenth century," in *By Things Seen*, ed. D. Jeffrey (Ottawa, 1979), 211–26; R. Gottfried, *The Black Death* (New York, 1983), 153–56; J. North, "1348 and all that: Science in late medieval Oxford," in *From Ancient Omens to Statistical Mechanics*, ed. J. Berggren and B. Goldstein (Copenhagen, 1987), 155–65.

84. P. d'Ailly, *On the Sentences* 1 q. 1, quoted in Michalski, *Philosophie au XIVe siècle*, 92; M. de Gandillac, "De l'usage et de la valeur des arguments probables dans les questions du Cardinal Pierre d'Ailly sur le 'Livre des Sentences,'" *Archives d'histoire doctrinale et littéraire du moyen âge* 8 (1933): 43–91, at 70; cf. 55, 65–66, 79, 84; lack of argumentative quality also in L. Smoller, *History, Prophecy and the Stars: The Christian Astrology of Pierre d'Ailly* (Princeton, 1994).

85. F. Petrarca, *On His Own Ignorance and That of Many Others*, trans. H. Nachod, in *The Renaissance Philosophy of Man*, ed. E. Cassirer et al. (Chicago, 1948), 47–133, at 126.

86. S. Toulmin, *Cosmopolis* (New York, 1990), 69–80.

87. L. Cohen, "Some historical remarks on the Baconian conception of probability," *Journal of the History of Ideas* 41 (1980): 219–31; P. Urbach, *Francis Bacon's Philosophy of Science* (La Salle, 1987), ch. 2.

88. F. Bacon, *De Augmentis Scientiarum* 5, ch. 2, in *The Works of Francis Bacon*, ed. J. Spedding (London, 1859), 4:410; cf. *Novum Organum* preface, in Spedding, *Works*, 4:42; *Novum Organum* 2, aph. 36, in Spedding, *Works*, 4:190.

89. J. Martin, *Francis Bacon, the State and the Reform of Natural Philosophy* (Cambridge, 1992), ch. 5.

90. Ibid., 166–68; H. Wheeler, "The invention of modern empiricism: Juridical foundations of Francis Bacon's philosophy of science," *Law Library Journal* 76 (1983): 78–120; R.-M. Sargent, "Scientific experiment and legal expertise: The way of experience in seventeenth-century England," *Studies in History and Philosophy of Science* 20 (1989): 19–46; for a modern example of this kind of reasoning see P. Winston, "Learning structural descriptions from examples," in *The Psychology of Computer Vision*, ed. P. Winston (New York, 1975), 157–209.

91. Bacon, *Novum Organum* 1 aph. 118, in Spedding, *Works*, 4:105; see Urbach, *Francis Bacon's Philosophy of Science*, 156–60.

92. Bacon, *Novum Organum* Plan of the Great Instauration, in Spedding, *Works* 1:141; further refs. in Martin, *Francis Bacon*, 217 n. 73.

93. Bacon, *Certain Observations Touching the Pacification of the Church*, in *The Works of Francis Bacon*, ed. B. Montagu (London, 1825–34), 2:425, quoted in Hall, "Some perspectives on the use of torture in Bacon's time," 315; J. Heath, *Torture and English Law* (Westport, 1982), 119–20, 134–35, 153; K. Cardwell, "Francis Bacon, inquisitor," in *Francis Bacon's Legacy of Texts*, ed. W. Sessions (New York, 1990), 269–89; Martin, *Francis Bacon, the State, and the Reform of Natural Philosophy*, 102–3.

94. Descartes, *Oeuvres*, ed. C. Adam and P. Tannery, rev. ed. (Paris, 1964–76), 10:362; *The Philosophical Writings of Descartes*, trans. J. Cottingham,

R. Stoothoff, and D. Murdoch (Cambridge, 1985), 1:10; cf. M. Fehér, "Galileo and the demonstrative ideal of science," *Studies in History and Philosophy of Science* 13 (1982): 87–110.

95. Descartes, *Rules for the Direction of the Mind* rule 3, in *Oeuvres*, 10:367, also in Cottingham, Stoothoff, and Murdoch, *Philosophical Writings of Descartes*, 1:13.

96. Descartes, *Oeuvres*, 10:367–68, also in Cottingham, Stoothoff, and Murdoch, *Philosophical Writings of Descartes*, 1:14; Descartes' early public statement of this position in Baillet, *Vie de Descartes*, 2: ch. 14, trans. as "Descartes' encounter with Chandoux," in N. Kemp Smith, *New Studies in the Philosophy of Descartes* (London, 1952), 40–46.

97. *Descartes' Conversation with Burman*, trans. J. Cottingham (Oxford, 1976), 48–49, see *Oeuvres* 4:177; cf. Descartes, *Discours de la Methode*, ed. E. Gilson (Paris, 1925), 128.

98. Descartes to Mersenne, Oct. 5, 1637, in Descartes, *Correspondance*, ed. C. Adam and G. Milhaud (Paris, 1936–), 2:44.

99. *Descartes' Discourse on the Method* pt. 6, see *Oeuvres* 6:64, also in Cottingham, Stoothoff, and Murdoch, *Philosophical Writings of Descartes*, 1:144; see D. Clarke, *Descartes' Philosophy of Science* (Manchester, 1982), §§ 18–19.

100. D. Garber, "Science and certainty in Descartes," in *Descartes: Critical and Interpretive Essays*, ed. M. Hooker (Baltimore, 1978), 114–51.

101. Descartes, *The Principles of Philosophy* pt. 3, princ. 43, in *Oeuvres* 8A:99, also in Cottingham, Stoothoff, and Murdoch, *Philosophical Writings of Descartes*, 1:255.

102. Descartes, *Principles of Philosophy* pt. 4, princ. 205, in *Oeuvres* 8A:327–28, also in Cottingham, Stoothoff, and Murdoch, *Philosophical Writings of Descartes*, 289–90; on codes and certainty, P. Pesic, "Secrets, symbols and systems," *Isis* 88 (1997): 674–92.

103. De Arriaga, *Cursus philosophicus*, quoted in E. Curley, "Certainty: Psychological, moral and metaphysical," in *Essays on the Philosophy and Science of René Descartes*, ed. S. Voss (New York, 1993), 16–17.

104. Descartes, *Principles of Philosophy* pt. 4 princ. 204, in *Oeuvres* 8A:327, also in Cottingham, Stoothoff, and Murdoch, *Philosophical Writings of Descartes*, 1:289; cf. pt. 3, princ. 44; see S. Voss, "Scientific and practical certainty in Descartes," *American Catholic Philosophical Quarterly* 67 (1993): 569–85; M. Morrison, "Hypotheses and certainty in Cartesian science," in *An Intimate Relation: Studies in History and Philosophy of Science*, ed. J. Brown and J. Mittelstrass (Dordrecht, 1989), 43–64; Clarke, *Descartes' Philosophy of Science*, §§ 14, 20. For later developments, L. Laudan, *Science and Hypotheses* (Boston, 1981), ch. 4, "The clock metaphor and hypotheses: The impact of Descartes on English methodological thought, 1650–1670."

105. Descartes, *Optics* disc. 1, in *Oeuvres* 6:83, also in Cottingham, Stoothoff, and Murdoch, *Philosophical Writings of Descartes*, 1:152–53; Descartes to Mersenne, May 17, 1638, in *Correspondance* 2:266.

106. Wallace, "The certitude of science in late medieval and Renaissance thought."

107. Kepler, *Mysterium Cosmographicum*, in *Keplers Gesammelte Werke* 1:15–16; see N. Jardine, *The Birth of History and Philosophy of Science: Kepler's Defense of Tycho against Ursus* (Cambridge, 1984), 216–17.

108. Cf. C. Hooker, H. Penfold, and R. Evans, "Control, connectionism and cognition," *British Journal for the Philosophy of Science* 43 (1992): 517–36.

109. C. Meinel, "Early seventeenth-century atomism: Theory, epistemology and the insufficiency of experiment," *Isis* 79 (1988): 68–103.

110. Descartes to Plempius, Oct. 3, 1637, in *Correspondance* 2:19; cf. J. Morris, "Descartes and probable knowledge," *Journal of the History of Philosophy* 8 (1970): 303–12.

111. Descartes to Morus, Feb. 5, 1649, in *Correspondance* 8:135–36; cf. Descartes to Elisabeth, Oct. 6, 1645, in *Correspondance* 6:319.

112. Descartes, *Rules for the Direction of the Mind* rule 12, in *Oeuvres* 10:424, also in Cottingham, Stoothoff, and Murdoch, *Philosophical Writings of Descartes*, 1:47–48; cf. *Replies to Objections V: Concerning the objections to the Fourth Meditation* 1, in *Oeuvres* 7:375, also in Cottingham, Stoothoff, and Murdoch, *Philosophical Writings*, 2:258; Descartes to (Mersenne?), May 27, 1630, in *Correspondance* 1:142; Clarke, *Descartes' Philosophy of Science*, ch. 6.

113. *Oeuvres Philosophiques de Descartes*, ed. F. Alquié (Paris, 1963), 1:151 n. 2; Cottingham, Stoothoff, and Murdoch, *Philosophical Writings of Descartes*, 1:48 n. 1.

114. *Discourse on the Method* pt. 3, in *Oeuvres* 6:25, also in Cottingham, Stoothoff, and Murdoch, *Philosophical Writings of Descartes*, 1:123; cf. Descartes to Elisabeth, Sept. 15, 1645, in *Correspondance* 6:303; cf. T. Carr, *Descartes and the Resilience of Rhetoric* (Carbondale, 1990), 32–33.

115. J. Caramuel, *Animadversiones in Meditationes Cartesianas, quibus demonstratur clarissime nihil demonstratur a Cartesio* (1644), in D. Pastine, "Caramuel contro Descartes: Obiezioni inedite alle Meditazioni," *Rivista critica di storia della filosofia* 27 (1972): 177–221; summary in J. Velarde Lombraña, *Juan Caramuel: vida y obra* (Oviedo, 1989), 120–35.

116. P. Gassendi, *Objections to the Meditations* 5, preface, in Cottingham, Stoothoff, and Murdoch, *Philosophical Writings of Descartes*, 2:179; on Gassendi's approval of probabilistic views in general, refs. in O. Bloch, *La Philosophie de Gassendi* (The Hague, 1971), 92–95; on the "probable syllogism," P. Gassendi, *Institutio Logica*, ed. and trans. H. Jones (Assen, 1981), 148–52; cf. *Syntagma: Logic* 2, ch. 2, in *Selected Works of Pierre Gassendi*, ed. and trans. C. Brush (New York, 1972), 293–94; Gassendi to Mersenne, in *Correspondance du P. Marin Mersenne* 2:185; B. Brundell, *Pierre Gassendi* (Dordrecht, 1987), 101–3; L. Joy, *Gassendi the Atomist* (Cambridge, 1987), 171–72; S. Fisher, "Gassendi's concepts of probability, inductive inference, and 'probabilistic' deductive inference" (unpublished).

117. J. de Lugo, *Disputationes Scholasticae et morales*, ed. J. Fournials (Paris,

1868), 2:240–41, quoted in S. Knebel, "Necessitas moralis ad optimum (III): Naturgesetz und Induktionsproblem in der Jesuitenscholastik während des zweiten Drittels des 17. Jahrhunderts," *Studia Leibnitiana* 24 (1992): 182–215, at 189 n. 33.

118. I. Derkennis, *De Deo uno, trino, creatore* (Brussels, 1655), 200, quoted in Knebel, "Necessitas moralis ad optimum (III)," 189 n. 34.

119. S. Knebel, "Die früheste Axiomatisierung des Induktionsprinzips: Pietro Sforza Pallavicino SJ (1607–1667)," *Salzburger Jahrbuch für Philosophie* 41 (1996): 97–128.

120. M. de Esparza, *Quaestiones disputandae de Deo uno et trino* (Rome, 1657), q. 28, in Esparza, *Cursus theologicus* (Lyons, 1666), 1: fols. 97a–105b, fols. 104a/b, quoted in Knebel, "Necessitas moralis ad optimum (III)," 200 n. 96.

121. R. Harrod, *Foundations of Inductive Logic* (London, 1956), ch. 3, summarized in J. Nicod, *Geometry and Induction* (London, 1970), 243–45.

122. T. Hobbes, *Human Nature* ch. 4, in *The English Works of Thomas Hobbes*, ed. W. Molesworth (London, 1839), 4:17–18; cf. ch. 6, in *Works* 4:29.

123. G. Polya, *Patterns of Plausible Inference*, 2d ed. (Princeton, 1968), 134–36; K. Popper, "On Carnap's version of Laplace's Rule of Succession," *Mind* 71 (1962): 69–73; but cf. S. Zabell, "The rule of succession," *Erkenntnis* 31 (1989): 283–321; M. Changizi and T. Barber, "A paradigm-based solution to the riddle of induction," *Synthese* 117 (1998): 419–84.

124. Pascal to M. Perier, Nov. 15, 1647, in *Récit de la grande expérience de l'équilibre des liqueurs*, in Pascal, *Oeuvres complètes*, ed. J. Mesnard (Paris, 1970), 2:679; in *The Physical Treatises of Pascal*, trans. I. Spiers and A. Spiers (New York, 1973), 100.

125. Pascal to Père Noël, Oct.–Nov. 1647, in Pascal, *Oeuvres complètes* 2:519; in *Great Shorter Works of Pascal*, trans. E. Cailliet and J. Blankenagel (Westport, 1974), 43.

126. Pascal, *Oeuvres complètes* 2:523–24; *Great Shorter Works*, 46–47.

127. Père Noël to Pascal, Oct.–Nov. 1647, in Pascal, *Oeuvres complètes* 2:532.

128. Pascal, *Oeuvres complètes* 2:538.

129. Pascal to Le Pailleur, Feb.–Mar. 1648, in Pascal, *Oeuvres complètes* 2:561; *Great Shorter Works*, 63.

130. Pascal, *Oeuvres complètes* 2:564; *Great Shorter Works*, 74–75; cf. Karl Popper's account of the origin of his philosophy of science in *Conjectures and Refutations*, 5th ed. (London, 1974), 34–35.

Chapter 9: Religion

1. Revelation 3:15–16.

2. 1 Corinthians 2:4.

3. Sextus Empiricus, *Adversus mathematicos* 9. 27, trans. R. Bury, in *Sextus Empiricus* (London, 1933–49), 3:15–16; similar in Chrysippus in Cicero, *De Natura Deorum* 2.v.15.

4. Xenophon, *Memorabilia* 1.4.6, 8, trans. E. Marchant (London, 1923),

55–58; also in Cicero, *De natura deorum* 2.vi.18; Sextus, *Adversus mathematicos* 9.93–94.

5. Sextus, *Adversus mathematicos* 9.111–12, in Bury, *Sextus Empiricus*, 3:61.

6. Cicero, *De natura deorum* 2.vi.17 and 3.x.26; cf. 2.xxii.57–58; Aëtius, *Placita* 1.6, in *Stoicorum veterum fragmenta*, ed. J. Arnim (Stuttgart, 1964), 2: § 1009.

7. Romans 1:20.

8. Philo, *On Rewards and Punishments* 41–43, trans. F. H. Colson, in *Philo* (London, 1929–62), 8:337; cf. *Special Laws* 1.33–35; *Allegorical Interpretation of Genesis* 3.97–99; on probabilistic reasoning in theology in general, *Special Laws* 1.38–40.

9. Lactantius, *Divinae Institutiones* 3.3.7–8; cf. L. Panizza, "Valla, Lactantius and oratorical scepticism," *Journal of the Warburg and Courtauld Institutes* 41 (1978): 76–107; Clement of Alexandria, *Stromateis* 2, 4 and 12; cf. 1 and 11; T. Brummel, "The Role of Reason in the Act of Faith in the Theologies of Origen, Basil and John Chrysostom" (Ph.D. diss., UCLA, 1988).

10. Marius Victorinus, *In De Inventione* 1.29, in *Rhetores Latini Minores*, ed. C. Halm (Frankfurt, 1964), 232; cf. Fortunatianus, *Artis rhetoricae libri tres* 1, excerpt in Miller, Prosser, and Benson, *Readings in Medieval Rhetoric*, 28–29.

11. St. Cyprian, *On the Unity of the Church* ch. 22, in *De Lapsis and De Ecclesiae Catholicae Unitate*, ed. and trans. M. Bevenot (Oxford, 1971), 93; Nicholas of Cusa, *Catholic Concordance* 1, ch. 8; 3, preface at 29, 206.

12. Augustine, *Contra Academicos* 3 chs. 15–16; cf. *De Sermone Domini in Monte* 1 ch. 9.

13. Gregory of Tours, *History of the Franks* 4 ch. 48; J. Høyrup, "Sixth-century intuitive probability: The statistical significance of a miracle," *Historia Mathematica* 10 (1983): 80–84.

14. *Averroes' "Tahafut al-Tahafut,"* trans. S. van den Bergh (Oxford, 1954), 1:316–17, 324; see M. Mamura, "Ghazali's attitude toward the secular sciences," in *Essays on Islamic Philosophy and Science*, ed. G. Hourani (Albany, 1975), 100–111; W. Watt, ed. and trans., *The Faith and Practice of al-Ghazali* (London, 1953), 21–22; earlier arguments undermining reason in W. Watt, *Islamic Philosophy and Theology* (Edinburgh, 1962), 67–68; E. Ormsby, *Theodicy in Islamic Thought* (Princeton, 1984), 23.

15. See Averroes, *On the Harmony of Religion and Philosophy*, trans. G. Hourani (London, 1961), 40.

16. John of Salisbury, *Metalogicon* 2 ch. 13, trans. D. McGarry (Berkeley, 1955), 104–5; cf. *Metalogicon* 3 ch. 9 at 188; ref. to Marius Victorinus, *In De Inventione* as above (n. 10), that to Augustine is *De Diversis Quaestionibus 83* q. 46.2.

17. John of Salisbury, *Metalogicon* 2 ch. 13 at 105; John of Salisbury, *Policraticus* 7 ch. 2.

18. John of Salisbury, *Metalogicon* 3 ch. 10 at 201; cf. *The Letters of John of Salisbury*, ed. W. Millor and C. Brooke (Oxford, 1979), 2:321; on the Academics and probability, *Policraticus* 7 chs. 6, 7, trans. J. Pike as *Frivolities of Courtiers*

and Footprints of Philosophers (Minneapolis, 1938), 236, 239; on the *Digest* and witnesses, *Policraticus* 5 ch. 14; cf. M. Pike, "Römisches und kirchliches Recht im *Policraticus* des Johannes von Salisbury," in *The World of John of Salisbury*, ed. M. Wilks (Oxford, 1984), 365–79; on the distinction between probable and rhetorical arguments, *Metalogicon* 2 chs. 3–5 at 78–83; cf. P. von Moos, "The use of *exempla* in the *Policraticus* of John of Salisbury," in Wilks, *World of John of Salisbury*, 207–61; P. von Moos, *Geschichte als Topik: Das rhetorische Exemplum von der Antike zur Neuzeit und die "historiae" im Policraticus Johanns von Salisbury* (Hildesheim, 1988); T. Stiefel, *Intellectual Revolution in Twelfth-Century Europe* (New York, 1985), 59; L. Cazzola Palazzo, "Il valore filosofico della probabilità nel pensiero di Giovanni di Salisbury," *Atti della Accademia delle scienze di Torino, Classe di scienze morali* 92 (1957–58): 96.

19. *Metalogicon* 2 ch. 14 at 106–7; cf. 2 ch. 15 at 109; the passage referred to in Aristotle, *Topics* 131b22–31, does not really make the same point; a modern treatment of probability dynamics in P. Forrest, *The Dynamics of Belief* (Oxford, 1986).

20. Hugh of St. Victor, *Didascalicon* 2 ch. 30, trans. J. Taylor (New York, 1961), 81.

21. Richard of St. Victor, *De Trinitate* 1 cap. 4, in Migne, *Patrologia Latina*, ed. J. Migne, 196: col. 892C; cf. Aureolus, *Scriptum super primum sententiarum* q. 2 at 1:135, 2:694; further refs. in P. Dronke, ed., *A History of Twelfth Century Philosophy* (Cambridge, 1988), 154; Hugh of St. Victor, *De sacramentis* 1 part 6 cap. 26, in Migne, *Patrologia Latina*, 176: col. 280B; on the need for a dialectical method even in theology, Abelard's *Sic et Non*, discussed in C. Radding, *A World Made by Men: Cognition and Society, 400–1200* (Chapel Hill, 1985), 205; cf. 156–72.

22. Maimonides, *Guide of the Perplexed* pt. 2 ch. 15 at 290–92.

23. Ibid., pt. 2 chs. 22–23 at 319–21; N. Rabinovitch, *Probability and Statistical Inference in Ancient and Medieval Jewish Literature* (Toronto, 1973), 138–39.

24. Maimonides, *Guide of the Perplexed* pt. 2 ch. 19 at 309–10; Rabinovitch, *Probability and Statistical Inference*, 122–23; H. Davidson, *Proofs for Eternity, Creation and the Existence of God in Medieval Islamic and Jewish Philosophy* (New York, 1987), 198–201.

25. Maimonides, *Guide of the Perplexed* pt. 3 ch. 26 at 509; Rabinovitch, *Probability and Statistical Inference*, 103.

26. T. Bukowski, "The eternity of the world according to Siger of Brabant: Probable or demonstrative?" *Recherches de théologie ancienne et médiévale* 36 (1969): 225–29; cf. A. Maurer, "Siger of Brabant and theology," *Mediaeval Studies* 59 (1988): 257–78, at 272; also Siger of Brabant, *Questions on de Anima* 3 q. 2, in F. van Steenberghen, *Siger de Brabant d'après les oeuvres inédites* (Louvain, 1931), 166.

27. Siger of Brabant, *Impossibilia*, in *Ecrits de logique, de morale et de physique*, ed. B. Bazán (Louvain, 1974), 81; further refs. in F. van Steenberghen, *Maître Siger de Brabant* (Louvain, 1977), 247 n. 42; cf. Maurer, "Siger of Brabant and theology," 268.

28. Siger of Brabant, *Quaestiones in metaphysicam* 4 q. 31, ed. A. Maurer (Louvain, 1983), 176.

29. J. Pecham, *Quaestiones de anima* q. 4, quoted in D. Douie, *Archbishop Pecham* (Oxford, 1952), 23.

30. H. Denifle and E. Chatelain, eds., *Chartularium Universitatis Parisiensis* (Paris, 1889–97), 1:543–55, trans. E. Fortrin and P. O'Neill, in *Medieval Political Philosophy: A Sourcebook*, ed. R. Lerner and M. Mahdi (New York, 1963), 335–54; further in R. Hissette, *Enquête sur les 219 articles condamnées à Paris le 7 mars 1277* (Louvain, 1977); L. Bianchi, *Il vescovo e i filosofi* (Bergamo, 1990); C. Normore, "Who was condemned in 1277?" *Modern Schoolman* 72 (1995): 273–81; J. Thijssen, "1277 revisited," *Vivarium* 35 (1997): 72–101; cf. A. Maurer, "Siger of Brabant on fables and falsehoods in religion," *Mediaeval Studies* 43 (1981): 515–30.

31. J. Wippel, "Thomas Aquinas and the condemnation of 1277," *Modern Schoolman* 72 (1995): 233–72.

32. E. Grant, "Medieval and Renaissance Scholastic conceptions of the influence of the celestial region on the terrestrial," *Journal of Medieval and Renaissance Studies* 17 (1987): 1–23, at 4.

33. Aristotle, *Metaphysics* 1017a19; Aquinas, *Summa contra gentiles* 4 ch. 62 § 12; see M. McCord Adams, "Aristotle and the sacrament of the altar," in *Aristotle and His Medieval Interpreters*, ed. R. Bosley and M. Tweedale, supp. *Canadian Journal of Philosophy* (1991), 195–249.

34. Aquinas, *Quaestiones quodlibetales* 9.l.3, ed. R. Spiazzi (Rome, 1949), 183–84.

35. Aquinas, *Summa theologiae* III q. 77 art. 1 ad 1.

36. Job 28:26; E. Zilsel, "The genesis of the concept of physical law," *Philosophical Review* 51 (1942): 245–79; R. Grant, *Miracle and Natural Law in Graeco-Roman and Early Christian Thought* (Amsterdam, 1952); H. Koester, "ΝΟΜΟΣ ΦΥΣΕΩΣ: The concept of natural law in Greek thought," in *Religions in Antiquity*, ed. J. Neusner (Leiden, 1968), 521–41.

37. Aquinas, *Commentary on "The Divine Names"* 10.1; Augustine, *De Genesi ad Litteram* 9 ch. 17; J. Ruby, "The origins of scientific 'law,'" *Journal of the History of Ideas* 47 (1986): 311–59.

38. F. Oakley, "Christian theology and the Newtonian science: The rise of the concept of the laws of nature," *Church History* 30 (1961): 433–57.

39. William of Ockham, *Quodlibeta* 6 q. 6, in *Philosophical Writings*, ed. and trans. P. Boehner (Edinburgh, 1957), 25–26; see E. Grant, "The condemnation of 1277, God's absolute power, and physical thought in the late Middle Ages," *Viator* 10 (1979): 211–44; and generally W. Courtenay, *Covenant and Causality in Medieval Thought* (London, 1984), esp. ch. 4; W. Courtenay, *Capacity and Volition: A History of the Distinction of Absolute and Ordained Power* (Bergamo, 1990); H. Klocker, *William of Ockham and the Divine Freedom* (Milwaukee, 1992); Aquinas, *Summa contra gentiles* 4 ch. 65 § 3; *Quaestiones quodlibetales* 3 q. 1 art. 2 40; the opposite asserted by no. 63 of the condemned propositions of 1277.

40. William of Ockham, *Dialogus de Imperio et Pontificia Potestate* pt. 1 bk. 4 cap. 2, in *Opera plurima* (repr. London, 1962), 1: fol. 22v.

41. *Dialogus* pt. 3 tract. 1, trans. in William of Ockham, *A Letter to the Friars Minor and Other Writings*, ed. and trans. A. McGrade and J. Kilcullen (Cambridge, 1995), 214.

42. *Dialogus* pt. 1 bk. 6 cap. 100, in *Opera Plurima* fol. 110r.

43. G. Mollat, *The Popes at Avignon*, trans. J. Love (Edinburgh, 1963), 210–12; cf. A. McGrade, *The Political Thought of William of Ockham* (Cambridge, 1974), ch. 2.

44. Aristotle, *Physics* 189a15; *On the Heavens* 271a33; W. Thorburn, "The myth of Occam's razor," *Mind* 27 (1918): 345–53; R. Ariew, "Did Ockham use his razor?" *Franciscan Studies* 37 (1977): 5–17; further refs. in A. Maurer, "Ockham's razor and Chatton's anti-razor," *Mediaeval Studies* 46 (1984): 463–75, at 463 n. 3.

45. Ockham, *Scriptum in librum primum Sententiarum Ordinatio* 1 dist. 14 q. 2; *Quodlibeta* 4 q. 30 in *Opera Philosophica et Theologica* (St. Bonaventure, 1967–86), 9:448–50.

46. L. Thorndike, *History of Magic and Experimental Science* (New York, 1934), 2:953–54, 959; cf. L. Thorndike, "Relations of the Inquisition to Peter of Abano and Cecco d'Ascoli," *Speculum* 1 (1926): 338–43, at 342; L. Thorndike, "Franciscus Florentinus, or Paduanus, an inquisitor of the fifteenth century, and his treatise on astrology and divination, magic and popular superstition," in *Mélanges Mandonnet* (Paris, 1930), 2:353–69, at 361, 363; H. Lemay, "The stars and human sexuality," *Isis* 71 (1980): 127–37, at 129.

47. E. le Roy Ladurie, *Montaillou*, trans. B. Bray (London, 1978), 141.

48. A. Dent, *The Plaine Man's Pathway to Heaven* (London, 1601), 118; see Sampson, "Laxity and liberty in seventeenth-century English political thought," 102.

49. T. Fuller, quoted in H. Henson, *Puritanism in England* (New York, 1972), 97–98.

50. N. Jones, *Faith by Statute* (London, 1982), ch. 6; W. Haugaard, *Elizabeth and the English Reformation* (Cambridge, 1968), 247–57.

51. A. Quinton, *The Politics of Imperfection: The Religious and Secular Traditions of Conservative Thought in England from Hooker to Oakeshott* (London, 1978); M. Perrott, "Hooker and the problem of authority in the Elizabethan church," *Journal of Ecclesiastical History* 49 (1998): 29–60.

52. R. Hooker, *Laws of Ecclesiastical Polity* (London, 1594, repr. Menston, 1969), 2 ch. 4, 102; cf. 2 ch. 7, 120.

53. J. Marshall, *Hooker and the Anglican Tradition* (London, 1963), chs. 7, 8; A. D'Entrèves, *The Medieval Contribution to Political Thought* (New York, 1959), ch. 6; P. Munz, *The Place of Hooker in the History of Thought* (London, 1952), 124–26, 175–93.

54. Hooker, preface to *Laws of Ecclesiastical Polity* at 16; cf. 1 ch. 8 at 63; 2 ch. 1 at 99.

55. Ibid., 2 ch. 7 at 117; cf. preface at 17.

56. Ibid., preface at 29; cf. 1 ch. 8 at 63; 2 ch. 1 at 99.

57. Ibid., preface at 23–24.

58. L. Lessius, *De providentia numinis, et animi immortalitate libri duo, adversus atheos et politicos* (Antwerp, 1613), 1, 21; summary and discussion in M. Buckley, *At the Origins of Modern Atheism* (New Haven, 1987), 42–55; cf. P. du Plessis Mornay, *De la verité de la religion chrestienne* (Antwerp, 1581), in *A Woorke Concerning the Trewnesse of the Christian Religion*, trans. P. Sidney and A. Golding (London 1587, repr. New York, 1976), 7, 12–13; on Lessius's anti-Anglican writings, J. Gillow, *A Literary and Biographical History, or Bibliographical Dictionary of the English Catholics* (New York, 1968), 5:365–66.

59. L. Lessius, *Rawleigh his Ghost*, trans. "A.B." (St. Omer, 1631, repr. Ilkley, 1977), 28–29.

60. Dutch original, Leiden, 1622.

61. P. Topliss, *The Rhetoric of Pascal* (Leicester, 1966), 148–49; J. ter Meulen and P. Diermanse, *Bibliographie des écrits sur Hugo Grotius imprimés au XVIIe siècle* (The Hague, 1961), nos. 20, 209, 413.

62. H. Grotius, *The Truth of the Christian Religion* 2 sec. 19, trans. J. Clarke (London, 1793), 139–40.

63. Grotius, *True Religion* 1 sec. 6 (London, 1632), 27–28; Grotius, *Truth of the Christian Religion*, 13.

64. H. Trevor-Roper, *Catholics, Anglicans and Puritans: Seventeenth Century Essays* (London, 1987), ch. 4.

65. W. Chillingworth to Sheldon, in *The Works of William Chillingworth* (Philadelphia, 1840), vii; *Additional Discourses* 8 at 737–38; connections with Grotius in van Leeuwen, *Problem of Certainty in English Thought*, 21 n. 18.

66. W. Chillingworth, *The Religion of Protestants a Safe Way to Salvation* ch. 7 par. 8 in *Works*, 500; cf. *The Whole Works of the Right Rev. Jeremy Taylor*, ed. R. Heber (London, 1822), 12:97.

67. Chillingworth, *Religion of Protestants* ch. 1 par. 8 at 81; the Scholastic origins of these ideas mentioned by the author: R. Orr, *Reason and Authority: The Thought of William Chillingworth* (Oxford, 1967), 51.

68. Chillingworth, *Religion of Protestants* ch. 2 par. 104 at 147.

69. Ibid., ch. 2 par. 154 at 167–68; cf. ch. 2 par. 118–20 at 152–53; ch. 3 par. 89 at 247–48; ch. 6 par. 7 at 433–34.

70. Ibid., ch. 3 par. 26 at 203–4.

71. Ibid., ch. 2 par. 67 at 131–32.

72. Ibid., ch. 4 par. 57 at 295–96.

73. J. Aubrey, *Brief Lives*, ed. O. Dick (London, 1958), 64; Chillingworth, *Works* vi.

74. Shapiro, *Probability and Certainty in Seventeenth-Century England*, esp. chs. 1–3; B. Shapiro, *John Wilkins, 1614–1672: An Intellectual Biography* (Berkeley, 1969), 228–34; M. Ferreira, *Scepticism and Reasonable Doubt: The British Naturalist Tradition in Wilkins, Hume, Reid and Newman* (Oxford, 1986), ch. 2; S. Shapin, *A Social History of Truth* (Chicago, 1994), 207–11.

75. K. Digby, *A Conference with a Lady about Choice of Religion* (Paris, 1638, repr. Menston, Eng., 1969), 70; cf. G. Tavard, *The Seventeenth Century Tradition: A Study in Recusant Thought* (Leiden, 1978), ch. 8.

76. Van Leeuwen, *Problem of Certainty in English Thought*, 17–19, 25–31.

77. *The Works of the Most Reverend Dr John Tillotson, Late Lord Archbishop of Canterbury*, 4th ed. (London, 1728), 3:125.

78. J. Ryan, "The argument of the wager in Pascal and others," *New Scholasticism* 19 (1945): 233–50; M. Asín Palacios, *Los precedentes musulmanes del Pari de Pascal* (Santander, 1920); review in P. Lønning, *Cet effrayant pari* (Paris, 1980), 129–35.

79. Arnobius, *Adversus Gentes* 2.4, in *Ante-Nicene Fathers*, trans. H. Bryce and H. Campbell (New York, 1907), 5: 434.

80. Chillingworth, *Religion of Protestants* ch. 6 par. 5 at 430–31.

81. R. Knox, *Enthusiasm* (Oxford, 1951), 185.

82. Further in J. de Silhon, *De la certitude des connaissances humaines* 5 (Paris, 1661).

83. J. de Silhon, *De l'Immortalité de l'Ame* (Paris, 1634), 227–29, in L. Blanchet, "L'attitude religieuse des jésuites et les sources du pari de Pascal," *Revue de métaphysique et de morale* 26 (1919): 477–516, 617–47, at 625.

84. A. Sirmond, *Démonstration de l'Immortalité de l'Ame, Tirée des Principes de la Nature, Fortifiée de ceux d'Aristote* (Paris, 1637), 456–61, quoted in Blanchet, "L'attitude religieuse des jésuites," 628–30; similar, with comparison to merchants' risks, in P. Hurtado de Mendoza, *De anima* disp. 18 sec. 1 at 6–7, in *Opera* (Lyons, 1617), in N. Meusnier, "L'émergence d'une mathématique du probable au XVIIe siècle," *Revue d'histoire des mathématiques* 2 (1996): 119–47, at 140.

85. A. Sirmond, *De immortalitate animae Demonstratio physica et Aristotelica Adversus Pomponatium et asseclas* (Paris, 1635), 390–92, in Blanchet, "L'attitude religieuse des jésuites," 633–34, selection in H. Busson, *La pensée religieuse française de Charron à Pascal* (Paris, 1933), 392–95.

86. A. Sirmond, *Défense de la Vertu* (Paris, 1641); Pascal, *Provincial Letter* 10, in Pascal, *Pensées and the Provincial Letters*; Blanchet, "L'attitude religieuse des jésuites," 636; further in H. Bremond, *La querelle du pur amour au temps de Louis XIII* (Paris, 1932).

87. J. Mesnard, *Pascal* (New York, 1952), 114.

88. See J. Pascal, *Pensées* nos. 167, 423, numbering following *Oeuvres complètes*, ed. L. Lafuma (Paris, 1963), and *Pensées*, trans. A. Krailsheimer (Harmondsworth, 1966), nos. 269, 277 in *Pensées*, trans. W. Trotter (London, 1931), numbering following *Oeuvres de Blaise Pascal*, ed. L. Brunschvicg and P. Boutroux (Paris, 1904–14).

89. Ibid., nos. 678/358, 116/398, 117/409, 430/431.

90. Ibid., nos. 429/229.

91. Ibid., nos. 149/430.

92. Ibid., nos. 835/564.

93. Ibid., nos. 157/225; cf. nos. 428/195, 165/210, 823/217; I. Hacking, *The Emergence of Probability* (Cambridge, 1975), ch. 8.

94. Pascal, *Pensées* nos. 418/233; see Hacking, *Emergence of Probability*, ch. 8; Lønning, *Cet effrayant pari*, ch. 4; L. Thirouin, *Le hasard et les règles: Le modèle du jeu dans la pensée de Pascal* (Paris, 1991), ch. 7; G. Brunet, *Le pari de Pascal* (Paris, 1956); P.-A. Cahné, *Pascal, ou le risque de l'espérance* (Paris, 1981), ch. 4; E. Seneta, "Pascal and probability," in *Interactive Statistics*, ed. D. McNeil (Amsterdam, 1979), 225–33, at 230–31; F. Chimenti, "Pascal's wager: A decision-theoretic approach," *Mathematics Magazine* 63 (1990): 321–25; objections to mathematical approaches in V. Carraud, *Pascal et la philosophie* (Paris, 1992), 434–50.

95. Hacking, *Emergence of Probability*, 66; also see R. Anderson, "Recent criticisms and defenses of Pascal's wager," *International Journal for Philosophy of Religion* 37 (1995): 45–56; J. Golding, "Pascal's wager (a critical examination based on recent discussion in Anglo-American literature)," *Modern Schoolman* 71 (1994): 115–43; J. Jordan, *Gambling on God* (Lanham, 1994); J. Franklin, "Two caricatures, I: Pascal's Wager," *International Journal for Philosophy of Religion* 44 (1998): 109–14.

96. Pascal, *Pensées* nos. 599/907, 721/916, 692/914.

97. Ibid., nos. 722/922.

98. Knox, *Enthusiasm*, 203.

Chapter 10: Aleatory Contracts

1. Aristotle, *Nicomachean Ethics* 1133a25–30.

2. A. Smith, *Wealth of Nations* 1.5.

3. [Demosthenes], *Private Orations* 34.33; see G. de Ste. Croix, "Ancient Greek and Roman maritime loans," in *Debits, Credits, Finance and Profits*, ed. H. Edey and B. Yamey (London, 1974), 41–59; P. Millett, "Maritime loans and the structure of credit in fourth-century Athens," in *Trade in the Ancient Economy*, ed. P. Garnsey, K. Hopkins, and C. Whittaker (Berkeley, 1983), 36–52.

4. *Private Orations* 35.10; G. Calhoun, "Risk in sea-loans in ancient Athens," *Journal of Economic and Business History* 2 (1930): 561–84, at 575, 579.

5. *Private Orations* 34.28.

6. Lysias, *Against Aeschines*, in Athenaeus, *Deipnosophists* 611d–12f; P. Millett, *Lending and Borrowing in Ancient Athens* (Cambridge, 1991), 1–2.

7. *Corpus iuris civilis, Digest* 18.1.8; cf. 18.4.7; 19.1.12; 19.1.11.18; H. Guitton, "Le droit romain en face de l'aléa," *Revue française de recherche opérationelle* 7 (1963): 194–95; R. Zimmermann, *The Law of Obligations* (Cape Town, 1990), 245–49.

8. *Digest* 35.2.73.1.

9. *Digest* 18.4.11; J. Thomas, "*Venditio hereditatis* and *emptio spei*," *Tulane Law Review* 33 (1958–59): 541–50.

10. Livy, *History* 23.49; 25.3; Suetonius, *Claudius* 18; *Theodosian Code* 13.9.1–2.

11. *Digest* 18.1.35.7; 18.6.1; 19.2.13.5; *Code* 4.48.2; F. de Zulueta, *The Roman*

Law of Sale (Oxford, 1945), 30–35; E. Betti, "Periculum: Problemi del rischio contrattuale in diritto romano classico e giustinianeo," *Jus* 5 (1954): 333–84; G. MacCormack, "Alfenus Varus and the law of risk in sale," *Law Quarterly Review* 101 (1985): 573–86; *Institutes* 3.23.3; cf. A. Hunt and C. Edgar, eds., *Select Papyri* (London, 1932), 1:121.

12. *Digest* 17.1.39; cf. D. Maffei, *Il caso fortuito nell'età dei glossatori* (Milan, 1957), 86–95.

13. *Digest* 22.2.4–5; *Code* 4.33.2; *Novels* 106; Paulus, *Sententiae* 2.14; P. Huvelin, *Etudes d'histoire du droit commercial romain* (Paris, 1929), 98–110; Zimmermann, *Law of Obligations*, 181–86.

14. *Digest* 35.2.68; M. Greenwood, "A statistical mare's nest?" *Journal of the Royal Statistical Society* 103 (1940): 246–48.

15. Talmud, Makkot 3a; cf. Baba Kama 89a; A. Hasofer, "The Talmud on expectation," *Milla wa-Milla (The Australian Bulletin of Comparative Religion),* no. 13 (Dec. 1973): 33–39; R. Ohrenstein and B. Gordon, "Risk, uncertainty and expectation in Talmudic literature," *International Journal of Social Economics* 18 (1991): 4–15.

16. H. Berman, *Law and Revolution* (Cambridge, Mass., 1983), 588 n. 85.

17. Rashi, quoted in Hasofer, "Talmud on expectation"; cf. N. Rabinovitch, *Probability and Statistical Inference in Ancient and Medieval Jewish Literature* (Toronto, 1973), 157 n. 22.

18. Maimonides, *Mishneh Torah*, Laws of Testimony 21.1, quoted in Hasofer, "Talmud on expectation," 36; cf. Isaac Alfasi, quoted in Ohrenstein and Gordon, "Risk, uncertainty and expectation in Talmudic literature," 13.

19. F. Rosenthal, *Gambling in Islam* (Leiden, 1975), ch. 3.

20. N. Saleh, *Unlawful Gain and Legitimate Profit in Islamic Law*, 2d ed. (London, 1992), 106, 64–65, 75, 100–101; Rosenthal, *Gambling in Islam*, 139; ancient debates on the saleability of uncertain future goods in E. Swan, "Futures and derivatives: From ancient Mesopotamia to the fall of Rome," in *The Development of the Law of Financial Services*, ed. E. Swan (London, 1993), 1–33, esp. 9, 24.

21. C. Hoover, "The sea loan in Genoa in the twelfth century," *Quarterly Journal of Economics* 40 (1925): 499–529; Islamic parallels in A. Udovitch, *Partnership and Profit in Medieval Islam* (Princeton, 1970), ch. 4.

22. R. Fraher, "Preventing crime in the High Middle Ages," in *Popes, Teachers, and Canon Law in the Middle Ages*, ed. J. Sweeney and S. Chodorow (Ithaca, 1989), 227–28.

23. J. Noonan, *The Scholastic Analysis of Usury* (Cambridge, Mass., 1957), 134–36, 141.

24. *Glossa ordinaria*, in ibid., 40–41, 135; cf. Thomas of Chobham, *Summa confessorum*, ed. F. Broomfield (Paris, 1968), 516; O. Langholm, *Economics in the Medieval Schools* (Leiden, 1992), 56.

25. Aquinas, *Summa theologiae* II-II q. 78 art. 2 ad 5; Noonan, *Scholastic Analysis of Usury*, 143–45.

26. Noonan, *Scholastic Analysis of Usury*, 90–91; *Decretals of Gregory IX* 5 tit. 19 cap. 6, in *Corpus iuris canonici*, ed. A. Friedberg (Leipzig, 1879) 2: col. 813; Langholm, *Economics in the Medieval Schools*, 87; A. Lapidus, "Information and risk in the medieval doctrine of usury during the thirteenth century," in *Perspectives on the History of Economic Thought*, ed. W. Barber (1991), 5:23–38.

27. Peter the Chanter, *Summa*, quoted in J. Baldwin, *Masters, Princes, and Merchants: The Social Views of Peter the Chanter and His Circle* (Princeton, 1970), 2:194 n. 52; cf. 2:286; cf. Albert the Great, *In IV Sententiarum* dist. 16B art. 46, in Albert the Great, *Opera omnia*, ed. A. Borgnet (Paris, 1890–99), 29:638; Raymond of Peñafort, *Summa de poenitentia* 2 tit. 7 n. 3 at col. 540; Thomas Aquinas [or Gilles de Lessines?], *De usuris in communi et de usurarum contractibus* cap. 6, in Aquinas, *Opera omnia* (Parma, 1864), 17:419–20, also quoted in A. Crombie, *Styles of Scientific Thinking in the European Tradition* (London, 1994), 2:1315; a further passage in cap. 7, quoted and discussed in I. Kantola, *Probability and Moral Uncertainty in Late Medieval and Early Modern Times* (Helsinki, 1994), 53.

28. Robert de Courçon, *De usura*, ed. and trans. G. Lefèvre (Lille, 1902), 61; see Baldwin, *Masters, Princes, and Merchants*, 275.

29. *Decretals of Gregory IX* 5 tit. 9 cap. 19, in Friedberg, *Corpus iuris canonici* 2: col. 816.

30. Noonan, *Scholastic Analysis of Usury*, 137; G. Coulton, "An episode in canon law," *History* 6 (1921): 67–76.

31. Noonan, *Scholastic Analysis of Usury*, 114, 129, 138–40, 148; Raymond of Peñafort, *Summa de poenitentia* 2 tit. 7 n. 7 at col. 544; cf. Antonius de Rosellis, *Tractatus de usuris* (Rome, ca. 1488; *Goldsmiths'-Kress Library of Economic Literature*, item 6.7), 4.

32. W. Cunningham, *The Growth of English Industry and Commerce* (Cambridge, 1890), 326 n. 1, 331; for the position of later English law, see the case of 1608 in *English Reports*, 79:182.

33. Hostiensis, *In Quintum decretalium librum commentaria* tit. 19 cap. 19 (Turin, 1965), fols. 59A–60.

34. D. Burr, "The apocalyptic element in Olivi's critique of Aristotle," *Church History* 40 (1971): 15–29.

35. D. Burr, "Olivi's apocalyptic timetable," *Journal of Medieval and Renaissance Studies* 11 (1981): 237–60, at 243; D. Burr, "Olivi, apocalyptic expectation, and visionary experience," *Traditio* 41 (1985): 273–88, at 285.

36. B. Tierney, *The Origins of Papal Infallibility, 1150–1350* (Leiden, 1972), 116; B. Tierney, "John Peter Olivi and papal inerrancy," in B. Tierney, *Rights, Laws and Infallibility in Medieval Thought* (Aldershot, 1997), 315–28; also J. Kilcullen, "Ockham and infallibility," *Journal of Religious History* 16 (1991): 387–409.

37. D. Burr, "The persecution of Peter Olivi," *Transactions of the American Philosophical Society* 66 (1976), pt. 5, 43, 45.

38. D. Burr, *Olivi and Franciscan Poverty* (Philadelphia, 1989); G. Bazzichi, "La proprietà secondo tre pensatori francescani del Medioevo: Pietro di Giovanni Olivi, Guglielmo Ockham e Alvaro Pelagio," *Rivista di politica economica*

75 (1985): 569–92; K. Madigan, "Aquinas and Olivi on evangelical poverty," *Thomist* 61 (1997): 567–86.

39. Langholm, *Economics in the Medieval Schools*, ch. 14; J. Kirshner and K. Lo Prete, "Peter John Olivi's treatises on contracts of sale, usury and restitution: Minorite economics or minor works?" *Quaderni fiorentini per la storia del pensiero giuridico moderno* 13 (1984): 233–86; D. Dixon, "Marketing as production: The development of a concept," *Journal of the Academy of Marketing Science* 18 (1990): 337–43.

40. G. Todeschini, ed., *Un trattato di economia politica francescana: il "De emptionibus et venditionibus, de usuris, de restitutionibus" di Pietro di Giovanni Olivi* (Rome, 1980), 82.

41. Ibid., 84; another subtraction at 112.

42. Ibid., 91.

43. Ibid., 109–10.

44. Duns Scotus, *In IV Sententiarum* dist. 15; J. Buridan, *Quaestiones et dubia in Aristotelis politica* 1, q. 13 (*Goldsmiths'-Kress Library of Economic Literature*, item 6.8), at fol. 18r.

45. P. Gerber, "Prometheus born: The High Middle Ages and the relationship between law and economic conduct," *St. Louis University Law Journal* 38 (1994): 673–738, esp. 711–13; "Usury and the medieval English courts," in R. Helmholz, *Canon Law and the Law of England* (London, 1987), 323–39; G. Seabourne, "Controlling commercial morality in late medieval London: The usury trials of 1421," *Journal of Legal History* 19 (1998): 116–42; B. Nelson, "The usurer and the merchant prince: Italian businessmen and the ecclesiastical law of restitution," supp., *Journal of Economic History* 7 (1947): 104–22; F. Galassi, "Buying a passport to heaven: Usury, restitution and the merchants of medieval Genoa," *Religion* 22 (1992): 313–26.

46. Survey in J. Kirshner, "Les travaux de Raymond de Roover sur la pensée économique des scolastiques," *Annales* 30 (1975): 318–38.

47. P. Milgrom, D. North, and B. Weingast, "The role of institutions in the revival of trade: The law merchant, private judges and the Champagne fairs," *Economics and Politics* 2 (1990): 1–23; A. Greif, P. Milgrom, and B. Weingast, "The merchant gild as a nexus of contracts," Working Paper 70 (Stanford Law School, John M. Olim Program in Law and Economics, 1990); A. Greif, "Institutions and international trade: Lessons from the commercial revolution," *American Economic Review* 82 (1992): 128–33; B. Benson, "The spontaneous evolution of commercial law," *Southern Economic Journal* 55 (1989): 644–61; more generally, F. Fukuyama, *Trust* (New York, 1995).

48. Todeschini, *Un trattato di economia politica francescana*, 71.

49. *Corpus iuris civilis, Institutes* 13.23.3; Todeschini, *Un trattato di economia politica francescana*, 110, 82.

50. *Digest* 13.4.2.8; 12.1.31; 18.6.20; *Code* 7.47.1.

51. K. Pribram, *A History of Economic Reasoning* (Baltimore, 1983), 639 n. 34.

52. H.-P. Baum, "Annuities in late medieval Hanse towns," *Business History*

Review 59 (1985): 24–48, at 28–30; F. Veraja, *Le origini della controversia teologica sull contratto di censo nel XIII secolo* (Rome, 1960), 22; cf. R. Necoechea, "Retirement made easy: Annuities in medieval England (Ph.D. diss., UCLA, 1989); W. Ogris, *Der mittelalterliche Leibrentenvertrag* (Vienna, 1961).

53. D. Nicholas, *The Metamorphosis of a Medieval City* (Lincoln, 1987), 214–18.

54. A. Jack, *An Introduction to the History of Life Assurance* (London, 1912), 175; M. Postan, ed., *The Cambridge Economic History of Europe* (Cambridge, 1963), 3: 531–32.

55. Veraja, *Le origini della controversia teologica*, 22; cf. J. Tracy, *A Financial Revolution in the Habsburg Netherlands* (Berkeley, 1985), 93 n. 58.

56. Veraja, *Le origini della controversia teologica*, 24–25.

57. Goffredo di Trani, *Summa in titulos decretalium* 5.19 (Venice, 1586): fol. 214v, quoted in Veraja, *Le origini della controversia teologica*, 30–31; cf. Langholm, *Economics in the Medieval Schools*, 97; further reasoning in William of Rennes, "Gloss on the *Summa* of Raymond of Peñafort," quoted in Veraja, *Le origini della controversia teologica*, 33.

58. Henry of Ghent, *Quaestiones quodlibetales* quod. 1 q. 39; quod. 2 q. 15, in *Opera omnia*, ed. R. Macken (Louvain, 1979–), 1.217–18, 6.97; Langholm, *Economics in the Medieval Schools*, 273–75.

59. A. Hamelin, *Un traité de morale économique au XIVe siècle: Le tractatus de usuris de maître Alexandre d'Alexandrie* (Louvain, 1962), 152–53; further refs. in B. Schnapper, "Les rentes chez les théologiens et les canonistes du XIIIe au XVIe siècle," *Etudes d'histoire du droit canonique dediée à Gabriel le Bras* (Paris, 1965), 965–95, at 971; discussion in Kantola, *Probability and Moral Uncertainty*, 54–56.

60. Hamelin, *Un traité de morale économique*, 154, 156, 157.

61. St. Antoninus, *Summa theologica* pt. 2 tit. 1 cap. 8 (Graz, 1959), 2: col. 139.

62. Jack, *Introduction to the History of Life Assurance*, 172; C. vom Hagen, *Tractatus de usu usurarum, et annuorum redituum ac interesse* (Wittenberg, 1631), cap. 11 at 147–58.

63. R. Swanson, *Church and Society in Late Medieval England* (Oxford, 1989), 236–37.

64. Baldus, *Consilia* 3: cons. 210 (Venice, 1575) 3: fol. 61r; 2: cons. 154 at 2: fol. 41r, with ref. to *Digest* 19.1.12 on the cast of a net.

65. Baldus, *Consilia* 5: cons. 292 at 5: fol. 73r; cf. Hostiensis, *In Quintum decretalium librum commentaria* tit. 19 cap. 6 at fol. 57r.

66. *Code* 4.44.2; cf. Aquinas, *Summa theologiae* II-II q. 77 art. 1 ad 1.

67. Laurentius de Ridolphis, *De usura*, in *Tractatus universi iuris* (Venice, 1584) 7: fols. 15r–50r at 43r.

68. J. Mundy, *Europe in the High Middle Ages, 1150–1309* (London, 1973), 182; Ghent's default on life annuities in Tracy, *Financial Revolution*, 14; laments on the diversion from productive investment in C. Cipolla, *The Monetary Policy of Fourteenth-Century Florence* (Berkeley, 1982), 60.

69. Tracy, *Financial Revolution*, 11; cf. J. Kirshner, "Conscience and public finance: A questio disputata of John of Lignano on the public debt of Genoa," in *Philosophy and Humanism*, ed. E. Mahoney (Leiden, 1976), 434–53; R. Mueller, *Money and Banking in Medieval and Renaissance Venice* (Baltimore, 1997), 2: chs. 11, 13, 14.

70. A. Molho, *Florentine Public Finances in the Early Renaissance* (Cambridge, Mass., 1971), 138–41.

71. I. Origo, *The World of San Bernardino* (London, 1963), 91; Bernardino's dependence on Olivi in Todeschini, *Un trattato di economia politica francescana*, 81 n. 42.

72. *Digest* 50.8.2.7; cf. ibid., 19.2.9.3; *Code* 4.24.6; Baldus, *Code* 4.34.1, in *Opera omnia* 6: fol. 97v; *Tractatus universi iuris* 7: fols. 195v–224v.

73. *Digest* 17.1.39.

74. Maffei, *Il caso fortuito nell'età dei glossatori*, 86–95.

75. F. Edler de Roover, "Early examples of marine insurance," *Journal of Economic History* 5 (1945): 172–200, at 181; F. Melis, *Origini e sviluppi delle assicurazione in Italia* (Rome, 1975), 1: table 1; cf. J. Ball, *Merchants and Merchandise* (London, 1977), 180; L. Boiteux, *La fortune de mer: Le besoin de sécurité et les débuts de l'assurance maritime* (Paris, 1968), chs. 4–5; the phrase *accipiebat risicum* in a 1336 document in G. Stefani, *Insurance in Venice from the Origins to the End of the Serenissima* (Trieste, 1958), 1:73; further on estimates of risk for economic purposes in L. Thorndike, *History of Magic and Experimental Science* (New York, 1934), 4:144; cf. J. Dotson, "Safety regulations for galleys in mid-fourteenth century Genoa: Some thoughts on medieval risk management," *Journal of Medieval History* 20 (1994): 327–36.

76. Noonan, *Scholastic Analysis of Usury*, 303; Melis, *Origini e sviluppi delle assicurazione in Italia*, 185–86; Boiteux, *La fortune de mer*, 80; Islamic parallels in V. Rispler, "Insurance in the world of Islam: origins and current practice" (Ph.D. diss., UC Berkeley, 1985), 29–30.

77. H. Nelli, "The earliest insurance contract: A new discovery," *Journal of Risk and Insurance* 39 (1972): 215–20; Melis, *Origini e sviluppi delle assicurazione in Italia*, 184–85.

78. Edler de Roover, "Early examples of marine insurance," 183.

79. Boiteux, *La fortune de mer*, 182, 190; I. Origo, *The Merchant of Prato* (London, 1963), 138–40.

80. A. Garcia i Sanz and M.-T. Ferrer i Mallol, *Assegurances i canvis marítims medievals a Barcelona* (Barcelona, 1983); J. Heers, "Le prix de l'assurance maritime à la fin du moyen âge," *Revue d'histoire économique et sociale* 37 (1959): 7–19; J. Heers, *Gênes au XVe siècle* (Paris, 1961), 206–14; M. del Treppo, "Assicurazioni e commercio internazionale a Barcellona nel 1428–1429," *Rivista storica italiana* 69 (1957): 508–41, 70 (1958): 44–81; M. del Treppo, *I mercanti catalani e l'espansione della corona aragonese nel secolo XV* (Naples, 1968), 403–522, 640–725; Boiteux, *La fortune de mer*; R. Lopez and W. Irving, *Medieval Trade in the*

Mediterranean World (New York, 1955), 169–70, 255–65; further refs. in B. Kedar, *Merchants in Crisis* (New Haven, 1976), 221 nn. 35–36.

81. Stefani, *Insurance in Venice*, 61, 83–84; Melis, *Origini e sviluppi delle assicurazione in Italia*, 271; K. Nehlsen-von Stryk, *L'assicurazione marittima a Venezia nel XV secolo* (Rome, 1988), 133–37.

82. M.-T. Ferrer i Mallol, "Intorno all'assicurazione sulla persona di Filipozzo Soldani, nel 1399, e alle attività dei Soldani, mercanti fiorentini, a Barcellona," in *Studi in Memoria di Federigo Melis* (Naples, 1978), 2:441–78, at 477–78; Melis, *Origini e sviluppi delle assicurazione in Italia*, 210–12; Jack, *Introduction to the History of Life Assurance*, 202; Stefani, *Insurance in Venice*, 81; R. Smith, "Life insurance in fifteenth century Barcelona," *Journal of Economic History* 1 (1941): 57–59; Heers, *Gênes au XVe siècle*, 214–15.

83. Stefani, *Insurance in Venice*, 76, 207.

84. Edler de Roover, "Early examples of marine insurance," 196; Stefani, *Insurance in Venice*, 88; Melis, *Origini e sviluppi delle assicurazione in Italia*, 215–17; cf. Nehlsen-von Stryk, *L'assicurazione marittima*, 178–79; S. Passamaneck, *Insurance in Rabbinic Law* (Edinburgh, 1974), ch. 2.

85. Heers, *Gênes au XVe siècle*, 216.

86. R. de Roover, *The Rise and Decline of the Medici Bank* (Cambridge, Mass., 1963), 151–52.

87. Baldus, *Code* 1.1.1 at fol. 6v.

88. Laurentius de Ridolphis, *De usura*, in *Tractatus universi iuris* 7: fols. 15r–50r at 38r; repeated in St. Antoninus, *Summa theologica*, pt. 2 tit. 1 cap. 7 § 46 at 2: cols. 121–22.

89. P. Santerna, *De assecurationibus et sponsionibus mercatorum* (Venice, 1552; Cologne, 1599); printed with B. Straccha, *De Assecurationibus* (*Goldsmiths'-Kress Library of Economic Literature*, item 185); repr. with notes by M. Amzalak and English, French, and Portuguese translations (Lisbon, 1971) (I am grateful to C. Pimenta for providing a copy); see D. Maffei, *Il giureconsulto portoghese Pedro de Santarém, autore del primo trattato sulle assicurazioni, 1488* (Separata do número especial do Boletim da Faculdade de Direito de Coimbra, Coimbra, 1983, 703–28).

90. Santerna, *Assecurationibus et sponsionibus mercatorum* pt. 1 §§ 7, 16; pt. 3 §§ 33 and 46 (Amzalak edition, 204, 207, 254, 261).

91. Ibid., pt. 3 § 55 (265).

92. Ibid., §§ 72–82 (274–82).

93. Ibid., pt. 5 §§ 3 and 4 (336).

94. Ibid., § 6 (337).

95. V. Barbour, *Capitalism in Amsterdam in the Seventeenth Century* (Ann Arbor, 1963), 33–35; E. Coornaert, *Les francais et le commerce international à Anvers* (Paris, 1961), 2:234–41; V. Vasquez de Prada, *Lettres marchandes d'Anvers* (Paris, 1960), 1:133–35, 243–44; J. Ashton, *The History of Gambling in England* (Montclair, N.J., 1969), 275–76.

96. *Select Pleas in the Court of Admiralty*, Publications of the Selden Society 11 (London, 1897), 2:47–57; H. Cockerell and E. Green, *The British Insurance Business, 1547–1970* (London, 1970), 4, 34–35; but cf. A. Thomas, ed., *Calendar of Plea and Memoranda Rolls, 1413–1427* (Cambridge, 1943), 208–10.

97. Nörr, "Bologna and the Court of Admiralty"; D. Coquillette, "Legal incorporation and ideology IV: The nature of civilian influence on modern Anglo-American commercial law," *Boston University Law Review* 67 (1987): 877–934; L. Trakman, *The Law Merchant: The Evolution of Commercial Law* (Littleton, 1983), ch. 2; D. Coquillette, *The Civilian Writers of Doctors Commons, London* (Berlin, 1988), pt. 2; J. Baker, "The Law Merchant and the Common Law before 1700," *Cambridge Law Journal* 38 (1979): 295–322; W. Vance, "The early history of insurance law," in *Select Essays in Anglo-American Legal History* (Boston, 1909), 3:98–116.

98. V. Barbour, "Marine risks and insurance in the seventeenth century," *Journal of Economic and Business History* 1 (1928–29): 561–96.

99. J. Nider, *De contractibus mercatorum*, sec. 3 (*Goldsmiths'-Kress Library of Economic Literature*, item 2; supp. item 6.0–1); trans. C. Reeves as *On the Contracts of Merchants* (Norman, 1966), 41–42; cf. 17, 35–36; the latter statement in later law in J. de Damhouder, *Sententiae selectae* (Antwerp, 1601; repr. Aalen, 1978), 162.

100. Nider, *De contractibus mercatorum*, sec. 2, 28; cf. sec. 2, 30, 32, 50.

101. Ibid., sec. 4, 62–63.

102. J. Nider, *Formicarius*, in Kramer and Sprenger, *Malleus Maleficarum* (Lyons, 1669), 1:339.

103. R. Ehrenberg, *Capital and Finance in the Age of the Renaissance*, trans. H. Lucas (New York, 1928), 234.

104. H. Pirenne, *Histoire de Belgique* (Brussels, 1953), 3:280–81, 284; H. Soly, "Grandspeculatie en kapitalisme te Antwerpen in de 16e eeuw," *Economisch en Sociaal Tijdschrift* 27 (1973); H. Soly, "Urbanisme en kapitalisme te Antwerpen in de 16de eeuw," *Historische Uitgaven Pro Civitate*, no. 47 (1977).

105. Barbour, *Capitalism in Amsterdam*, 74.

106. Azo, *Summa codicis* 3.43–44, in I. Caccialupus, *De ludo*, in *Tractatus universi iuris* (Venice, 1584), 7: fols. 155r–161v at 157r; cf. Paris de Puteo, *De ludo*, in *Tractatus universi iuris*, 7: fols. 151r–155r, at 153v.

107. H. van der Wee, *The Growth of the Antwerp Market and the European Economy* (The Hague, 1963), 2:364–65.

108. Ehrenberg, *Capital and Finance*, 240–42; R. Ehrenberg, *Le siècle des Fugger* (Paris, 1955), 208–9; cf. C. Bühler, "Sixteenth century prognostications," *Isis* 33 (1941–42): 609–20.

109. P. Jeannin, *Merchants of the Sixteenth Century*, trans. P. Fittingoff (New York, 1972), 107–8.

110. H. Hauser, "The European financial crisis of 1559," *Journal of Economic and Business History* 2 (1929–30): 241–55.

111. F. García, *Tratado utilísmo y muy general de todos los contractos* (Valencia,

1583), selections in M. Grice-Hutchinson, *The School of Salamanca* (Oxford, 1952), 107; further on risk and prices in L. Sarvia de la Calle, *Instrucción de mercaderes* (Medina del Campo, 1544), ch. 3, selections in Grice-Hutchinson, *School of Salamanca*, 81–82; cf. D. Soto, *De Iustitia et Iure* 6, q. 2, art. 3, in Grice-Hutchinson, *School of Salamanca*, 86.

112. Shakespeare, *Henry IV, Part II* 1.1.178–84; further refs. to "ten to one" in C. Crawford, *Marlowe Concordance* (Vaduz, 1963), 2:1270; [Kyd], *First Part of Ieronimo* 2.3.122; "a thousand to one" in G. Babington, *A Very Fruitfull Exposition of the Commandments* (1583), in G. Babington, *Works* (London, 1637), 7:68; see D. Bellhouse and J. Franklin, "The language of chance," *International Statistical Review* 65 (1997): 73–85.

113. J. Palsgrave, *L'éclaircissement de la langue francaise* (London, 1530; Paris, 1852), 712.

114. T. Howell, ed., *A Complete Collection of State Trials* (London, 1816), 1: cols. 393–94; the argument but not the precise numbers appear in Roper's account, *Two Early Tudor Lives*, ed. R. Sylvester and D. Harding (New Haven, 1962), 249, but the numbers are probably original: J. Derrett, "The trial of Sir Thomas More," in *Essential Articles for the Study of Thomas More*, ed. R. Sylvester and G. Marc'hadour (Hamden, 1977), 72; other examples of proportions in a reference class in Anon., *Arden of Feversham* (1592), line 1256, in L. Ule, *A Concordance to the Shakespeare Apocrypha* (Hildesheim, 1987), 1:24; J. Marston, *The Malcontent* (1604), 3.1.85–91.

115. Shakespeare, *Henry V* 4.1.307–9; cf. Marlowe, *Tamburlaine the Great, Part I* 1.2.122, 143; Shakespeare, *Two Gentlemen of Verona* 1.1.72; Shakespeare, *Twelfth Night* 1.3.112.

116. R. Trexler, "Une table florentine d'espérance de vie," *Annales* 26 (1971): 137–39; J. Dupâquier, "Sur une table (prétendument) florentine d'espérance de vie," *Annales* 28 (1973): 1066–70.

117. *Oeuvres Complètes de Huygens* (La Haye, 1897), 7:95–98; G. Alter and J. Riley, "How to bet on lives: a guide to life-contingent contracts in early modern Europe," *Research in Economic History* 10 (1986): 1–53, at 29; G. Alter, "Plague and the Amsterdam annuitant," *Population Studies* 37 (1983): 23–41; see also D. Houtzager, *Hollands lijf- en losrenteleningen voor 1672* (Schiedam, 1950).

118. J.-L. Thireau, *Charles du Moulin (1500–1566). Etude sur les sources, la méthode, les idées politiques et économiques d'un juriste de la Renaissance* (Geneva, 1980), 355–59.

119. C. du Moulin, or Molinaeus, *Tractatus commerciorum, et usurarum, redituumque pecunia constitutorum, et monetarum* (Paris, 1546; *Goldsmiths'-Kress Library of Economic Literature*, item 38.1), 210–11.

120. *Mitchell v. Mulholland* [1972], 1 Queen's Bench 65, at 85–86, quoted in D. Wickens, "Actuarial assistance in assessing damages," *Australian Law Journal* 48 (1974): 286–94, at 293.

121. Jack, *Introduction to the History of Life Assurance*, 206–7.

122. E. Krelage, *Bloemenspeculatie in Nederland* (Amsterdam, 1942);

N. Posthumus, "The tulip mania in Holland in the years 1636 and 1637," *Journal of Economic and Business History* 1 (1929): 434–66; W. Blunt, *Tulips and Tulipomania* (London, 1977); P. Garber, "Who put the mania in tulipmania?" *Journal of Portfolio Management* 16 (1) (Fall 1989): 53–60.

123. J. van Dillen, "Isaac le Maire et la commerce des actions de la compagnie des Indes Orientales," *Revue d'histoire moderne* 10 (1935): 5–21, 121–37; N. De Marchi and P. Harrison, "Trading 'in the wind' and with guile: The troublesome matter of short selling of shares in 17th-century Holland," supp. *History of Political Economy* 26 (1994): 47–65; S. Schama, *The Embarrassment of Riches* (New York, 1987), 347–50.

124. J. Nef, *Cultural Foundations of Industrial Civilization* (Cambridge, 1958), 12–14; later history in J. Klein, *Statistical Visions in Time: A History of Time Series Analysis, 1662–1938* (Cambridge, 1997).

125. F. Endemann, *Beiträge zur Geschichte der Lotterie und zum Heutigen Lotterierechte* (Bonn, 1882), 34–36; J. Handelsman, *La loterie d'état en Pologne et dans les autres pays d'Europe* (Paris, 1933), 5–6; see R. Kinsey, "The Role of Lotteries in Public Finance"(Ph.D. diss., Columbia Univ., 1959), 12–13; possible Islamic forerunners in Rosenthal, *Gambling in Islam*, 81–83.

126. C. Ewen, *Lotteries and Sweepstakes* (London, 1932), 25–28, 30; Kinsey, "Role of Lotteries," 13–14; J. van Houtte, "Anvers aux XVe et XVIe siècles," *Annales* 16 (1961): 248–78, at 277–78.

127. G. Fokker, *Geschiedenis der Loterijen in de Nederlanden* (Amsterdam, 1862), summarized in Schama, *Embarrassment of Riches*, 306–10.

128. Ewen, *Lotteries and Sweepstakes*, 34–63; R. Woodhall, "The English state lotteries," *History Today* 14 (July 1964): 497–504; P. Hughes and J. Larkin, eds., *Tudor Royal Proclamations* (New Haven, 1969), 2:291–92, 294–95, 298, 306–7.

129. Shakespeare, *Coriolanus* 5.2.10.

130. *Three Proclamations Concerning the Lottery for Virginia* (Providence, 1907); R. Johnson, "The lotteries of the Virginia Company, 1612–1621," *Virginia Magazine of History and Biography* 74 (July 1966): 259–90; Ewen, *Lotteries and Sweepstakes*, 70–88; J. Ezell, *Fortune's Merry Wheel* (Cambridge, Mass., 1960), ch. 1; J. Findlay, *People of Chance* (New York, 1986), 11–15.

131. Proverbs 18:18; Mesopotamian parallels in A. Oppenheim, *Ancient Mesopotamia*, rev. ed. (Chicago, 1977), 99–100, 208–9.

132. Talmud, Yoma 22a; A. Hasofer, "Some aspects of Talmudic probabilistic thought," *Proceedings of the Association of Orthodox Jewish Scientists* 2 (1969): 63–80, at 64.

133. Talmud, Yoma 39a; Rabinovitch, *Probability and Statistical Inference in Ancient and Medieval Jewish Literature*, 27; cf. 30; see also A. Hasofer, "Random mechanisms in Talmudic literature," *Biometrika* 54 (1967): 316–21, also in *Studies in the History of Statistics and Probability*, ed. E. Pearson and M. Kendall (London, 1970), 1:39–43.

134. T. Gataker, *Of the Nature and Use of Lots* (London, 1619), 146–47, in

D. Bellhouse, "Probability in the sixteenth and seventeenth centuries: An analysis of Puritan casuistry," *International Statistical Review* 56 (1988): 63–74, at 69; also N. Rescher, "The philosophers of gambling," in Brown and Mittelstrass, *An Intimate Relation*, 203–20.

135. Gataker, *Nature and Use of Lots*, 159; Bellhouse, "Probability in the sixteenth and seventeenth centuries," 70.

136. P. Clark, s.v. "Lots," in *Dictionary of the Apostolic Church*, ed. J. Hastings (New York, 1916), 1:710–13.

137. W. Ames, *Medulla SS. Theologiae*, 3d ed. (London, 1629), trans. as *The Marrow of Sacred Divinity* (London, 1638), 262; Bellhouse, "Probability in the sixteenth and seventeenth centuries," 71; cf. G. Mosse, *The Holy Pretence: A Study in Christianity and Reason of State from William Perkins to John Winthrop* (New York, 1968), 77, 80.

138. T. Gataker, *T. Gatakeri Londinatis Antithesis, partim G. Amessi, partim G. Voetii de sorte thesibus reposita* (London, 1638), 12–13; Bellhouse, "Probability in the sixteenth and seventeenth centuries," 72.

139. Noonan, *Scholastic Analysis of Usury*, 209–17; G. O'Brien, *An Essay on Medieval Economic Teaching* (New York, 1967), 210–11; J. Brodrick, *The Economic Morals of the Jesuits* (New York, 1972), ch. 6; R. Savelli, "Between law and morals: interest in the dispute on exchanges during the 16th century," in *The Courts and the Development of Commercial Law*, ed. V. Piergiovanni (Berlin, 1987), 39–102.

140. *Jacobi Lainez disputationes Tridentinae*, quoted in Brodrick, *Economic Morals of the Jesuits*, 124.

141. T. Wilson, *A Discourse upon Usury* (1572), ed. R. Tawney (London, 1925), 306–7; cf. 246–48; N. Jones, *God and the Moneylenders: Usury and Law in Early Modern England* (Oxford, 1989), ch. 1; the issue also in Grice-Hutchinson, *School of Salamanca*, 124; cf. Cajetan, *De cambiis* ch. 2, in Thomas de Vio, Cardinal Cajetan, *Scripta philosophica: opuscula oeconomico-socialia*, ed. P. Zammit (Rome, 1934), 99; N. Sanders, *A Brief Treatise of Usury* (Louvain, 1568; *Goldsmiths'-Kress Library of Economic Literature*, item 133), 66–67; details of how to arrange contracts to avoid risk in C. de Villalon, *Provechoso tratado de cambios y contrataciones de mercaderes* (Valladolid, 1542), in Ehrenberg, *Capital and Finance*, 243–44; cf. H. Doneau, or Donellius, *De usuris, et nautico foenore* (Paris, 1556; *Goldsmiths'-Kress Library of Economic Literature*, item 59), 32–34.

142. W. Wallace, "The enigma of Domingo Soto: *uniformiter difformis* and falling bodies in late medieval physics," *Isis* 59 (1968): 384–401; W. Wallace, "Duhem and Koyré on Domingo de Soto," *Synthese* 83 (1990): 239–60; C. Calderón, "The 16th-century Iberian calculatores," *Revista de la Union Matematica Argentina* 35 (1989): 245–58; R. Hernández, "The internationalization of Francisco de Vitoria and Domingo de Soto," *Fordham International Law Journal* 15 (1992): 1031–59; D. Soto, *De dubio et opinione*, discussed in Kantola, *Probability and Moral Uncertainty*, 117–23; A. Chafuen, *Christians for Freedom: Late Scholastic Economics* (San Francisco, 1986), 126, 135, 146.

143. D. Soto, *De Iustitia et Iure* 6 q. 7 (Lyons, 1569), 207; cf. B. Straccha, *De*

Assecurationibus (Venice, 1569), preface, sec. 13, 46, 79–80 (*Goldsmiths'-Kress Library of Economic Literature*, item 145.4), on which see L. Franchi, *Benvenuto Straccha: note bio-bibliographice* (Florence, 1975), 153–58; P. de Navarra, *De ablatorum restitutione in foro conscientiae*, 2d ed. (Lyons, 1593), 2:261–62, quoted in Crombie, *Styles of Scientific Thinking in the European Tradition*, 2:1320; R. de Roover, "The Scholastics, usury and foreign exchange," *Business History Review* 41 (1967): 257–71, at 269.

144. B. Gordon, *Economic Analysis before Adam Smith: Hesiod to Lessius* (London, 1975), ch. 9; J. Schumpeter, *History of Economic Analysis* (London, 1954), 94–107, 116; J. Olsem, *Deus in Oeconomico: Lessius, Molina, Lugo, artisans de la révolution Galileo-Cartésienne dans la théorie de la valeur* (Besançon, 1980); R. Beutels, *Leonardus Lessius (1554–1623): Portret van een Zuidnederlandse laat-scholastieke econoom* (Wommelgem, 1987); T. van Houdt, "Tradition and renewal in late Scholastic economic thought: The case of Leonardus Lessius," *Journal of Medieval and Early Modern Studies* 28 (1998): 51–73.

145. L. Lessius, *De Iustitia et Iure*, quoted in Gordon, *Economic Analysis before Adam Smith*, 252; cf. Brodrick, *Economic Morals of the Jesuits*, 14.

146. Lessius, *De Iustitia et Iure* 2 cap. 22 dub. 6 (Lyons, 1622), 273; cf. dub. 9–10 at 275–77; C. vom Hagen, *Tractatus de usu usurarum* cap. (Wittenberg, 1631), at 147–58.

147. Lessius, *De Iustitia et Iure* 2 cap. 25 dub. 2–3 at 304–7, partly quoted in E. Coumet, "La théorie du hasard est-elle née par hasard?" *Annales* 25 (1970): 574–98, at 592; 2 cap. 28 disp. 4 at 321.

148. J. de Lugo, *De Iustitia et Iure* disp. 31 sec. 7 (Lyons, 1652), 2:447.

149. Ibid.; cf. H. Grotius, *The Jurisprudence of Holland* 3, ch. 24.5, trans. R. Lee (Oxford, 1926), 1:421.

150. Noonan, *Scholastic Analysis of Usury*, 289.

151. P. Bauny, *Somme des péchez qui se commettent en tous estats, de leurs conditions et qualitez* (Paris, 1653), 227, quoted in Coumet, "La théorie du hasard," 592 n. 1; cf. 591.

Chapter 11: Dice

1. G. Lenski and J. Lenski, *Human Societies*, 3d ed. (New York, 1978), 136.

2. M. Ascher, *Ethnomathematics* (Pacific Grove, Calif., 1991), 87–89, referring to S. Culin, *Games of the North American Indians* (New York, 1975); further in E. Tylor, "On American lot-games, as evidence of Asiatic intercourse before the time of Columbus," supp., *International Archives for Ethnography* 9 (1896): 56–66, repr. in *The Study of Games*, ed. E. Avedon and B. Sutton-Smith (New York, 1971), 77–93.

3. K.-H. Hagstroem, *Les préludes antiques de la théorie des probabilités* (Stockholm, 1932), chs. 7–11, based on L. Becq de Fouquières, *Les jeux des anciens*, 2d ed. (Paris, 1873).

4. F. David, *Games, Gods and Gambling* (London, 1962), 15–16; S. Kunoff and S. Pines, "Teaching elementary probability through its history," *College Mathematics Journal* 17 (1986): 210–19.

5. Aristotle, *On the Heavens* 292a30.

6. Cicero, *De divinatione* 1.xii.23–1.xiv.25; 1.lix.121.

7. Aristotle, *Rhetoric* 1407a37–b5, in *The Complete Works of Aristotle*, ed. J. Barnes (Princeton, 1984).

8. Aristotle, *On Prophecy in Sleep* 463b19–23, trans. W. Ross, in Barnes, *Complete Works of Aristotle*.

9. H. Lamer, s.v. "Lusoria tabula," in *Paulys Realencyclopädie der Classischen Altertumswissenschaft* (Stuttgart, 1927), 13, 2: cols. 1900–2029; H. Lüders, *Das Würfelspiel im alten Indien* (Göttingen, 1907); K. de Vreese, "The game of dice in ancient India," *Orientalia Neerlandica* (Leiden, 1948): 349–62; C. Panduranga Bhatta, *Dice-Play in Sanskrit Literature* (Delhi, 1985); W. Tauber, *Das Würfelspiel im Mittelalter und in der frühen Neuzeit* (Frankfurt, 1987), Einleitung; F. Rosenthal, *Gambling in Islam* (Leiden, 1975), 34–37.

10. David, *Games, Gods and Gambling*, 22; Lamer, "Lusoria tabula," col. 2023; L. Maistrov, *Probability Theory: A Historical Sketch* (New York, 1974), 10–11; Avedon and Sutton-Smith, *Study of Games*, 123; D. Bennett, *Randomness* (Cambridge, Mass., 1998), 18–27; illustrations in O. Dilke, *Mathematics and Measurement* (London, 1987), 54.

11. Xenophon, *Hellenica* 6.3.16.

12. *Corpus iuris civilis, Code* 3.43.1; *Digest* 11.5.

13. *Decretals of Gregory IX* 3 tit. 1 cap. 15, in Friedberg, *Corpus* 2:454; Raymond of Peñafort, *Summa de poenitentia* 2 tit. 8 n. 11 at col. 572.

14. Aquinas, *Summa theologiae* II-II q. 32 art. 7 ad 3; cf. Nikolaus von Dinkelsbühl, *Predigt über das Spiel* (1414), in Tauber, *Würfelspiel im Mittelalter*, 93–99; D. Soto, *De Iustitia et Iure* 4, q. 5 art. 2; quoted in Crombie, *Styles of Scientific Thinking*, 2:1319.

15. Baldus, *Code* 3.43 and 44, quoted in T. Malvetius, *De sortibus*, pt. 2, in *Tractatus universi iuris*, 11: pt. 2, fols. 398r–402r, at 400r; cf. Stephanus Costa, *De ludo*, in *Tractatus universi iuris* (Venice, 1584), 7: fols. 161v–168v, at 166v; J. Ashton, *The History of Gambling in England* (Montclair, N.J., 1969), 13–15.

16. Azo, *Summa codicis* 3.43; see L. Zdekauer, *Il gioco d'azzardo nel medioevo italiano*, ed. G. Ortalli (Florence, 1993), 98.

17. Baldus, *Digest* 11.5, quoted in Caccialupus, *De ludo* in *Tractatus universi iuris* 7:fols. 155r–161v at 158r–v; brief information on Costa and Caccialupus in Zdekauer, *Il gioco d'azzardo nel medioevo italiano*, 31.

18. Tauber, *Würfelspiel im Mittelalter*, 12–15; K. Meadows, *Backgammon, Its History and Practice* (New York, 1931).

19. I. Butrigarius, *Code* 1.1.1, in Malvetius, *De sortibus* at 400r; cf. Butrigarius, *Lectura super codice* 4.33 (Bologna, 1973), fol. 135v.

20. V. Flint, *The Rise of Magic in Early Medieval Europe* (Princeton, 1991), 220–21; W. Braekman, *Fortune-Telling by the Casting of Dice: A Middle English Poem and Its Background* (Brussels, 1981); Tauber, *Würfelspiel im Mittelalter*, 30–34.

21. Tauber, *Würfelspiel im Mittelalter*, 11, 50.

22. Alfonso X el Sabio, *Libro de los dados*, in *Obras*, ed. A. Solalinde (Madrid,

1922), 2:130–32; cf. R. Ineichen, "Der schlechte Würfel—Ein selten behandeltes Problem in der Geschichte der Stochastik," *Historia Mathematica* 18 (1991): 253–61; Statute of Vicenza on the regulation of dicing, in Zdekauer, *Il gioco d'azzardo nel medioevo italiano*, 107; F. Semrau, *Würfel und Würfelspiel im alten Frankreich* (Halle, 1910); J. Norwich, *The Kingdom in the Sun* (New York, 1970), 258.

23. Tauber, *Würfelspiel im Mittelalter*, 42–45.

24. Rabelais, *Pantagruel* 3 chs. 39 and 44, trans. in *Gargantua and Pantagruel* (London, 1929), 2:40, 55; Rabelais' legal studies in J. Plattard, *The Life of François Rabelais* (London, 1930), chs. 5–6.

25. A. Martinelli, "Notes on the origin of double entry bookkeeping," *Abacus* 13 (1977): 1–27.

26. F. Pegolotti, *La pratica della mercatura*, ed. A. Evans (Cambridge, Mass., 1936), 301–2.

27. R. Franci and L. Toti Rigatelli, *Introduzione all'aritmetica mercantile del medioevo e del renascimento* (Urbino, 1982); W. van Egmond, "The Commercial Revolution and the Beginnings of Western Mathematics in Renaissance Florence, 1300–1500" (Ph.D. diss., Indiana Univ., 1976), summarized in P. Grendler, *Schooling in Renaissance Italy* (Baltimore, 1989), ch. 11; F. Swetz, *Capitalism and Arithmetic* (La Salle, Ill., 1987); R. Hadden, *On the Shoulders of Merchants: Exchange and the Mathematical Conception of Nature in Early Modern Europe* (Albany, 1994), ch. 4; J. Williams, "Mathematics and the alloying of coinage, 1202–1700," *Annals of Science* 52 (1995): 213–63.

28. R. Franci and L. Toti Rigatelli, "Towards a history of algebra from Leonardo of Pisa to Luca Pacioli," *Janus* 72 (1985): 17–82; W. van Egmond, "The earliest vernacular treatment of algebra: The Libro di Ragioni of Paolo Gerardi (1328)," *Physis* 20 (1978): 155–89.

29. B. Hughes, "An early 15th-century algebra codex: A description," *Historia Mathematica* 14 (1987): 167–72, at 169.

30. G. Sarton, *Introduction to the History of Science*, 3: pt. 2 (Baltimore, 1948), 1813; I. Todhunter, *A History of the Mathematical Theory of Probability* (Cambridge, 1865), 1; David, *Games, Gods and Gambling*, 35; R. Ineichen, "Dante—Kommentare und die Vorgeschichte der Stochastik," *Historia Mathematica* 15 (1988): 264–69; Zdekauer, *Il gioco d'azzardo nel medioevo italiano*, 22.

31. David, *Games, Gods and Gambling*, 31–34, and plates 6, 7; Tauber, *Würfelspiel im Mittelalter*, 28–29; also in M. Ayrer, *Würfelbuch* (Bamberg, 1483), in Tauber, *Würfelspiel im Mittelalter*, 146–62.

32. David, *Games, Gods and Gambling*, 35.

33. Pseudo-Ovid, *De vetula* (French, 13th century?); editions of Cologne, 1479; Wolfenbüttel, 1662; Leiden, 1967 (ed. P. Klopsch); Amsterdam, 1968 (ed. D. Robothan); relevant part in M. Kendall, "The beginnings of a probability calculus," *Biometrika* 43 (1956): 1–14, also in Pearson and Kendall, *Studies in the History of Statistics and Probability* 1:19–34; David, *Games, Gods and Gambling*, 32–34.

34. L. Toti Rigatelli, "Il 'problema delle parti' in manoscritti del XIV e XV

secolo," in *Mathemata: Festschrift für Helmuth Gericke*, ed. M. Folkerts and U. Lindgren (Stuttgart, 1985), 229–36, at 234–35 (I am grateful to M. Cowling for help in understanding this passage); also in I. Schneider, "The market place and games of chance in the fifteenth and sixteenth centuries," in *Mathematics from Manuscript to Print, 1300–1600*, ed. C. Hay (Oxford, 1988), 220–35, at 227–29.

35. Contrary to Toti Rigatelli, "Il 'problema delle parti' in manoscritti," 233.

36. Schneider, "The market place and games of chance," 229.

37. H. Murray, *A History of Chess* (Oxford, 1913), 408–11, 458 n. 12; Alfonso X el Sabio, *Libro de las tablas*, in *Obras*, 2:133–38, at 136–37.

38. Toti Rigatelli, "Il 'problema delle parti' in manoscritti," 230–32; cf. H. Kurath, ed., *Middle English Dictionary* (Ann Arbor, 1954–), s.v. "juparti."

39. E. Coumet, "Le problème des partis avant Pascal," *Archives internationales d'histoire des sciences* 17 (1965): 244–72, at 248–53; David, *Games, Gods and Gambling*, 37–38; I. Schneider, "Luca Pacioli und das Teilungsproblem: Hintergrund und Lösungsversuch," in Folkerts and Lindgren, *Mathemata*, 237–46.

40. Coumet, "Le problème des partis avant Pascal," 254.

41. O. Ore, *Cardano, the Gambling Scholar* (Princeton, 1953), 36; generally, I. Maclean, "The interpretation of natural signs: Cardano's *De subtilitate* versus Scaliger's *Exercitationes*," in *Occult and Scientific Mentalities in the Renaissance*, ed. B. Vickers (Cambridge, 1984), 231–49.

42. A. Grafton, "Girolamo Cardano and the tradition of classical astrology," *Proceedings of the American Philosophical Society* 142 (1998): 323–54, at 338.

43. M. Fierz, *Girolamo Cardano*, trans. H. Niman (Boston, 1983), 127–28; G. Cardano, *Commentarium in Ptolomaeum de Astrorum iudiciis* preface, in Cardano, *Opera omnia* (Lyons, 1663), 5:93; generally in N. Siraisi, *The Clock and the Mirror: Girolamo Cardano and Renaissance Medicine* (Princeton, 1997), pt. 4; cf. N. Campion, "Astrological historiography in the Renaissance," in *History of Astrology*, ed. A. Kitson (London, 1989), 89–136, at 102–3.

44. Fierz, *Girolamo Cardano*, 134, 136, 131; cf. D. Allen, *Doubt's Boundless Sea* (Baltimore, 1964), 49–52.

45. Ore, *Cardano, the Gambling Scholar*, 22; W. Shumaker, ed., *Renaissance Curiosa: John Dee's Conversation with Angels, Girolamo Cardano's Horoscope of Christ, Johannes Trithemius and Cryptography, George Dalgarno's Universal Language* (Binghamton, 1982).

46. Ore, *Cardano, the Gambling Scholar*, ch. 3.

47. G. Cardano, *Book of My Life* chs. 19, 13, 25, trans. J. Stoner (New York, 1962), 73, 54, 87.

48. G. Cardano, *Practica arithmetica et mensurandi singularis* (Milan, 1539), ch. 69, in *Opera omnia* 4:214; Coumet, "Le problème des partis avant Pascal," 261.

49. Cardano, *Practica arithmetica* ch. 61, in *Opera omnia* 4:112; Coumet, "Le problème des partis avant Pascal," 262–64.

50. Ore, *Cardano, the Gambling Scholar*, 122.

51. G. Cardano, *Liber de ludo aleae*, in *Opera omnia* 1:262–76, trans. S. Gould,

in Ore, *Cardano, the Gambling Scholar,* ch. 9 (*Opera omnia* 1:264, Ore, *Cardano, the Gambling Scholar,* 193–94); David, *Games, Gods and Gambling,* 58–59.

52. Cardano, *De ludo aleae* ch. 11, in *Opera omnia* 1:264–65, Ore, *Cardano, the Gambling Scholar,* 195–96.

53. Cardano, *De ludo aleae* ch. 12, in *Opera omnia* 1:265, Ore, *Cardano, the Gambling Scholar,* 197.

54. Cardano, *De ludo aleae* ch. 14, in *Opera omnia* 1:266, Ore, *Cardano, the Gambling Scholar,* 202.

55. Cardano, *De ludo aleae* ch. 15, in *Opera omnia* 1:267, Ore, *Cardano, the Gambling Scholar,* 204–5.

56. Cardano, *De ludo aleae* ch. 30, in *Opera omnia* 1:274, Ore, *Cardano, the Gambling Scholar,* 233.

57. G. Walker [?], *A Manifest Detection of the Most Vile and Detestable Use of Diceplay* (London, 1552), in A. Judges, *The Elizabethan Underworld* (London, 1930), 26–50, at 33; cf. 44.

58. Judges, *Elizabethan Underworld,* 39; see D. Bellhouse, "The role of roguery in the history of probability," *Statistical Science* 8 (1993): 410–20.

59. R. Greene, *A Notable Discouerie of Coosenage* (London, 1591), in Judges, *Elizabethan Underworld,* 130.

60. A. Cotton, *The Compleat Gamester* (London, 1674), repr. in *Games and Gamesters of the Restoration,* ed. C. Hartmann (London, 1930), ch. 34; discussion in Bellhouse, "Role of roguery."

61. Lessius, *De Iustitia et Iure* (Lyons, 1662) 2, cap. 26, dub. 5 (314–15); de Escobar y Mendoza, *Liber theologiae moralis* (Lyons, 1644) tract. 3 ex. 12 cap. 1 (400–401); cf. J. Caramuel Lobkowitz, *Mathesis Biceps* (Campania, 1670), 998.

62. A. Diana, *Summa Diana,* ed. A. Cotonio (Madrid, 1646; Antwerp, 1656, etc.), quoted in Caramuel, *Mathesis Biceps,* 999; the material is from the author's *Resolutiones morales* (Palermo, 1629; Lyons, 1633; Venice, 1636, etc.); cf. Jonsen and Toulmin, *Abuse of Casuistry,* 156.

63. Galileo, "Sopra le scoperte dei dadi," in *Opere,* ed. A. Favaro (Florence, 1890–1909), 8:591–94; trans. in David, *Games, Gods and Gambling,* 192–95, commentary 65–67.

64. Maistrov, *Probability Theory,* 30.

65. S.-D. Poisson, *Recherches sur la probabilité des jugements en matière criminelle et en matière civile* (Paris, 1837), 1.

66. Pascal to Fermat, July 29,1654, in Pascal, *Oeuvres complètes,* ed. J. Mesnard (Paris, 1970), 2:1137; trans. in D. Smith, *A Source Book in Mathematics* (New York, 1959), 2:547–48; also in David, *Games, Gods and Gambling,* 229–53.

67. Pascal, *Oeuvres complètes* 2:1142; Smith, *Source Book in Mathematics,* 552.

68. O. Ore, "Pascal and the invention of probability theory," *American Mathematical Monthly* 67 (1960): 409–19, at 412.

69. *Oeuvres de Blaise Pascal,* ed. L. Brunschvicg, P. Boutroux, and F. Gazier (Paris, 1914), 9:215–23, at 217–18; Pascal, *Oeuvres complètes,* ed. J. Mesnard (Paris, 1991), 3:348–59, at 355.

70. J. Mesnard, *Pascal et les Roannez* (Paris, 1965), 1:370.

71. Brunschvicg, Boutroux, and Gazier, *Oeuvres de Blaise Pascal*, 9:225–26; cf. Leibniz, *Philosophischen Schriften*, ed. C. Gerhardt (Hildesheim, 1978), 5:447.

72. Mesnard, *Pascal et les Roannez*, 1:372.

73. Pascal, *Oeuvres complètes* 2:1142; Smith, *Source Book in Mathematics*, 552.

74. J. Franklin, "Achievements and fallacies in Hume's account of infinite divisibility," *Hume Studies* 20 (1994): 85–101.

75. L. Auger, *Un savant méconnu: Gilles Personne de Roberval* (Paris, 1962), 146–47; cf. Huygens to Roberval, Apr. 18, 1656, in *Oeuvres complètes de Huygens* 1:404; *Oeuvres de Fermat*, ed. P. Tannery and C. Henry (Paris, 1894), 2:290, 302, 310; P. Dupont, "Concetti probabilistici in Roberval, Pascal e Fermat," *Rendiconti del Seminario Matematico dell'Università e del Politecnico di Torino* 34 (1975–76): 235–45.

76. Mesnard, *Pascal et les Roannez*, 1:370; Brunschvicg, Boutroux, and Gazier, *Oeuvres de Blaise Pascal*, 9:226.

77. D. Wetsel, "Pascal and Mitton: Theological objections to 'l'honnêteté' in the *Pensées*," *French Studies* 47 (1993): 404–11; Mesnard, *Pascal et les Roannez*, 1:376.

78. Pascal, *Oeuvres complètes* 2:1136; Smith, *Source Book in Mathematics*, 546–47.

79. Seneta, "Pascal and probability," 225–33, at 227–28; Thirouin, *Le hasard et les règles*, ch. 6.

80. Pascal to Fermat, Aug. 24, 1654, in Pascal, *Oeuvres complètes* 2:1147–48; Smith, *Source Book in Mathematics*, 555.

81. Pascal, *Oeuvres complètes* 2:1148–49; Smith, *Source Book in Mathematics*, 556–57.

82. Fermat to Pascal, Sept. 25, 1654, in Pascal, *Oeuvres complètes* 2:1155; Smith, *Source Book in Mathematics*, 562.

83. Pascal, *Oeuvres complètes* 2:1150–57; Smith, *Source Book in Mathematics*, 557–63; analysis in A. Edwards, "Pascal and the problem of points," *International Statistical Review* 50 (1982): 259–66.

84. Pascal to Fermat, July 29, 1654, in Pascal, *Oeuvres complètes* 2:1138; Smith, *Source Book in Mathematics*, 548.

85. Pascal, *Oeuvres complètes* 2:1139; Smith, *Source Book in Mathematics*, 549.

86. Pascal, *Oeuvres complètes* 2:1314–17; see A. Edwards, *Pascal's Arithmetical Triangle* (London, 1987), 76; A. Hald, *History of Probability and Statistics and Their Applications before 1750* (New York, 1990), 45–63.

87. Pascal, *Usage du triangle arithmétique pour déterminer les partis qu'on doit faire entre deux joueurs qui jouent en plusieurs parties*, in Pascal, *Oeuvres complètes* 2:1308.

88. Pascal, *Oeuvres complètes* 2:1034–35; see R. Nelson, *Pascal, Adversary and Advocate* (Cambridge, Mass., 1981), 109–10.

89. Mesnard, *Pascal et les Roannez*, 1:370.

90. Fermat to Pascal, Aug. 29, 1654, in Pascal, *Oeuvres complètes* 2:1154–55; Smith, *Source Book in Mathematics*, 561.

91. Mesnard, *Pascal et les Roannez*, 1:210.

92. Carcavy to Huygens, Sept. 28, 1656, in *Oeuvres complètes de Huygens* 1:492–94; and Pascal, *Oeuvres complètes* 3:860–63; cf. Huygens to Carcavy, Oct. 12, 1656, in *Oeuvres complètes de Huygens* 1:505–7; Pascal, *Oeuvres complètes* 3:863–65; A. Edwards, "Pascal's problem: The 'gambler's ruin,'" *International Statistical Review* 51 (1983): 73–79.

93. J. Mesnard, "Sur le chemin de l'Académie des Sciences: le cercle du mathématicien Claude Mylon (1654–1660)," *Revue d'histoire des sciences* 44 (1991): 211–51; Huygens to Mylon, Feb. 1, 1657, in *Oeuvres complètes de Huygens* 2:7.

94. Huygens to Wallis, July 21, 1656, in *Oeuvres complètes de Huygens* 1:459–60.

95. Huygens to Roberval, Apr. 18, 1656, in *Oeuvres complètes de Huygens* 1:404; Huygens to Mylon, late Apr. 1656, in *Oeuvres complètes de Huygens* 1:399; cf. E. Coumet, "Sur 'le calcul ès jeux de hasard' de Huygens: dialogues avec les mathématiciens français (1655–1657)," in *Huygens et la France*, foreword by R. Taton (Paris, 1982), 123–37; ed. intro. in Pascal, *Oeuvres complètes* 3:837–43.

96. Huygens to Mylon, June 1, 1656, in *Oeuvres complètes de Huygens* 1:426–27; Mylon to Huygens, June 23, 1656, in *Oeuvres complètes de Huygens* 1:438–39.

97. Carcavy to Huygens, June 22, 1656, in *Oeuvres complètes de Huygens* 1:431–34.

98. Huygens to Carcavy, July 6, 1656, in *Oeuvres complètes de Huygens* 1:442–47, at 442.

99. Carcavy to Huygens, Sept. 28, 1656, in *Oeuvres complètes de Huygens* 1:492–94; Huygens to Carcavy, Oct. 12, 1656, in *Oeuvres complètes de Huygens* 1:505–6.

100. *Oeuvres complètes de Huygens* 14:52–95, preceded by editor's analysis; Dutch original of intro. sent to van Schooten Apr. 20, 1656, in *Oeuvres complètes de Huygens* 1:406–7; partial English trans. in H. Freudenthal, "Huygens' foundations of probability," *Historia Mathematica* 7 (1980): 113–17; summary in Todhunter, *History of the Mathematical Theory of Probability*, ch. 3.

101. *Oeuvres complètes de Huygens* 14:60–61; Hacking, *Emergence of Probability*, ch. 11; Hald, *History of Probability and Statistics*, ch. 6; David, *Games, Gods and Gambling*, ch. 11; I. Schneider, "Christiaan Huygens's contribution to the development of a calculus of probabilities," *Janus* 67 (1980): 269–79.

102. Huygens to van Schooten, Apr. 20, 1656, in *Oeuvres complètes de Huygens* 1:404–5; ibid., 14:5 n. 13.

103. Aquinas, *Summa theologiae* I-II q. 40 art. 2 obj. 1; cf. P. Dupont and C. Silvia Roero, "Il trattato 'De ratiociniis in ludo aleae' di Christiaan Huygens con le 'Annotationes' di Jakob Bernoulli ('Ars Conjectandi,' parte 1) presentati in traduzione italiana, con commento storico-critico e risoluzioni moderne," *Memorie della Accademia delle Scienze di Torino, Classe di Scienze Fisiche, Matematiche e Naturali*, s. 5, 8, fasc. 1–2 (1984), esp. "Nota filologica," at 19–27; also

E. Shoesmith, "Expectation in the early probabilists," *Historia Mathematica* 10 (1983): 78–80.

104. Huygens to Carcavy, July 6, 1656, in *Oeuvres complètes de Huygens* 1:442–47, at 442; *Reckoning*, prop. 3, in *Oeuvres complètes de Huygens* 14:64.

105. *Reckoning*, prop. 4, in *Oeuvres complètes de Huygens* 14:68–69.

106. D. Pastine, *Probabilismo ed. enciclopedia* (Florence, 1975), 140; Caramuel, *Mathesis Biceps*, syn. 6, ch. 23, at 923–72; see P. Barbieri, "Juan Caramuel Lobkowitz (1606–1682): Über die musikalischen Logarithmen und das Problem der musikalischen Temperatur," *Musiktheorie* 2 (1987): 145–68.

107. Caramuel, *Mathesis Biceps* syn. 6, ch. 24, at 972–1002; Huygens' book at 986–93; partial summary in Todhunter, *History of the Mathematical Theory of Probability*, 44–46; comments in Velarde Lombraña, *Juan Caramuel: vida y obra*, 158–61.

108. Caramuel, *Mathesis Biceps* syn. 6 ch. 24 at 973.

109. Ibid., 974.

110. Ibid., 975; cf. 999.

111. Ibid., 975.

112. Ibid., 976–77, 980; Todhunter, *History of the Mathematical Theory of Probability*, 46.

113. Caramuel, *Mathesis Biceps* syn. 6 ch. 24 at 981.

114. Ibid., 997–98, 1013; D. Bellhouse, "The Genoese lottery," *Statistical Science* 6 (1991): 141–48; for possible earlier origins of lotto games, see Zdekauer, *Il gioco d'azzardo nel medioevo italiano*, 133.

115. Caramuel, *Mathesis Biceps* syn. 6 ch. 24 at 999; a Scholastic treatment of dice of slightly later date in S. Knebel, "Agustin de Herrera: A treatise on aleatory probability," *Modern Schoolman* 73 (1996): 199–264.

Chapter 12: Conclusion

1. Plato, *Meno* 80–86.

2. Galileo, *Dialogue Concerning the Two Chief World Systems*, trans. S. Drake (Berkeley, 1953), 190–91; J. Franklin, "Diagrammatic reasoning and modelling in the imagination: The secret weapons of the scientific revolution," in *1543 and All That: Image and Word, Change and Continuity in the Proto-Scientific Revolution*, ed. G. Freeland and A. Corones (Dordrecht, 2000), 53–115, at 96–99.

3. A. Karmiloff-Smith, *Beyond Modularity: A Developmental Perspective on Cognitive Science* (Cambridge, Mass., 1992); cf. M. Polanyi, "The logic of tacit inference," *Philosophy* 41 (1966): 1–18; recent reviews in A. Reber, *Implicit Learning and Tacit Knowledge: An Essay on the Cognitive Unconscious* (New York, 1993); A. Cleeremans, A. Destrebecqz, and M. Boyer, "Implicit learning: News from the front," *Trends in Cognitive Science* 2 (1998): 406–16; Z. Dienes and J. Perner, "A theory of implicit and explicit knowledge," *Behavioral and Brain Sciences* 22 (1999): 735–808.

4. Review in J. Holland, K. Holyoak, R. Nisbett, and P. Thagard, *Induction* (Cambridge, Mass., 1986), sec. 5.2.

5. S. Lima and L. Dill, "Behavioral decisions made under the risk of predation: A review and prospectus," *Canadian Journal of Zoology* 68 (1990): 619–40; M. Lawes and M. Perrin, "Risk sensitive foraging behaviour of the round-eared elephant shrew," *Behavioral Ecology and Sociobiology* 37 (1995): 31–37; R. Templeton and J. Franklin, "Adaptive information and animal behaviour," *Evolutionary Theory* 10 (1992): 145–55.

6. M. Richard and R. Lippmann, "Neural net classifiers estimate Bayesian *a posteriori* probabilities," *Neural Computation* 3 (1991): 461–83; R. Neal, *Bayesian Learning for Neural Networks* (New York, 1996); D. Shanks, "Connectionism and the learning of probabilistic concepts," *Quarterly Journal of Experimental Psychology A* 42 (1990): 209–37.

7. W. Cooper, "Decision theory as a branch of evolutionary theory: A biological derivation of the Savage axioms," *Psychological Review* 94 (1987): 395–411; A. Robson, "A biological basis for expected and non-expected utility," *Journal of Economic Theory* 68 (1996): 399–424.

8. G. Gigerenzer and D. Murray, *Cognition as Intuitive Statistics* (Hillsdale, N.J., 1987); C. Peterson and L. Beach, "Man as an intuitive statistician," *Psychological Bulletin* 68 (1967): 29–46; G. Keren and C. Lewis, eds., *A Handbook of Data Analysis in the Behavioral Sciences: Methodological Issues* (Hillsdale, N.J., 1993), pt. 3; L. Cosmides and J. Tooby, "Are humans good intuitive statisticians after all?" *Cognition* 58 (1996): 1–73.

9. A. Hoerl and H. Fallin, "Reliability of subjective evaluations in a high incentive situation," *Journal of the Royal Statistical Society A* 137 (1974): 227–30; G. Gigerenzer and U. Hoffrage, "How to improve Bayesian reasoning without instruction—frequency formats," *Psychological Review* 102 (1995): 684–704; P. Juslin, P. Wennerholm, and H. Olsson, "Format dependence in subjective probability calibration," *Journal of Experimental Psychology: Learning Memory and Cognition* 25 (1999): 1038–52.

10. M. Lawrence and M. O'Connor, "Exploring judgemental forecasting," *International Journal of Forecasting* 8 (1992): 15–26; E. Welch, S. Bretschneider, and J. Rohrbaugh, "Accuracy of judgemental extrapolation of time series data," *International Journal of Forecasting* 14 (1998): 95–110.

11. J. Piaget and B. Inhelder, *The Origin of the Idea of Chance in the Child*, trans. L. Leake et al. (London, 1975); E. Fischbein, *The Intuitive Sources of Probabilistic Thinking in Children* (Boston, 1975); C. Acredolo, J. O'Connor, L. Banks, and K. Horobin, "Children's ability to make probability estimates," *Child Development* 60 (1989): 933–45; R. Falk and F. Wilkening, "Children's construction of fair chances," *Developmental Psychology* 34 (1998): 1340–57.

12. Y.-K. Ng, "The paradox of the adventurous young and the cautious old: Natural selection vs. rational calculation," *Journal of Theoretical Biology* 152 (1991): 339–52.

13. P. Johnson-Laird, P. Legrenzi, V. Girotto, M. Legrenzi, and J. Caverni, "Naive probability: A mental model theory of extensional reasoning," *Psycho-*

logical Review 106 (1999): 62–88; D. Kahneman, P. Slovic, and A. Tversky, eds., *Judgement Under Uncertainty: Heuristics and Biases* (Cambridge, 1982).

14. C. Howarth, "The relationship between objective risk, subjective risk and behaviour," *Ergonomics* 31 (1988): 527–35; I. Erev and T. Wallsten, "The effect of explicit probabilities on decision weights and on the reflection effect," *Journal of Behavioral Decision Making* 6 (1993): 221–41.

15. J. Wright and G. Murphy, "The utility of theories in intuitive statistics," *Journal of Experimental Psychology: General* 113 (1984): 301–22.

16. R. Reagan, F. Mosteller, and C. Youtz, "Quantitative meanings of verbal probability expressions," *Journal of Applied Psychology* 74 (1989): 433–42; A. Kong, G. Barnett, F. Mosteller, and C. Youtz, "How medical professionals evaluate expressions of probability," *New England Journal of Medicine* 315 (1986): 740–44; K. Teigen and W. Brun, "The directionality of verbal probability expressions: effects on decisions, predictions and probabilistic reasoning," *Organizational Behavior and Human Decision Processes* 80 (1999): 155–90.

17. Review in G. Gigerenzer et al., *The Empire of Chance: How Probability Changed Science and Ordinary Life* (Cambridge, 1989), ch. 6.

18. A. Edwards, *Pascal's Arithmetical Triangle* (London, 1987); B. Datta and A. Singh, "Use of permutations and combinations in India," *Indian Journal of History of Science* 27 (1992): 231–49; S. Koelblen, "Une pratique de la composition des raisons dans un exercice de combinatoire," *Revue d'histoire des sciences* 47 (1994): 209–47; R. Ceñal, intro. to *La combinatoria de Sebastian Izquierdo* (Madrid, 1974).

19. J. Kassler, "The emergence of probability reconsidered," *Archives internationales d'histoire des sciences* 36 (1986): 17–44, at 29–33; E. Knobloch, "Marin Mersennes Beiträge zur Kombinatorik," *Sudhoffs Archiv* 58 (1974): 356–79; E. Knobloch, "Musurgia universalis: unknown combinatorial studies in the age of baroque absolutism," *History of Science* 17 (1979): 258–75; E. Coumet, "Mersenne: Dénombrements, répertoires, numérotations de permutations," *Mathématiques et sciences humaines* 10 (1972): 5–37; P. Dear, *Mersenne and the Learning of the Schools* (Ithaca, 1988), 193–96.

20. R. Eggleston, *Evidence, Proof, and Probability*, 2d ed. (London, 1983), ch. 9.

21. P. Bonissone, s.v. "Reasoning, plausible," in *Encyclopedia of Artificial Intelligence*, 2d ed., ed. S. Shapiro (New York, 1992), 2:1307–22.

22. See P. Forrest, *The Dynamics of Belief* (Oxford, 1986), ch. 8; C. Hempel, *Aspects of Scientific Explanation* (New York, 1965), ch. 2.

23. E. Jaynes, "Information theory and statistical mechanics," *Physical Review* 106 (1957): 620–30, at 627.

24. G. Polya, *Patterns of Plausible Inference*, 2d ed. (Princeton, 1968), 4.

25. Cf. P. Raymond, *De la combinatoire aux probabilités* (Paris, 1975), chs. 3, 5; C. Howson, "The prehistory of chance," *British Journal for the Philosophy of Science* 29 (1978): 274–80.

26. Franklin, "Diagrammatic reasoning"; A. Crombie, "Quantification in medieval science," in *Science, Optics and Music in Medieval and Early Modern Thought* (London, 1990); A. Maier, *On the Threshold of Exact Science*, ed. and trans. S. Sargent (Philadelphia, 1982), 168–70.

27. *Oeuvres complètes de Huygens* 14:18 n. 4; 11:93–94; cf. *Correspondance du P. Marin Mersenne*, 6:131.

28. E. Pascal and Roberval to Fermat, Aug. 16, 1636, and Fermat's reply, Aug. 23, in Pascal, *Oeuvres complètes* 2:129–46.

29. P. Holgate, "The influence of Huygens' work in dynamics on his contribution to probability," *International Statistical Review* 52 (1984): 137–40.

30. J. Yoder, *Unrolling Time: Christiaan Huygens and the Mathematization of Nature* (Cambridge, 1988), ix.

31. *Oeuvres complètes de Huygens* 14:57; cf. I. Schneider, "Why do we find the origin of a calculus of probabilities in the seventeenth century?" in *Probabilistic Thinking, Thermodynamics and the Interaction of the History and Philosophy of Science: Proceedings of the 1978 Pisa Conference on the History and Philosophy of Science*, ed. J. Hintikka, D. Gruender, and E. Agazzi (Dordrecht, 1981), 3–24, at 4; Huygens to Carcavy, June 1, 1656, in *Oeuvres complètes de Huygens* 1:427–29.

32. Pascal, *Traité de l'equilibre des liqueurs* ch. 2, in *The Physical Treatises of Pascal*, trans. I. Spiers and A. Spiers (New York, 1973), 8; legal quote in Lambarde, *Eirenarcha* 1, ch. 16, at 127; cf. Chaucer, *Boece* 5 prosa 3, in *The Works of Geoffrey Chaucer*, ed. F. Robinson (London, 1933), 376; H. Kurath, ed., *Middle English Dictionary* (Ann Arbor, 1954–), s.v. "indifference," "indifferentli."

33. Pascal, "Avis nécessaire à ceux qui auront curiosité de voir la machine arithmétique, et de s'en servir," in Pascal, *Oeuvres complètes* 2:336.

34. *Oeuvres complètes de Huygens* 14:57–59; Leibniz, "Du jeu de quinquenove" (1678), in S. de Mora-Charles, "Quelques jeux de hazard selon Leibniz," *Historia Mathematica* 19 (1992): 125–57, at 132.

35. J. Mill, *System of Logic*, 8th ed. (London, 1872), 353.

36. Schneider, "Why do we find. . . ?" 5–6; cf. J. van Brakel, "Some remarks on the prehistory of the concept of statistical probability," *Archive for History of Exact Sciences* 16 (1976–77): 119–36.

37. Aristotle, *Rhetoric* 1362a6–11; cf. *Metaphysics* 1025a14–18, 1065a33–35; *Magna Moralia* 1206b38–1207a4.

38. *Physics* 197a32–33; cf. 196b10–13.

39. *Posterior Analytics* 87b19–25; cf. *Physics* 197a18–21; *On Generation and Corruption* 333b5–7; *Metaphysics* 1026b33–36, 1064b30–1065a5; see further J. Junkersfeld, *The Aristotelian-Thomistic Conception of Chance* (Notre Dame, 1945); J. Lennox, "Aristotle on chance," *Archiv für Geschichte der Philosophie* 66 (1984): 52–60; M. Boeri, "Chance and teleology in Aristotle's *Physics*," *International Philosophical Quarterly* 35 (1995): 87–96; O. Sheynin, "On the prehistory of the theory of probability," *Archive for History of Exact Sciences* 12 (1974): 97–141.

40. *On Prophecy in Sleep* 463b23–31.

41. *Generation of Animals* 767b7–12, 767a35; cf. 768a2–b33.

42. Plutarch, *De stoicorum repugnantiis* c. 23 (*Moralia* 1045C).

43. Pascal, *Pensées* no. 413, numbering following *Oeuvres complètes*, ed. L. Lafuma (Paris, 1963).

44. Pascal, *Trois discours sur la condition des grands*, in Cailliet and Blankenagel, *Great Shorter Works of Pascal*, 211–17, at 212–13; see Thirouin, *Le hasard et les règles*, ch. 1; *Pensées* no. 194; Krailsheimer, *Pensées*, 87.

45. Pascal to Perier, Oct. 17, 1651, in Pascal, *Oeuvres complètes* 2:852; also in Cailliet and Blankenagel, *Great Shorter Works of Pascal*, 83; see Thirouin, *Le hasard et les règles*, ch. 8.

46. H. Patch, *The Goddess Fortuna in Medieval Literature* (Cambridge, Mass., 1927); V. Cioffari, *Fortune and Fate from Democritus to Saint Thomas Aquinas* (New York, 1935); J. Frakes, *The Fate of Fortune in the Early Middle Ages* (Leiden, 1988).

47. J. Roberts and B. Sutton-Smith, "Cross-cultural correlates of games of chance," *Behavior Science Notes* 1 (1966): 131–44; H. Barry and J. Roberts, "Infant socialization and games of chance," *Ethnology* 11 (1972): 296–308; B. Sutton-Smith, "Games as models of power," in *The Content of Culture: Constants and Variants*, ed. B. Bolton (New Haven, 1989), 3–18.

48. J. Tobacyck and L. Wilson, "Paranormal beliefs and preference for games of chance," *Psychological Reports* 68 (1991): 1088–90.

49. The first four are taken from M. Kendall, "The beginnings of a probability calculus," *Biometrika* 43 (1956): 1–14, also in *Studies in the History of Statistics and Probability*, ed. E. Pearson and M. Kendall (London, 1970), 1:19–34, at 30; cf. Garber and Zabell, "On the emergence of probability," 49; J. Tiago de Oliveira, review of M. de Mora Charles, *Los inicios de la teoria de la probabilidad: siglos XVI y XVII*, *Annals of Science* 50 (1993): 396–97.

50. L. Maistrov, *Probability Theory: A Historical Sketch* (New York, 1974), 3–4, 7.

51. I. Hacking, *The Emergence of Probability* (Cambridge, 1975), esp. chs. 4–5.

52. F. David, *Games, Gods and Gambling* (London, 1962), 26, 36.

53. D. Garber and S. Zabell, "On the emergence of probability," *Archive for History of Exact Sciences* 21 (1979): 33–53, at 49.

54. R. Plackett, "The principle of the arithmetic mean," *Biometrika* 45 (1958): 130–35; also in *Studies in the History of Statistics and Probability*, ed. E. Pearson and M. Kendall (London, 1970), 1:121–26.

55. Pascal, *Celeberrimae Matheseos Academiae Parisiensi*, in Pascal, *Oeuvres complètes* 2:1034.

56. L. Daston, *Classical Probability in the Enlightenment* (Princeton, 1988), 138–41, 163–82.

57. S. Sambursky, *The Physical World of the Greeks*, trans. M. Dagat (London, 1956), 180, 237–38.

58. Hoerl and Fallin, "Reliability of subjective evaluations."

59. Piaget and Inhelder, *Origin of the Idea of Chance in the Child*; Fischbein, *Intuitive Sources of Probabilistic Thinking in Children*; Acredolo et al., "Children's ability to make probability estimates"; Falk and Wilkening, "Children's construction of fair chances."

60. M. Caveing, "La proportionnalité des grandeurs dans la doctrine de la nature d'Aristote," *Revue d'histoire des sciences* 47 (1994): 163–88.

61. Maier, *On the Threshold of Exact Science*, 82–85; M. Wolff, *Geschichte der Impetustheorie* (Frankfurt, 1978), 174–91; Grant, *Source Book in Medieval Science*, 316–19.

62. C. Boyer, *The History of the Calculus and its Conceptual Development* (New York, 1949), chs. 3–5.

63. K. Pedersen, "Roberval's method of tangents," *Centaurus* 13 (1968): 151–82.

64. K. Pedersen, "Roberval's comparison of the arclength of a spiral and a parabola," *Centaurus* 15 (1970): 26–43; *Correspondance du P. Marin Mersenne*, 12:54–56.

65. Yoder, *Unrolling Time*, 9.

66. Refs. in J. de Kleer, "A view on qualitative physics," *Artificial Intelligence* 59 (1993): 105–14.

67. M. McCloskey, "Intuitive physics," *Scientific American* 248 (4) (Apr. 1983): 114–22; M. McCloskey and D. Kohl, "The curvilinear impetus principle and its role in interacting with moving objects," *Journal of Experimental Psychology: Learning, Memory, and Cognition* 9 (1983): 146–56; M. Kaiser, M. McCloskey, and D. Proffitt, "Development of intuitive theories of motion—curvilinear motion in the absence of external forces," *Developmental Psychology* 22 (1986): 67–71.

68. C. Wright, *Perspective in Perspective* (London, 1983), 38; cf. A. Trendall and T. Webster, *Illustrations of Greek Drama* (London, 1971), 3:3, 43; R. Tobin, "Ancient perspective and Euclid's *Optics*," *Journal of the Warburg and Courtauld Institutes* 53 (1990): 14–41; K. Andersen, "Ancient roots of linear perspective," in Berggren and Goldstein, *From Ancient Omens to Statistical Mechanics*, 75–89.

69. Grant, *Source Book in Medieval Science*, 423–26; Lindberg, *Theories of Vision from al-Kindi to Kepler*, ch. 6.

70. Ghiberti, *Second Commentary*, selection in E. Holt, *A Documentary History of Art* (Garden City, N.Y., 1957), 1:154.

71. C. Cennini, *The Craftsman's Handbook*, trans. D. Thompson (New York, 1960), ch. 67; M. Kemp, *The Science of Art* (New Haven, 1990), 10–11.

72. L. Alberti, *On Painting*, trans. J. Spencer, rev. ed. (London, 1966), 45.

73. Original in *Lorenzo Ghibertis Denkwürdigkeiten*, ed. and trans. J. von Schlosser (Berlin, 1912); discussion in G. ten Doesschate, *De derde commentaar van Lorenzo Ghiberti in verband met de middeleeuwsche optiek* (Utrecht, 1940); G. Federici Vescovini, *Studi sulla prospettiva medievale* (Turin, 1987), esp. chs. 11–12; S. Edgerton, *The Renaissance Rediscovery of Linear Perspective* (New York, 1975), esp. ch. 5.

74. H. Lang, *Aristotle's Physics and Its Medieval Varieties* (Albany, 1992), 284; Chillingworth, preface to *Religion of Protestants*, § 19.

75. J. Urmson, *Aristotle's Ethics* (Oxford, 1988), 71; some connections between the two in W. Elton, "Aristotle's *Nicomachean Ethics* and Shakespeare's *Troilus and Cressida*," *Journal of the History of Ideas* 58 (1997): 331–37.

76. Byrne, *Probability and Opinion*; Breny, "Les origines de la théorie des probabilités."

77. Boyer, *History of the Calculus*, chs. 3–5; Grant, *Source Book in Medieval Science*, esp. §§ 26, 41, 51; G. Molland, *Mathematics and the Medieval Ancestry of Physics* (Aldershot, 1995).

78. Albert of Saxony, *Commentary on De Caelo*, summarized in J. Sesiano, "On an algorithm for the approximation of surds from a Provençal treatise," in *Mathematics from Manuscript to Print, 1300–1600*, ed. C. Hay (Oxford, 1988), 30–56, at 42–47.

79. S. Siegel, "The Aristotelian basis of English law, 1450–1800," *New York University Law Review* 56 (1981): 18–59.

80. J. Gordley, *The Philosophical Origins of Modern Contract Doctrine* (Oxford, 1991); P. Grossi, ed., *La seconda scolastica nella formazione del diritto privato moderno* (Milan, 1973).

81. A. Nussbaum, *A Concise History of the Law of Nations* (New York, 1962), ch. 2; "Bartolo on the conflict of laws," trans. J. Smith, *American Journal of Legal History* 14 (1970): 157–83, 247–75; J. Muldoon, *Canon Law, the Expansion of Europe, and World Order* (Aldershot, 1998); F. de Vitoria, *De Indis et De Iure Belli Relectiones*, ed. E. Nys (Washington, D.C., 1917); F. Suarez, *Selections from Three Works* (Oxford, 1944); see Scott, *Spanish Origin of International Law*; and to the contrary, Nussbaum, *Concise History of the Law of Nations*, app. 2; D. Kennedy, "Primitive legal scholarship," *Harvard International Law Journal* 27 (1986): 1–98; Stein, *Roman Law and European History*, 94–97.

82. A. Black, *Political Thought in Europe, 1250–1450* (Cambridge, 1992); J. Blythe, *Ideal Government and the Mixed Constitution in the Middle Ages* (Princeton, 1992); Q. Skinner, *The Foundations of Modern Political Thought* (Cambridge, 1978), 2: ch. 5; Hamilton, *Political Thought in Sixteenth Century Spain*; D. Luscombe, "The state of nature and the origin of the state," in Kretzmann, Kenny, Pinborg, *Cambridge History of Later Medieval Philosophy*; J. Burns, "The Scholastics," in *Cambridge History of Political Thought, 1450–1700*, ed. J. Burns (Cambridge, 1991); H. Lloyd, "Constitutionalism," in Burns, *Cambridge History*; de Vitoria, *Political Writings*.

83. J. Doyle, "Francisco Suárez on preaching the Gospel to people like the American Indians," *Fordham International Law Review* 15 (1992): 879–951.

84. Langholm, *Economics in the Medieval Schools*; O. Langholm, *The Legacy of Scholasticism in Economic Thought* (Cambridge, 1998); Gordon, *Economic Analysis before Adam Smith*, chs. 6–9; Noonan, *Scholastic Analysis of Usury*; M. Grice-Hutchinson, *The School of Salamanca* (Oxford, 1952); de Roover, "The Scholastics, usury and foreign exchange"; J. Schumpeter, *History of Economic Analysis*

(London, 1972), ch. 2; J. Viner, "The economic doctrines of the Scholastics," *History of Political Economy* 10 (1978): 46–113; L. Rivas, "Business ethics and the history of economics in Spain: 'The School of Salamanca,' a bibliography," *Journal of Business Ethics* 22 (1999): 191–202; Chafuen, *Christians for Freedom*; overview in Colish, *Medieval Foundations of the Western Intellectual Tradition*, ch. 25; later connections in R. Prasch, "The origins of the a priori method in classical political economy," *Journal of Economic Issues* 30 (1996): 1105–25.

85. J. Kaye, *Economy and Nature in the Fourteenth Century: Money, Market Exchange and the Emergence of Scientific Thought* (New York, 1998); Hadden, *On the Shoulders of Merchants*, ch. 4.

86. H. Schüling, *Bibliographie der psychologischen Literatur des 16. Jahrhunderts* (Hildesheim, 1967); S. Kemp, *Medieval Psychology* (New York, 1990); J. Zupko, "What is the science of the soul?" *Synthese* 110 (1997): 297–334; S. Knebel, "Scotists vs. Thomists: What seventeenth-century Scholastic psychology was about," *Modern Schoolman* 74 (1997): 219–26.

87. A. Kenny, *Aquinas* (Oxford, 1980), ch. 3; E. Harvey, *The Inward Wits* (London, 1975); E. Mahoney, "Sense, intellect, and imagination in Albert, Thomas, and Siger," in *The Cambridge History of Later Medieval Philosophy* 602–22; N. Steneck, "Albert on the psychology of sense perception," in *Albertus Magnus and the Sciences*, ed. J. Weisheipl (Toronto, 1980), 263–90; refs. to Avicenna's originals in Black, *Logic and Aristotle's Rhetoric and Poetics*, 202 n. 66.

88. S. Stich, *From Folk Psychology to Cognitive Science* (Cambridge, Mass., 1983); T. Greenwood, ed., *The Future of Folk Psychology* (Cambridge, 1991).

89. J. Fodor, *The Modularity of Mind: An Essay in Faculty Psychology* (Cambridge, Mass., 1983); M. Minsky, *Society of Mind* (London, 1987).

90. E. Gilson, *Index scolastico-cartésien* (Paris, 1912); E. Gilson, *Etudes sur le rôle de la pensée médiévale dans la formation du système cartésien* (Paris, 1951); F. van der Pitte, "Some of Descartes' debts to Eustachius a Sancto Paulo," *Monist* 71 (1988): 487–97; R. Ariew, "Descartes and Scholasticism," in *Cambridge Companion to Descartes*, ed. J. Cottingham (Cambridge, 1992), 58–90; A. Tellkamp, *Das Verhältnis John Lockes zur Scholastik* (Münster, 1927); W. Kenney, "John Locke and the Oxford Training in Logic and Metaphysics" (Ph.D. diss., St. Louis Univ., 1959); E. Ashworth, "'Do words signify ideas or things?' The Scholastic sources of Locke's theory of language," *Journal of the History of Philosophy* 19 (1981): 299–326; D. Connell, *The Vision in God: Malebranche's Scholastic Sources* (Louvain, 1967); P. Reif, "The textbook tradition in natural philosophy, 1600–1650," *Journal of the History of Ideas* 30 (1969): 17–32.

91. M. Curtis, *Oxford and Cambridge in Transition* (Oxford, 1959), 111; W. Costello, *The Scholastic Curriculum at Early Seventeenth-Century Cambridge* (Cambridge, Mass., 1958), 49–50; H. Fletcher, *The Intellectual Development of John Milton* (Urbana, 1961), 2: ch. 7; R. Ariew and A. Gabbey, "The scholastic background," in *Cambridge History of Seventeenth Century Philosophy*, ed. D. Garber and M. Ayers (Cambridge, 1998), ch. 15.

92. P. Gassendi, *Disquisitio Metaphysica* med. 3, dub. 9, in *Opera omnia*, 3:348.

93. E. Vineis and A. Maierù, "Medieval linguistics," in *History of Linguistics*, ed. G. Lepschy (London, 1994), 2: ch. 2; G. Bursill-Hall, *Speculative Grammars of the Middle Ages* (The Hague, 1971); U. Eco and C. Marmo, eds., *On the Medieval Theory of Signs* (Amsterdam, 1989); L. Kaczmarek, "The age of the sign: New light on the role of the fourteenth century in the history of semiotics," *Dialogue* (Canada) 31 (1992): 509–15.

94. Franklin, "Mental furniture from the philosophers," *Et Cetera* 40 (1983): 177–91.

95. P. Duhem, *Etudes sur Léonard de Vinci* (Paris, 1906–13); R. Ariew and P. Barker, eds., "Duhem as Historian of Science," *Synthese* 83 (2) (1990); S. Jaki, *Uneasy Genius: The Life and Work of Pierre Duhem* (The Hague, 1984), ch. 10; R. Martin, *Pierre Duhem* (La Salle, Ill., 1991), chs. 8–9; J. Murdoch, "Pierre Duhem and the history of late medieval science and philosophy in the Latin west," in *Gli studi di filosofia medievale fra otto e novecento*, ed. R. Imbach and A. Maierù (Rome, 1991), 253–302; overview in E. Grant, *The Foundations of Modern Science in the Middle Ages* (Cambridge, 1996), ch. 8.

96. Wallace, *Galileo's Early Notebooks*; W. Wallace, *Galileo and His Sources* (Princeton, 1984); E. Reitan, "Thomistic natural philosophy and the Scientific Revolution," *Modern Schoolman* 73 (1996): 265–81.

97. Many articles in E. Grant and J. Murdoch, eds., *Mathematics and Its Applications to Science and Natural Philosophy in the Middle Ages* (Cambridge, 1987); A. Crombie, "Quantification in medieval physics," *Isis* 52 (1961): 143–60; also Colish, *Medieval Foundations of the Western Intellectual Tradition*, ch. 24.

98. P. Duhem, *The Origins of Statics*, trans. G. Leneaux, V. Vagliente, and G. Wagener (Dordrecht, 1991).

99. *Digest* 22.6.9; *English Reports* 76:386, 393; S. Davies, "The jurisprudence of willfulness: An evolving theory of excusable ignorance," *Duke Law Journal* 48 (1998): 341–427.

100. P. Stein, *Regulae Iuris* (Edinburgh, 1966), 28–29.

101. Review in J. Boose, s.v. "Knowledge acquisition," in *Encyclopedia of Artificial Intelligence*, ed. S. Shapiro, 2d ed. (New York, 1992), 1:719–42.

102. Howell, *Complete Collection of State Trials*, 27:1060, quoted in M. McNamara, *2000 Famous Legal Quotations*, 2d ed. (Rochester, 1967), 149.

103. F. Guicciardini, *Ricordi*, in *Maxims and Reflections of a Renaissance Statesman*, trans. M. Domandi (New York, 1965), 69.

104. See generally W. Bouwsma, "Lawyers and early modern culture," *American Historical Review* 78 (1973): 303–27; further refs in Brundage, *Medieval Canon Law*, 65–67.

105. E. Ives, *The Common Lawyers of Pre-Reformation England* (Cambridge, 1983), ch. 1; in France, K. Bezemer, "The law school of Orleans as school of public administration," *Legal History Review* 66 (1998): 247–77.

106. Aston, "Oxford's medieval alumni," 28–30; cf. D. Owen, *The Medieval Canon Law: Teaching, Literature and Transmission* (Cambridge, 1990), ch. 2; for an earlier period, Southern, *Scholastic Humanism and the Unification of Europe*, 1:ch. 5.

107. See A. Murray, *Reason and Society in the Middle Ages* (Oxford, 1978), 222–23.

108. E. Kittell, *From Ad Hoc to Routine: A Case Study in Medieval Bureaucracy* (Philadelphia, 1991); P. Partner, *The Pope's Men: The Papal Civil Service in the Renaissance* (Oxford, 1990).

109. B. Tierney, *Medieval Poor Law: A Sketch of Canonical Theory and Its Application in England* (Berkeley, 1959); J. Brundage, "Legal aid for the poor and the professionalization of law in the Middle Ages," *Journal of Legal History* 9 (1988): 169–79.

110. C. Cheney, *Notaries Public in England in the Thirteenth and Fourteenth Centuries* (Oxford, 1972); M. Giansante, *Retorica e politica nel duecento: notai bolognese e l'ideologia comunale* (Rome, 1999); more generally M. Bellomo, *The Common Legal Past of Europe, 1000–1800* (Washington, D.C., 1995).

111. W. Jones, *The Elizabethan Court of Chancery* (Oxford, 1967), 1, 238; cf. Ives, *Common Lawyers of Pre-Reformation England*, 7.

112. G. Keeton, *Shakespeare's Legal and Political Background* (New York, 1968); O. Phillips, *Shakespeare and the Lawyers* (London, 1972); D. Barton, *Shakespeare and the Law* (New York, 1971).

113. J. Hornsby, *Chaucer and the Law* (Norman, 1988), 108, 56–66, and ch. 3; on other authors, J. Alford and D. Seniff, *Literature and Law in the Middle Ages: A Bibliography of Scholarship* (New York, 1984); M. Bloch, *Medieval French Literature and Law* (Berkeley, 1977); R. Kay, "Roman law in Dante's *Monarchia*," in King and Ridyard, *Law in Medieval Life and Thought*, 259–68.

114. G. de Ste. Croix, "Greek and Roman accounting," in *Studies in the History of Accounting*, ed. A. Littleton and B. Yamey (London, 1956), 14–74; G. Costouros, *Accounting in the Golden Age of Greece* (Urbana, 1979), ch. 2.

115. S. Epstein, "Business cycles and the sense of time in medieval Genoa," *Business History Review* 62 (1988): 238–60.

116. S. Kuttner, "Johannes Andreae on the style of dating papal documents," *Jurist* 48 (1988): 448–53.

117. A. Youschkevitch, s.v. "Abu'l-Wafa," in *Dictionary of Scientific Biography*, ed. C. Gillispie, 1:39–43; K. Vogel, s.v. "Fibonacci," in *Dictionary of Scientific Biography*, 4:604–13.

118. Hunt and Edgar, *Select Papyri*, 1:121–22.

119. *Oxford English Dictionary* (2d ed.), s.vv. "per," "hundred," "cent."

120. Herodotus, *Histories* 2, ch. 109; later developments in R. Kain and E. Baigert, *The Cadastral Map in the Service of the State: A History of Property Mapping* (Chicago, 1992).

121. *La Tiberiade di Bartole da Sasoferato del modo di dividere l'alluuione, l'isole and l'aluei* (Rome, 1579); see J. van Maanen, "Teaching geometry to 11 year old 'medieval lawyers,'" *Mathematical Gazette* 76 (1992): 37–45.

122. Franklin, "Diagrammatic reasoning and modelling in the imagination," 59–61; J. Murdoch, *Album of Science: Antiquity and the Middle Ages* (New York, 1984), illus. 37–39; A. Errera, *"Arbor actionum": genere letterario*

e forma di classificazione delle azioni nella dottrina dei glossatori (Bologna, 1995), chs. 1, 2.

123. K. Pennington, *The Prince and the Law, 1200–1600: Sovereignty and Rights in the Western Legal Tradition* (Berkeley, 1993); B. Tierney, *Church Law and Constitutional Thought in the Middle Ages* (London, 1979), esp. ch. 15; A. Monahan, *Consent, Coercion and Control: The Medieval Origins of Parliamentary Democracy* (Kingston, Ont., 1987); D. Kelley, *Cambridge History of Political Thought, 1450–1700*, ch. 3; P. Riga, "The influence of Roman law on state theory in the eleventh and twelfth centuries," *American Journal of Jurisprudence* 35 (1990): 121–86; R. Helmholz, "Magna Carta and the *ius commune*," *University of Chicago Law Review* 66 (1999): 297–371; overview in Colish, *Medieval Foundations of the Western Intellectual Tradition*, ch. 26.

124. R. Tuck, *Natural Rights Theories* (Cambridge, 1979), ch. 1; B. Tierney, *The Idea of Natural Rights: Studies on Natural Rights, Natural Law and Church Law, 1150–1625* (Atlanta, 1997); C. Reid, "The canonistic contribution to the Western rights tradition," *Boston College Law Review* 33 (1991): 37–92; A. Monahan, *From Personal Duties towards Personal Rights: Late Medieval and Early Modern Political Thought, 1300–1600* (Montreal, 1994).

125. References in J. Kelly, *A Short History of Western Legal Theory* (Oxford, 1992), 105, 146.

126. *Digest* 17.2.

127. *Digest* 5.1.76; cf. 41.3.30, 45.1.83.5; the ship example is in Plutarch, *Theseus* 23.1; see Jones, *Law and Legal Theory of the Greeks*, ch. 8; L. Schnorr von Carolsfeld, *Geschichte der juristischen Person* (Aalen, 1969); M. Radin, *The Legislation of the Greeks and Romans on Corporations* (New York, 1910); Jolowicz, *Roman Foundations of Modern Law*, ch. 11.

128. Huff, *The Rise of Early Modern Science*, ch. 4; based on P. Gillet, *La personnalité juridique en droit ecclésiastique* (Malines, 1927); P. Michaud-Quantin, *Universitas: Expressions du mouvement communautaire dans le moyen-âge latin* (Paris, 1970); also J. Canning, "Law, sovereignty and corporation theory," in *Cambridge History of Medieval Political Thought*, ed. J. Burns (Cambridge, 1988), ch. 15, pt. 2; J. Schacht, "Islamic religious law," in *The Legacy of Islam*, 2d ed., ed. J. Schacht and C. Bosworth (New York, 1974), 392–403, at 398.

129. *Digest* 50.16.6.1.

130. Plato, *Laws* 861a; cf. Kelly, *Short History of Western Legal Theory*, 32–34.

131. *Digest* 50.17.40; R. Pickett, *Mental Affliction and Church Law* (Ottawa, 1952); N. Walker, *Crime and Insanity*, vol. 1 (Edinburgh, 1967); M. Dols, *Majnun: The Madman in Medieval Islamic Society* (Oxford, 1992), ch. 14; J.-M. Fritz, *Le discours du fou au moyen âge: XIIe–XIIIe siècles: étude comparée des discours littéraire, medical, juridique et théologique de la folie* (Paris, 1992); C. Lanza, *Ricerche su "Furiosus" in diritto romano* (Rome, 1990); D. Roffe and C. Roffe, "Madness and care in the community: a medieval perspective," *British Medical Journal* 311 (Dec. 23, 1995): 1708–12.

132. *Corpus iuris civilis, Institutes* 4.3; *Digest* 50.17.76.

133. *Digest* 30.1.17; 50.17.30; J. Perry, "The canonical concept of marital consent: Roman law influences," *Catholic Lawyer* 25 (1978): 228–36; C. Donahue, "The case of the man who fell into the Tiber: The Roman law of marriage in the time of the Glossators," *American Journal of Legal History* 22 (1978): 1–53.

134. Gordley, *Philosophical Origins of Modern Contract Doctrine*, 32.

135. Kelly, *Short History of Western Legal Theory*, 28–29, 52–57, 153–54, 189–90; Jolowicz, *Roman Foundations*, ch. 7; *Digest* 50.17.90 and 50.17.183.

136. Kelly, *Short History of Western Legal Theory*, 57–63.

137. *Digest* 1.1.1.

138. *Digest* 50.17.50; 50.17.197; 50.17.206.

139. Vitoria, *De Indis*, sec. 1 § 23 at 127; B. de Las Casas, *The Defense of the Indians*, ch. 4, trans. S. Poole (De Kalb, 1974); further refs. in E. Sevilla-Casas, "Notes on Las Casas' ideological and political practice," in *Western Expansion and Indigenous Peoples*, ed. E. Sevilla-Casas (The Hague, 1977), 15–29.

140. R. Williams, "The medieval and Renaissance origins of the status of the American Indian in Western legal thought," *Southern California Law Review* 57 (1983): 1–99; L. Hanke, *Aristotle and the American Indians* (London, 1959).

141. *Digest* 50.16 *De verborum significatione*; I. MacLean, *Interpretation and Meaning in the Renaissance: The Case of Law* (Cambridge, 1992); J. Faur, "Law and hermeneutics in rabbinic jurisprudence: A Maimonidean perspective," *Cardozo Law Review* 14 (1993): 1657–78.

142. *Digest* 34.5.3; cf. 50.17.96.

143. *Digest* 1.5.1; Gaius, *Institutes* 1.8; also *Digest* 1.8, *De divisione rerum et qualitate*; D. Kelley, *History, Law and the Human Sciences* (London, 1984), 621; Jolowicz, *Roman Foundations*, ch. 8.

144. *Digest* 50.17.24.

145. J. Given, *Inquisition and Medieval Society* (Ithaca, 1997), sec. 1.

146. Brundage, *Law, Sex, and Society in Medieval Europe*; T. Kuehn, *Law, Family and Women: Towards a Legal Anthropology of Renaissance Italy* (Chicago, 1991); J. Bossy, ed., *Disputes and Settlements: Law and Human Relations in the West* (Cambridge, 1983); M. Sheehan, *Marriage, Family and Law in Medieval Europe* (Toronto, 1996); J. Pluss, "Reading legal doctrine historically: Three 14th-century jurists on dowries and social standing," *Historian* 51 (1989): 283–310.

147. J. Gardner, *Women in Roman Law and Society* (London, 1986); A. Arjava, *Women and Law in Late Antiquity* (Oxford, 1996); K. Gravdal, *Ravishing Maidens: Writing Rape in Medieval French Literature and Law* (Philadelphia, 1991); D. Nicholas, *The Domestic Life of a Medieval City: Women, Children and the Family in Fourteenth-Century Ghent* (Lincoln, 1985); R. Karras, *Common Women: Prostitution and Sexuality in Medieval England* (New York, 1996).

148. P. Hyams, *King, Lord and Peasants in Medieval England* (Oxford, 1980); K. Stow, *Alienated Minority: The Jews of Medieval Europe* (Cambridge, Mass., 1992); A. Linder, ed., *The Jews in the Legal Sources of the Early Middle Ages* (Detroit, 1997).

149. Plutarch, *Moralia* 386C–387D.

150. F. Jackson, ed., *Conditionals* (Oxford, 1991); V. Dudman, "On conditionals," *Journal of Philosophy* 91 (1994): 113–28; B. Slater, "Non-conditional 'Ifs,'" *Ratio* 9 (1996): 47–55, etc.

151. P. Cheng and K. Holyoak, "Pragmatic reasoning schemas," *Cognitive Psychology* 17 (1985): 391–416.

152. L. Cosmides, "The logic of social exchange: Has natural selection shaped how humans reason?" *Cognition* 31 (1989): 187–276.

153. For example, *Digest* 46.3.98.5.

154. *Digest* 34.5.13.3.

155. *Digest* 28.3.16; Gaius, *Institutes* 3.98; *Digest* 50.17.185.

156. *Digest* 13.4.2.8; 12.1.31; 18.6.20; *Code* 7.47.1; also J. Kortmann, "*Ab alio ictu(s):* Misconceptions about Julian's view of causation," *Journal of Legal History* 20 (1999): 95–103.

157. *Digest* preface, *De conceptione digestorum*, 8.

158. Plutarch, "On the E at Delphi," *Moralia* 386F.

159. *Digest* 50.17.81; 50.17.147; 50.17.65.

160. F. Schulz, *History of Roman Legal Science* (Oxford, 1953), 62–75; M. Frost, "Greco-Roman legal analysis: The topics of invention," *St. Johns Law Quarterly* 66 (1992): 107–28; J. Miquel, "Stoische Logik und römische Jurisprudenz," *Zeitschrift der Savigny-Stiftung für Rechtsgeschichte. Romanistische Abteilung* 87 (1970): 85–122; G. Wright, "Stoic midwives at the birth of jurisprudence," *American Journal of Jurisprudence* 28 (1983): 169–88.

161. Leibniz, *Study for Universal Characteristic*, in *Philosophischen Schriften*, 7:167, quoted in Crombie, *Styles of Scientific Thinking in the European Tradition*, 2:1338.

162. *Digest* 50.17.185.

163. M. Krygier, "Julius Stone: Leeways of choice, legal tradition and the declaratory theory of law," *University of New South Wales Law Journal* 9 (1986): 26–38.

164. Twining, *Rethinking Evidence*, ch. 2, "Taking facts seriously."

Epilogue

1. I. Hacking, *The Emergence of Probability* (Cambridge, 1975), ch. 9.

2. A. Arnauld and P. Nicole, *La logique, ou l'art de penser* (Paris, 1662) 4, ch. 15, trans. in *The Art of Thinking*, by J. Dickoff and P. James (Indianapolis, 1964), 351; some true figures on fraudulent notaries in C. Cheney, "Notaries public in Italy and England," in *The English Church and Its Laws* (London, 1982), 173–88, at 186; cf. Cheney, *Notaries Public in England*, 130–34.

3. Arnauld and Nicole, *La logique*, 4, ch. 16, at 354–56.

4. Leibniz, *New Essays* 4, ch. 2, in *Leibniz, Selections*, ed. P. Wiener (New York, 1951), 82–83.

5. Ibid., 4, ch. 16 at 85–87; further in *Opuscules et fragments inédits de Leibniz*, ed. L. Couturat (Paris, 1903), 210–14; Leibniz, *Philosophischen Schriften* 7:477; see L. Couturat, *La logique de Leibniz d'après des documents inédits*

(Hildesheim, 1969), 238–55; I. Schneider, "Leibniz on the probable," in *Mathematical Perspectives*, ed. J. Dauben (New York, 1981), 201–19; M. Parmentier, "Concepts juridiques et probabilistes chez Leibniz," *Revue d'histoire des sciences* 46 (1993): 439–85; L. Krüger, "Probability in Leibniz: On the internal coherence of a dual concept," *Archiv für Geschichte der Philosophie* 63 (1981): 47–60; further refs. in notes to Crombie, *Styles of Scientific Thinking in the European Tradition*, 2:1335.

6. Leibniz, *Textes inédits* 2:548.

7. Leibniz, *Philosophical Letters and Papers*, trans. L. Loemker (Chicago, 1956), 1:133.

8. Leibniz, *Sämtliche Schriften und Briefe* (Darmstadt, 1923–), ser. 6, 1:496; ser. 2, 1:38; cf. *Textes inédits* 1:43 n. 165, 1:292.

9. Leibniz, *Study for Universal Characteristic*, in *Philosophische Schriften*, ed. C. Gerhardt (Hildesheim, 1978), 7:201.

10. K. Biermann and M. Faak, "G. W. Leibniz, 'De incerti aestimatione,'" *Forschungen und Fortschritte* 31 (1957): 45–50; see Schneider, "Leibniz on the probable," 207.

11. Leibniz to Wagner, 1698; Schneider, "Leibniz on the probable," 203.

12. Schneider, "Leibniz on the probable," 210; further in Crombie, *Styles of Scientific Thinking in the European Tradition*, 2:1353–64.

13. See O. Sheynin, "Early history of the theory of probability," *Archive for History of Exact Sciences* 17 (1977): 201–59, at 205–6 (on N. Bernoulli); Voltaire, "Essai sur les probabilités en fait de justice," 1772, in *Oeuvres complètes de Voltaire*, ed. L. Maland (Paris, 1879), 38:495–516; Condorcet, *Essai sur l'application de l'analyse à la probabilité des décisions rendues à la pluralité des voix* (Paris, 1785); see Daston, *Classical Probability in the Enlightenment*, § 6.4; J. Bentham, *Traité des preuves judiciaires* (Paris, 1823); Poisson, *Recherches sur la probabilité*.

14. W. Blackstone, *Commentaries on the Laws of England*, 3, ch. 23 (Oxford, 1768, repr. London, 1966, 3:370–72).

15. W. Best, *A Treastise on Presumptions of Law and Fact, with the Theory and Rules of Presumptive or Circumstantial Proof in Criminal Cases* (London, 1844, Philadelphia, 1845, repr. Littleton, Colo., 1981), esp. nn. to ch. 1; cf. W. Best, *A Treatise on the Principles of Evidence* (London, 1849, repr. New York, 1978; 12th ed., London, 1922), nn. to §§ 35 and 43.

16. E. Morgan, "Presumptions," *Washington Law Review* 12 (1937): 255–81, at 255; for recent developments see Symposium on Presumptions and Burdens of Proof, *Harvard Journal of Law and Public Policy* 17 (3) (1994); more generally, R. Gaskins, *Burdens of Proof in Modern Discourse* (New Haven, 1992).

17. B. Shapiro, *"Beyond Reasonable Doubt" and "Probable Cause"* (Berkeley, 1991), ch. 1, esp. 22, 33, 35, 273–74; Waldman, "Origins of the legal doctrine of reasonable doubt."

18. *Miller v. Minister for Pensions* [1947] 2 *All England Reports* 372 at 373–74; cf. *Cross on Evidence*, 6th ed., ed. R. Cross and C. Tapper (London, 1985), 141, also 149–51.

19. Cross and Tapper, *Cross on Evidence*, 150.

20. M. Finkelstein, *Quantitative Methods in Law* (New York, 1978), esp. app., 288–320; R. Eggleston, *Evidence, Proof, and Probability*, 2d ed. (London, 1983), esp. ch. 12; R. Wright, "Causation, risk, probability, naked statistics and proof," *Iowa Law Review* 73 (1988): 1001–77, § 5; D. Kaye, "Apples and oranges: Confidence coefficients and the burden of persuasion," *Cornell Law Review* 73 (1987): 54–77; N. Cohen, "Conceptualizing proof and calculating probabilities," *Cornell Law Review* 73 (1987): 78–95; C. Aitken, *Statistics and the Evaluation of Evidence for Forensic Scientists* (Chichester, 1995).

21. F. Pollock, *Law of Torts*, 4th ed. (London, 1895), 36–37; cf. *Hammer v. Slive*, Appellate Court of Illinois, 1960, *North Eastern Reporter*, 2d ser. 169:400.

22. A. Herbert, *Uncommon Law* (London, 1935), 2.

23. T. Deman, s.v. "Probabilisme," *Dictionnaire de théologie catholique*, 13: pt. 1 (Paris, 1936), cols. 417–619, at cols. 520–21; F. Deininger, *Johannes Sinnich. Der Kampf der Löwener Universität gegen den Laxismus* (Düsseldorf, 1928).

24. E. Amann, s.v. "Laxisme," *Dictionnaire de théologie catholique*, 9: pt. 1 (1926), cols. 42–86, at col. 55.

25. Deman, "Probabilisme," col. 532; *Dictionnaire de théologie catholique*, 1: col. 741.

26. Amann, "Laxisme," cols. 74, 79, 80.

27. Deman, "Probabilisme," col. 539; see A. Eberle, "Das 'Probabile' bei Thyrsus Gonzales als Grundlage Seines Moralsystems," *Theologische Quartalschrift* 127 (1947): 295–331; Kantola, *Probability and Moral Uncertainty in Late Medieval and Early Modern Times*, ch. 4.

28. D. Concina, *Storia del probabilismo e rigorismo* (Venice, 1743); cf. I. de Camargo, *Saggio sulla storia del probabilismo nella descrizione delli cangiamenti di sei insigni probabilisti in probabilioristi* (Verona, 1736); Alphonsus Liguori, *Dell'uso moderato dell'opinione probabile*, see D. Capone, "Dissertazione e note di S. Alfonso sulla probabilità e la conscienza," *Studia Moralia* 1 (1963): 265–343, 2 (1964): 89–155, 3 (1965): 82–149.

29. W. Gladstone, *Studies Subsidiary to the Works of Bishop Butler* (Oxford, 1896), 365.

30. M. Hunter, "Casuistry in action: Robert Boyle's confessional interviews with Gilbert Burnet and Edward Stillingfleet, 1691," *Journal of Ecclesiastical History* 14 (1993): 80–98, at 97–98.

31. R. Mortimer, *The Elements of Moral Theology* (Oxford, 1947), 92, 95.

32. *Codex Iuris Canonici* (Vatican City, 1933), canons 1825, 1829, 1869; cf. *Code of Canon Law* (London, 1983), canons 1573, 1584; *Acta Apostolicae Sedis* (1942), 339; cf. L. Robitaille, "Evaluating proofs: Is it becoming a lost art?" *Jurist* 57 (1997): 541–59.

33. C. Huygens, preface to *Treatise on Light*, trans. S. Thompson (London, 1912), vi–vii.

34. I. Newton, *Opticks* query 31 (New York, 1952), 376, 404.

35. C. Darwin, *The Origin of Species*, 1st ed. (London, 1859), 1, 402, 18; cf.

156, 160, 354, 379, 437, 444; R. Curtis, "Darwin as an epistemologist," *Annals of Science* 44 (1987): 379–408.

36. G. Polya, *Mathematics and Plausible Reasoning* (Princeton, 1954); Franklin, "Non-deductive logic in mathematics," *British Journal for the Philosophy of Science* 38 (1987): 1–18.

37. Refs. in Lawrence and O'Connor, "Exploring judgemental forecasting"; P. Goodwin and G. Wright, "Improving judgmental time series forecasting," *International Journal of Forecasting* 9 (1993): 147–61.

38. C. Jones, *Assessment and Control of Software Risks* (Englewood Cliffs, 1994); B. Marick, *The Craft of Software Testing* (Englewood Cliffs, 1995).

39. R. Cooke, *Experts in Uncertainty: Opinion and Subjective Probability in Science* (New York, 1991).

40. K. Schlechta, *Nonmonotonic Logics* (Berlin, 1997); survey in P. Bonissone, s.v. "Reasoning, plausible," in *Encyclopedia of Artificial Intelligence*, 2d ed., ed. S. Shapiro (New York, 1992), 2:1307–22; E. Charniak, "Bayesian networks without tears," *AI Magazine* 12 (4) (1991): 50–63; F. Bacchus, *Representing and Reasoning with Probabilistic Knowledge: A Logical Approach to Probabilities* (Cambridge, Mass., 1990); psychological comparisons in T. Wallsten, D. Budescu, I. Erev, and A. Diederich, "Evaluating and combining subjective probability estimates," *Journal of Behavioral Decision Making* 10 (1997): 243–68.

41. E. Gibbon, *Memoirs of My Life*, ed. G. Bonnard (London, 1966), 78.

42. C. McCullagh, *The Truth of History* (London, 1998); K. Windschuttle, *Killing of History: How a Discipline Is Being Murdered by Literary Critics and Social Theorists* (New York, 1997), ch. 7; C. Coady, *Testimony: A Philosophical Study* (Oxford, 1992), ch. 13.

43. J. Austen, *Pride and Prejudice*, ch. 36 (Harmondsworth, 1972), 234, 237; cf. *Northanger Abbey*, in *The Novels of Jane Austen*, ed. R. Chapman, 3d ed. (London, 1933), 5:197; W. Hancock, "Jane Austen, historian," *Historical Studies* 10 (1963): 422–30.

44. S. LeRoy and L. Singell, "Knight on risk and uncertainty," *Journal of Political Economy* 95 (1987): 394–406.

45. J. Locke, *Essay Concerning Human Understanding* 4, chs. 15.1, 15.6; P. Romanell, *John Locke and Medicine* (Buffalo, 1984), 132, 191, 200–202; D. Owen, "Locke on reason, probable reasoning, and opinion," *Locke Newsletter* 24 (1993): 35–79.

46. J. Butler, intro. to *Analogy of Religion*; cf. T. Penelhum, *Butler* (London, 1985), 91–93, 102, 177, 202–3.

47. Cf. H. Trowbridge, "Scattered atoms of probability," *Eighteenth Century Studies* 5 (1971): 1–31; *Encyclopaedia Britannica*, 1st ed. (1771), 3:531.

48. J. Keynes, *Treatise on Probability* (London, 1921), ch. 3; cf. D. Stove, s.v. "Keynes," in *Encyclopedia of Philosophy*, ed. P. Edwards (New York, 1967); for related developments see T. Hailperin, "The development of probability logic from Leibniz to MacColl," *History and Philosophy of Logic* 9 (1988): 131–91; T. Hailperin, "Probability logic in the twentieth century," *History and Philoso-*

phy of Logic 12 (1991): 71–110; a purely comparative system of probability in B. Koopman, "The axioms and algebra of intuitive probability," *Annals of Mathematics* 41 (1940): 269–92; B. Koopman, "The bases of probability," *Bulletin of the American Mathematical Society* 46 (1940): 763–74.

49. R. Carnap, *Logical Foundations of Probability* (London, 1950); J. Earman, *Bayes or Bust?* (Cambridge, Mass., 1992); C. Howson and P. Urbach, *Scientific Reasoning: The Bayesian Approach*, 2d ed. (Chicago, 1993); E. Jaynes, *Probability Theory: The Logic of Science*, in press.

50. D. Williams, *The Ground of Induction* (Cambridge, Mass., 1947), 16, 18.

51. D. Stove, *Popper and After: Four Modern Irrationalists* (Oxford, 1982); repr. as *Anything Goes* (Sydney, 1998).

52. P. Gross and N. Levitt, *Higher Superstition: The Academic Left and Its Quarrels with Science* (Baltimore, 1994); N. Koertge, ed., *A House Built on Sand: Exposing Postmodernist Myths about Science* (New York, 1998).

53. D. Bloor and D. Edge, "Knowing reality through knowing society," *Physics World* 11 (3) (Mar. 1998): 23, repr. in *Social Studies of Science* 30 (2000): 158–60.

54. Windschuttle, *Killing of History*.

Index